Mathematical Statistics

Mathematical Statistics

An Introduction to Likelihood Based Inference

Richard J. Rossi
Montana Tech of the University of Montana

Registered Office(s)
John Wiley & Sons, Inc., 111 River Street, Hoboken, NJ 07030, USA

Editorial Office
111 River Street, Hoboken, NJ 07030, USA

For details of our global editorial offices, customer services, and more information about Wiley products visit us at www.wiley.com.

Wiley also publishes its books in a variety of electronic formats and by print-on-demand. Some content that appears in standard print versions of this book may not be available in other formats.

Library of Congress Cataloging-in-Publication Data
Names: Rossi, Richard J., 1956- author.
Title: Mathematical statistics : an introduction to likelihood based
   inference / by Richard J. Rossi.
Description: 1st edition. | Hoboken, NJ : John Wiley & Sons, 2018. | Includes
   bibliographical references and index. |
Identifiers: LCCN 2018003972 (print) | LCCN 2018010628 (ebook) | ISBN
   9781118770979 (pdf) | ISBN 9781118771167 (epub) | ISBN 9781118771044
   (cloth)
Subjects: LCSH: Mathematical statistics.
Classification: LCC QA276.A2 (ebook) | LCC QA276.A2 R67 2018 (print) | DDC
   519.5/4–dc23
LC record available at https://lccn.loc.gov/2018003972

Cover Design: Wiley
Cover Image: ©VikaSuh/Shutterstock

Set in 10/12pt WarnockPro by SPi Global, Chennai, India

Printed in the United States of America

V10003572_081318

This book is dedicated to
Lloyd Gavin.

Contents

Preface

Mathematical Statistics: An Introduction to Likelihood-Based Inference is intended to be used in a two-semester or three-quarter sequence on mathematical statistics. This textbook is primarily written for senior-level statistics or mathematics majors, but is also appropriate for a first-year graduate sequence on mathematical statistics. The purpose of this book is to provide students with rigorous introduction to statistical theory behind point estimation, hypothesis testing, and statistical modeling. The primary emphasis in the chapters on mathematical statistics is likelihood inference. The prerequisites for a course taught from this textbook include univariate calculus, multivariate calculus, and linear algebra.

Topic Coverage

The standard mathematical statistics topics are covered in this book. Throughout the chapters on statistics, the primary emphasis of *Mathematical Statistics: An Introduction to Likelihood-Based Inference* is placed on likelihood-based inference; however, Bayesian methods are also included in this textbook. The organization of the topics was chosen for statistical and pedagogical reasons, which have been tried, tested, updated, and continually modified over the last 30 years. *Mathematical Statistics: An Introduction to Likelihood-Based Inference* is organized into two parts with Chapters 1–3 covering probability theory, random variables/vectors, and probability models and Chapters 4–7 covering mathematical statistics. In particular, the following topics are presented in this textbook.

Chapter 1 – Probability: Chapter 1 provides a rigorous introduction to probability theory covering the foundations of probability, probability rules, probability assignment, conditional probability, and counting methods.

Chapter 2 – Random Variables and Random Vectors: Chapter 2 covers random variables and random vectors including transformations, marginal and conditional distributions, independence, mathematical expectation, variance, moment generation functions, covariance, correlation, and sums of random variables.

Chapter 3 – Probability Models: Commonly used probability models are introduced in Chapter 3. The discrete models included in Chapter 3 are the binomial, Poisson, hypergeometric, geometric, negative binomial, and multinomial; the continuous models included in Chapter 3 are the uniform, exponential, gamma, normal, log-normal, and beta distributions. The T and F distributions are also introduced in Chapter 3.

Chapter 4 – Parametric Point Estimation: Statistics, sampling distributions, properties of statistics, bias, consistency, sufficiency, exponential family distributions, UMVUEs, Cramér–Rao lower bound, and order statistics are covered in Chapter 4.

Chapter 5 – Likelihood-Based Estimation: Likelihood and log-likelihood functions, maximum likelihood estimation, properties of MLEs, Bayesian estimation, and exact, large sample, and Bayesian interval estimation are covered in Chapter 5.

Chapter 6 – Hypothesis Testing: Chapter 6 covers hypothesis testing including components of a hypothesis test, most powerful tests, uniformly most powerful tests, generalized likelihood ratio tests, and large sample tests including Wald and Score tests.

Chapter 7 – Generalized Linear Models: Generalized linear models are introduced in Chapter 7, including model fitting, parameter estimation and testing, asymptotic properties of the MLEs, models for a normal response variable, models for a binary response variable, and models for a Poisson response variable.

Special Features

Special features in this book include

- a relevant case study at the end of each chapter;
- emphasis on likelihood-based inference;
- emphasis on properties of the exponential family distributions;
- an introduction to Bayesian point and interval estimation;
- a separate chapter on general linear models;
- the use of the statistical computing package R throughout the text.

Teaching from This Book

In teaching a two-semester mathematical statistic sequence of courses from this textbook, it is possible to cover all of Chapters 1–3 in the first semester and all of Chapters 4–7 in the second semester. While I have not taught a course from this book on the quarter system, I believe that all of Chapters 1 and 2 can be covered in the first quarter, Chapters 3 and 4 in the second quarter, and Chapters 5–7 in the third quarter. However, there are many different ways to teach from this book, and I leave that to the discretion and goals of the instructor.

A large number of exercises have been included in each chapter of this textbook, and the solutions to many of the exercises are included at the back of this book. The exercises accompanying this textbook do cover a wide range of topics with differing levels of difficulty. While some exercises require the use of R, this textbook has been written with the expectation that students will have access to the statistical computing package R. Moreover, since vast majority of the examples and exercises do not depend on the use of R, *Mathematical Statistics: An Introduction to Likelihood-Based Inference* can easily be used in courses where R will not be used.

A Web page containing the data used in *Mathematical Statistics: An Introduction to Likelihood-Based Inference* is housed on my personal Web page at the Montana Tech of the University of Montana Web page; however, because the Web page addresses change frequently at Montana Tech, no Web address is provided here. The easiest route to this web page is to go to www.mtech.edu and then search for Rick Rossi, which will provide a link to my Web page.

Acknowledgments

First and foremost, I wish to thank my good friend and mentor Lloyd Gavin, who first introduced me to mathematical statistics. I am forever indebted to Lloyd for his guidance and support throughout my career.

I would also like to thank H. Dan Brunk, Glen Meeden, David Birkes, Fred Ramsey, Ray Carroll, and David Ruppert, who motivated me to write this book, and especially my editor Susanne Steitz-Filler for her help and guidance on this book and my previous two books. Last but not least, I wish to thank my family and friends for their support while pursuing this endeavor.

Silverbow, Montana or somewhere on the Big Hole River   *Richard J. Rossi*
March 5, 2018

1

Probability

The mathematical foundation upon which mathematical statistics and likelihood inference are built is probability theory. Modern probability theory is primarily due to the foundational work of the Russian mathematician Andrei Nikolaevich Kolmogorov (1903–1987). Kolmogorov published his treatise on probability in 1933 [1], which framed probability theory in a rigorous mathematical framework. Kolmogorov's work provided probability theory with an axiomatic mathematical structure that produces a consistent and coherent theory of probability. Specifically, Kolmogorov's structure is based on measure theory, which deals with assigning numerical values to sets (i.e. measuring a set) and the theory of integration and differentiation.

1.1   Sample Spaces, Events, and σ-Algebras

The structure under which probabilities are relevant and can be assigned in a consistent and coherent fashion requires a probability model consisting of a chance experiment, the collection of all possible outcomes of the chance experiment, and a function that assigns probabilities to collections of outcomes of the chance experiment.

Definition 1.1   A chance experiment is any task for which the outcome of the task is unknown until the task is actually performed.

Experiments where the outcome is known before the experiment is actually performed are called *deterministic experiments* and are not interesting with regard to probability. Probability theory, probability assignments, and statistics only apply to chance experiments. The set of possible outcomes of a chance experiment is called the *sample space*, and the sample space defines one component of a probability model.

Mathematical Statistics: An Introduction to Likelihood Based Inference, First Edition. Richard J. Rossi.
© 2018 John Wiley & Sons, Inc. Published 2018 by John Wiley & Sons, Inc.

Definition 1.2 The sample space associated with a chance experiment is the set of all possible outcomes of a chance experiment. The sample space will be denoted by Ω.

The sample space consists of the outcomes that are considered feasible and interesting. Probabilities can only be assigned to outcomes or subsets of outcomes in the sample space.

Example 1.1 Suppose that a chance experiment consists of flipping a two-sided coin with heads on one side and tails on the other side. The most commonly used sample space is $\Omega = \{$Heads, Tails$\}$; however, another possible sample space that could be used is $\Omega' = \{$heads, tails, edge$\}$; these two sample spaces produce two different probability models for the same chance experiment. As long as Kolmogorov's measure theoretic approach is used, both probability models will produce consistent and coherent probability assignments.

Chance experiments cover a wide range of everyday tasks such as dealing a hand of cards, forecasting weather, driving in excess of the speed limit at the risk of getting a speeding ticket, and buying a lottery ticket. In each of these cases there is a chance experiment where the outcome is unknown until the experiment is actually completed.

Example 1.2 Suppose that a chance experiment consists of weighing a brown trout randomly selected from the Big Hole River in Montana. A reasonable sample space for this chance experiment is $\Omega = (0, 50]$ since the largest known brown trout to come from the river is less than 50 lb. If the upper limit on the weight of a Big Hole brown trout is unknown, it would also be reasonable to use $\Omega = (0, \infty)$ for the sample space. Choosing the probability assignment takes care of probabilities for likely and unlikely values of the weight of a Big Hole River brown trout.

Note that in many chance experiments, the limits of the sample space will be unknown, and in this case, the sample space can be taken to be an infinite length subset of $\mathbb{R}$. The sample space is only the list of possible outcomes, while the choice of the function used to make the probability assignments controls the probabilities of the values in the sample space. The three components required of a probability model are the sample space, a collection of subsets of the sample space for which probabilities will be assigned, and the function used to assign the probabilities to subsets of the sample space.

Under Kolmogorov's probability structure, not all subsets of the sample space can be assigned probabilities. The collection of subsets of the sample space that can be assigned probabilities must have a particular structure so that the probability assignments are coherent and consistent. In particular, the collection of subsets of Ω that can be assigned probabilities must be a *σ-algebra*.

Definition 1.3 Let $\mathcal{A}$ be a collection of events of Ω. $\mathcal{A}$ is said to be a σ-algebra of events if and only if

  i) $\Omega \in \mathcal{A}$.
  ii) $A^c \in \mathcal{A}$ whenever $A \in \mathcal{A}$.
  iii) $\bigcup_{i=1}^{\infty} A_i \in \mathcal{A}$ whenever $A_i \in \mathcal{A}$, $\forall i$.

A subset of Ω that is in a σ-algebra associated with Ω is called an *event*.

Definition 1.4 An event A of a sample space Ω is any subset in a σ-algebra associated with Ω. An event A is said to have occurred when the chance experiment results in an outcome in A.

A σ-algebra associated with a sample space Ω contains the only events that probabilities can be assigned to. There are many σ-algebras of subsets associated with a sample space (see Example 1.3); however, the appropriate σ-algebra must be chosen so that it is large enough to contain all of the relevant events to be considered. It is important to note that in order to have a consistent and coherent probability assignment, not all events of Ω can be assigned probabilities.

Example 1.3 Examples of σ-algebras associated with a sample space Ω include the following:

1) The trivial σ-algebra $\mathcal{A}_0 = \{\emptyset, \Omega\}$. This is the smallest σ-algebra possible and not very useful for a probability model.
2) $\mathcal{A}_1 = \{\emptyset, A, A^c, \Omega\}$, where A is a subset of Ω. This is the smallest σ-algebra that includes the event A.
3) The Borel σ-algebra, which is the smallest σ-algebra containing all of the open intervals of $\mathbb{R}$. The Borel σ-algebra can only be used when the elements of Ω are real numbers, and in this case, it is a commonly used σ-algebra.

The σ-algebra of events associated with Ω will also include all of the compound events that can constructed using the basic set operations intersection, union, and complementation. The definitions for the compound events are given in Definitions 1.5–1.7.

Definition 1.5 Let A and B be events of Ω. The event formed by the intersection of the events A and B is denoted by $A \cap B$ and is defined to be $A \cap B = \{\omega \in \Omega : \omega \in A \text{ and } \omega \in B\}$.

Definition 1.6 Let A and B be events of Ω. The event formed by the union of the events A and B is denoted by $A \cup B$ and is defined to be $A \cup B = \{\omega \in \Omega : \omega \in A \text{ or } \omega \in B\}$.

Definition 1.7 Let A be an event of Ω. The event that is the complement of the event A is denoted by A^c and is defined to be $A^c = \{\omega \in \Omega : \omega \notin A\}$.

Note that union and intersection are commutative operations. That is, $A \cup B = B \cup A$ and $A \cap B = B \cap A$. Also, with complementation, $(A^c)^c = A$. Another set operation that is used to create a compound event is the *set difference*. The set difference between two sets A and B consists of the elements of A that are not elements of B.

Definition 1.8 Let A and B be events of Ω. The set difference $A - B$ is defined to be $A - B = \{\omega \in \Omega : \omega \in A \text{ and } \omega \notin B\}$.

Set difference is not a commutative operation, and $A - B$ can also be written as $A \cap B^c$. The following example illustrates how compound events can be created using the set operations union, intersection, complementation, and set difference.

Example 1.4 Suppose that a card will be drawn from a standard deck of 52 playing cards. Then, the sample space is

$$\Omega = \{AH, \ldots, KH, AD, \ldots, KD, AC, \ldots, KC, AS, \ldots, KS\},$$

where in the outcome XY, X is the denomination of the card $(A, 2, \ldots, K)$ and Y is the suit of the card (H, D, C, S). Let A be the event that a heart is selected, and let B be the event that an ace is selected. Then, $A = \{AH, \ldots, KH\}$, $B = \{AH, AD, AC, AS\}$, and

$$A \cap B = \{AH\},$$
$$A \cup B = \{AH, \ldots, KH, AD, AC, AS\},$$
$$A^c = \{AD, \ldots, KD, AC, \ldots, KC, AS, \ldots, KS\},$$
$$B^c = \{2H, \ldots, KH, 2D, \ldots, KD, 2C, \ldots, KC, 2S, \ldots, KS\},$$
$$A - B = \{2H, \ldots, KH\},$$
$$B - A = \{AD, AC, AS\}.$$

Events that share no common elements are called *disjoint events* or *mutually exclusive events*.

Definition 1.9 Two events A and B of Ω are said to be disjoint when $A \cap B = \emptyset$.

When two events A and B are disjoint, the chance experiment cannot result in an outcome where both the events A and B occur. The events A and A^c are always disjoint events as are $A - B$ and $B - A$.

Compound events can also be constructed using the set operations intersection, union, and complementation on a family of sets, say $\mathcal{F} = \{A_i : i \in \Delta\}$, where the index set Δ is a finite or countably infinite set. In most cases, Δ will be taken to be a subset of $\mathbb{N}$, and the compound events created using intersection and union are

$$\omega \in \bigcap_{i \in \Delta} A_i, \qquad \text{if and only if } \omega \in A_i, \ \forall i \in \Delta,$$

$$\omega \in \bigcup_{i \in \Delta} A_i, \quad \text{if and only if } \omega \in A_i \text{ for some } i \in \Delta.$$

Example 1.5 Suppose that a two-sided coin will be flipped until the first head appears. Let A_i be the event that the first head appears on the ith flip, B be the event that it takes at least two flips of the coin to observe the first head, and let C be the event that it takes less than 10 flips to observe the first head. Then,

$$B = \bigcup_{i=2}^{\infty} A_i$$

and

$$C = \bigcup_{i=1}^{9} A_i.$$

The set laws given in Theorems 1.1 and 1.2 can often be used to simplify the computation of the probability of a compound event.

Theorem 1.1 (De Morgan's Laws) *If $\{A_i : i \in \Delta\}$ is a family of events of Ω, Δ is a subset of $\mathbb{N}$, and $\mathcal{D} \subset \Delta$, then*

i) $\left(\displaystyle\bigcup_{i \in \mathcal{D}} A_i \right)^c = \displaystyle\bigcap_{i \in \mathcal{D}} A_i^c.$

ii) $\left(\displaystyle\bigcap_{i \in \mathcal{D}} A_i \right)^c = \displaystyle\bigcup_{i \in \mathcal{D}} A_i^c.$

Corollary 1.1 *If A and B are events of Ω, then*

i) $(A \cup B)^c = A^c \cap B^c.$
ii) $(A \cap B)^c = A^c \cup B^c.$

Note that the complement of a union is the intersection of the complements, and the complement of an intersection is the union of the complements.

Theorem 1.2 (Distributive Laws) *If $\{A_i : i \in \Delta\}$ is a family of events of Ω, Δ is a subset of $\mathbb{N}$, and $\mathcal{D} \subset \Delta$, then*

i) *for any event B of Ω,* $B \cap \left(\bigcup_{i \in \mathcal{D}} A_i \right) = \bigcup_{i \in \mathcal{D}} (B \cap A_i).$

ii) *for any event B of Ω,* $B \cup \left(\bigcap_{i \in \mathcal{D}} A_i \right) = \bigcap_{i \in \mathcal{D}} (B \cup A_i).$

In particular, Theorem 1.1 holds for $\mathcal{D} = \{1, 2, \dots, n\}$ or $\mathcal{D} = \mathbb{N}$. That is,

$$\left(\bigcup_{i=1}^{n} A_i \right)^c = \bigcup_{i=1}^{n} A_i^c \quad \text{and} \quad \left(\bigcup_{i=1}^{\infty} A_i \right)^c = \bigcup_{i=1}^{\infty} A_i^c$$

and

$$\left(\bigcap_{i=1}^{n} A_i \right)^c = \bigcap_{i=1}^{n} A_i^c \quad \text{and} \quad \left(\bigcap_{i=1}^{\infty} A_i \right)^c = \bigcap_{i=1}^{\infty} A_i^c$$

Similarly, Theorem 1.2 also holds for the finite set $\mathcal{D} = \{1, 2, \dots, n\}$ and the infinite set $\mathcal{D} = \mathbb{N}$. The simplest version of the *Distributive Laws* is given in Corollary 1.2

Corollary 1.2 *If A, B, and C are events of Ω, then*

i) $A \cap (B \cup C) = (A \cap B) \cup (A \cap C).$
ii) $A \cup (B \cap C) = (A \cup B) \cup (A \cup C).$

A family $\mathcal{P}$ of disjoint events of a sample space Ω whose union is Ω is called a *partition*. Partitions also can be used to simplify the computation of the probability of an event.

Definition 1.10 A collection of events $P = \{A_i : i \in \mathbb{N}\}$ is said to be a partition of a sample space Ω if and only if

i) $\bigcup_{i=1}^{\infty} A_i = \Omega.$

ii) $A_i \cap A_j = \emptyset$, whenever $i \neq j$.

A partition may consist of a finite number of events. That is, if $\{A_1, A_2, \dots, A_n\}$ is a collection of disjoint events whose union is Ω, a partition $P = \{A_j : j \in \mathbb{N}\}$ is formed by letting $A_j = \emptyset$ for $j > n$.

Example 1.6 Let A be an event in Ω. The simplest partition of a sample space Ω is $P = \{A, A^c\}$ since $A \cup A^c = \Omega$ and $A \cap A^c = \emptyset$.

Theorem 1.3 shows that a partition $\mathcal{P}$ can be used to partition an event B into disjoint events whose union is the event B.

Theorem 1.3 *If $P = \{A_i : i \in \mathbb{N}\}$ is a partition of Ω, then for any event B of Ω.*

$$B = \bigcup_{i=1}^{\infty}(B \cap A_i).$$

Proof. Let $\{A_i : i \in \mathbb{N}\}$ be a partition of Ω and B an event of Ω. Then,

$$B = B \cap \Omega = B \cap \left(\bigcup_{i=1}^{\infty} A_i \right) = \bigcup_{i=1}^{\infty}(B \cap A_i).$$

■

Corollary 1.3 shows that an arbitrary event A of Ω can be used to partition any other event B of Ω.

Corollary 1.3 *If A and B are events of Ω, then $B = (B \cap A) \cup (B \cap A^c)$.*

Problems

1.1.1 Determine a reasonable sample space when the chance experiment involves
a) selecting a student at random and recording their GPA.
b) selecting an adult male at random and measuring their weight.
c) selecting an adult female at random and measuring their weight.
d) selecting a student at random and recording the color of their hair.
e) rolling two standard six-sided dice and summing the outcomes on each die.
f) selecting a person at random and recording their birthday.
g) selecting a student at random and recording how many credits they are enrolled in.

1.1.2 Two numbers will be drawn at random and without replacement from the numbers $1, 2, 3, 4, 5$. Let A be the event that at least one even number is drawn in the two draws, and let B be the event that the sum of the draws is equal to 4. Determine the outcomes in

a) sample space.   b) A.
c) B.   d) $A \cap B$.
e) $A \cup B$.   f) $B - A$.

1.1.3 Flip a two-sided coin four times. Let A be the event exactly two heads are flipped and let B be the event at least one tail is flipped. Determine the outcomes in

a) the sample space.   b) A.
c) A^c.   d) B.
e) $A \cap B$.   f) $A \cup B$.
g) $A - B$.   h) $B - A$.

1.1.4 Draw a card at random from a standard deck of 52 playing cards. Let A be the event that an ace is drawn, and let B be the event that a diamond is drawn. Determine the outcomes in

a) the sample space.
b) A.
c) B.
d) B^c.
e) $A \cap B$.
f) $A \cup B$.
g) $A - B$.
h) $B - A$.

1.1.5 Flip a two-sided coin until a head appears. Let A be the event the first head is flipped on the fourth flip and let B be the event the first head is flipped in less than five flips. Determine the outcomes in

a) A.
b) B.
c) $A \cap B$.
d) $A \cup B$.

1.1.6 Let A be an event of Ω. Show that $(A^c)^c = A$.

1.1.7 Let A and B be events of Ω. If $A \subset B$, show that $B = A \cup (B - A)$.

1.1.8 Let A and B be events of Ω. Show that $A \cup B = A \cup (B - A)$.

1.1.9 Let A and B be events of Ω. Show that $A - B$ and $B - A$ are disjoint events.

1.1.10 Let A and B be disjoint events of Ω, and let C be any other event of Ω. Show that $A \cap C$ and $B \cap C$ are disjoint events.

1.1.11 Let $\{A_i\}$ be a countable collection of events of Ω with $\bigcup_{i=1}^{\infty} A_i = \Omega$, and let $B_1 = A_1$ and $B_k = A_k - \left(\bigcup_{i=1}^{k-1} A_i \right)$ for $n \geq 2$. Show that $\{B_k\}$ is a partition of Ω.

1.1.12 Let $\Omega = \mathbb{R}$ and define $A_i = [-\frac{1}{n}, n)$ for $i \in \mathbb{N}$. Determine

a) $\bigcup_{i=1}^{10} A_i$.
b) $\bigcap_{i=1}^{10} A_i$.
c) $\bigcup_{i=1}^{\infty} A_i$.
d) $\bigcap_{i=1}^{\infty} A_i$.

1.1.13 Let $\Omega = \mathbb{R}^+$ and define $A_i = (\frac{1}{n}, 1 + \frac{1}{n})$ for $i \in \mathbb{N}$. Determine

a) $\bigcup_{i=1}^{20} A_i$.

b) $\bigcap_{i=1}^{20} A_i$.

c) $\bigcup_{i=1}^{\infty} A_i$.

d) $\bigcap_{i=1}^{\infty} A_i$.

1.1.14 Show that $\{\emptyset, A, A^c, \Omega\}$ is a σ-algebra.

1.1.15 Show that if $\mathcal{A}$ is a σ-algebra and $A, B \in \mathcal{A}$, then $A \cap B \in \mathcal{A}$.

1.1.16 Let $\mathcal{A}$ be a σ-algebra and let $A_i \in \mathcal{A}$ for $i \in \mathbb{N}$. Show that $\bigcap_{i=1}^{\infty} A_i \in \mathcal{A}$.

1.1.17 Determine the smallest σ-algebra that contains the events A and B.

1.2 Probability Axioms and Rules

The third component of a probability model is the *probability function*, which is a set function whose domain $\mathcal{A}$ is a σ-algebra of events of Ω. Kolmogorov's measure theoretic approach to probability requires a probability function satisfying the properties given in Definition 1.11.

Definition 1.11 (Kolmogorov's Probability Function) Let Ω be the sample space associated with a chance experiment, and let $\mathcal{A}$ be a σ-algebra of events of Ω. A set function P on $\mathcal{A}$ satisfying the following three properties is called a probability function or a probability measure.

A1: $P(\Omega) = 1$.
A2: $P(A) \geq 0$ for every event $A \in \mathcal{A}$.
A3: If $\{A_i : i \in \mathbb{N}\} \subset \mathcal{A}$ is a collection of disjoint events, then

$$P\left(\bigcup_{i=1}^{\infty} A_i\right) = \sum_{i=1}^{\infty} P(A_i).$$

The triple $(\Omega, \mathcal{A}, P)$ is called a probability space.

Conditions A1–A3 are known as *Kolmogorov's Axioms of Probability*, and the sets in the σ-algebra $\mathcal{A}$ are the measurable sets and the only sets that can be assigned probabilities. Theorem 1.4 reveals some of the basic consequences of Kolmogorov's Axioms.

Theorem 1.4 *Let $(\Omega, \mathcal{A}, P)$ be a probability space and let $A, B \in \mathcal{A}$.*

i) $P(\emptyset) = 0$.

ii) *If $A \subset B$, then $P(A) \leq P(B)$.*

iii) $0 \leq P(A) \leq 1$.

iv) $P\left(\bigcup_{i=1}^{n} A_i\right) = \sum_{i=1}^{n} P(A_i)$ *when $A_1, A_2, \ldots, A_n$ are disjoint events in $\mathcal{A}$.*

Proof. Let $(\Omega, \mathcal{A}, P)$ be a probability space and let $A, B \in \mathcal{A}$.

i) Since $\Omega = \Omega \bigcup_{i=1}^{\infty} \emptyset$, let $A_1 = \Omega$ and $A_{i+1} = \emptyset$ for $i \in \mathbb{N}$. Then, $\{A_i\}$ is a collection of disjoint sets and

$$1 = P(\Omega) = P\left(\bigcup_{i=1}^{\infty} A_i\right) = \underbrace{\sum_{i=1}^{\infty} P(A_i)}_{\text{Axiom 3}}$$

$$= P(A_1) + \sum_{i=2}^{\infty} P(A_i) = 1 + \sum_{i=2}^{\infty} P(A_i).$$

Thus, $\sum_{i=2}^{\infty} P(A_i) = 0$ and since $P(A_i) \geq 0$, $\forall i \in \mathbb{N}$. Hence, it follows that $P(A_{i+1}) = P(\emptyset) = 0$, $\forall i \in \mathbb{N}$.

ii) Since $B = A \cup (B - A)$ and $A \cap (B - A) = \emptyset$, it follows that

$$P(B) = P(A \cup (B - A)) = P(A) + \underbrace{P(B - A)}_{P(B-A) \geq 0} \geq P(A).$$

iii) This follows directly from parts (i) and (ii), since $A \subset \Omega$ and $\emptyset \subset A$, it follows that $0 = P(\emptyset) \leq P(A) \leq P(\Omega) = 1$.

iv) The proof of part (iv) is left as an exercise. ∎

Further consequences of Kolmogorov's Axioms are given in Theorem 1.5. In particular, the results given in Theorem 1.5 provide several useful rules for computing probabilities.

Theorem 1.5 *Let $(\Omega, \mathcal{A}, P)$ be a probability space and let $A, B \in \mathcal{A}$. Then,*

i) $P(A^c) = 1 - P(A)$.

ii) $P(B) = P(B \cap A) + P(B \cap A^c)$.

iii) $P(A - B) = P(A) - P(A \cap B)$.

iv) if $A \subset B$, then $P(B - A) = P(B) - P(A)$.
v) $P(A \cup B) = P(A) + P(B) - P(A \cap B)$.
vi) if A is countable, say $A = \{\omega_i : i \in \mathbb{N}\}$, then $P(A) = \sum_{i=1}^{\infty} P(\{\omega_i\})$.

Proof. Let $(\Omega, \mathcal{A}, P)$ be a probability space and let $A, B \in \mathcal{A}$.

i) Note that $\Omega = A \cup A^c$ and A and A^c are disjoint events. Thus,

$$1 = P(\Omega) = P(A \cup A^c) = \underbrace{P(A) + P(A^c)}_{\text{since disjoint}}.$$

Therefore, $P(A^c) = 1 - P(A)$.

ii) From Corollary 1.3, $B = (A \cap B) \cup (A^c \cap B)$ and the events $A \cap B$ and $A^c \cap B$ are disjoint. Thus,

$$P(B) = P[(A \cap B) \cup (A^c \cap B)] = P(A \cap B) + P(A^c \cap B).$$

iii) The proof of part (iii) is left as an exercise.
iv) The proof of part (iv) is left as an exercise.
v) Since $A \cup B$ can be written as the disjoint union of $A \cap B^c$, $A \cap B$, and $A^c \cap B$, it follows that

$$P(A \cup B) = P(A \cap B^c) + P(A \cap B) + P(A^c \cap B)$$
$$= P(A - B) + P(A \cap B) + P(B - A).$$

Hence, by Theorem 1.5 part (iii), it follows that

$$P(A \cup B) = P(A - B) + P(A \cap B) + P(B - A)$$
$$= P(A) - P(A \cap B) + P(A \cap B) + P(B) - P(A \cap B)$$
$$= P(A) + P(B) - P(A \cap B).$$

vi) The proof of part (vi) is left as an exercise. ∎

Examples 1.7 and 1.8 illustrate the use of Theorem 1.5 for computing probabilities.

Example 1.7 Let $(\Omega, \mathcal{A}, P)$ be a probability space and let $A, B \in \mathcal{A}$. Suppose that A and B are events of Ω with $P(A) = 0.6$, $P(B) = 0.75$, and $P(A \cap B) = 0.55$. Then,

1) $P(A^c) = 1 - P(A) = 1 - 0.6 = 0.4$.
2) $P(A \cup B) = P(A) + P(B) - P(A \cap B) = 0.6 + 0.75 - 0.55 = 0.8$.
3) $P(A - B) = P(A) - P(A \cap B) = 0.6 - 0.55 = 0.05$.
4) $P(A^c \cap B^c) = P[(A \cup B)^c] = 1 - P(A \cup B) = 1 - 0.8 = 0.2$.
5) $P(A^c \cup B) = P(A^c) + P(B) - P(A^c \cap B) = 0.4 + 0.75 - (0.75 - 0.55) = 0.95$.

Example 1.8 Suppose that 70% of the fishermen in Montana use fly fishing gear, 40% use spin fishing gear, and 10% use both fly and spin fishing gear. The probability that a fisherman in Montana uses fly fishing gear (F) but not spin fishing gear (S) is

$$P(F - S) = P(F) - P(F \cap S) = 0.70 - 0.10 = 0.60,$$

and the probability that a fisherman in Montana uses spin fishing gear but not fly fishing gear is

$$P(S - F) = P(S) - P(S \cap F) = 0.40 - 0.10 = 0.30.$$

Theorem 1.5 part (ii) is known as the *Law of Total Probability*. The Law of Total Probability can be used to solve many probability problems, and in particular, it is useful when there are two or more cases that must be considered when computing the probability of an event. A generalized version of the Law of Total Probability is given in Theorem 1.6.

Theorem 1.6 (General Law of Total Probability) *Let $(\Omega, \mathcal{A}, P)$ be a probability space and let $A \in \mathcal{A}$. If $B_i \in \mathcal{A}$, $\forall i \in \mathbb{N}$ and $\{B_i\}$ is a partition of Ω, then,*

$$P(A) = \sum_{i=1}^{\infty} P(A \cap B_i).$$

Proof. The proof of Theorem 1.6 follows directly from Theorem 1.3 and axiom A3. ∎

Note that Theorem 1.6 also works with a finite partition $\{B_1, B_2, \dots, B_n\}$ in which case $P(A) = \sum_{i=1}^{n} P(A \cap B_i)$.

Example 1.9 Suppose that a game is played where two dice are rolled until either a total of 6 or 7 is rolled. Let A be the event that a 6 is rolled before a 7. Note that $A = \bigcup_{i=1}^{\infty}(A \cap B_i)$, where B_i = the game terminates on the ith roll.

Thus, the probability of event A can be computed using the General Law of Total Probability with

$$P(A) = \sum_{i=1}^{\infty} P(A \cap B_i).$$

This example will be revisited in a later section after further development of the rules of probability.

Example 1.10 Suppose that two marbles are drawn at random and without replacement from an urn containing 5 white marbles, 9 red marbles, and

11 black marbles. Let A be the event that two marbles of the same color are selected, W be the event that the first marble selected is white, R be the event that the first marble selected is red, and B be the event that the first marble selected is black. Then, W, R, and B partition the sample space, and the probability of drawing two marbles of the same color is

$$P(A) = P(A \cap W) + P(A \cap R) + P(A \cap B).$$

Theorem 1.7 (Boole's Inequality) *If $\{A_i\}$ is a collection of events in $\mathcal{A}$, then*

$$P\left(\bigcup_{i=1}^{\infty} A_i\right) \leq \sum_{i=1}^{\infty} P(A_i).$$

Proof. Let $\{A_i\}$ be a collection of events in $\mathcal{A}$ and let

$$B_i = A_i - \left(\bigcup_{j=1}^{i-1} A_i\right)$$

Then, $\displaystyle\bigcup_{i=1}^{\infty} A_i = \bigcup_{i=1}^{\infty} B_i$, $B_i \cap B_j = \emptyset$ for $i \neq j$, and $B_i \subset A_i$, $\forall i$. Hence,

$$P\left(\bigcup_{i=1}^{\infty} A_i\right) = P\left(\bigcup_{i=1}^{\infty} B_i\right) = \underbrace{\sum_{i=1}^{\infty} P(B_i) \leq \sum_{i=1}^{\infty} P(A_i)}_{\text{since } B_i \subset A_i}. \qquad \blacksquare$$

Boole's Inequality is also often referred to as *Bonferroni's Inequality* and is sometimes used when performing multiple comparisons in a hypothesis testing scenario.

Example 1.11 Suppose that five hypothesis tests will be carried out with each test having a probability of false rejection equal to 0.01. Let A_i be the event that the ith test makes a false rejection. Then, by Boole's Inequality, the probability of making at least one false rejection in the five hypothesis tests is

$$P\left(\bigcup_{i=1}^{5} A_i\right) \leq \sum_{i=1}^{5} P(A_i) = 0.05.$$

Using Boole's Inequality is one way to protect against making false rejections in a multiple comparison setting and can be generalized to handle any number of hypothesis tests. For example, if the goal is to have the overall probability of making at least one false rejection less than α when n hypothesis tests are performed, then by taking $P(A_i) = \frac{\alpha}{n}$, the overall chance of making one or more false rejections is less than α. This procedure is referred to as the *Bonferroni Multiple Comparison Procedure*.

Problems

1.2.1 Suppose that $P(A) = 0.6$, $P(B) = 0.75$, and $P(A \cap B) = 0.55$. Determine

    a) $P(A^c)$.
            b) $P(B^c)$.
    c) $P(A \cup B)$.
            d) $P(A - B)$.
    e) $P(A^c \cap B^c)$.
            f) $P(A^c \cup B^c)$.
    g) $P(A^c \cup B)$.
            h) $P[(A - B) \cup (B - A)]$.

1.2.2 Suppose that $A \subset B$, $P(A) = 0.4$, and $P(B) = 0.6$. Determine

    a) $P(A^c)$.
            b) $P(B^c)$.
    c) $P(A \cap B)$.
            d) $P(A \cup B)$.
    e) $P(B - A)$.
            f) $P(A - B)$.

1.2.3 Suppose that A and B are disjoint events with $P(A) = 0.5$ and $P(B) = 0.35$. Determine

    a) $P(A \cap B)$.
            b) $P(A \cup B)$.
    c) $P(B - A)$.
            d) $P(A - B)$.
    e) $P[(A \cup B)^c]$.
       f) $P(A^c \cap B^c)$.

1.2.4 Suppose that $P(B) = 0.6$, $P(A \cap B) = 0.45$, and $P(A - B) = 0.25$. Determine

    a) $P(B - A)$.
            b) $P(A)$.
    c) $P(A \cup B)$.
            d) $P(A^c \cap B^c)$.

1.2.5 Suppose that $\Omega = \{1, 2, 3, 4, 5, 6, 7, 8, 9, 10\}$ and $P(\{i\}) = \frac{i}{55}$ for $i \in \Omega$. If a number is drawn at random, determine the probability that
    a) an even number is drawn.
    b) a multiple of 3 is drawn.
    c) a number less than 5 is drawn.
    d) a prime number is drawn.

1.2.6 A large computer sales company reports that 80% of their computers are sold with a DVD drive, 95% with a CD drive, and 75% with both. Determine the probability that
    a) a computer without a DVD drive is sold.
    b) a computer with a DVD drive or a CD drive is sold.
    c) a computer with a DVD drive but not a CD drive is sold.
    d) a computer with a DVD drive or a CD drive but not both is sold.

1.2.7 Mr. Jones watches the 6 p.m. news 50% of the time, he watches the 11 p.m. news 75% of the time, and watches both the 6 p.m. and 11 p.m. news 28% of the time. Determine the probability that
   a) Mr. Jones watches either the 6 p.m. or 11 p.m. news.
   b) Mr. Jones watches either the 6 p.m. or 11 p.m. news, but not both.
   c) Mr. Jones watches neither the 6 p.m. nor 11 p.m. news.
   d) Mr. Jones watches the 6 p.m. but not the 11 p.m. news.
   e) Mr. Jones watches the 11 p.m. but not the 6 p.m. news.

1.2.8 Suppose that the probability of a two-child family having two male children is 0.23, having a male child first followed by a female child is 0.25, having a female child first followed by a male child is 0.25, and having two female children is 0.27. Determine the probability that a two-child family has
   a) at least one male child.
   b) at least one female child.
   c) a male first child.
   d) a female second child.

1.2.9 Airlines A and B have 9 a.m. flights from San Francisco to Seattle. Suppose that the probability that airline A's flight is fully booked is 0.80, the probability that airline B's flight is fully booked is 0.75, and the probability that both airlines 9 a.m. flights to Seattle are fully booked is 0.68. Determine the probability that
   a) airline A or airline B has a fully booked flight.
   b) neither airline A nor airline B has a fully booked flight.
   c) airline A has a fully booked flight but airline B does not.

1.2.10 Suppose that A, B, and C are disjoint events with $P(A) = 0.1$, $P(B) = 0.25$, and $P(C) = 0.6$. Determine
   a) $P(A \cap B \cap C)$.   b) $P(A \cup B \cup C)$.

1.2.11 Suppose that A, B, and C are events with $A \subset B \subset C$. If $P(A) = 0.1, P(B) = 0.25$, and $P(C) = 0.6$, determine
   a) $P(A \cap B \cap C)$.   b) $P(A \cup B \cup C)$.
   c) $P(C - A)$.   d) $P(C - B)$.

1.2.12 Suppose that $P(A) = 0.25$, $P(B - A) = 0.3$, and $P[C - (A \cup B)] = 0.1$. Determine
   a) $P(A \cup B)$.   b) $P(A \cup B \cup C)$.

1.2.13 Prove: If A and B are events, then
  a) the probability that event A or event B occurs, but not both, is $P(A \cup B) - P(A \cap B)$.
  b) the probability that exactly one of the events A or B occurs is $P(A) + P(B) - 2P(A \cap B)$.
  c) $P(A \cap B) \leq P(A \cup B) \leq P(A) + P(B)$.

1.2.14 Prove: If A, B, and C are events, then
  a) $P(A \cup B \cup C) \leq P(A) + P(B) + P(C)$.
  b) $P(A \cup B \cup C) = P(A) + P(B) + P(C) - P(A \cap B) - P(A \cap C) - P(B \cap C) + P(A \cap B \cap C)$.

1.2.15 Prove Theorem 1.4 part (iv).

1.2.16 Prove
  a) Theorem 1.5 part (iii).
  b) Theorem 1.5 part (iv).
  c) Theorem 1.5 part (vi).

1.2.17 Suppose that $\Omega = \mathbb{N}$ and for $i \in \mathbb{N}$, $P(\{i\}) = \frac{k}{3^i}$, where k is an unknown constant. Determine the value of
  a) k.   b) $P(\{\omega \in \mathbb{N} : \omega \leq 5\})$.

1.2.18 Let $\Omega = \mathbb{N}$ and for $i \in \mathbb{N}$, let $A_i = \{\omega \in \mathbb{N} : \omega \leq i\}$. If $P(A_i) = 1 - \left(\frac{1}{2}\right)^i$, determine
  a) $P(A_i - A_j)$ for $i, j \in \mathbb{N}$ and $i > j$.
  b) $P(\{i\})$ for $i \in \mathbb{N}$.

1.2.19 Let $\Omega = (0, \infty)$ and for $t \in (0, \infty)$, let $A_t = \{\omega \in \mathbb{N} : \omega \geq t\}$. If $P(A_t) = e^{-t}$, determine $P(A_s - A_t)$ for $t, s \in (0, \infty)$ and $t > s$.

1.3 Probability with Equally Likely Outcomes

When Ω is a finite set and each of the outcomes in Ω has the same chance of occurring, then the outcomes in Ω are said to be *equally likely outcomes*. When the outcomes of the chance experiment are equally likely, computing the probability of any event A is often simple. In particular, if $N(A)$ is the number of outcomes in event A and $N(\Omega) = N$, then $P(A) = \frac{N(A)}{N}$.

Theorem 1.8 *If Ω is a finite sample space with N possible outcomes, say $\Omega = \{\omega_1, \omega_2, \ldots, \omega_N\}$, and the outcomes in Ω are equally likely to occur, then the probability of an event A is $P(A) = \frac{N(A)}{N}$.*

Proof. Follows directly from Theorem 1.5 part (vi) with each outcome having equal probability $\frac{1}{N(\Omega)}$. ∎

Equally likely outcomes often arise when sampling at random from a finite population. For example, when drawing a card at random from a standard deck of 52 playing cards, each of the 52 cards has an equal chance of being selected. Other common chance experiments in which equally likely outcomes arise include flipping a fair coin n times, rolling n fair dice, selecting n balls at random from an urn, randomly drawing lottery numbers, dealing cards from a well shuffled deck, and choosing names at random out of a hat.

Examples 1.12–1.14 illustrate probability computations where the sample space contains a finite number of equally likely outcomes.

Example 1.12 A card will be drawn at random from a standard deck of 52 playing cards. The sample space associated with this chance experiment is

$$\Omega = \{\underbrace{AH, 2H, \ldots, KH, AD, 2D, \ldots, KD}_{\text{red cards}}, \underbrace{AS, 2S, \ldots, KS, AC, 2C, \ldots, KC}_{\text{black cards}}\}.$$

Thus, $N(\Omega) = 52$. Let A be the event that an ace is drawn and B be the event that a black card is drawn. Then, $P(A) = \frac{4}{52}$, $P(B) = \frac{26}{52}$, $P(A \cap B) = \frac{2}{52}$, and $P(A \cup B) = P(A) + P(B) - P(A \cap B) = \frac{4}{52} + \frac{26}{52} - \frac{2}{52} = \frac{28}{52}$.

Theorem 1.9 *If a chance experiment consists of repeating a task that has N possible outcomes n times, then $N(\Omega) = N^n$.*

Thus, when a chance experiment consists of drawing twice, at random and with replacement, from a collection of N objects, there are N^2 possible outcomes in Ω. For example, if two cards are drawn at random and with replacement from a standard deck of 52 playing cards, then there are $52^2 = 2704$ possible outcomes in Ω.

Example 1.13 If a chance experiment consists of flipping a fair coin 10 times, then there are $2^{10} = 1024$ possible outcomes in Ω. If A is the event that at least one of the flips is a head, then

$$P(A) = 1 - P(A^c) = 1 - P(0 \text{ heads are flipped}) = 1 - \frac{1}{1024} = \frac{1023}{1024}.$$

Theorem 1.10 *When a chance experiment consists of drawing at random and without replacement twice from a collection of N objects, there are $N(N-1)$ outcomes in Ω.*

For example, if two cards are to be drawn at random and without replacement from a standard deck of 52 playing cards, then $N(\Omega) = 52 \times 51 = 2652$. Theorem 1.10 can be generalized to a chance experiment consisting of drawing n times, at random and without replacement, from a collection of N objects.

In this case, $N(\Omega) = N(N-1)\cdots(N-n+1)$. For example, when three objects are drawn at random and without replacement from a collection of 10 objects, the sample space for this chance experiment will contain $10 \times 9 \times 8 = 720$ outcomes.

Example 1.14 Draw two numbers at random and without replacement from the numbers 1, 2, 3, 4, and 5. The sample space associated with this chance experiment contains 20 equally likely outcomes. The event that at least one number greater than or equal to 4 is chosen comprised the outcomes (4,1), (1,4), (4,2), (2,4), (4,3), (3,4),(4,5), (5,4), (5,3), (3,5), (5,2), (2,5), (5,1), and (1,5). Thus,

$$P(\text{at least one number greater than or equal to 4 is chosen}) = \frac{14}{20}.$$

Probabilities concerning chance experiments with equally likely outcomes will be considered in more detail in Section 1.6.

Problems

1.3.1 Draw a card at random from a standard deck of 52 playing cards. Let A be the event that an ace is drawn, B be the event that a black card is drawn, C be the event that a club is drawn, and K be the event that a king is drawn. Determine

a) $P(A \cap B)$.   b) $P(A \cup B)$.
c) $P(A \cap C)$.   d) $P(A \cup K)$.
e) $P(A \cap (B \cup C))$.   f) $P(A \cup B \cup K)$.

1.3.2 A fair six-sided die is to be rolled twice. Determine
a) the sample space associated with this chance experiment.
b) the probability that the total is 5.
c) the probability that the absolute value of the difference between the two rolls is 2.
d) the probability that the sum of the two rolls is even.

1.3.3 A fair die will be rolled and then a fair coin will be flipped twice. Determine the
a) probability that an even number is rolled and two heads are flipped.
b) probability that a 1 is rolled and two tails are flipped.
c) probability that a 1 is rolled and at least one tail is flipped.

1.3.4 Three fair dice will be rolled. Determine the
a) number of outcomes in Ω.
b) probability that exactly two sixes will be rolled.
c) probability that no sixes will be rolled.

1.3.5 A fair coin will be flipped five times. Determine the
    a) number of outcomes in Ω.
    b) probability that exactly two heads will be flipped.
    c) probability that fewer than two heads will be flipped.
    d) probability that at least two heads will be flipped.

1.3.6 Two marbles will be drawn at random and without replacement from an urn containing three white marbles, one blue marble, and one red marble. Determine the
    a) number of outcomes in Ω.
    b) probability that only white marbles are drawn.
    c) probability that no white marbles are drawn.
    d) probability that the blue marble is drawn.

1.3.7 Determine the number of outcomes in Ω when the chance experiment consists of drawing
    a) three objects at random and without replacement from a collection of 12 objects.
    b) three objects at random and with replacement from a collection of 12 objects.

1.3.8 Determine the number of outcomes in Ω when the chance experiment consists of drawing
    a) four cards at random and with replacement from a standard deck of 52 playing cards.
    b) four cards at random and without replacement from a standard deck of 52 playing cards.

1.3.9 Let Ω contain N equally likely outcomes. Show that
    a) $P(A^c) = \frac{N - N(A)}{N}$.
                                        b) $P(A - B) = \frac{N(A) - N(A \cap B)}{N}$.

1.4 Conditional Probability

When probability theory is applied to real-world problems, it is often the case that an event A being studied is dependent on several other factors. For example, when a two-sided coin is flipped, the probability of flipping a head depends on whether or not the coin is a fair coin or a coin biased in favor of heads or tails. Without the knowledge of the type of coin being flipped, the probability of heads must be computed unconditionally. On the other hand, when the type of coin being flipped is known, the probability of flipping a head can be conditioned on this knowledge.

In general, a statistical model is built to explain how a set of explanatory variables $X_1, X_2, \ldots, X_p$ are related to a response variable Y, and conditional probability models form the foundation of statistical modeling. For example, a researcher studying the incidence of lung cancer may be interested in the incidence of lung cancer among individuals who are long-term smokers or have been exposed to asbestos, rather than the unconditional rate of lung cancer among the general population, because the incidence of lung cancer would be expected to be higher for long-term smokers than it would be for the general population.

A probability computed utilizing known information about a chance experiment is called a *conditional probability*.

Definition 1.12   Let $(\Omega, \mathcal{A}, P)$ be a probability space, and let A and B be events in $\mathcal{A}$. The conditional probability of the event A given the event B, denoted by $P(A|B)$, is defined to be

$$P(A|B) = \frac{P(A \cap B)}{P(B)}$$

provided $P(B) > 0$. The conditional probability $P(A|B)$ is said to be undefined when $P(B) = 0$.

$P(A|B)$ is stated as "the probability of the event A, given the event B has or will occur" or simply as "the conditional probability of event A given the event B." The event B is the known condition upon which the probability of A is computed, and the event B serves as the conditional sample space. That is, since B is assumed to have or will occur, only the outcomes in B are relevant to the chance experiment and the conditional probability that event A occurs. Thus, given the event B, the event A occurs only when the chance experiment results in an outcome in $A \cap B$.

Example 1.15   Suppose that a fair coin is flipped twice with $\Omega = \{HH, HT, TH, TT\}$ and each of the outcomes in Ω is equally likely. Let B be the event that at least one head has been flipped, and let A be the event that exactly two heads are flipped. Consider $P(A|B) = \frac{P(A \cap B)}{P(B)}$.

Since there is only one outcome in $A \cap B$, namely HH, and B contains the outcomes HH, HT, and TH, it follows that

$$P(A|B) = \frac{P(A \cap B)}{P(B)} = \frac{\frac{1}{4}}{\frac{3}{4}} = \frac{1}{3}.$$

Note that $P(A)$ unconditionally is $\frac{1}{4}$; however, given the knowledge that at least one head was flipped, there is a larger chance that A will occur.

The probability rules for conditional probability are similar probability rules given in Section 1.2. Several conditional probability rules are given in Theorem 1.11.

Theorem 1.11 *Let* $(\Omega, \mathcal{A}, P)$ *be a probability space and let* $A, B, C \in \mathcal{A}$. *Then,*

i) $0 \le P(A|B) \le 1$.
ii) $P(B|B) = 1$.
iii) $P(A^c|B) = 1 - P(A|B)$.
iv) $P(A \cup B|C) = P(A|C) + P(B|C) - P(A \cap B|C)$.
v) $P(A \cap B) = P(A)P(B|A) = P(B)P(A|B)$.

Proof. Let $(\Omega, \mathcal{A}, P)$ be a probability space and let $A, B, C \in \mathcal{A}$.

i) First, $P(A|B) = \dfrac{P(A \cap B)}{P(B)}$ and $A \cap B \subset B$. Thus, it follows that $0 \le P(A \cap B) \le P(B)$, and hence, $0 \le P(A|B) \le 1$.

ii) $P(B|B) = \dfrac{P(B \cap B)}{P(B)} = \dfrac{P(B)}{P(B)} = 1$.

iii) Note that,

$$P(A^c|B) = \frac{P(A^c \cap B)}{P(B)} = \frac{P(B) - P(A \cap B)}{P(B)}$$
$$= \frac{P(B)}{P(B)} - \frac{P(A \cap B)}{P(B)} = 1 - P(A|B).$$

iv) The proof of (iv) is left as an exercise.
v) The proof of (v) is left as an exercise. ∎

The result given in Theorem 1.11(v) is known as the *Multiplication Law* and can be generalized to conditional probabilities involving than more than two events. For example, with three events A, B, and C, the Multiplication Law is $P(A \cap B \cap C) = P(A)P(B|A)P(C|A \cap B)$.

The Multiplication Law is often used in computing probabilities in chance experiments involving sampling without replacement from a collection of N objects, such as the chance experiment in Example 1.16.

Example 1.16 Suppose that an urn contains 8 red, 10 blue, and 7 green marbles. Three marbles will be drawn from the urn at random and without replacement. The probability of selecting three red marbles consists of selecting a red marble on the first draw, say event $R1$; a red marble on the second draw, say event $R2$; and selecting a red marble on the third draw, say event $R3$. Thus, by the Multiplication Law,

$$P(R1 \cap R2 \cap R3) = P(R1) \times P(R2|R1) \times P(R2|R1 \cap R2)$$
$$= \frac{8}{25} \times \frac{7}{24} \times \frac{6}{23} = \frac{336}{13\,800}.$$

The Multiplication Law also provides an alternative form for the Law of Total Probability, which is given in Theorem 1.12.

Theorem 1.12 (Law of Total Probability) *Let A and B be events. Then,*

$$P(B) = P(B|A)P(A) + P(B|A^c)P(A^c).$$

Proof. Let A and B be events. From Theorem 1.4,

$$P(B) = P(B \cap A) + P(B \cap A^c),$$

and by the Multiplication Law, $P(B \cap A) = P(B|A)P(A)$ and $P(B \cap A^c) = P(B|A^c)P(A^c)$. Thus,

$$P(B) = P(B \cap A) + P(B \cap A^c) = P(B|A)P(A) + P(B|A^c)P(A^c). \qquad \blacksquare$$

Example 1.17 Draw two cards at random and without replacement from a standard deck of 52 playing cards. Because the second draw is dependent on the outcome of the first draw, the Law of Total Probability will be used to determine the probability that an ace is drawn on the second draw, say event $A2$. Now, let $A1$ be the event that an ace was selected on the first draw. Then, $A1$ and $A1^c$ partition Ω, and by conditioning on the outcome of the first draw, it follows that $P(A2|A1) = \frac{3}{51}$ and $P(A2|A1^c) = \frac{4}{51}$. Thus, by Theorem 1.12,

$$P(A2) = P(A1)P(A2|A1) + P(A1^c)P(A2|A1^c)$$
$$= \frac{4}{52} \times \frac{3}{51} + \frac{48}{52} \times \frac{4}{51} = \frac{4}{52}.$$

Theorem 1.13 generalizes the conditional form of the Law of Total Probability to partitions consisting of more than two events.

Theorem 1.13 (Generalized Law of Total Probability) *Let $\{B_k : k \in \mathbb{N}\}$ be a partition of Ω and let $A \in \mathcal{A}$. Then, $P(A) = \sum_{k=1}^{\infty} P(A|B_k)P(B_k)$.*

Proof. The proof follows directly from Theorem 1.6 and the Multiplication Law. $\qquad \blacksquare$

Note that the Generalized Law of Total Probability also applies to a finite partition, say $\{B_1, B_2, \ldots, B_n\}$, $\forall n \in \mathbb{N}$, and in this case, $P(A) = \sum_{k=1}^{n} P(A|B_k)P(B_k)$.

Example 1.18 Suppose that a town T is in a region of high tornado activity. Let A be the event that a tornado hits town T, and let B_i be the event that there are $i \in \mathbb{W}$ tornadoes in town T's region during a tornado season. Suppose that $P(B_i) = \frac{e^{-3}3^i}{i!}$ and $P(A|B_i) = 1 - \frac{1}{2^i}$. The probability that a tornado hits town T is

$$P(A) = \sum_{i=0}^{\infty} P(A|B_i)P(B_i) = \sum_{i=0}^{\infty} \left(1 - \frac{1}{2^i}\right) \times \frac{e^{-3}3^i}{i!}$$

$$= \sum_{i=0}^{\infty} \frac{e^{-3}3^i}{i!} - \sum_{i=0}^{\infty} \frac{e^{-3}3^i}{2^i i!} = 1 - e^{-3} \sum_{i=1}^{\infty} \frac{\left(\frac{3}{2}\right)^i}{i!}$$

$$= 1 - e^{-3}e^{\frac{3}{2}} = 1 - e^{-\frac{3}{2}} = 0.777.$$

The Generalized Law of Total Probability shows that $P(A)$ can be computed when a partition $\{B_i\}$ is available and $P(B_i)$ and $P(A|B_i)$ are known for each of the events in the partition. Theorem 1.14, *Bayes' Theorem*, provides the solution to the inverse problem that concerns the probability of $B_i|A$.

Theorem 1.14 (Bayes' Theorem) *Let $\{B_k : k \in \mathbb{N}\}$ be a partition of Ω and let $A \in \mathcal{A}$, then for $i \in \mathbb{N}$*

$$P(B_i|A) = \frac{P(A|B_i)P(B_i)}{\sum_{k=1}^{\infty} P(A|B_k)P(B_k)}.$$

Proof. Let $\{B_k\}$ be a partition of Ω and let $A \in \mathcal{A}$. Let $i \in \mathbb{N}$ be arbitrary but fixed, then

$$P(B_i|A) = \frac{P(B_i \cap A)}{P(A)} = \overbrace{\frac{P(A|B_i)P(B_i)}{P(A)}}^{\text{by the Multiplication Law}}$$

$$= \underbrace{\frac{P(A|B_i)P(B_i)}{\sum_{k=1}^{\infty} P(A|B_k)P(B_k)}}_{\text{by the Generalized Law of Total Probability}}.$$ ∎

Note that the Generalized Law of Total Probability and Bayes' Theorem also work with finite partitions. For example, using B and B^c to partition Ω, Bayes' Theorem becomes $P(B|A) = \frac{P(A|B)P(B)}{P(A|B)P(A)+P(A|B^c)P(B^c)}$.

Example 1.19 Suppose that a new and rare infectious disease has been diagnosed. It is known that this new disease is contracted with probability $P(D) = 0.0001$, which is referred to as the *prevalence probability*. Suppose that a diagnostic test has been developed to diagnose this disease, and given that a person has the disease, the probability that the test is positive is $P(+|D) = 0.99$; $P(+|D)$ is called the *sensitivity* of the test and measures the ability of the test to correctly diagnose that an individual does have the disease. Also, suppose that given a person does not have the disease, the probability that the test is positive is $P(+|D^c) = 0.02$; the *specificity* of the test is $P(-|D^c) = 1 - P(+|D^c)$ and measures the ability of the test to correctly diagnose that an individual does not have the disease.

The probability that a person has the disease given a positive test result is

$$P(D|+) = \frac{P(+|D)P(D)}{P(+|D)P(D) + P(+|D^c)P(D^c)}$$

$$= \frac{0.0001 \times 0.99}{0.0001 \times 0.99 + 0.9999 \times 0.02} = 0.004\,926.$$

Example 1.20 Suppose that three silicon wafer plants produce blank DVDs with plant A manufacturing 45%, plant B manufacturing 30%, and plant C manufacturing 25%. The probability of a defective DVD given plant A produced the DVD is 0.01, the probability of a defective DVD given plant B produced the DVD is 0.02, and the probability of a defective DVD given plant C produced the DVD is 0.05. If a defective DVD was found, the probability that it was manufactured by plant A is

$$P(A|D) = \frac{P(D|A)P(A)}{P(D|A)P(A) + P(D|B)P(B) + P(D|C)P(C)}$$

$$= \frac{0.01 \times 0.45}{0.01 \times 0.45 + 0.02 \times 0.30 + 0.05 \times 0.25} = 0.196.$$

Similarly, if a defective DVD was found, the probability that it was manufactured by plant B is

$$P(B|D) = \frac{P(D|B)P(B)}{P(D|A)P(A) + P(D|B)P(B) + P(D|C)P(C)}$$

$$= \frac{0.02 \times 0.30}{0.01 \times 0.45 + 0.02 \times 0.30 + 0.05 \times 0.25} = 0.261.$$

Finally, for manufacturer C, the probability is

$$P(C|D) = \frac{P(D|C)P(C)}{P(D|A)P(A) + P(D|B)P(B) + P(D|C)P(C)}$$

$$= \frac{0.05 \times 0.25}{0.01 \times 0.45 + 0.02 \times 0.30 + 0.05 \times 0.25} = 0.543.$$

Thus, given a defective DVD is found, it was most likely produced by manufacturer C.

Example 1.21 Suppose that in Example 1.18, town T was hit by a tornado. The probability that there were three tornadoes in town T's region during a tornado season given that town T was hit by a tornado is

$$P(B_3|A) = \frac{P(B_3)P(A|B_3)}{\sum_{i=0}^{\infty} P(A|B_i)P(B_i)} = \frac{\left(1 - \frac{1}{2^3}\right)\frac{e^{-3}3^3}{3!}}{\sum_{i=0}^{\infty}\left(1 - \frac{1}{2^i}\right) \times \frac{e^{-3}3^i}{i!}}$$

$$= \frac{0.196}{0.777} = 0.252.$$

Problems

1.4.1 Suppose that an urn contains 10 red marbles and 5 black marbles. If two marbles are drawn from the urn at random and without replacement, determine the probability that
a) a red marble is drawn on the second draw given a black marble was drawn on the first draw.
b) a red marble is drawn on the second draw and a black marble was drawn on the first draw.
c) a red marble is drawn on the second draw.
d) two marbles of different colors are selected.
e) two marbles of the same color are selected.

1.4.2 Suppose that an urn contains 12 red marbles, 8 white marbles, and 5 black marbles. If three marbles are drawn from the urn at random and without replacement, determine the probability that
a) a white marble is drawn on the first draw, a red marble is drawn on the second draw, and a white marble is drawn on the third draw.
b) three marbles of the same color are selected.
c) a white marble is selected on the second draw.
d) a white marble is selected on the third draw.

1.4.3 Two marbles will be drawn at random and with replacement from an urn having 8 red marbles and 12 black marbles. Determine
a) the number of possible outcomes in Ω.
b) the probability of drawing marbles of different colors.
c) the probability of drawing marbles of the same color.

1.4.4 Two marbles will be drawn at random and with replacement from an urn having 8 red marbles, 4 black marbles, and 5 white marbles. Determine
a) the number of possible outcomes in Ω.
b) the probability of drawing two white marbles.
c) the probability of drawing one white marble.
d) the probability of drawing marbles of different colors.
e) the probability of drawing marbles of the same color.

1.4.5 Suppose that two cards are drawn at random and without replacement from a standard deck of 52 playing cards, determine the probability that
a) an ace is drawn on the second draw given a king was selected on the first draw.

b) an ace is drawn on the second draw and a king was selected on the first draw.

c) a club is drawn on the second draw.

d) cards of different colors are selected.

1.4.6 Suppose that urn A contains 5 red marbles and 10 black marbles and urn B contains 8 red marbles and 4 black marbles. If a marble is drawn at random from urn A and placed in urn B and then a marble is drawn from urn B at random, determine the probability that

a) a black marble is drawn from urn A and a red marble is drawn from urn B.

b) a red marble is drawn from urn B.

c) marbles of the same color are drawn from urns A and B.

d) marbles of the opposite colors are drawn from urns A and B.

1.4.7 Using the information in Example 1.19 with $P(+|D^c) = p$. Determine the value of p so that $P(D|+) = 0.75$.

1.4.8 A computer company has two assembly locations with 84% of their computers assembled in location A and 16% assembled in location B. Given that a computer is assembled in location A, the probability that a computer works perfectly is 0.98; given that a computer is assembled in location B, the probability that the computer works perfectly is 0.92. Determine the probability that

a) one of the computers supplied by this company works perfectly.

b) one of the computers supplied by this company was assembled in location A, given that it works perfectly.

1.4.9 A university club holds dances at three different bars. The club uses bar A 25% of the time, bar B 60% of the time, and bar C 15% of the time. Suppose that the probability that a fight breaks out given that bar A was used is 0.30, 0.10 given that bar B was used, and 0.50 given that bar C was used. Determine

a) the probability that a fight breaks out a club dance.

b) the probability bar B was used given that a fight broke out.

c) the probability bar C was used given that a fight broke out.

d) the bar most likely to have been used given that a fight broke out.

1.4.10 Suppose that two cards are drawn at random and without replacement from a standard deck of 52 playing cards, determine the probability that

a) a king, queen, or a jack is selected on the second draw.

b) a king was selected on the first draw given a king, queen, or jack was selected on the second draw.

1.4.11 A fishing fleet has three different captains it uses with a particular ship during salmon season. The fleet uses Captain I 65% of the time, Captain II 25% of the time, and Captain III 10% of the time. Suppose that the probability that the ship reaches its quota given that Captain I was used is 0.72, 0.43 given that Captain II was used, and 0.83 given that Captain III was used. Determine
a) the probability that the ship fulfills its quota.
b) the probability that Captain I was used given that the ship fulfills its quota.
c) the probability Captain III was used given that the ship fulfills its quota.
d) the captain most likely to have been used given the ship fulfills its quota.

1.4.12 A gold mining company uses four different remediation techniques, say R1, R2, R3, and R4. The company uses remediation technique R1 20% of the time, remediation technique R2 40% of the time, remediation technique R3 30% of the time, and remediation technique R4 10% of the time. Suppose that the probability of a successful remediation given that technique R1 was used is 0.5, 0.3 given that R2 was used 0.2, given that technique R3 was used, and 0.25 given technique R4 was used. Determine the probability
a) of a successful remediation.
b) remediation technique R2 was used given that the remediation was successful.
c) remediation technique R1 or technique R3 was used given that the remediation was successful.

1.4.13 In a small town, three lawyers (A, B, and C) serve as public defenders. Court records indicate that lawyer A handles 40% of the cases, lawyer B 30% of the cases, and lawyer C 30% of the cases. Furthermore, the probability that a defendant is acquitted given that lawyer A handled the case is 0.25, 0.20 given that lawyer B handled the case, and 0.30 given that lawyer C handled the case. Determine the probability that
a) a defendant using a public defender is acquitted.
b) lawyer B handled the case given that a defendant using a public defender is acquitted.
c) lawyer C handled the case given that a defendant using a public defender is acquitted.
d) lawyer A handled the case given that a defendant using a public defender is acquitted.

1.4.14 In Example 1.21, given a tornado hit town T, determine the most likely number of tornadoes there were in town T's region. *Hint:* Determine the largest value of $P(B_i|A)$.

1.4.15 *Prove:* If $P(A|B) > P(A)$, then $P(B|A) > P(B)$.

1.4.16 Prove
a) Theorem 1.11 part (iv).   b) Theorem 1.11 part (v).

1.5 Independence

In some cases, the conditional probability of an event A given that the event B is the same as the unconditional probability of the event A. In this case, knowledge of the event B occurring does not provide any information about the probability that event A occurs, and the events A and B are said to be *independent events*.

Definition 1.13 Let $(\Omega, \mathcal{A}, P)$ be a probability space and let $A, B \in \mathcal{A}$. The events A and B are said to be independent events if and only if any one of the following three conditions is satisfied.
i) $P(A \cap B) = P(A)P(B)$.
ii) $P(A|B) = P(A)$, provided $P(B) > 0$.
iii) $P(B|A) = P(B)$, provided $P(A) > 0$.

Example 1.22 Suppose that a fair coin is flipped twice. Let A be the event that a head is flipped on the first flip, and let B be the event that a tail is flipped on the second flip. There are four equally likely outcomes in Ω and the event $A = \{HT, HH\}$ and the event $B = \{TT, HT\}$. Thus, $P(A) = \frac{1}{2}$, $P(B) = \frac{1}{2}$, and since $P(A \cap B) = P(\{HT\}) = \frac{1}{4}$, it follows that A and B are independent events.

Note that in a chance experiment where n objects are drawn with replacement, the outcome of each draw is independent of the outcome of any other draw; however, this is not the case when the draws are made without replacement.

Theorem 1.15 shows that when A and B are independent events, then so are the pairs of events A and B^c, A^c and B, and A^c and B^c.

Theorem 1.15 *If A and B are independent events, then*
i) A^c and B are independent.
ii) A and B^c are independent.
iii) A^c and B^c are independent.

Proof. Let A and B be independent events.

i) Consider $P(A^c \cap B)$.

$$P(A^c \cap B) = P(B) - P(A \cap B) = P(B) - P(A)P(B)$$
$$= P(B) \times [1 - P(A)] = P(B)P(A^c).$$

Thus, A^c and B are independent whenever A and B are independent.

ii) The proof of part (ii) is left as an exercise.

iii) The proof of part (iii) is left as an exercise. ∎

The definition of independence can be generalized to a family of events, say $\{A_i : i \in \mathbb{N}\}$. For example, if every pair of events in $\mathcal{A}$ is a pair of independent events, then the events in $\{A_i : i \in \mathbb{N}\}$ are called a *pairwise independent* events. Definition 1.14 gives the definition of a family of *mutually independent* events.

Definition 1.14 A family of events $\{A_i : i \in \mathbb{N}\}$ is said to be a family of mutually independent events if and only if

$$P\left(\bigcap_{i \in J} A_i\right) = \prod_{i \in J} P(A_i)$$

for all subsets J of $\mathbb{N}$.

Note that, in a family of mutually independent events, the probability of the intersection of any subcollection of the family of events is the product on the events in the subcollection. For example, three events A, B, and C are mutually independent only when $P(A \cap B) = P(A)P(B)$, $P(A \cap C) = P(A)P(C)$, $P(B \cap C) = P(B)P(C)$, and $P(A \cap B \cap C) = P(A)P(B)P(C)$. Furthermore, it is possible for a family to be a family of pairwise independent events but not a family of mutually independent events as illustrated in Example 1.23.

Example 1.23 Flip a fair coin twice so that the four outcomes in Ω are equally likely. Let $A = \{HH, HT\}, B = \{HT, TH\}$, and $C = \{HH, TH\}$. Then, $P(A) = P(B) = P(C) = \frac{1}{2}$ and

$$P(A \cap B) = P(\{HT\}) = \frac{1}{4} = P(A)P(B)$$

$$P(A \cap C) = P(\{HH\}) = \frac{1}{4} = P(A)P(C)$$

$$P(B \cap C) = P(\{TH\}) = \frac{1}{4} = P(B)P(C)$$

but

$$P(A \cap B \cap C) = P(\emptyset) = 0 \neq \frac{1}{8} = P(A)P(B)P(C).$$

Thus, A, B, and C are pairwise independent events but not mutually independent events.

In chance experiments consisting of n trials where the outcome of each trial is independent of the outcomes of the other trials, the outcomes of the trials are mutually independent events. Moreover, the probability of a particular outcome of the chance experiment is simply the product of the probabilities of the corresponding trial outcomes.

Example 1.24 A basketball practice will end when a particular player makes a shot from half court. If the outcome of each shot, hit (H) or miss (M), is independent, and the probability that a half court shot will be made is 0.05, then the probability that it will take five shots to end the practice is

$$P(\text{M and M and M and M and H}) = P(\text{M})P(\text{M})P(\text{M})P(\text{M})P(\text{H})$$
$$= 0.95^4 \times 0.05 = 0.041.$$

In general, the probability that it takes k shots to end the practice is $0.95^{k-1} \times 0.05$.

Example 1.25 Suppose that two fair dice will be rolled repeatedly and independently until a total of 6 or a total of 7 appears. To determine the probability that a total of 6 is rolled before a total of 7 is rolled, let A_i be the event that the first total of 6 is rolled before a total of 7 is rolled occurs on the ith roll. Then, the collection of events $\{A_i\}$ is a partition of the event that a 6 is rolled before a 7 and

$$P(6 \text{ before } 7) = P\left(\overset{\infty}{\underset{i=1}{\bigcup}} A_i\right) \overset{\text{disjoint}}{=} \sum_{i=1}^{\infty} P(A_i).$$

Now, the event A_i occurs when the first $i - 1$ rolls result in totals that are not 6 or 7, and the ith roll is a total of 6. Thus, since the rolls of the dice are independent,

$$P(A_i) = P\left(\overset{i-1}{\underset{i=1}{\bigcap}} (6 \cup 7)^c \cap 6\right) \overset{\text{ind.}}{=} \left(\frac{25}{36}\right)^{i-1} \times \frac{5}{36}.$$

Hence,

$$P(6 \text{ before } 7) = \sum_{i=1}^{\infty} P(A_i) = \sum_{i=1}^{\infty} \left(\frac{25}{36}\right)^{i-1} \times \frac{5}{36}$$

$$= \frac{5}{36} \underbrace{\sum_{i=1}^{\infty} \left(\frac{25}{36}\right)^{i-1}}_{\text{geometric series}} = \frac{5}{36} \times \frac{1}{1 - \frac{11}{36}} = \frac{5}{11}.$$

Problems

1.5.1 Suppose that $P(A) = 0.3$, $P(B) = 0.4$, and A and B are independent events. Determine

    a) $P(A \cup B)$.
    b) $P(A - B)$.
    c) $P(A^c \cup B^c)$.
    d) $P((A - B) \cup (B - A))$.

1.5.2 Two cards will be drawn at random and with replacement from a standard deck of playing cards. Determine the probability of drawing

    a) two aces.
    b) two cards of different suits.
    c) two cards of the same suit.

1.5.3 Three geological exploration companies are searching independently for new shale deposits in Eastern Montana. Suppose that the probability that company A finds new shale in Eastern Montana is 0.4, the probability that company B finds new shale in Eastern Montana is 0.6, and the probability that company C finds new shale in Eastern Montana is 0.5. Determine the probability that

    a) exactly two companies find new shale deposits in Eastern Montana.
    b) at least two of the companies find new shale deposits in Eastern Montana.
    c) at least one of the companies finds a new shale deposit in Eastern Montana.

1.5.4 The goal of a military operation is to destroy a strategic target by firing missiles at the target. The target will be destroyed when two missiles have hit the target. Suppose that each missile is fired at the target independently, and the probability that a missile hits the target is 0.6. If missiles will be fired until the second missile hits the target, determine

    a) the probability that only two missiles will have to be fired in order to destroy the target.
    b) the probability that four missiles will have to be fired in order to destroy the target.
    c) the probability that at most four missiles will have to be fired in order to destroy the target.

1.5.5 Suppose that the probability of winning any money at all on a single play of the Lucky Dollar poker machine is 0.14. If each play of the Lucky Dollar poker machine is independent, determine the probability that

    a) a player wins on five consecutive plays.
    b) a player wins at least once in five consecutive plays.
    c) a player's first win occurs on the kth play.
    d) a player's first win occurs in 10 or fewer plays.

1.5.6 A basketball practice will end when a player hits a shot from half court. Suppose that each shot is independent of the others, and the probability that a half court shot will be made is 0.05. Determine the probability that it takes more than 20 shots to end practice.

1.5.7 Let A, B, C be mutually independent events. Show that
a) A and $B \cap C$ are independent events.
b) A and $B \cup C$ are independent events.
c) A and $B - C$ are independent events.

1.5.8 Let $\Omega = \{ABB, BAB, BBA, AAA\}$ and suppose that the outcomes in Ω are equally likely. Let A_i be the event that A occurs in the ith position for $i = 1, 2, 3$. Show that A_1, A_2, and A_3 are pairwise independent but not mutually independent.

1.5.9 Suppose that $P(A) = 0.25$, $P(B) = 0.3$, and $P(C) = 0.16$. Determine $P(A \cup B \cup C)$ when A, B, and C are
a) disjoint events.
b) mutually independent events.

1.5.10 Suppose that a multiple-choice test has n questions and each question has four possible choices. A student will guess at random on each question making each guess independent of the others, and the probability of guessing the correct answer on any question is 0.25. Determine the number of questions (n) that would have to be guessed so that the probability of guessing the correct answer on at least one question is at least 0.98.

1.5.11 An academic integrity committee consists of three members, two faculty representatives and one student representative. Suppose that each of the committee members votes independently on each student appeal case, and at least two committee members must agree with the student appeal for the committee to rule in favor of the student. Furthermore, based on the past records, the faculty rule in favor of a student appeal with probability 0.10 and the student rules in favor with probability 0.3. Determine the probability that the committee rules in favor of a student appeal.

1.5.12 Two fair dice will be rolled repeatedly and independently until a total of 6 or a total of 5 appears. Determine the probability that a total of 6 is rolled before a total of 5 is rolled.

1.5.13 Players A and B will alternate rolling a fair die independently until a 6 appears. The first player to roll a 6 wins the game. Determine the probability that player A wins the game
a) when player A rolls first.
b) when player B rolls first.
c) when the players randomly chose who goes first.

1.5.14 If $\{A_i : i \in \mathbb{N}\}$ are mutually independent events, show that

$$P\left(\bigcup_{i=1}^{n} A_i\right) = 1 - \prod_{i=1}^{n}[1 - P(A_i)].$$

1.5.15 Let A and B be independent events. Show that
a) A and B^c are independent events.
b) A^c and B^c are independent events.

1.5.16 If $P(A) = 0$, show that A is independent of every other event B of Ω.

1.5.17 Let A and B be events and let A be a subset of B. Show that A and B are independent events when $P(A) = 0$ or $P(B) = 0$.

1.5.18 Let A and B be events, and let A be a nonempty subset of B with $P(A) > 0$. If $P(B) < 1$, show that A and B cannot be independent events.

1.6 Counting Methods

Probability computations often involve the enumeration of the possible outcomes of a chance experiment, enumeration of the outcomes in an event, or the enumeration of the ways an event can occur. The *Fundamental Principal of Counting*, given in Theorem 1.16, is one of the most important tools for enumeration and can be used on chance experiments with or without equally likely outcomes.

Theorem 1.16 (Fundamental Principle of Counting) *Let a chance experiment $\mathcal{E}$ be carried out by carrying out the subexperiments $\mathcal{T}_1, \mathcal{T}_2, \mathcal{T}_3, \ldots, \mathcal{T}_k$. If there are n_i possible outcomes to subexperiment $\mathcal{T}_i$, then the number of possible outcomes for the chance experiment $\mathcal{E}$ is $N(\Omega) = \prod_{i=1}^{k} n_i$.*

Example 1.26 Suppose that a pizza parlor offers pizzas based on three different types of crusts, two different sauces, and 15 different toppings. Making a three distinct topping pizza is based on five subexperiments, namely,

- $\mathcal{T}_1$ – choose the type of crust with $n_1 = 3$,
- $\mathcal{T}_2$ – choose the type of sauce with $n_2 = 2$,
- $\mathcal{T}_3$ – choose the first topping with $n_3 = 15$,
- $\mathcal{T}_4$ – choose the second topping with $n_4 = 14$ since one topping has already been chosen,
- and $\mathcal{T}_5$ – choose the third topping with $n_5 = 13$ since two toppings have already been chosen.

Thus, the number of possible pizzas having three different toppings is $N(\Omega) = 3 \times 2 \times 15 \times 14 \times 13 = 16\,380$.

Theorem 1.17 *If a chance experiment consists of drawing n times*

i) *with replacement from N objects, then $N(\Omega) = N^n$.*
ii) *without replacement from N objects, then*
   $N(\Omega) = N(N-1) \times \cdots \times (N-n+1).$

Proof. This theorem follows directly from the Fundamental Principle of Counting. ■

Example 1.27 Suppose that a chance experiment involves rolling a fair six-sided die five times. Since there are six possible outcomes to each roll, $N(\Omega) = 6^5 = 7776$.

Example 1.28 Suppose that a chance experiment involves selecting three people at random and without replacement from a group of seven people to serve as club president, vice president, and secretary. Then, $N(\Omega) = 7 \times 6 \times 5 = 210$.

The Fundamental Principle of Counting can be very useful when computing probabilities associated with a chance experiment having equally likely outcomes. To use the Fundamental Principle of Counting to compute $P(A)$ when the chance experiment consists of several tasks and equally likely outcomes:

1) Define each of the k tasks involved in the chance experiment.
2) Determine the number of possible outcomes to task T_k, say n_k.
3) $N(\Omega) = \prod_{i=1}^{n} n_i$.
4) Determine the number of favorable outcomes to task T_k, say f_k.
5) $N(A) = \prod_{i=1}^{n} f_i$.
6) $P(A) = \frac{N(A)}{N(\Omega)}$.

Example 1.29 Suppose that a standard state license plate is made of three letters selected from $\{A, B, C, \dots, Z\}$ followed by three digits selected from $\{0, 1, 2, \dots, 9\}$. Then, the number of possible distinct license plates that can be made is $N(\Omega) = 26 \times 26 \times 26 \times 10 \times 10 \times 10 = 17\,576\,000$.

Now, if a license plate is selected at random, the probability that a license plate will have three different letters is

$$P(\text{three different letters}) = \frac{N(\text{three different letters})}{N(\Omega)}$$
$$= \frac{26 \times 25 \times 24 \times 10 \times 10 \times 10}{17\,576\,000} = \frac{15\,600\,000}{17\,576\,000}$$
$$= 0.888.$$

The probability that all three letters on a license plate match while none of the digits on a license plate match is

$$\frac{26 \times 1 \times 1 \times 10 \times 9 \times 8}{17\,576\,000} = \frac{18\,720}{17\,576\,000} = 0.0011.$$

The two most important considerations when computing probabilities in a chance experiment that involves sampling from N distinct objects are (i) whether or not the sampling is with replacement and (ii) whether or not the order in which the objects are sampled is important. For example, when dealing five cards without replacement to form a hand of cards, the order in which the five cards are dealt does not matter; that is, a hand of cards consisting of $AH, KS, 5D, 4C$, and $2C$ is the same hand, no matter in which order the cards were dealt. On the other hand, the order in which the objects are selected may be important such as in the case where nine digits are chosen to form a social security number.

Definition 1.15 A permutation of n distinct objects is an ordered sequence of the n objects. A partial permutation of size r is an ordered sequence of r of the n objects.

For example, abc and bca are two different permutations of the letters a, b, and c, and ab and cb are different partial permutations of size 2 made from the letters a, b, and c.

Theorem 1.18 *The number of partial permutations of size r made from n distinct objects when sampling without replacement is $\dfrac{n!}{(n-r)!}$.*

Proof. This theorem follows directly from the Fundamental Principle of Counting. ∎

Example 1.30 There are $\dfrac{12!}{(12-4)!} = 11\,880$ partial permutations of size 4 when sampling with replacement from 12 distinct objects, and there are $12! = 479\,001\,600$ permutations of all 12 objects.

Example 1.31 Five couples are arranged in a row of 10 chairs. Let A be the event that five couples are seated in a fashion so that a husband is seated next to his wife. Now, the number of ways to seat 10 people in the 10 chairs is $N(\Omega) = 10! = 3\,628\,800$, and the event A comprised the following six tasks:

- $\mathcal{T}_1$ – Choose the order in which the 5 couples will be seated. $n_1 = 5! = 120$.
- $\mathcal{T}_2$ – Seat the first couple. $n_2 = 2$.
- $\mathcal{T}_3$ – Seat the second couple. $n_2 = 2$.
- $\mathcal{T}_4$ – Seat the third couple. $n_2 = 2$.
- $\mathcal{T}_5$ – Seat the fourth couple. $n_2 = 2$.
- $\mathcal{T}_6$ – Seat the fifth couple. $n_2 = 2$.

Thus, $N(A) = 5! \times 2^5 = 3840$. Hence, $P(A) = \dfrac{3840}{3\,628\,800} = 0.0011$.

When sampling without replacement, and the order the objects are selected is not important, the objects selected are called a *combination*.

Definition 1.16 A combination of size r is an unordered subset of r distinct objects selected from a set of n distinct objects.

For example, *abc* and *cab* are the same combination of the letters a, b, and c. The four combinations of size 3 for the letters a, b, c, and d are abc, abd, acd, bcd.

Theorem 1.19 *The number of combinations of size r that can be formed by selecting r objects without replacement from n distinct objects is*
$$\binom{n}{r} = \frac{n!}{r!(n-r)!}.$$

Proof. The proof of Theorem 1.19 is left as an exercise. ∎

Note that $\binom{n}{k}$ is the kth term in the nth row of Pascal's Triangle, and $\binom{n}{k}$ is also the kth coefficient in the *binomial expansion* of $(x+y)^n = \sum_{k=0}^{n} \binom{n}{k} x^k y^{n-k}$.

Example 1.32 The number of possible five-card poker hands that can be dealt from a standard deck of 52 playing cards is $\binom{52}{5} = 2\,598\,960$. Let A be the event a five-card hand contains four aces. Then, A is comprised of two tasks $\mathcal{T}_1$ – deal all four aces from the four aces in the deck to the hand and $\mathcal{T}_2$ – deal the fifth card from the remaining 48 cards in the deck to the hand. Thus,

$$P(A) = \frac{\binom{4}{4}\binom{48}{1}}{\binom{52}{5}} = \frac{1 \cdot 48}{2\,598\,960} = 0.000\,018.$$

The number of combinations of size r can also be used to count the number of ways n objects can be assigned to two cells with one cell receiving r objects and the other cell $n - r$ objects.

Example 1.33 The number of possible equally likely outcomes when a fair coin is flipped 20 times is $2^{20} = 1\,048\,576$. The number of outcomes resulting in 12 heads and eight tails is $\binom{20}{12} = 125\,970$, and thus, the probability of flipping 12 heads when a fair coin is flipped 20 times is $\dfrac{125\,970}{1\,048\,576} = 0.12$.

Example 1.34 A state lottery generally consists of selecting six numbers from 1 to n without replacement, and the order of selection is unimportant. A lottery player will select their own six numbers from 1 to n, and then the state will pick the six winning lottery numbers without replacement. Because order is unimportant, the number of possible lottery combinations is $\binom{n}{6}$. For example, when $n = 51$, there are $\binom{51}{6} = 18\,009\,460$ possible lottery combinations.

A player will usually win prize money for matching three, four, five, or six of the state's numbers. Now, for a player to match x of the winning numbers, x of the player's numbers must be match the six winning numbers and $6 - x$ of the player's numbers must match the 45 numbers that were not chosen. Thus, the probability that a player will match x of the state's six numbers is

$$P(x \text{ matched numbers}) = \frac{\binom{6}{x}\binom{n-6}{6-x}}{\binom{n}{6}}.$$

With $n = 51$ and $x = 3$, the probability that a player matches 3 of the winning numbers is $\dfrac{\binom{6}{3}\binom{45}{3}}{\binom{45}{6}} = 0.0158$.

Theorem 1.20 generalizes the number of ways n distinct objects can be assigned to more than two cells.

Theorem 1.20 *The number of ways n distinct objects can be distributed into k distinct cells is*

$$\frac{n!}{n_1! n_2! \cdots n_k!} = \binom{n}{n_1, n_2, n_3, \ldots, n_k},$$

where $n_i \geq 0$ for $i = 1, 2, 3, \ldots, k$ and $\sum_{i=1}^{k} n_i = n$.

Example 1.35 The number of ways that a standard deck of 52 playing cards can be dealt so that each player receives 13 cards is $\binom{52}{13,13,13,13} = \frac{52!}{13!^4}$. Let A be the event that one of the four players is dealt all four aces. Then, A is comprised the tasks $\mathcal{T}_1$ – choose one of the players to receive the four aces, $\mathcal{T}_2$ – distribute the four aces to the player, and $\mathcal{T}_3$ – distribute the remaining 48 cards so that each player has been dealt 13 cards. Thus,

$$P(A) = \frac{N(A)}{N(\Omega)} = \frac{\binom{4}{1}\binom{4}{4}\binom{48}{9, 13, 13, 13}}{\binom{52}{13, 13, 13, 13}} = 0.0106.$$

Counting methods can also be useful for enumerating the possible outcomes in chance experiments where the outcomes are not equally likely.

Example 1.36 Suppose that a multiple-choice test consists of 10 questions, with each question having four options of which only one option is the correct answer. If a student guesses at random on each question, then the probability that a student guesses correctly on any particular question is $\frac{1}{4}$. If a student guesses independently on each of the questions, the probability of answering seven questions correctly is

$$\binom{10}{7}\left(\frac{1}{4}\right)^7\left(\frac{3}{4}\right)^3 = 0.0031$$

since (1) the probability of any sequence of seven correct answers and three incorrect answers is $\left(\frac{1}{4}\right)^7\left(\frac{3}{4}\right)^3$ and (2) there are $\binom{10}{7}$ ways to assign the 10 questions to seven correct answers and three incorrect answers.

Problems

1.6.1 Phone numbers in a particular state begin with one of three area codes, followed by one of seven prefixes and finally four digits (i.e. xxx-xxx-xxxx). Determine how many possible distinct phone numbers are possible

a) without further restriction.

b) if a phone number must begin with the prefix 491.

c) if the last four digits cannot be 0000.

d) if the last three digits must be 000.

1.6.2 A particular state's license plates have the form LCCDDDD, where L is either an A for auto or a T for truck, CC is one of 56 county identifiers, and D is a digit. Determine how many distinct license plates there are
a) for automobiles.
b) for a particular county.
c) with no repeated digits.
d) where the product of the digits is even.

1.6.3 A pizza parlor makes pizzas with three different types of crusts, two types of sauces, 10 different toppings, and all pizzas come with cheese. Determine the number of distinct types of
a) one-topping pizzas that can be made.
b) two-topping pizzas that can be made.
c) pizzas that can be made.

1.6.4 Suppose that a club consists of six men and three women. From the nine club members, three members will be selected at random and without replacement to serve as club officers. If, the first member selected will serve as President, the second as Vice President, and the third as Treasurer, determine
a) the number of possible outcomes for selecting this club's officers.
b) the probability that all of the club's officers are female.
c) the probability that the club's president is female.

1.6.5 Three numbers are chosen at random and without replacement from a pool of 15 numbers consisting of 0, six positive numbers, and eight negative numbers. Determine the probability that
a) no negative numbers are chosen.
b) at least one negative number is chosen.
c) the product is positive.
d) the product is nonnegative.
e) the product is 0.
f) the product is negative.

1.6.6 A bridge hand consists of dealing 52 playing cards to four players such that each player receives 13 cards. Determine the probability that
a) each player has one ace.
b) one player has all four aces and all four kings.

1.6.7 From a standard deck of 52 playing cards, five cards are dealt at random and without replacement to four players so that each player receives five cards and 32 are left undealt.
a) How many possible ways are there to deal the four hands?
b) What is the probability that each player is dealt one ace?
c) What is the probability that one player receives all four aces?

1.6.8 A company has 25 trucks of which 20 are in working order and 5 are in the shop for repair. If four trucks are selected at random and without replacement, determine the probability that
a) all four of the selected trucks are in working order.
b) all four of the selected trucks are in the shop for repair.
c) at least one of the selected trucks is in working order.
d) at least two of the selected trucks are in working order.

1.6.9 A wardrobe consists of five pairs of pants of which three pairs are blue, 12 shirts of which 4 are white, seven pairs of socks, and black, brown, and tan shoes. Determine the number of distinct ensembles (i.e. pants, shirt, socks, and shoes) possible
a) with no restrictions.
b) with blue pants.
c) with blue pants, a white shirt, and black or brown shoes.

1.6.10 Suppose that four cards are drawn at random and without replacement from a standard deck of 52 playing cards. Determine the probability that
a) two aces are drawn.
b) two hearts are drawn.
c) at least one ace is drawn.
d) one ace and one king are drawn.

1.6.11 Suppose that four digits are to be selected at random with replacement. Determine the probability that
a) the product is even.
b) the product is odd.
c) the sum is even.
d) the product is positive.

1.6.12 Suppose that five male/female couples are to be seated in a row of chairs. Determine the probability that
a) the seating arrangement alternates MFMFMFMFMF.
b) the seating arrangement is MMMMMFFFFF.
c) no couple is seated together.

1.6.13 For the state lottery given in Example 1.34 with $n = 51$, compute the probability of matching
a) 0 numbers.   b) 1 number.
c) at least 3 numbers.   d) at most 1 number.

1.6.14 For the state lottery given in Example 1.34 with $n = 56$, compute the probability of matching

a) 3 numbers.
c) 5 numbers.

b) 4 numbers.
d) at least 3 numbers.

1.6.15 Prove Theorem 1.19.

1.6.16 Using the binomial expansion of $(1 + 1)^n$, show that $\sum_{k=0}^{n} \binom{n}{k} = 2^n$.

1.6.17 Show that

a) $\binom{n}{k} = \binom{n}{n-k}$.

b) $\binom{n}{k} = \binom{n-1}{k} + \binom{n-1}{k-1}$.

1.6.18 Show that $\binom{n}{n_1, n_2 \ldots, n_k} = \binom{n}{n_1}\binom{n-n_1}{n_2} \cdots \binom{n-n_1-n_2-\cdots-n_{k-1}}{n_k}$ where $\sum_{i=1}^{k} n_k = n$.

1.6.19 Evaluate the following sums:

a) $\sum_{k=0}^{n} \binom{n}{k}(-1)^k$.

b) $\sum_{k=0}^{n} \binom{n}{k} 2^k$.

c) $\sum_{k=0}^{n} \binom{n}{k}(\alpha - 1)^k$.

1.6.20 Show that $\sum_{j=0}^{k} \binom{n}{j}\binom{m}{k-j} = \binom{n+m}{k}$ for $n, m, k \in \mathbb{N}$. *Hint:* Consider the binomial expansion of $(x + 1)^{n+m}$.

1.7 Case Study – The Birthday Problem

A famous probability problem posed sometime in the twentieth century is called the *Birthday Problem*. The Birthday Problem deals with the probability of people sharing common birthdays, which turns out to be a fairly common coincidence.

Statement of the Birthday Problem: What is the probability that two or more people in a room of n randomly assembled people share a common birthday?

The solution to the Birthday Problem depends on the assumptions being made on the assumed probability model, and different models will yield different solutions. The probability model being used to solve the Birthday Problem here is one of the simplest models possible; more complicated models may yield a more accurate probability for the solution to the Birthday Problem, but they are seldom used.

The Assumptions of the Probability Model:

- Because the people in the room are randomly assembled, assume that the birthdays of the individuals in the room are independent events.
- Because leap-year birthdays (i.e. February 29) are rare, assume that the birthdays all occur in nonleap years, leaving 365 possible birthdays.
- Assume that birthdays are uniformly distributed over the 365 possible birthdays so that each of the 365 birthdays is equally likely.

With these assumptions, the birthday problem is reduced to a repeated sampling problem where the chance experiment consists of choosing n birthdays with replacement from the 365 possible birthdays. Thus, under this model, $N(\Omega) = 365^n$.

Let A be the event that two or more people share the same birthday. While simple in statement, the event A is actually a fairly complex event. For example, a small sampling of the possible outcomes in A includes that only two people share the same birthday, three people share the same birthday, two people share the same birthday and another two share a different birthday, all but one person in the room share the same birthday, and all n people share the same birthday. To compute the probability of A directly, one would need to determine all of the possible ways at least two people in a room of n people share the same birthday and then sum the probabilities of these outcomes.

On the other hand, complementary event A^c is the event that no one in the room shares a common birthday. Because $P(A^c)$ is easier to compute, the probability of A will be found using the complement rule, $P(A) = 1 - P(A^c)$.

Now, the number of ways none of the individuals in the room share a common birthday can be modeled as sampling n birthdays without replacement from the 365 possible birthdays. In other words, the first person can have any of the 365 possible birthdays, the second person can have any of the 364 remaining birthdays, and so on, until the last person can have any of the $365 - n + 1$ remaining birthdays. Thus,

$$N(A^c) = 365 \times 364 \times 363 \times \cdots \times (365 - n + 1)$$

and

$$P(A^c) = \frac{365 \times 364 \times 363 \times \cdots \times (365 - n + 1)}{365^n}.$$

Finally, the probability that two or more people in a room of n randomly assembled people share a common birthday is

$$P(A) = 1 - P(A^c) = 1 - \frac{365 \times 364 \times 363 \times \cdots \times (365 - n + 1)}{365^n}$$

$$= 1 - \frac{\frac{365!}{(365-n)!}}{365^n}.$$

For example, for $n = 10$, the probability that two or more people in a room of 10 randomly assigned people share a common birthday is

$$P(A) = 1 - \frac{365 \times 364 \times 363 \times \cdots \times 356}{365^{10}} = 0.1169,$$

and for $n = 23$, $P(A) = 1 - \frac{365 \times 364 \times 363 \times \cdots \times 344}{365^{22}} = 0.5073$. Thus, there is better than a 50% chance that at least two people in a room of 23 randomly assembled people will share a common birthday.

The probability of two or more people in a room of n randomly assembled people sharing the same birthday plotted as a function of n is shown in Figure 1.1. Note that the probability at least two people in a room of n randomly assembled people will share a common birthday is greater than 0.90 when $n \geq 41$.

Finally, other probability models could have been used for modeling the probability that two or more people in a room of n randomly assembled people share a common birthday. For example, a probability model that includes February 29 as possible birthday could be considered; however, in this model, it would not

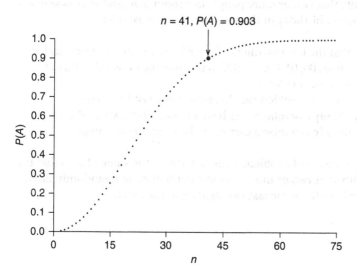

Figure 1.1 A plot of $P(A)$ for $n = 2, 3, \ldots, 75$.

be reasonable to assume that the 366 birthdays are equally likely. An alternative probability model could also be built using unequal empirical probabilities for each of the possible birthdays.

Problems

1.7.1 Determine the probability that in a room of 40 people, at least two people share the same birthday.

1.7.2 Determine the probability that in a room of 50 people, at least two people share the same birthday.

1.7.3 Using a probability model analogous to the one used in solving the Birthday Problem, determine the probability that two or more people in a room of n randomly assembled people have a birthday in the same week
a) assuming that there are 52 equally likely weeks in a year.
b) assuming that there are 52 equally likely weeks in a year and $n = 10$.
c) assuming that there are 52 equally likely weeks in a year and $n = 15$.
d) assuming that there are 52 equally likely weeks in a year and $n = 25$.

1.7.4 Using the setting of Problem 1.7.3, determine the value of n so that the probability that two or more people in a room of n randomly assembled people have a birthday in the same week is at least 0.50.

1.7.5 Suppose that the last two digits of a car license plate are numerical with possible values $00, 01, 02, \ldots, 99$. Determine the probability that at least two cars in a parking lot of
a) n randomly assembled cars have the same last two digits.
b) 10 randomly assembled cars have the same last two digits.
c) 20 randomly assembled cars have the same last two digits.

1.7.6 Using the setting of Problem 1.7.5, determine the value of n so that the probability that two or more cars in a parking lot of n randomly assembled cars have the same last two digits is at least 0.75.

2

Random Variables and Random Vectors

2.1 Random Variables

In a statistical analysis, a sample of units is selected from a well-defined population, preferably at random, and then measurements are recorded on each of
the units in the sample. In this setting, the selection of a particular unit from
the population can be modeled as a chance experiment where the sample space
consists of all of the units in the population. The characteristics being measured
on each unit are the variables in the study, and when a quantitative variable is
collected in a random sample, it is called a *random variable*. The study of random variables forms the foundation on which modern statistics has been built.

Definition 2.1 Let $(\Omega, \mathcal{A}, P)$ be a probability space. A random variable X is
a function defined on Ω such that $\{\omega \in \Omega : X(\omega) \leq \alpha\} \in \mathcal{A}$, $\forall \alpha \in \mathbb{R}$. The set
of possible values of X is $\mathcal{S} = \{x \in \mathbb{R} : \exists\ \omega \in \Omega$ such that $X(\omega) = x\}$, which is
called the support of the random variable X.

Example 2.1 Consider a study of the grade point average (GPA) of the students currently enrolled at a particular university. The sample space is the collection of students currently enrolled at the university. The primary variable of
interest is GPA, but other variables of interest that might be of importance in
this study include the age, gender, number of credits earned, and the major of a
student. Because GPA, age, and the number of credits earned are quantitative
variables, they are random variables; the qualitative variables gender and major
are called *random quantities*.

Note that a random variable is a function that maps a sample space Ω to a subset
of $\mathbb{R}$. The support $\mathcal{S}$ of a random variable X is the set of possible values the
random variable X can take on, and for any subset B of $\mathbb{R}$, the probability that
the random variable X is in B is $P(X \in B) = P(\{\omega \in \Omega : X(\omega) \in B\})$.

Mathematical Statistics: An Introduction to Likelihood Based Inference, First Edition. Richard J. Rossi.
© 2018 John Wiley & Sons, Inc. Published 2018 by John Wiley & Sons, Inc.

Example 2.2 Roll a fair die twice. Then, the sample space is $\Omega = \{(x, y) : x = 1, \ldots, 6; y = 1, \ldots, 6\}$, and each of the 36 outcomes is equally likely. Let D be the difference of the outcomes of rolls one and two. Then, the support of the random variable D is $\mathcal{S} = \{-5, -4, -3, -2, 1, -0, 1, 2, 3, 4, 5\}$, and the probability that $D = 1$ is $P(D = 1) = P(\{(6, 5), (5, 4), (4, 3), (3, 2), (2, 1)\}) = \frac{5}{36}$.

2.1.1 Properties of Random Variables

The two types of random variables that will be discussed in this chapter are the *discrete* and the *continuous* random variables, which are distinguished from one another by the structure of their supports.

Definition 2.2 A random variable X is said to be a discrete random variable when the support $\mathcal{S}$ of X is a finite or a countable subset of $\mathbb{R}$. For $x \in \mathcal{S}$, the function $f(x) = P(X = x)$ is called the probability density function or pdf.

For example, the number of heads in 10 tosses of a coin, the number of errors made in a baseball game, the number of students attending a talk, the number of tornadoes occurring in a tornado season, and the number of points scored in basketball game are all discrete random variables.

Example 2.3 A fair die will be rolled independently until a 6 appears. Let X be the number of rolls it takes to observe the first 6. Then, $\mathcal{S} = \mathbb{N}$ and

$$f(x) = P(X = x)$$
$$= P(\text{the first } x - 1 \text{ rolls are not sixes and the } x\text{th roll is a six})$$
$$= \left(\frac{5}{6}\right)^{x-1} \frac{1}{6}.$$

Definition 2.3 A random variable X is said to be a continuous random variable when the support of X is the union of one or more intervals, and there exists a nonnegative real-valued function $f(x)$ such that $P(X \leq x) = \int_{-\infty}^{x} f(t) \, dt$. The function $f(x)$ is called the probability density function or pdf.

The time between two consecutive phone calls, the top speed of a cheetah, and the weight of a quarter are all examples of continuous random variables.

Example 2.4 The time to failure (in days) of an electronic component might have the pdf $f(x) = 0.02 \, e^{-0.02x}$ on $\mathcal{S} = (0, \infty)$ In this case, the probability that the time to failure (X) of the electronic component is no more than 30 days is

$$P(X \leq 30) = \int_{0}^{30} 0.02 \, e^{-0.02x} \, dx = 1 - e^{-0.6} = 0.4512.$$

The pdf and support of a random variable contain complete information about a random variable, and therefore, all of the important characteristics of a random variable can be derived from the pdf. Note that for a discrete random variable X with pdf $f(x)$ on support $\mathcal{S}$,

1) $f(x) \geq 0$, $\forall x \in \mathcal{S}$ and $f(x) = 0$, $\forall x \notin \mathcal{S}$.
2) $\sum_{x \in \mathcal{S}} f(x) = 1$.
3) $P(X \in B) = \sum_{x \in B} f(x)$.

Analogously, for a continuous random variable X with pdf $f(x)$ on support $\mathcal{S}$,

1) $f(x) \geq 0$, $\forall x \in \mathcal{S}$ and $f(x) = 0$, $\forall x \notin \mathcal{S}$.
2) $\int_{\mathcal{S}} f(x)\, dx = 1$.
3) $P(X \in B) = \int_{B} f(x)\, dx$.

The pdf of any random variable can be defined on $\mathbb{R}$ by writing the pdf in terms of an *indicator function* on the set $\mathcal{S}$. In particular, let

$$I_{\mathcal{S}}(x) = \begin{cases} 0 & x \notin \mathcal{S}, \\ 1 & x \in \mathcal{S}. \end{cases}$$

Then, $f(x) = f(x) \times I_{\mathcal{S}}(x)$. Using pdfs defined on $\mathbb{R}$ plays an important role in Chapter 5 on likelihood estimation and Chapter 6 on hypothesis testing.

Example 2.5 In Example 2.4, the pdf $f(x)$ defined over $\mathbb{R}$ is

$$f(x) = 0.02\, e^{-0.02x} \times I_{(0,\infty)(x)}, x \in \mathbb{R}.$$

One of the key differences between the discrete and continuous pdfs is that the pdf of a discrete random variable gives the probability of a value of $x \in \mathcal{S}$, whereas the pdf of a continuous random variable does not. In fact, for a continuous random variable $P(X = x) = 0$, $\forall x \in \mathcal{S}$. A second probability function that also completely describes a random variable is the *cumulative distribution function* (CDF).

Definition 2.4 The cumulative distribution function or CDF of a random variable X is defined to be $F(x) = P(X \leq x)$, $\forall x \in \mathbb{R}$.

For a discrete random variable, the CDF is a step function where the jump at $X = x_0 \in \mathcal{S}$ is $f(x_0) = P(X = x_0)$; the CDF of a discrete random variable is constant between two consecutive values in the support. An example of a discrete CDF is shown in Figure 2.1. The discontinuity points of the CDF of a discrete random variable occur at points in S, and the CDF of a discrete random variable is a right-continuous function on $\mathbb{R}$.

Example 2.6 Let X be a discrete random variable with pdf

$$f(x) = \frac{x}{21}, \quad \text{for} \quad x = 1, 2, 3, 4, 5, 6.$$

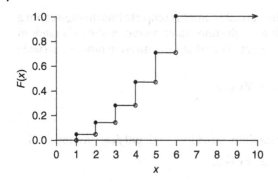

Figure 2.1 The CDF of the discrete random variable X in Example 2.6.

Then,

$$F(x) = P(X \leq x) = \begin{cases} 0 & x < 1, \\ \frac{1}{21} & 1 \leq x < 2, \\ \frac{3}{21} & 2 \leq x < 3, \\ \frac{6}{21} & 3 \leq x < 4, \\ \frac{10}{21} & 4 \leq x < 5, \\ \frac{15}{21} & 5 \leq x < 6, \\ 1 & x \geq 6. \end{cases}$$

Unlike the CDF of a discrete random variable, the CDF of a continuous random variable is a continuous function on $\mathbb{R}$. An example of the CDF of a continuous random variable is shown in Figure 2.2.

Example 2.7 Suppose that the lifetime of a light bulb in hours is a continuous random variable with pdf $f(x) = \frac{1}{750}e^{-\frac{x}{750}}$ for $x > 0$. Then,

$$F(x) = \begin{cases} 0 & x \leq 0, \\ 1 - e^{-\frac{x}{750}} & x > 0. \end{cases}$$

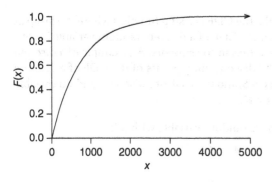

Figure 2.2 The CDF of the continuous random variable X in Example 2.7.

Theorem 2.1 summarizes several of the important properties of a CDF.

Theorem 2.1 *If X is a random variable with CDF F(x), then*

i) $\lim_{x \to -\infty} F(x) = 0$ *and* $\lim_{x \to \infty} F(x) = 1$.
ii) $F(x)$ *is nondecreasing; that is* $F(x) \leq F(y)$ *whenever* $x < y$.
iii) $F(x)$ *is continuous from the right.*
iv) $P(a < X \leq b) = F(b) - F(a)$, $\forall a, b \in \mathbb{R}$.

Proof. The proof of Theorem 2.1 is left as an exercise. ∎

Theorems 2.2 and 2.3 show how to determine the pdf from the CDF of a random variable.

Theorem 2.2 *If X is a discrete random variable with CDF F(x) and support* $S = \{x_0, x_1, x_2, \ldots\}$, *where* $x_0 < x_1 < x_2 < \cdots$, *then for* $x_k \in S$,

$$f(x_k) = F(x_k) - F(x_{k-1}).$$

Proof. Let X be a discrete random variable with CDF $F(x)$ and support $S = \{x_0, x_1, x_2, \ldots\}$ with $x_0 < x_1 < x_2 < \cdots$. Then, for $k \geq 1$,

$$f(x_k) = P(X = x_k) = P(x_{k-1} < X \leq x_k) = F(x_k) - F(x_{k-1}).$$ ∎

Theorem 2.3 *For a continuous random variable* $f(x) = \frac{d}{dx} F(x)$, $\forall x \in \mathbb{R}$.

Proof. Theorem 2.3 follows directly from the Fundamental Theorem of Calculus. ∎

Note that Theorems 2.2 and 2.3 show the existence of a one-to-one correspondence between the pdf and the CDF of a random variable. Thus, both the CDF and the pdf of a random variable contain complete information about the random variable.

Inverting the CDF of a continuous random variable produces a *quantile* of a random variable.

Definition 2.5 For $p \in (0, 1)$, the $100 \cdot p$th quantile of a continuous random variable X is the value of $x \in S$ such that $F(x) = p$. The $100 \cdot p$th quantile of a random variable will be denoted by x_p.

The commonly used quantiles include the following:

- *percentiles,* $x_{0.01}, x_{0.02}, x_{0.03}, \ldots, x_{0.99}$
- *deciles,* $x_{0.10}, x_{0.20}, x_{0.30}, \ldots, x_{0.90}$
- *quintiles,* $x_{0.20}, x_{0.40}, x_{0.60}, x_{0.80}$
- *quartiles,* $x_{0.25}, x_{0.50}, x_{0.75}$
- *median,* $x_{0.50}$, which will be denoted by $\tilde{\mu}$.

The median is often used as a measure of the typical value of a random variable since 50% of the distribution lies below the median and 50% lies above the median.

Example 2.8   Standardized tests often report a raw score and its corresponding percentile. For example, according to the College Board, a raw score of 700 on the SAT Math Exam corresponds to the 93rd percentile. That is, 93% of the individuals taking the SAT Math Exam scored 700 or lower (i.e. $F(700) = 0.93$).

Example 2.9   Suppose that X is a continuous random variable with pdf $f(x) = \frac{2}{x^3}$ for $x \in (1, \infty)$. Then, for $x \in (1, \infty)$, the CDF is

$$F(x) = \int_1^x \frac{2}{t^3}\, dt = 1 - \frac{1}{x^2}.$$

Thus, the $100 \cdot p$th quantile is found by solving $p = 1 - \frac{1}{x^2}$, which yields $x_p = \sqrt{\frac{1}{1-p}}$ for $p \in (0, 1)$. Hence, the median is $\tilde{\mu} = \sqrt{\frac{1}{1-0.5}} = \sqrt{2}$, and the 90th percentile is $x_{0.90} = \sqrt{\frac{1}{1-0.9}} = \sqrt{10}$.

Because the CDF of a discrete random variable is discontinuous, there may be values of $p \in (0, 1)$ where there is no solution to the equation $p = F(x_p)$. Quantiles can still be defined for discrete random variables by defining x_p to be the smallest value of $x \in \mathcal{S}$ such that $F(x_p) \geq p$. That is, for a discrete random variable and $p \in (0, 1)$, $x_p = \inf_{x \in \mathcal{S}} P(X \leq x_p) \geq p$. In general, the quantiles of a discrete random variable are not very useful.

Problems

2.1.1   Suppose that a fair coin is flipped four times. Determine the support and pdf of the random variable
a) $H =$ the number of heads flipped.
b) $D =$ the difference in number of heads and tails flipped.
c) $A = |D|$.

2.1.2   Suppose that a biased coin with $P(H) = 0.6$ will be flipped independently until a tail is flipped. Let X be the number of flips required to flip the first tail and Y be the number of heads flipped before the first tail is flipped. Determine

a) the support of X.   b) the support of Y.
c) the pdf of X.   d) the pdf of Y.
e) $P(X = 5)$.   f) $P(Y = 5)$.

2.1.3 Let X be a discrete random variable with pdf $f(x) = \frac{x}{k}$, $x = 1, 2, 3, 4, 5, 6$. Determine
  a) k.   b) $F(x)$.
  c) $P(X > 2)$.   d) $P(2 \leq X < 5)$.

2.1.4 Let X be a discrete random variable with pdf $f(x) = k \cdot 3^{-k}$ for $x \in \mathbb{N}$. Determine
  a) k.   b) $F(x)$.
  c) $P(X > 2)$.   d) $P(2 \leq X < 5)$.

2.1.5 Let X be a discrete random variable with pdf $f(x) = k(x + 1)$ for $x = 0, 1, 2, 3, 4, 5$. Determine
  a) k.   b) $F(x)$.
  c) $P(X > 3)$.   d) $P(2 \leq X < 5)$.

2.1.6 Let X be a continuous random variable with pdf $f(x) = kx^2$ for $x \in (-1, 1)$. Determine
  a) k.   b) $F(x)$.
  c) $P(X > 0.2)$.   d) $\tilde{\mu}$.

2.1.7 Let X be a continuous random variable with pdf $f(x) = \frac{k}{x^2}$ for $x \in (1, 2)$. Determine
  a) k.   b) $F(x)$.
  c) $P(X > 1.2)$.   d) $\tilde{\mu}$.

2.1.8 Let X be a continuous random variable with pdf $f(x) = \alpha x^{\alpha-1}$ for $x \in (0, 1)$ and $\alpha > 0$. Determine
  a) x_p.   b) $\tilde{\mu}$.

2.1.9 Let X be a continuous random variable with pdf $f(x) = kx(1 - x)$ for $x \in (0, 1)$. Determine
  a) k.   b) $F(x)$.
  c) $P(X > 0.4)$.   d) $P(0.1 < X < 0.9)$.

2.1.10 Let X be a continuous random variable with pdf $f(x) = e^{-(x-5)}$ for $x \in (5, \infty)$. Determine
  a) $F(x)$.   b) $P(X \geq 7.5)$.
  c) x_p.   d) $x_{0.95}$.

2.1.11 Let X be a continuous random variable with pdf $f(x) = \frac{1}{x^2}$ for $x \geq 1$.
   a) Use an indicator function to write $f(x)$ so that it is defined over $\mathbb{R}$.
   b) Determine $F(x)$, $\forall x \in \mathbb{R}$.
   c) Determine $P(2 \leq X \leq 5)$.

2.1.12 Let X be a continuous random variable with pdf $f(x) = \frac{1}{3}e^{-\frac{x-2}{3}}$ for $x \geq 2$.
   a) Use an indicator function to write $f(x)$ so that it is defined over $\mathbb{R}$.
   b) Determine $F(x)$, $\forall x \in \mathbb{R}$.
   c) Determine $P(3 \leq X \leq 6)$.

2.1.13 Let X be a random variable with CDF $F(x) = \begin{cases} 0 & x < 1 \\ 1 - \frac{1}{x^4} & x \geq 1 \end{cases}$.
   Determine
   a) the support of X.   b) the pdf of X.
   c) $x_{0.25}$.   d) $\tilde{\mu}$.

2.1.14 Let X be a random variable with CDF $F(x) = \begin{cases} 0 & x < 0 \\ 1 - e^{-\frac{x^2}{8}} & x \geq 0 \end{cases}$.
   Determine
   a) the support of X.   b) the pdf of X.
   c) $P(X \geq 4)$.   d) $P(1 \leq X < 4)$.
   e) $x_{0.95}$.   f) $\tilde{\mu}$.

2.1.15 Let X be a random variable with CDF $F(x) = \begin{cases} 0 & x < 0 \\ 1 - e^{-\left(\frac{x}{4}\right)^3} & x \geq 0 \end{cases}$.
   Determine
   a) the support of X.   b) the pdf of X.
   c) $P(X \leq 2)$.   d) $P(3 < X \leq 5)$.
   e) $x_{0.05}$.   f) $\tilde{\mu}$.

2.1.16 Let X be a continuous random variable with CDF $F(x)$. Show that
   a) $P(a < X \leq b) = F(b) - F(a)$, $\forall a, b \in \mathbb{R}$.
   b) $P(a < X \leq b) = P(a \leq X \leq b) = P(a < X < b) = P(a \leq X < b)$
   $= F(b) - F(a)$, $\forall a, b \in \mathbb{R}$.

2.1.17 Prove Theorem 2.1.

2.2 Random Vectors

In most statistical analyses, more than one variable will be measured on each unit that is sampled. For example, in a cardiovascular study, a researcher might include variables for systolic and diastolic blood pressure, age, weight, cholesterol level, and body mass index. Each observation in the study is a vector of observations on a sampled unit called a *random vector*.

Definition 2.6 A random vector $\vec{X} = (X_1, X_2, \ldots, X_k)$ is a k-dimensional vector, where $X_1, X_2, \ldots, X_k$ are random variables. A random vector is called a discrete random vector when each of its component random variables is discrete and called a continuous random vector when each of its component random variables is continuous.

Definition 2.7 A two-dimensional random vector $\vec{X} = (X_1, X_2)$ is called a bivariate random variable.

2.2.1 Properties of Random Vectors

Similar to random variables, random vectors have a pdf, a support, and a CDF. The support of a k-dimensional random vector $\vec{X}$ is the set of values it can take on, which is denoted by $\mathcal{S}_{\vec{X}}$, and $\mathcal{S}_{\vec{X}}$ is a subset of $\mathbb{R}^k$. A k-dimensional random vector $\vec{X}$ has a *joint pdf* and a *joint CDF*, which contain complete information about the random vector and are used for computing probabilities with regard to $\vec{X}$. The definitions of the joint pdf for discrete and continuous random vectors are given in Definitions 2.8 and 2.9, respectively.

Definition 2.8 Let $\vec{X}$ be a k-dimensional discrete random vector. The joint pdf of $\vec{X}$ is defined to be $f(\vec{x}) = f(x_1, \ldots, x_k) = P(X_1 = x_1, X_2 = x_2, \ldots, X_k = x_k)$ for $\vec{x} = (x_1, x_2, \ldots, x_k) \in \mathcal{S}_{\vec{X}}$.

When $\vec{X}$ is a k-dimensional discrete random vector with joint pdf $f(x_1, \ldots, x_k)$, then

- $0 \leq f(x_1, \ldots, x_k) \leq 1$, $\forall \vec{x} \in \mathcal{S}_{\vec{X}}$.
- $\sum_{\vec{x} \in \mathcal{S}_{\vec{X}}} f(x_1, \ldots, x_k) = 1$.
- for any event $B \subset \mathcal{S}_{\vec{X}}$, $P(\vec{X} \in B) = \sum_{\{\vec{x} \in \mathcal{S}_{\vec{X}} : \, \vec{x} \in B\}} f(x_1, \ldots, x_k)$.

Definition 2.9 Let $\vec{X}$ be a k-dimensional continuous random vector. The joint pdf of $\vec{X}$ is defined to be any nonnegative function such that

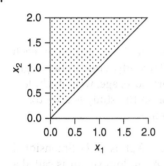

Figure 2.3 The support of the bivariate continuous random vector $\vec{X} = (X_1, X_2)$ in Example 2.11.

1) $f(x_1, \ldots, x_k) > 0$, $\forall \vec{x} \in S_{\vec{X}}$.
2) $\int_{S_{\vec{X}}} f(x_1, \ldots, x_k) \, dx_1 \cdots dx_k = 1$.
3) for any event $B \subset S_{\vec{X}}$, $P(\vec{X} \in B) = \int_B f(x_1, \ldots, x_k) \, dx_1 \cdots dx_k$.

A random vector also has a CDF, which is called the *joint cumulative distribution function* or *joint CDF*. The definition of the joint CDF of a random vector $\vec{X}$ is given in Definition 2.10.

Definition 2.10 Let $\vec{X} = (X_1, \ldots, X_k)$ be a k-dimensional random vector. The joint CDF of $\vec{X}$ is defined to be $F(x_1, \ldots, x_k) = P(X_1 \leq x_1, \ldots, X_k \leq X_k)$, $\forall (x_1, \ldots, x_k) \in \mathbb{R}^k$.

Examples 2.10 and 2.11 provide examples of bivariate discrete and continuous random vectors and their joint pdfs.

Example 2.10 Roll a fair die twice. Let X_1 be the outcome of roll one and X_2 be the outcome of roll two. Then, $\vec{X} = (X_1, X_2)$ is a discrete bivariate random variable with $S_{\vec{X}} = \{(x_1, x_2) : x_i = 1, 2, 3, 4, 5, 6; i = 1, 2\}$. Assuming that the outcomes of roll one and roll two are independent, the joint pdf of $\vec{X}$ is $f(x_1, x_2) = \frac{1}{36}$ for $(x_1, x_2) \in S_{\vec{X}}$. The joint CDF of $\vec{X}$ is

$$F(x_1, x_2) = P(X_1 \leq x_1, X_2 \leq x_2) = \sum_{i=1}^{x_2} \sum_{j=1}^{x_1} \frac{1}{36} = \frac{x_1 x_2}{36}$$

for $\vec{x} \in S_{\vec{X}}$.

Example 2.11 Let $\vec{X} = (X_1, X_2)$ be a bivariate continuous random variable with joint pdf given by $f(x_1, x_2) = \frac{x_1 x_2}{2}$, $0 < x_1 < x_2 < 2$. The support of $\vec{X}$ is shown in Figure 2.3.

Now, $F(1.2, 1.5) = P(X_1 < 1.2, X_2 < 1.5)$ is found by integrating over the two regions shown in Figure 2.4, which yields

$$F(1.2, 1.5) = \int_0^{1.5} \int_0^{x_2} \frac{x_1 x_2}{2} \, dx_1 \, dx_2 + \int_{1.2}^{1.5} \int_0^{1.2} \frac{x_1 x_2}{2} \, dx_1 \, dx_2 = 0.4622.$$

Figure 2.4 The values of the random vector $\vec{X} = (X_1, X_2)$ in Example 2.11 where $P(X_1 < 1.2, X_2 < 1.5)$.

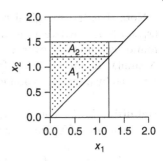

Several important properties of a joint CDF are given in Theorem 2.4.

Theorem 2.4 *If $F(x_1, x_2)$ is the joint CDF of $\vec{X} = (X_1, X_2)$, then*

i) $\lim\limits_{x_1 \to -\infty} F(x_1, x_2) = F(-\infty, x_2) = 0, \ \forall x_2 \in \mathbb{R}$.

ii) $\lim\limits_{x_2 \to -\infty} F(x_1, x_2) = F(x_1, -\infty) = 0, \ \forall x_1 \in \mathbb{R}$.

iii) $\lim\limits_{\substack{x_1 \to \infty \\ x_2 \to \infty}} F(x_1, x_2) = F(\infty, \infty) = 1$.

iv) $\lim\limits_{h \to 0^+} F(x_1 + h, x_2) = \lim\limits_{h \to 0^+} F(x_1, x_2 + h) = F(x_1, x_2), \ \forall x_1, x_2 \in \mathbb{R}$.

v) $F(b, d) - F(b, c) - F(a, d) + F(a, c) \geq 0$ *whenever* $a < b$ *and* $c < d$.

The first three properties of Theorem 2.4 are intuitive, the fourth property shows that a joint CDF is right continuous in each of its variable, and the last property is a monotonicity condition that ensures that $P(a < X_1 \leq b, c < X_2 \leq d) \geq 0$ whenever $a < b$ and $c < d$.

Since the components of a random vector $\vec{X}$ are random variables, the pdf of each component random variable X_i in $\vec{X}$ can be derived from the joint pdf of $\vec{X}$. The pdf of a component random variable is called a *marginal pdf*. The marginal pdf of the random variable X_i will be denoted by $f_i(x_i)$. The definitions of the marginal pdfs for a bivariate random vector $\vec{X} = (X_1, X_2)$ are given in Definitions 2.11 and 2.12.

Definition 2.11 Let $\vec{X} = (X_1, X_2)$ be a discrete random vector. The marginal pdfs of X_1 and X_2 are given by

$$f_1(x_1) = \sum_{x_2 \in \mathcal{S}_2'} f(x_1, x_2)$$

and

$$f_2(x_2) = \sum_{x_1 \in \mathcal{S}_1'} f(x_1, x_2),$$

where $\mathcal{S}_1' = \{x_1 : (x_1, x_2) \in \mathcal{S}_{\vec{X}}\}$ and $\mathcal{S}_2' = \{x_2 : (x_1, x_2) \in \mathcal{S}_{\vec{X}}\}$ for fixed values of x_2 and x_1, respectively.

Note that the marginal pdf of X_1 is found by summing over $x_2 \in \{x_2 : (x_1, x_2) \in \mathcal{S}_{\vec{X}}\}$, where x_1 is arbitrary but fixed, and the marginal pdf of X_2 is found by summing over $x_1 \in \{x_1 : (x_1, x_2) \in \mathcal{S}_{\vec{X}}\}$ for arbitrary but fixed x_2. A similar definition holds for continuous random vectors with summation replaced by integration and is given in Definition 2.12.

Definition 2.12 Let $\vec{X} = (X_1, X_2)$ be a continuous random vector. The marginal pdfs of X_1 and X_2 are given by

$$f_1(x_1) = \int_{x_2 \in \mathcal{S}_2'} f(x_1, x_2) \, dx_2$$

and

$$f_2(x_2) = \int_{x_1 \in \mathcal{S}_1'} f(x_1, x_2) \, dx_1,$$

where $\mathcal{S}_1' = \{x_1 : (x_1, x_2) \in \mathcal{S}_{\vec{X}}\}$ and $\mathcal{S}_2' = \{x_2 : (x_1, x_2) \in \mathcal{S}_{\vec{X}}\}$ for fixed values of x_2 and x_1, respectively.

Examples 2.12 and 2.13 illustrate the determination of the marginal pdfs from the joint pdf for discrete and continuous bivariate random vectors, respectively.

Example 2.12 Let $\vec{X} = (X_1, X_2)$ be a discrete random vector with joint pdf $f(x_1, x_2) = \frac{2}{n(n+1)}$ with $\mathcal{S} = \{(x_1, x_2) : x_1, x_2 \in \mathbb{N} \text{ and } 1 \leq x_1 \leq x_2 \leq n\}$. The marginal pdf of X_1 is found by summing over all values of x_2 such that $x_1, x_2 \in \mathbb{N}$ and $x_1 \leq x_2 \leq n$. Thus, for an arbitrary but fixed value of x_1,

$$f_1(x_1) = \sum_{x_2 = x_1}^{n} \frac{2}{n(n+1)} = \frac{2(n - x_1 + 1)}{n(n+1)}, \quad x_1 = 1, 2, 3, \dots, n.$$

The marginal pdf of X_2 is found by summing over all values of x_1 such that $x_1, x_2 \in \mathbb{N}$ and $x_2 \geq x_1 \geq 1$. Thus, for an arbitrary but fixed value of x_2,

$$f_2(x_2) = \sum_{x_1 = 1}^{x_2} \frac{2}{n(n+1)} = \frac{2x_2}{n(n+1)}, \quad x_2 = 1, 2, 3, \dots, n.$$

Example 2.13 Let $\vec{X} = (X_1, X_2)$ be a continuous random vector with joint pdf $f(x_1, x_2) = e^{-x_2}$ with $\mathcal{S} = \{(x_1, x_2) : 0 < x_1 < x_2 < \infty\}$. The marginal pdf of X_1 is found by integrating over all values of x_2 such that $0 < x_1 < x_2 < \infty$. Thus, for an arbitrary but fixed value of x_1,

$$f_1(x_1) = \int_{x_1}^{\infty} e^{-x_2} \, dx_2 = e^{-x_1}, \ x_1 \in (0, \infty).$$

The marginal pdf of X_2 is found by integrating over all values of x_1 such that $0 < x_1 < x_2$. Thus, for an arbitrary but fixed value of x_2,

$$f_2(x_2) = \int_0^{x_2} e^{-x_2} \, dx_1 = x_2 e^{-x_2}, \, x_2 \in (0, \infty).$$

The marginal CDFs of X_1 and X_2 are denoted by $F_1(x_1)$ and $F_2(x_2)$ and are given by $F_i(x_i) = P(X_i \leq x_i)$ for $i = 1, 2$.

Example 2.14 In Example 2.13, the marginal pdf of X_1 is $f_1(x_1) = e^{-x_1}$ for $x_1 > 0$. Thus,

$$F_1(x_1) = \begin{cases} 0 & x \leq 0, \\ \int_0^{x_1} e^{-u} \, du = 1 - e^{-x_1} & x > 0. \end{cases}$$

Theorem 2.5 shows that the marginal CDFs can also be computed directly from the joint CDF.

Theorem 2.5 *If* $\vec{X} = (X_1, X_2)$ *is a random vector with joint CDF* $F(x_1, x_2)$, *then*

i) $F_1(x_1) = \lim\limits_{x_2 \to \infty} F(x_1, x_2) = F(x_1, \infty).$

ii) $F_2(x_2) = \lim\limits_{x_1 \to \infty} F(x_1, x_2) = F(\infty, x_2).$

Proof. Let $\vec{X} = (X_1, X_2)$ be a random vector with joint CDF $F(x_1, x_2)$. WLOG consider $F_1(x_1)$.

$$F_1(x_1) = P(X_1 \leq x_1) = P(X_1 \leq x_1, X_2 \leq \infty) = F(x_1, \infty).$$

The proof for $F_2(x_2)$ follows similarly mutatis mutandis. ∎

For a bivariate random vector $\vec{X} = (X_1, X_2)$, the definition of the *conditional distribution* of $X_1 | X_2$ is similar to the definition of the conditional probability of an event A given the event B. In particular, the conditional pdf of $X_1 | X_2$ is defined by substituting the joint pdf of (X_1, X_2) and the marginal pdf of X_2 for $P(A \cap B)$ and $P(B)$ in the conditional probability definition. The definitions of the conditional distribution of $X_1 | X_2$ and $X_2 | X_1$ are given in Definition 2.13.

Definition 2.13 Let $\vec{X} = (X_1, X_2)$ be a random vector. For a fixed value of x_2 where $f_2(x_2) > 0$, the conditional pdf for $X_1 | X_2 = x_2$ is defined to be

$$f(x_1 | x_2) = \frac{f(x_1, x_2)}{f_2(x_2)} \quad \text{for} \quad x_1 \in \mathcal{S}_{X_1 | X_2 = x_2} = \{x_1 : (x_1, x_2) \in \mathcal{S}_{\vec{X}}\}.$$

For a fixed value of x_1 where $f_1(x_1) > 0$, the conditional pdf of $X_2|X_1 = x_1$ is defined to be

$$f(x_2|x_1) = \frac{f(x_1, x_2)}{f_1(x_1)} \quad \text{for} \quad x_2 \in \mathcal{S}_{X_2|X_1 = x_1} = \{x_2 : (x_1, x_2) \in \mathcal{S}_{\vec{X}}\}.$$

Note that when $\vec{X}$ is a bivariate discrete random vector, the pdf of $X_1|X_2$ produces conditional probabilities. In particular,

$$f(x_1|x_2) = \frac{f(x_1, x_2)}{f_2(x_2)} = \frac{P(X_1 = x_1, X_2 = x_2)}{P(X_2 = x_2)} = P(X_1 = x_1|X_2 = x_2).$$

However, for a bivariate continuous random vector, $f(x_1|x_2)$ is not a probability, and conditional probabilities are found by integrating over a set of X_1 values, say B. That is,

$$P(X_1 \in B|X_2 = x_2) = \int_B f(x_1|x_2) \, dx_1.$$

Example 2.15 For the discrete random vector in Example 2.12 with joint pdf $f(x_1, x_2) = \frac{2}{n(n+1)}$ and marginal pdf $f_1(x_1) = \frac{2(n-x_1+1)}{n(n+1)}$, the conditional pdf of $X_2|X_1$ is

$$f(x_2|x_1) = \frac{\frac{2}{n(n+1)}}{\frac{2(n-x_1+1)}{n(n+1)}} = \frac{1}{n - x_1 + 1}$$

for $x_2 \in \{x_2 \in \mathbb{N} : x_1 \le x_2 \le n\}$.

Thus, for $X_1 = 4$, the conditional pdf of $X_2|X_1 = 4$ is $f(x_2|4) = \frac{1}{n-3}$ for $x_2 = 4, 5, \ldots, n$.

Example 2.16 For the continuous random vector in Example 2.13 with joint pdf $f(x_1, x_2) = e^{-x_2}$ and marginal pdf $f_1(x_1) = e^{-x_1}$, the conditional pdf of $X_2|X_1$ is

$$f(x_2|x_1) = \frac{e^{-x_2}}{e^{-x_1}} = e^{-(x_2 - x_1)}$$

for $x_2 \in (x_1, \infty)$.

In particular, for $X_1 = 10$, the conditional pdf of $X_2|X_1 = 10$ is $f(x_2|10) = e^{-(x_2-10)}$ for $x_2 \in (10, \infty)$.

For $X_1 = 10$, the conditional CDF of $X_2|X_1 = 10$ is

$$F_{X_2|X_1}(x_2|10) = \begin{cases} 0 & x \le 10, \\ \int_{10}^{x_2} e^{-(u-10)} \, du = 1 - e^{-(x_2-10)} & x > 10. \end{cases}$$

Theorem 2.6 is an analog of the Law of Total Probability given in Chapter 1 applied to random variables X_1 and X_2.

Theorem 2.6 (Law of Total Probability) *Let $\vec{X}$ be a bivariate random vector with joint pdf $f(x_1, x_2)$, then*

i) for a continuous random vector $\vec{X}$, $f_1(x_1) = \int_{x_2 \in S_2'} f(x_1 | x_2) f_2(x_2)\, dx_2$, where $S_2' = \{x_2 : (x_1, x_2) \in S_{\vec{X}}\}$.

ii) for a discrete random vector $\vec{X}$, $f_1(x_1) = \sum_{x_2 \in S_2'} f(x_1 | x_2) f_2(x_2)$, where $S_2' = \{x_2 : (x_1, x_2) \in S_{\vec{X}}\}$.

Proof. Let $\vec{X}$ be a bivariate random vector with joint pdf $f(x_1, x_2)$ defined on support $S_{\vec{X}}$.

i) Suppose that $\vec{X}$ is a continuous random vector. Then, $f(x_1, x_2) = f(x_1 | x_2) f_2(x_2)$. Thus,

$$f_1(x_1) = \int_{x_2 \in S'} f(x_1, x_2)\, dx_2 = \int_{x_2 \in S'} f(x_1 | x_2) f_2(x_2)\, dx_2.$$

ii) The proof of part (ii) is similar to the proof of part (i) with integration replaced by summation. ∎

Example 2.17 Suppose that $\vec{X} = (X_1, X_2)$ is a continuous random vector where the marginal pdf of X_2 is $f_2(X_2) = \frac{1}{2} x_2^2\, e^{-x_2}$, $x_2 > 0$ and the conditional pdf of $X_1 | X_2$ is $f(x_1 | x_2) = \frac{1}{x_2}$, $0 < x_1 < x_2$. Then, the support of $X_1 | X_2$ is $0 < x_1 < x_2$, and by the Law of Total Probability,

$$f_1(x) = \int_{x_1}^{\infty} \underbrace{\frac{1}{2} x_2^2 e^{-x_2} \times \frac{1}{x_2}}_{f(x_1 | x_2) f(x_2)}\, dx_2 = \int_{x_1}^{\infty} \frac{1}{2} x_2 e^{-x_2}\, dx_2 = \frac{x_1 + 1}{2} e^{-x_1}$$

for $x_1 \in (0, \infty)$.

The definitions for the marginal and conditional distributions for a bivariate random vector can easily be extended to higher dimensional random vectors. For example, if $\vec{X} = (X_1, \dots, X_k)$, then the marginal pdf of the random variable X_i is

$$f_i(x_i) = \begin{cases} \underbrace{\sum \sum \cdots \sum}_{x_j, j \neq i} f(x_1, \dots, x_k) & \text{when } \vec{X} \text{ is discrete,} \\[2em] \underbrace{\int \int \cdots \int}_{x_j, j \neq i} f(x_1, \dots, x_k) \prod_{j \neq i} dx_j & \text{when } \vec{X} \text{ is continuous.} \end{cases}$$

Example 2.18 Let $\vec{X} = (X_1, X_2, X_3)$ be a continuous random vector with joint pdf given by $f(x_1, x_2, x_3) = \frac{3}{4}$ for $0 < x_1 < x_2 < x_3 < 2$. Then, for $x_1 \in (0, 2)$, it follows that $x_1 < x_2 < x_3 < 2$, and hence, the marginal pdf of X_1 is

$$f_1(x_1) = \int_{x_1}^{2} \int_{x_1}^{x_3} \frac{3}{4} \, dx_2 \, dx_3 = \frac{3}{8}(2 - x_1)^2, \; x_1 \in (0, 2).$$

For $x_2 \in (0, 2)$, it follows that $0 < x_1 < x_2 < x_3 < 2$. Thus, the marginal pdf of X_2 is

$$f_2(x_2) = \int_{x_2}^{2} \int_{0}^{x_2} \frac{3}{4} \, dx_1 \, dx_3 = \frac{3}{4}x_2(2 - x_2), \; x_2 \in (0, 2).$$

And finally, for $x_3 \in (0, 2)$, it follows that $0 < x_1 < x_2 < x_3$. Thus, the marginal pdf of X_3 is

$$f_3(x_3) = \int_{0}^{x_3} \int_{0}^{x_2} \frac{3}{4} \, dx_1 \, dx_2 = \frac{3}{8}x_3^2, \; x_3 \in (0, 2).$$

Now, for $x_1 \in (0, 2)$, the conditional pdf of $X_2, X_3 | X_1 = x_1$ is

$$f_{X_2, X_3 | X_1}(x_2, x_3 | x_1) = \frac{f(x_1, x_2, x_3)}{f_1(x_1)} = \frac{\frac{3}{4}}{\frac{3}{8}(2 - x_1)^2} = \frac{2}{(2 - x_1)^2}$$

for $x_1 < x_2 < x_3 < 2$.

Problems

2.2.1 Let $\vec{X}$ be a bivariate discrete random vector with joint pdf $f(x_1, x_2)$ given in Table 2.1. Determine
a) $P(X_1 = 3, X_2 < 2)$.
b) $f_1(x_1)$ and $\mathcal{S}_{X_1}$.
c) $f_2(x_2)$ and $\mathcal{S}_{X_2}$.
d) $f(x_1 | x_2 = 2)$ and $\mathcal{S}_{X_1 | X_2 = 2}$.

Table 2.1 The values of $f(x_1, x_2)$ for Problem 2.2.1.

X_1	X_2			
	0	1	2	3
1	0.1	0.1	0.05	0.01
2	0.2	0.1	0.08	0.02
3	0.3	0.02	0.01	0.01

2.2.2 Let $\vec{X}$ be a bivariate discrete random vector with joint pdf $f(x_1, x_2) = k$ for $\mathcal{S}_{\vec{X}} = \{(x_1, x_2) \in \mathbb{N} \times \mathbb{N} : 1 \le x_1 \le x_2 \le 10\}$. Determine
a) the value of k so that $f(x_1, x_2)$ is a joint pdf.
b) $P(X_1 \le 5, X_2 = 4)$.
c) $P(X_1 \le 5, X_2 \ge 4)$.
d) $f_1(x_1)$ and $\mathcal{S}_{X_1}$.
e) $f_2(x_2)$ and $\mathcal{S}_{X_2}$.
f) $f(x_1|x_2 = 6)$ and $\mathcal{S}_{X_1|X_2=6}$.

2.2.3 Let $\vec{X}$ be a bivariate discrete random vector with joint pdf given by $f(x_1, x_2) = p^{x_2}(1-p)^2$ for $\mathcal{S}_{\vec{X}} = \{(x_1, x_2) \in \mathbb{W} \times \mathbb{W} : x_1 \le x_2\}$ for $p \in (0, 1)$. Determine
a) $f_1(x_1)$ and $\mathcal{S}_{X_1}$.
b) $f_2(x_2)$ and $\mathcal{S}_{X_2}$.
c) $f(x_1|x_2)$ and $\mathcal{S}_{X_1|X_2}$.
d) $f(x_2|x_1)$ and $\mathcal{S}_{X_2|X_1}$.

2.2.4 Let $\vec{X}$ be a bivariate random vector with joint pdf $f(x_1, x_2) = kx_1x_2$ for $0 < x_1 < x_2 < 1$. Determine
a) the value of k so that $f(x_1, x_2)$ is a joint pdf.
b) $P(X_1 \le .5, X_2 > 0.6)$.
c) $f_1(x_1)$ and $\mathcal{S}_{X_1}$.
d) $f_2(x_2)$ and $\mathcal{S}_{X_2}$.
e) $f(x_1|x_2)$ and $\mathcal{S}_{X_1|X_2}$.
f) $P(X_1 \le 0.5|X_2 = 0.6)$.

2.2.5 Let $\vec{X}$ be a bivariate random vector with joint pdf $f(x_1, x_2) = k(x_1 - x_2)$ for $0 < x_2 < x_1 < 1$. Determine
a) the value of k so that $f(x_1, x_2)$ is a joint pdf.
b) $P(X_1 \ge 0.5, X_2 \le 0.4)$.
c) $f_1(x_1)$ and $\mathcal{S}_{X_1}$.
d) $f_2(x_2)$ and $\mathcal{S}_{X_2}$.
e) $f(x_2|x_1)$ and $\mathcal{S}_{X_2|X_1}$.
f) $F(x_2|x_1)$.
g) $P(X_2 \ge 0.4|X_1 = 0.8)$.

2.2.6 Let $\vec{X}$ be a bivariate random vector with joint pdf $f(x_1, x_2) = kx_1x_2$ for $0 < x_1 + x_2 < 2$. Determine
a) the value of k so that $f(x_1, x_2)$ is a joint pdf.
b) $f_1(x_1)$ and $\mathcal{S}_{X_1}$.
c) $f_2(x_2)$ and $\mathcal{S}_{X_2}$.
d) $f(x_1|x_2)$ and $\mathcal{S}_{X_1|X_2}$.
e) $P(0.25 < X_1 \le 1.5|X_2 = 0.6)$.

2.2.7 Let $\vec{X}$ be a bivariate random vector with joint pdf $f(x_1, x_2) = 2e^{-x_1-x_2}$ for $0 < x_1 < x_2 < \infty$. Determine
a) $P(X_1 > 2, X_2 < 4)$.
b) $f_1(x_1)$ and S_{X_1}.
c) $f_2(x_2)$ and S_{X_2}.
d) $f(x_1|x_2)$ and $S_{X_1|X_2}$.
e) $f(x_2|x_1)$ and $S_{X_2|X_1}$.

2.2.8 Let $\vec{X}$ be a bivariate random vector with joint pdf $f(x_1, x_2) = e^{-x_1-x_2}$ for $0 < x_1 < \infty, 0 < x_2 < \infty$. Determine
a) $P(X_1 < X_2)$.
b) $P(X_1 + X_2 < 5)$.

2.2.9 Let $\vec{X}$ be a three-dimensional random vector with $f(x_1, x_2, x_3) = 6$ for $0 < x_1 < x_2 < x_3 < 1$. Determine
a) $f_1(x_1)$ and S_{X_1}.
b) $f_2(x_2)$ and S_{X_2}.
c) $f_3(x_3)$ and S_{X_3}.
d) $f(x_1, x_2|x_3)$ and $S_{X_1,X_2|X_3}$.

2.2.10 Let $\vec{X}$ be a three-dimensional random vector with $f(x_1, x_2, x_3) = 24x_1$ for $0 < x_1 < x_2 < x_3 < 1$. Determine
a) $f_1(x_1)$ and S_{X_1}.
b) $f_2(x_2)$ and S_{X_2}.
c) $f_3(x_3)$ and S_{X_3}.
d) $f(x_1, x_3|x_2)$ and $S_{X_1,X_3|X_2}$.

2.2.11 Let X_1 and X_2 be random variables with $f(x_2|x_1) = \frac{1}{3}\left(\frac{2}{3}\right)^{x_2-x_1}$ for $x_2 \geq x_1, x_2 \in \mathbb{N}$ and $f_1(x_1) = \frac{1}{2^{x_1}}$ for $x_1 \in \mathbb{N}$. Determine the marginal pdf of X_2 and S_{X_2}.

2.2.12 Let X_1 and X_2 be random variables with $f(x_1|x_2) = \frac{2x_1}{x_2^2}$ for $0 < x_1 < x_2$ and $f_2(x_2) = 3x_2^2$ for $0 < x_2 < 1$. Determine the marginal pdf of X_1 and S_{X_1}.

2.2.13 Let X_1 and X_2 be random variables with $f(x_2|x_1) = \frac{3x_2^2}{1-x_1^3}$ for $x_1 < x_2 < 1$ and $f_1(x_1) = \frac{4}{3}(1-x_1^3)$ for $0 < x_1 < 1$. Determine the marginal pdf of X_2 and S_{X_2}.

2.2.14 Let X_1 and X_2 be random variables with $f(x_2|x_1) = \dfrac{2(x_1 + x_2)}{(1 - x_1)(1 + 3x_1)}$ for $x_1 < x_2 < 1$ and $f_1(x_1) = (1 + 2x_1 - 3x_1^2)$ for $0 < x_1 < 1$. Determine the marginal pdf of X_2 and $\mathcal{S}_{X_2}$.

2.2.15 Let X_1 and X_2 be random variables with $f(x_1|x_2) = x_2 e^{-x_2 x_1}$ for $x_1 > 0$ and $f_2(x_2) = e^{-x_2}$ for $0 < x_2 < \infty$.
a) Determine the marginal pdf of X_1 and $\mathcal{S}_{X_1}$.
b) Determine $F_1(x_1)$, $\forall x_1 \in \mathbb{R}$.

2.2.16 Let $\vec{X}$ be a bivariate random vector with joint CDF $F(x_1, x_2)$. Show that
a) $F(x_1, x_2) \leq F(x_1 + \epsilon, x_2)$, $\forall \epsilon > 0$.
b) $F(x_1, x_2) \leq F(x_1, x_2 + \epsilon)$, $\forall \epsilon > 0$.
c) $P(a < X_1 \leq b, c < X_2 \leq d) = F(b, d) - F(a, d) - F(b, c) + F(a, c)$ whenever $a < b$ and $c < d$.

2.3 Independent Random Variables

Recall that two events A and B are independent when $P(A \cap B) = P(A)P(B)$, and a collection of events are mutually independent events when the probability of the intersection of any subset of the collection of events is equal to the product of the event probabilities. The definition of independent events can be extended to two or more random variables in an analogous fashion. The definition of independent random variables is given in Definition 2.14.

Definition 2.14 A collection of random variables $\{X_1, X_2, \ldots, X_k\}$ is said to be a collection of independent random variables when

$$F(x_1, x_2, \ldots, x_k) = \prod_{i=1}^{k} F_i(x_i), \ \forall (x_1, x_2, \ldots, x_k) \in \mathbb{R}^k,$$

where $F_i(x_i)$ is the marginal CDF of the random variable X_i.

Now, because a pdf contains complete information about a random variable, an equivalent definition of independent random variables based on pdfs is given in Definition 2.15.

Definition 2.15 A collection of random variables $X_1, X_2, \ldots, X_n$ are said to be independent random variables if and only if

$$f(x_1, x_2, \ldots, x_k) = \prod_{i=1}^{k} f_i(x_i), \ \forall (x_1, x_2, \ldots, x_k) \in \mathcal{S}_{(X_1, X_2, \ldots, X_n)}.$$

Table 2.2 The joint pdf of (X, Y) for Example 2.19.

X	Y 0	1
0	$\frac{1}{3}$	$\frac{1}{6}$
1	$\frac{1}{4}$	$\frac{1}{4}$

Note that for $X_1, \ldots, X_n$ to be independent random variables, (i) the factorization of the joint pdf and joint CDF must hold for every $(x_1, \ldots, x_k)$ and (ii) the support must be Cartesian product of the supports of the component random variable. Thus, for the random variables $X_1, X_2, \ldots, X_k$ in $\vec{X}$ to be independent, it must be the case that $\mathcal{S}_{\vec{X}}$ factors into $\mathcal{S}_{X_1} \times \mathcal{S}_{X_2} \times \cdots \times \mathcal{S}_{X_k}$, where $\mathcal{S}_{X_i}$ is the support of X_i. Random variables that are not independent are called *dependent random variables*.

Example 2.19 Let X and Y be random variables with joint pdf given in Table 2.2. If X and Y are to be independent, then $f(x, y) = f_1(x) f_2(y)$, $\forall (x\,y) \in \mathcal{S}_{\vec{X}}$. The marginal pdf of X is found by summing across a row, and the marginal pdf for Y is found by summing down the column. Thus, $f_1(0) = \frac{1}{3} + \frac{1}{6} = \frac{1}{2}$ and $f_2(0) = \frac{1}{3} + \frac{1}{4} = \frac{7}{12}$, but $f(0,0) = \frac{1}{3} \neq \frac{1}{2} \times \frac{7}{12}$. Hence, X and Y are dependent random variables.

Example 2.20 Let $\vec{X}$ be a continuous random vector with joint pdf $f(x_1, x_2) = e^{-x_1 - x_2}$ for $(x_1, x_2) \in \mathbb{R}^+ \times \mathbb{R}^+$. Then,

$$f_1(x_1) = \int_0^\infty e^{-x_1 - x_2} \, dx_2 = e^{-x_1} \int_0^\infty e^{-x_2} \, dx_2 = e^{-x_1}$$

for $x_1 \in \mathbb{R}^+$ and by symmetry, $f_2(x_2) = e^{-x_2}$ for $x_2 \in \mathbb{R}^+$.

Thus, $f(x_1, x_2) = f_1(x_1) f_2(x_2)$, $\forall (x_1, x_2) \in \mathcal{S}_{\vec{X}}$ and $\mathcal{S}_{\vec{X}} = \mathcal{S}_{X_1} \times \mathcal{S}_{X_2}$, and hence, X_1 and X_2 are independent random variables.

Note that independence of X_1 and X_2 does not simply follow from having the joint pdf of (X_1, X_2) factor into the product of functions of X_1 and X_2 alone. To see this, consider the random vector $\vec{X}$ given in Example 2.21.

Example 2.21 Let $\vec{X}$ be a continuous random vector with joint pdf $f(x_1, x_2) = 2e^{-x_1 - x_2}$ for $0 < x_1 < x_2 < \infty$. Note that the joint pdf factors into

$$f(x_1, x_2) = 2e^{-x_1} \times e^{-x_2} = g(x_1) \times h(x_2);$$

however, the joint support of (X_1, X_2) is not of $\mathcal{S}_{X_1} \times \mathcal{S}_{X_2}$. Hence, the random variables in Example 2.21 are not independent random variables.

Definition 2.16 A collection of random variables $X_1, X_2, \ldots, X_n$ is said to be a collection of independently identically distributed(iid) random variables if and only if $X_1, X_2, \ldots, X_n$ are independent random variables and the pdf of each random variable is identical.

Note that iid random variables have common pdf, CDF, and support. Thus, when $X_1, X_2, \ldots, X_n$ are iid, only the common pdf needs to be specified to completely describe $X_1, \ldots, X_n$. Furthermore, when $X_1, X_2, \ldots, X_n$ are iid random variables with common pdf $f(x_1, x_2, \ldots, x_n)$ and common CDF $F(x_1, x_2, \ldots, x_n)$,

i) $f(x_1, x_2, \ldots, x_n) = \prod_{i=1}^{n} f(x_i)$.
ii) $F(x_1, x_2, \ldots, x_n) = \prod_{i=1}^{n} F(x_i)$.

Example 2.22 Suppose that $X_1, X_2, \ldots, X_n$ are iid random variables with common pdf $f(x) = \frac{1}{\theta} e^{-\frac{x}{\theta}}$ for $x > 0$. Then, the joint pdf of $X_1, X_2, \ldots, X_n$ is

$$f(x_1, x_2, \ldots, x_n) = \prod_{i=1}^{n} \frac{1}{\theta} e^{-\frac{x_i}{\theta}} = \frac{1}{\theta^n} e^{-\frac{1}{\theta}\sum_{i=1}^{n} x_i}.$$

For $x > 0$, the common CDF of $X_1, X_2, \ldots, X_n$ is $F(x) = 1 - e^{-\frac{x}{\theta}}$, and therefore, the joint CDF of $X_1, X_2, \ldots, X_n$ is

$$F(x_1, x_2, \ldots, x_n) = \prod_{i=1}^{n} (1 - e^{-\frac{x_i}{\theta}})$$

provided $x_i > 0$ for $i = 1, 2, \ldots, n$.

An important property of independent random variables is given in Theorem 2.7. In particular, when functions are applied to independent random variables $X_1, \ldots, X_n$, the transformed random variables are also independent.

Theorem 2.7 *If $X_1, X_2, \ldots, X_n$ are independent random variables and g_i is a function for $i = 1, \ldots, n$, then $g_1(X_1), g_2(X_2), \ldots, g_n(X_n)$ are independent random variables.*

Thus, if $X_1, X_2, \ldots, X_n$ are independent, then so are $X_1^2, X_2^2, \ldots, X_n^2$. Furthermore, when $\vec{Y}$ and $\vec{Z}$ are random vectors based on disjoint collections of the random variables $X_1, X_2, \ldots, X_n$ and $g_1(y)$ and $g_2(z)$ are functions mapping into $\mathbb{R}$, then $g_1(Y)$ and $g_2(Z)$ are also independent random variables.

Example 2.23 If $X_1, X_2, \ldots, X_n$ are independent random variables, then X_1 and $\sum_{i=2}^{n} X_i$ are independent random variables. Also, $X_1 + X_2$ and $X_3 - X_4$ are independent random variables.

Problems

2.3.1 Let $\vec{X} = (X_1, X_2)$ be a discrete random vector with joint pdf given in Table 2.3. Determine
a) $f_1(x_1)$ and $\mathcal{S}_{X_1}$.
b) $f_2(x_2)$ and $\mathcal{S}_{X_2}$.
c) whether or not X_1 and X_2 are independent random variables.

2.3.2 Let $\vec{X} = (X_1, X_2)$ be a discrete random vector with joint pdf given in Table 2.4. Determine
a) $f_1(x_1)$ and $\mathcal{S}_{X_1}$.
b) $f_2(x_2)$ and $\mathcal{S}_{X_2}$.
c) whether or not X_1 and X_2 are independent random variables.

2.3.3 Let X_1 and X_2 be random variables with joint pdf $f(x_1, x_2) = \frac{e^{-1}}{2^{x_1} x_2!}$ for $x_1 \in \mathbb{N}$ and $x_2 \in \mathbb{W}$. Determine
a) the marginal pdf of X_1 and $\mathcal{S}_{X_1}$.
b) the marginal pdf of X_2 and $\mathcal{S}_{X_2}$.
c) whether or not X_1 and X_2 are independent random variables.

2.3.4 Let X_1 and X_2 be random variables with joint pdf $f(x_1, x_2) = \left(\frac{1}{3}\right)^{x_1 + x_2} \left(\frac{2}{3}\right)^{2 - x_1 - x_2}$ for $x_1 = 0, 1$ and $x_2 = 0, 1$. Determine

Table 2.3 The joint pdf for Problem 2.3.1.

		X_2	
X_1	1	2	3
0	0.10	0.20	0.10
1	0.15	0.15	0.10
2	0.05	0.1	0.05

Table 2.4 The joint pdf for Problem 2.3.2.

		X_2	
X_1	−1	0	1
1	0.10	0.20	0.10
2	0.15	0.30	0.15

a) the marginal pdf of X_1 and S_{X_1}.
b) the marginal pdf of X_2 and S_{X_2}.
c) whether or not X_1 and X_2 are independent random variables.

2.3.5 Let X_1 and X_2 be random variables with joint pdf $f(x_1, x_2) = 2x_1$ for $0 < x_1 < 1$ and $0 < x_2 < 1$. Determine
a) the marginal pdf of X_1 and S_{X_1}.
b) the marginal pdf of X_2 and S_{X_2}.
c) whether or not X_1 and X_2 are independent random variables.

2.3.6 Let X_1 and X_2 be random variables with joint pdf $f(x_1, x_2) = 6x_2$ for $x_1 > 0, x_2 > 0$ and $0 < x_1 + x_2 < 1$. Determine
a) the marginal pdf of X_1 and S_{X_1}.
b) the marginal pdf of X_2 and S_{X_2}.
c) whether or not X_1 and X_2 are independent random variables.

2.3.7 Let X_1, X_2, and X_3 be random variables with joint pdf $f(x_1, x_2, x_3) = 48x_1x_2x_3$ for $0 < x_1 < x_2 < x_3 < 1$. Determine
a) the marginal pdf of X_1 and S_{X_1}.
b) the marginal pdf of X_2 and S_{X_2}.
c) the marginal pdf of X_3 and S_{X_3}.
d) whether or not X_1, X_2, and X_3 are independent random variables.

2.3.8 Let X_1, X_2, and X_3 be random variables with joint pdf $f(x_1, x_2, x_3) = \frac{2}{3}(x_1 + x_2 + x_3)$ for $0 < x_1 < 1, 0 < x_2 < 1$ and $0 < x_3 < 1$. Determine
a) the marginal pdf of X_1 and S_{X_1}.
b) the marginal pdf of X_2 and S_{X_2}.
c) the marginal pdf of X_3 and S_{X_3}.
d) whether or not X_1, X_2, and X_3 are independent random variables.

2.3.9 Let $X_1, X_2, \ldots, X_n$ be independent random variables. Does Theorem 2.7 imply
a) $\sum_{i=1}^{5} X_i$ and $\sum_{i=6}^{n} X_i$ are independent?
b) $\sum_{i=1}^{n} X_i$ and $\sum_{i=1}^{n} X_i^2$ are independent?
c) $\sum_{i=1}^{5} X_i$ and $\sum_{i=6}^{n} X_i^2$ are independent?

2.3.10 Let $X_1, X_2, \ldots, X_n$ be independent random variables with pdf $f_i(x_i) = \frac{1}{a_i}e^{-\frac{x_i}{a_i}}$, for $x_i > 0$ and $a_i > 0$. Determine
a) the joint pdf of $X_1, X_2, \ldots, X_n$.
b) the joint CDF of $X_1, X_2, \ldots, X_n$.

2.3.11 Let $X_1, X_2, \ldots, X_n$ be independent random variables with the pdf of X_i given by $f_i(x_i) = \frac{1}{a_i}$, for $0 < x_i < a_i$. Determine
a) the joint pdf of $X_1, X_2, \ldots, X_n$.
b) the joint CDF of $X_1, X_2, \ldots, X_n$.

2.3.12 Let $X_1, X_2, \ldots, X_n$ be independent random variables with the pdf of X_i given by $f_i(x_i) = \left(1 - \frac{1}{2^i}\right) \times \left(\frac{1}{2^i}\right)^{x_i}$, for $x_i \in \mathbb{W}$. Determine
a) the joint pdf of $X_1, X_2, \ldots, X_n$.
b) the joint CDF of $X_1, X_2, \ldots, X_n$.

2.3.13 Let $X_1, X_2, \ldots, X_n$ be iid random variables with common pdf $f(x) = \theta e^{-\theta x}$ for $x > 0$. Determine
a) the joint pdf of $X_1, X_2, \ldots, X_n$.
b) the joint CDF of $X_1, X_2, \ldots, X_n$.

2.3.14 Let $X_1, X_2, \ldots, X_n$ be iid random variables with common pdf $f(x) = \theta x^{\theta-1}$ for $0 < x < 1$. Determine
a) the joint pdf of $X_1, X_2, \ldots, X_n$.
b) the joint CDF of $X_1, X_2, \ldots, X_n$.

2.4 Transformations of Random Variables

Many of the basic ideas of the theory of statistics are based on a collection of iid random variables $X_1, \ldots, X_n$ with probability model $f(x; \theta)$. In this case, the collection of random variables $X_1, \ldots, X_n$, which is called a *random sample*. Also, the parameter θ in the probability model is unknown, and the information in the random sample is used for estimating and testing claims about θ. The formulas derived for estimating θ are often based on transformations of the random variables $X_1, \ldots, X_n$. For example, common transformations that arise in statistics are $|X|, X^2$, and $\ln(X)$.

In this section, transformations of random variables and bivariate random vectors are studied. In particular, the focus of this section is to determine the distribution of a transformed random variable or transformations of a bivariate random vector.

2.4.1 Transformations of Discrete Random Variables

When X is a discrete random variable, $g(x)$ is a function defined on $\mathcal{S}_X$, and $Y = g(X)$ is a transformation of X, the pdf of Y is $f_Y(y) = P(Y = y)$ can be converted into a probability statement about the random variable X. In particular,

Table 2.5 The probability distribution of X in Example 2.24.

X	-2	-1	0	1	2
$f(x)$	0.10	0.15	0.20	0.25	0.30

for $y \in \mathcal{S}_Y = \{y : \exists\, x \in \mathcal{S}_X \text{ such that } g(x) = y\}$, the pdf of Y is

$$f_Y(y) = P(Y = y) = P(g(X) = y) = P(X \in B) = \sum_{x \in B} f(x),$$

where $B = \{x \in \mathcal{S}_X : g(x) = y\}$.

Example 2.24 Let X be a discrete random variable with pdf given in Table 2.5 and let $Y = |X|$.

Now, $Y = |X|$ maps $\mathcal{S}_X$ to $\mathcal{S}_Y = \{0, 1, 2\}$. For $y = 0$,

$$f_y(0) = P(Y = 0) = P(|X| = 0) = P(X = 0) = f(0) = 0.20;$$

for $y = 1$

$$f_y(1) = P(Y = 1) = P(|X| = 1) = P(X = \pm 1) = f(-1) + f(1) = 0.40;$$

and for $y = 2$

$$f_y(2) = P(Y = 2) = P(|X| = 2) = P(X = \pm 2) = f(-2) + f(2) = 0.40.$$

Thus, the pdf of $Y = |X|$ is

$$f_Y(y) = \begin{cases} 0.20 & y = 0, \\ 0.40 & y = 1, \\ 0.40 & y = 2. \end{cases}$$

Example 2.25 Let X be a discrete random variable with pdf $f(x) = \binom{n}{x} p^x (1-p)^{n-x}$ for $x = 0, 1, \ldots, n$ and $Y = \frac{X}{n}$. Then, $\mathcal{S}_Y = \{0, \frac{1}{n}, \frac{2}{n}, \ldots, \frac{n-1}{n}, 1\}$, and for $y \in \mathcal{S}_Y$,

$$f_Y(y) = P(Y = y) = P\left(\frac{X}{n} = y\right) = P(X = ny) = \binom{n}{ny} p^{ny} (1-p)^{n-ny}.$$

2.4.2 Transformations of Continuous Random Variables

When X is a continuous random variable with pdf $f(x)$, $g(x)$ is a function defined on $\mathcal{S}_X$, and $Y = g(X)$ is a transformation of X, the CDF of Y can be found by converting it into a probability statement about the random variable X. In particular, for $y \in \mathcal{S}_Y = \{y : \exists\, x \in \mathcal{S}_X \text{ such that } g(x) = y\}$, the CDF of Y is

$$F_Y(y) = P(Y \leq y) = P(g(X) \leq y) = P(X \in B) = \int_{x \in B} f(x)\, dx,$$

where $B = \{x \in \mathcal{S}_X : g(x) \leq y\}$.

Note that when X is a continuous random variable, the pdf of Y, $f_Y(y)$, is the derivative of the CDF of Y. That is, $f_Y(y) = F_Y'(y)$. This method of finding the pdf of Y is called the *CDF method*.

Examples 2.26–2.28 illustrate the CDF method for determining the pdf of a transformation $Y = g(X)$.

Example 2.26 Let X be a continuous random variable with pdf $f(x) = 3x^2$ for $0 < x < 1$ and $Y = -\ln(X)$. First, Y maps $\mathcal{S}_X = (0, 1)$ to $\mathcal{S}_Y = (0, \infty)$. Now, consider the CDF of Y for $y \in (0, \infty)$.

$$\begin{aligned}
F_Y(y) &= P(Y \leq y) = P(-\ln(X) \leq y) = P(X > e^{-y}) \\
&= \int_{e^{-y}}^{1} 3x^2\, dx = x^3 \Big|_{e^{-y}}^{1} \\
&= 1 - e^{-3y}.
\end{aligned}$$

Hence, the pdf of Y is $f_Y(y) = \frac{d}{dy} F_Y(y) = 3e^{-3y}$, $y \in (0, \infty)$.

Example 2.27 Let X be a continuous random variable with pdf $f(x) = \frac{1}{2}$ for $-1 < x < 1$ and $Y = X^2$. First, Y maps $\mathcal{S}_X = (-1, 1)$ to $\mathcal{S}_Y = (0, 1)$. Consider the CDF of Y for $y \in (0, 1)$.

$$\begin{aligned}
F_Y(y) &= P(Y \leq y) = P(X^2 \leq y) = P(-\sqrt{y} < X < \sqrt{y}) \\
&= \int_{-\sqrt{y}}^{\sqrt{y}} \frac{1}{2}\, dx = \frac{x}{2} \Big|_{-\sqrt{y}}^{\sqrt{y}} \\
&= \sqrt{y}.
\end{aligned}$$

Thus, the pdf of Y is $f_Y(y) = \frac{dF_Y(y)}{dy} = \frac{1}{2\sqrt{y}}$, $y \in (0, 1)$.

Example 2.28 Let X be a continuous random variable with pdf $f(x) = \frac{1}{4}$ for $-3 < x < 1$ and $Y = |X|$. Then, Y maps $\mathcal{S}_X = (-3, 1)$ to $\mathcal{S}_Y = (0, 3)$, and for $y \in (0, 3)$, the CDF of Y is

$$F_Y(y) = P(Y \leq y) = P(|x| \leq y) = \begin{cases} P(-y \leq X \leq y) & \text{when } y \in (0, 1) \\ P(-1 \leq X \leq y) & \text{when } y \in (1, 3) \end{cases}$$

since $\mathcal{S}_X = (-3, 1)$. Thus,

$$F_Y(y) = \begin{cases} \int_{-y}^{y} \frac{1}{4}\, dy = \frac{y}{2} & \text{when } 0 < y < 1, \\ \int_{-1}^{y} \frac{1}{4}\, dy = \frac{1+y}{4} & \text{when } 1 \leq y < 3. \end{cases}$$

Figure 2.5 The pdf of
$Y = |X|$ in Example 2.28.

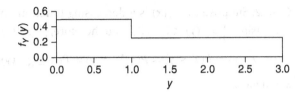

Hence, the pdf of $Y = |X|$ is

$$f_Y(y) = \begin{cases} \frac{1}{2} & \text{when } 0 < y < 1, \\ \frac{1}{4} & \text{when } 1 \leq y < 3, \end{cases}$$

which is shown in Figure 2.5.

When $Y = g(X)$ is a one-to-one transformation on $\mathcal{S}_X$, such as in Example 2.26, Theorem 2.8 shows how the pdf of Y is found by applying the CDF method.

Theorem 2.8 *Let X be a continuous random variable with pdf $f(x)$ on support $\mathcal{S}_X = \{x : f(x) > 0\}$, $g(x)$ a one-to-one transformation mapping $\mathcal{S}_X$ onto $\mathcal{S}_Y$, and $g^{-1}(y)$ the inverse function associated with $g(x)$ on $\mathcal{S}_X$. If the derivative of $g^{-1}(y)$ with respect to y is continuous and nonzero on $\mathcal{S}_Y$, then the pdf of Y is given by*

$$f_Y(y) = f(g^{-1}(y)) \left| \frac{dg^{-1}(y)}{dy} \right|$$

for $y \in \mathcal{S}_Y$.

Proof. Let X be a continuous random variable with pdf $f(x)$ on support $\mathcal{S}_X = \{x : f(x) > 0\}$, $g(x)$ a one-to-one transformation mapping $\mathcal{S}_X$ onto $\mathcal{S}_Y$, and let $g^{-1}(y)$ be the inverse function associated with $g(x)$ on $\mathcal{S}_X$ with continuous nonzero derivative with respect to y on $\mathcal{S}_Y$. Let $y \in \mathcal{S}_Y$ and consider $F_Y(y)$.

$$F_Y(y) = P(Y \leq y) = P(g(X) \leq y).$$

Now, since $g(x)$ is a one-to-one function on $\mathcal{S}_X$, it follows that either $g(x)$ is increasing or $g(x)$ is decreasing on $\mathcal{S}_X$.

Case 1: Suppose that $g(x)$ is an increasing function on $\mathcal{S}_X$. Then, $g^{-1}(y)$ is also increasing and $\frac{d}{dy}g^{-1}(y) > 0$ on $\mathcal{S}_Y$. Thus,

$$F_Y(y) = P(Y \leq y) = P(g(X) \leq y) = P(X \leq g^{-1}(y)) = F_X(g^{-1}(y)),$$

where $F_X(x)$ is the CDF associated with the random variable X. Hence,

$$f_Y(y) = \frac{d}{dy}F_Y(y) = \frac{d}{dy}F_X(g^{-1}(y)) = f(g^{-1}(y))\frac{d}{dy}g^{-1}(y)$$

and since $\frac{d}{dy}g^{-1}(y) > 0$ when $g(x)$ is increasing, $f_Y(y) = f(g^{-1}(y)) \left| \frac{d}{dy}g^{-1}(y) \right|$.

Case 2: Suppose that $g(x)$ is a decreasing function on $\mathcal{S}_X$. Then, $g^{-1}(y)$ is also decreasing, $\frac{d}{dy}g^{-1}(y) < 0$ on $\mathcal{S}_Y$, and therefore, $\left|\frac{d}{dy}g^{-1}(y) < 0\right| = -\frac{d}{dy}g^{-1}(y)$. Thus,

$$F_Y(y) = P(Y \le y) = P(g(X) \le y) = P(X \ge g^{-1}(y)) = 1 - F_X(g^{-1}(y)),$$

and hence,

$$f_Y(y) = \frac{d}{dy}F_Y(y) = \frac{d}{dy}[1 - F_X(g^{-1}(y))]$$

$$= -f(g^{-1}(y))\frac{d}{dy}g^{-1}(y) = f(g^{-1}(y))\left(-\frac{d}{dy}g^{-1}(y)\right)$$

$$= f(g^{-1}(y))\left|\frac{d}{dy}g^{-1}(y)\right|.$$

Therefore, in either case, $f_Y(y) = f(g^{-1}(y))\left|\frac{dg^{-1}(y)}{dy}\right|$. ∎

Example 2.29 Let X be a continuous random variable with pdf $f(x) = \frac{1}{\sqrt{2\pi}}e^{-\frac{x^2}{2}}$ for $x \in \mathbb{R}$ and $Y = e^X$. Then, $Y = g(X) = e^X$ is a one-to-one transformation mapping $\mathbb{R}$ to $\mathbb{R}^+$, and $g^{-1}(x) = \ln(x)$ has continuous nonzero derivative on $\mathbb{R}^+$. Hence, for $y > 0$, the pdf of $Y = e^X$ is

$$f_Y(y) = f(\ln(y))\left|\frac{d}{dy}\ln(y)\right| = \frac{1}{\sqrt{2\pi}}e^{-\frac{\ln(y)^2}{2}}\left|\frac{1}{y}\right| = \frac{1}{y\sqrt{2\pi}}e^{-\frac{\ln(y)^2}{2}}$$

for $y \in \mathcal{S}_Y$.

An important transformation of a continuous random variable X is the transformation $Y = F(X)$, where $F(X)$ is the CDF of X. This transformation is called the *Probability Integral Transformation* and results in the same distribution for each continuous random variable. In particular, Theorem 2.9 shows that $Y = F(X)$ follows a *uniform distribution* over the interval $(0, 1)$. The definition of a uniform distribution will be given before stating the Probability Integral Transformation, Theorem 2.9.

Definition 2.17 A random variable X is said to follow a uniform distribution over the interval $(0, 1)$, denoted by $X \sim U(0, 1)$, if and only if the CDF of X is $F(x) = x$ for $x \in (0, 1)$.

Theorem 2.9 (Probability Integral Transform) *If X is a continuous random variable with CDF $F(x)$, then $Y = F(X)$ is distributed as a $U(0, 1)$ random variable.*

Proof. Let X be a continuous random variable with CDF $F(x)$. Then, $Y = F(X)$ maps $\mathcal{S}_X$ onto $(0, 1)$. Since $F(x)$ is a continuous, it follows that $F(x)$ is a one-to-one increasing function with inverse $F^{-1}(y)$. Let $y \in (0,1)$ and consider $F_Y(y)$.

$$F_Y(y) = P(Y \leq y) = P(F(X) \leq y) = P(X \leq F^{-1}(y)) = F(F^{-1}(y)) = y.$$

Thus, the CDF of Y is the CDF of a $U(0,1)$ random variable, and therefore, $Y = F(X) \sim U(0,1)$. ∎

Example 2.30 Let X be a continuous random variable with pdf $f(x) = 4x^3$ for $x \in (0,1)$. Then, $F(x) = x^4$ for $x \in (0,1)$, and hence, $Y = X^4 \sim U(0,1)$.

Example 2.31 Let X be a continuous random variable with pdf $f(x) = 4e^{-4x}$ for $x \in \mathbb{R}^+$. Then, $F(x) = 1 - e^{-4x}$ for $x \in \mathbb{R}^+$, and hence, $Y = 1 - e^{-4X} \sim U(0,1)$.

2.4.3 Transformations of Continuous Bivariate Random Vectors

Because general transformations of k-dimensional random vectors can be complicated, only one-to-one transformations of bivariate continuous random vectors will be discussed in this section; specific transformations of random vectors, such as the sums and linear combinations of random variables, will be dealt with in later sections of Chapters 2 and 3.

When $\vec{X} = (X_1, X_2)$ is a bivariate continuous random vector, and $\vec{Y} = (g_1(x_1, x_2), g_2(x_1, x_2))$ is a one-to-one transformation from $\mathcal{S}_{\vec{X}}$ onto $\mathcal{S}_{\vec{Y}}$, then there exist functions $g_1^{-1}(y_1, y_2)$ and $g_2^{-1}(y_1, y_2)$ on $\mathcal{S}_{\vec{Y}}$ such that $x_1 = g_1^{-1}(y_1, y_2)$ and $x_2 = g_2^{-1}(y_1, y_2)$. The Jacobian of the transformation is the determinant of the 2×2 matrix of partial derivatives of $\frac{\partial x_i}{\partial y_j}$, $i, j = 1, 2$. In particular, the Jacobian is

$$|J| = \begin{vmatrix} \frac{\partial x_1}{\partial y_1} & \frac{\partial x_1}{\partial y_2} \\ \frac{\partial x_2}{\partial y_1} & \frac{\partial x_2}{\partial y_2} \end{vmatrix} = \begin{vmatrix} \frac{\partial g_1^{-1}}{\partial y_1} & \frac{\partial g_1^{-1}}{\partial y_2} \\ \frac{\partial g_2^{-1}}{\partial y_1} & \frac{\partial g_2^{-1}}{\partial y_2} \end{vmatrix}.$$

The Jacobian is used in determining the joint pdf of $\vec{Y}$ in place of $|\frac{d}{dy}g^{-1}(y)|$ in Theorem 2.8.

The joint pdf of a one-to-one transformation of a continuous bivariate random vector is given in Theorem 2.10.

Theorem 2.10 *Let $\vec{X} = (X_1, X_2)$ be a continuous bivariate random vector with joint pdf $f_{\vec{X}}(x_1, x_2)$ on $\mathcal{S}_{\vec{X}}$ and $\vec{Y} = (g_1(X_1, X_2), g_2(X_1, X_2))$ be a transformation of $\vec{X}$. If $y_1 = g_1(x_2, x_2)$ and $y_2 = g_2(x_1, x_2)$ are a one-to-one transformation mapping $\mathcal{S}_{\vec{X}}$ onto $\mathcal{S}_{\vec{Y}}$, the first partial derivatives of $x_1 = g_1^{-1}(y_1, y_2)$ and $x_2 = g_2^{-1}(y_1, y_2)$ are continuous over $\mathcal{S}_{\vec{Y}}$, and the Jacobian $|J|$ of the transformation is nonzero on $\mathcal{S}_{\vec{X}}$,* *then the joint density of* $\vec{Y} = (Y_1, Y_2)$ *is*

$$f_{\vec{Y}}(y_1, y_2) = f_{\vec{X}}(g_1^{-1}(y_1, y_2), g_2^{-1}(y_1, y_2))|J|, \quad \text{for } (y_1, y_2) \in S_{\vec{Y}}.$$

After applying Theorem 2.10 to find the joint pdf of the transformation $\vec{Y}$, the marginal pdf of Y_1 can be found by integrating over Y_2. Similarly, the marginal pdf of Y_2 is found by integrating over Y_1.

Example 2.32 Let X_1 and X_2 be iid random variables with common pdf $f(x) = 3e^{-3x}$ for $x > 0$. Let $Y_1 = X_1 + X_2$ and $Y_2 = X_2$. Then, $g(x_1, x_2) = (x_1 + x_2, x_2)$ is a one-to-one transformation mapping $S_{\vec{X}} = \mathbb{R}^+ \times \mathbb{R}^+$ to $S_{\vec{Y}} = \{(y_1, y_2) : y_1 > 0, 0 < y_2 < y_1\}$. Note that $y_2 \leq y_1$ since $y_1 = x_1 + x_2 = x_1 + y_2$ and $x_1 > 0$.

Thus, $(x_1, x_2) = (g_1^{-1}(y_1, y_2), g_2^{-1}(y_1, y_2)) = (y_1 - y_2, y_2)$, and the Jacobian for this transformation is

$$|J| = \begin{vmatrix} \frac{\partial x_1}{\partial y_1} & \frac{\partial x_1}{\partial y_2} \\ \frac{\partial x_2}{\partial y_1} & \frac{\partial x_2}{\partial y_2} \end{vmatrix} = \begin{vmatrix} 1 & -1 \\ 0 & 1 \end{vmatrix} = 1.$$

Hence, $f_{\vec{Y}}(y_1, y_2) = f_{\vec{X}}(y_1 - y_2, y_2)|J| = 9e^{-3(y_1 - y_2) - 3y_2} = 9e^{-3y_1}$. The marginal pdf of Y_1 is

$$f_{Y_1}(y_1) = \int_0^{y_1} 9e^{-y_1}\, dy_2 = 9y_1 e^{-3y_1}, y_1 > 0.$$

Example 2.33 Suppose that X_1 and X_2 are iid random variables with common pdf $f(x) = e^{-x}$ for $x > 0$. Let $Y_1 = \frac{X_1}{X_2}$ and $Y_2 = X_2$. Then, $g(x_1, x_2) = \left(\frac{x_1}{x_2}, x_2\right)$ is a one-to-one transformation mapping $S_{\vec{X}} = \mathbb{R}^+ \times \mathbb{R}^+$ to $S_{\vec{Y}} = \mathbb{R}^+ \times \mathbb{R}^+$.

Now, $(x_1, x_2) = g^{-1}(y_1, y_2) = (y_1 y_2, y_2)$, and the Jacobian for this transformation is

$$|J| = \begin{vmatrix} \frac{\partial x_1}{\partial y_1} & \frac{\partial x_1}{\partial y_2} \\ \frac{\partial x_2}{\partial y_1} & \frac{\partial x_2}{\partial y_2} \end{vmatrix} = \begin{vmatrix} y_2 & y_1 \\ 0 & 1 \end{vmatrix} = |y_2| = y_2$$

since $y_2 = x_2 > 0$.

Hence, $f_{\vec{Y}}(y_1, y_2) = f_{\vec{X}}(y_1 y_2, y_2)|J| = y_2 e^{-y_1 y_2 - y_2}$, and the marginal pdf of Y_1 is

$$f_{Y_1}(y_1) = \int_0^\infty e^{-y_1 y_2 - y_2} y_2\, dy_2 = \int_0^\infty y_2 e^{-y_2(1 + y_1)}\, dy_2 = \frac{1}{(1 + y_1)^2}$$

for $y_1 > 0$.

Table 2.6 The pdf for Problem 2.4.1.

X	−1	0	1
$f(x)$	$\dfrac{1}{3}$	$\dfrac{1}{6}$	$\dfrac{1}{2}$

Problems

2.4.1 Let X be a discrete random variable with pdf given in Table 2.6. Determine the pdf and support of

a) $Y = X + 1$.   b) $Y = |X|$.

c) $Y = X^2$.

2.4.2 Let X be a discrete random variable with pdf $f(x) = \frac{1}{2^x}$ for $x \in \mathbb{N}$. Determine the pdf and support of

a) $Y = X - 1$.   b) $Y = 2X$.

c) $Y = \ln(X)$.

2.4.3 Let X be a continuous random variable with pdf $f(x) = \frac{1}{2}$ for $0 < x < 2$. Determine the pdf and support of

a) $Y = \frac{X}{2}$.   b) $Y = -\ln\left(\frac{X}{2}\right)$.

c) $Y = \frac{1}{X}$.

2.4.4 Let X be a continuous random variable with pdf $f(x) = \frac{1}{3}$ for $-1 < x < 2$. Determine the pdf and support of

a) $Y = X + 1$.   b) $Y = |X|$.

c) $Y = X^2$.

2.4.5 Let X be a continuous random variable with pdf $f(x) = 4x^3$ for $0 < x < 1$. Determine the pdf and support of

a) $Y = X^4$.   b) $Y = -\ln(X)$.

c) $Y = \frac{1}{X}$.

2.4.6 Let X be a continuous random variable with pdf $f(x) = e^{-(x+3)}$ for $x > -3$. Determine the pdf and support of

a) $Y = |X|$.   b) $Y = X^2$.

2.4.7 Let X be a continuous random variable with pdf $f(x) = 6x(1-x)$ for $0 < x < 1$. Determine the pdf and support of

a) $Y = 1 - X$.
b) $Y = 4X$.
c) $Y = -\ln(X)$.

2.4.8 Let X be a continuous random variable with pdf $f(x) = \frac{3}{2}x^2$ for $-1 < x < 1$. Determine the pdf and support of

a) $Y = |X|$.
b) $Y = |X|^3$.
c) $Y = X^4$.
d) $Y = X + 1$

2.4.9 Let X be a continuous random variable with pdf $f(x) = \frac{x}{2}$ for $0 < x < 2$. Determine the pdf and support of

a) $Y = \frac{X}{2}$.
b) $Y = X^3$.
c) $Y = (X-1)^2$.

2.4.10 Let X be a continuous random variable with pdf $f(x) = \frac{3}{x^4}$ for $1 < x < \infty$. Determine the pdf and support of

a) $Y = \frac{1}{X}$.
b) $Y = \ln(X)$.
c) $Y = 1 - \frac{1}{X^3}$.

2.4.11 Let X be a continuous random variable with pdf $f(x) = \theta x^{\theta - 1}$ for $0 < x < 1$ and $\theta > 0$. Determine the pdf and support of

a) $Y = X + 2$.
b) $Y = -\ln(X)$.
c) $Y = \sqrt{X}$.

2.4.12 Let X be a continuous random variable with pdf $f(x) = \frac{1}{\theta}e^{-\frac{x}{\theta}}$ for $x > 0$ and $\theta > 0$. Determine the pdf and support of

a) $Y = X + 5$
b) $Y = \frac{2}{\theta}X$.
c) $Y = X^2$.
d) $Y = e^X$.

2.4.13 Let X_1 and X_2 be random variables with joint pdf $f(x_1, x_2) = e^{-x_1 - x_2}$ for $x_1 > 0$ and $x_2 > 0$ and $Y_1 = X_1 + X_2$ and $Y_2 = X_2$. Determine the

a) $|J|$.
b) joint pdf of (Y_1, Y_2).
c) pdf of Y_1.
d) pdf of Y_2.

2.4.14 Let X_1 and X_2 be iid random variables with common pdf $f(x) = 1$ for $0 < x < 1$. Let $Y_1 = X_1 X_2$ and $Y_2 = X_1$. Determine the

a) joint pdf of (X_1, X_2).
b) $|J|$.
c) the joint pdf of (Y_1, Y_2).
d) pdf of Y_1.

2.4.15 Let X_1 and X_2 be random variables with joint pdf $f(x_1, x_2) = \frac{e^{-\frac{1}{2}(x_1^2 + x_2^2)}}{2\pi}$ for $x_1 \in \mathbb{R}$ and $x_2 \in \mathbb{R}$ and $Y_1 = X_1 + X_2$ and $Y_2 = X_1 - X_2$. Determine the

a) $|J|$.
b) joint pdf of (Y_1, Y_2).
c) the pdf of Y_1.
d) pdf of Y_2.

2.4.16 Let X_1 and X_2 be random variables with joint pdf $f(x_1, x_2) = \frac{e^{-\frac{1}{2}(x_1^2 + x_2^2)}}{2\pi}$ for $x_1 \in \mathbb{R}$ and $x_2 \in \mathbb{R}$ and $Y_1 = 4X_1 + 3X_2$ and $Y_2 = X_2$. Determine the

a) $|J|$.
b) joint pdf of (Y_1, Y_2).
c) pdf of Y_1.

2.4.17 Let X_1 and X_2 be iid random variables with common pdf $f(x) = 4x^3$ for $0 < x < 1$ and $Y_1 = -\ln(X_1) - \ln(X_2)$ and $Y_2 = -\ln(X_2)$. Determine the

a) joint pdf of (X_1, X_2).
b) $|J|$.
c) pdf of Y_1.
d) pdf of Y_2.

2.4.18 Let X be a continuous random variable with pdf $f(x) = \frac{2}{x^3}$ for $x > 1$. Determine

a) the CDF associated with X.
b) the distribution of $Y = 1 - \frac{1}{X^2}$.

2.4.19 Let X be a continuous random variable with pdf $f(x) = 2e^{-2x}$ for $x > 0$. Use Theorem 2.9 to determine the distribution of $Y = 1 - e^{-2X}$.

2.4.20 Let X be a continuous random variable with pdf $f(x)$ on $\mathcal{S}_X$ and $Y = aX + b$ for $a, \in \mathbb{R}$ with $a \neq 0$. Show that $f_Y(y) = \frac{1}{|a|} f\left(\frac{y-b}{a}\right)$.

2.5 Expected Values for Random Variables

Several useful characteristics of a random variable are based on the *moments* of a random variable. In particular, the moments of a random variable are used to summarize important characteristics such as the location or the spread of a random variable; however, unlike pdfs and CDFs, which contain complete information about a random variable, moments only provide summary information about a random variable. In this section, *expected values* and moments of random variables are introduced and their properties discussed.

2.5.1 Expected Values and Moments of Random Variables

Expected values and moments of a random variable often contain important information about a random variable; however, it is possible that a random will not have any moments.

Definition 2.18 Let X be a random variable with pdf $f(x)$ on $\mathcal{S}_X$. The expected value of the random variable X, denoted by $E(X)$, is defined to be

$$E(X) = \sum_{x \in S} x f(x)$$

when X is a discrete random variable provided the series converges, and

$$E(X) = \int_S x f(x)\, dx$$

when X is a continuous random variable provided the integral converges.

The expected value of a random variable X is also referred to as the *mean* or the *first moment* and is a weighted average of the values in $\mathcal{S}_X$. Commonly used notation for the mean of a random variable includes $E(X)$, μ, and μ_X.

Examples 2.34 and 2.35 show how to compute the expected value for a discrete random variable and a continuous random variable, respectively. An example of a random variable without a mean is given in Example 2.36.

Example 2.34 Let X be a discrete random variable with pdf $f(x) = \frac{1}{2^x}$ for $x \in \mathbb{N}$. Then,

$$E(X) = \sum_{x=1}^{\infty} x \cdot \frac{1}{2^x} = \sum_{x=1}^{\infty} x \cdot \frac{1}{2^{x-1}} \cdot \frac{1}{2}$$

$$= \frac{1}{2} \sum_{x=1}^{\infty} x \cdot \frac{1}{2^{x-1}} = \frac{1}{2} \frac{1}{\left(1 - \frac{1}{2}\right)^2} = 2$$

since

$$\sum_{x=1}^{\infty} x p^{x-1} = \frac{d}{dp}\left[\sum_{x=1}^{\infty} p^x\right] = \frac{d}{dp}\left[\frac{1}{1-p}\right] = \frac{1}{(1-p)^2}$$

for $|p| < 1$.

Example 2.35 Let X be a continuous random variable with pdf $f(x) = 3e^{-3x}$ for $x > 0$. Then,

$$E(X) = \int_0^{\infty} x \cdot 3e^{-3x}\, dx = \underbrace{-xe^{-3x}\Big|_0^{\infty} - \int_0^{\infty} -e^{-3x}\, dx}_{\text{int. by parts with } u=x \text{ and } dv=3e^{-3x}}$$

$$= 0 + \int_0^{\infty} e^{-3x}\, dx = \frac{1}{3}.$$

Example 2.36   Let X be a continuous random variable with pdf $f(x) = \frac{1}{x^2}$ for $x > 1$. Then,

$$E(X) = \int_1^\infty x \cdot \frac{1}{x^2}\, dx = \int_1^\infty \frac{1}{x}\, dx$$

$$= \ln |x| \Big|_1^\infty = \infty,$$

which does not converge. Hence, $E(X)$ does not exist.

Theorem 2.11 *If X is a random variable and $E(|X|) < \infty$, then $E(X)$ exists.*

Proof. Because the proofs are similar for discrete and continuous random variables, only the proof for continuous random variables will be given; the proof when X is a discrete random variable follows by replacing integration with summation.

Let X be a continuous random variable with pdf $f(x)$ on S_X with $E(|X|) < \infty$. Consider $|E(X)|$.

$$|E(X)| = \left| \int_{S_x} x f(x)\, dx \right| \le \int_{S_x} |x f(x)|\, dx = \int_{S_x} |x| f(x)\, dx = E(|X|) < \infty.$$

Therefore, $E(X)$ exists whenever $E(|X|) < \infty$ for a continuous random variable X. ∎

The expected value of a transformed random variable can be found using definition 2.19 and does not require the pdf of the transformed random variable.

Definition 2.19   Let X be a random variable with pdf $f(x)$ on S_X, and let $g(x)$ be a function defined on of S_X. The expected value of $g(X)$, denoted by $E[g(X)]$, is defined to be

$$E[g(X)] = \sum_{x \in S} g(x) f(x)$$

when X is a discrete random variable provided the series converges, and

$$E[g(X)] = \int_S g(x) f(x)\, dx$$

when X is a continuous random variable provided the integral converges.

Example 2.37   Let X be a random variable with pdf $f(x) = \theta e^{-\theta x}$ for $x > 0$. Let $g(X) = X^2$, then

$$E(X^2) = \int_0^\infty x^2 \theta e^{-\theta x}\, dx = \theta \underbrace{\int_0^\infty \left(\frac{u}{\theta}\right)^2 e^{-u}\, \frac{du}{\theta}}_{u = \theta x}$$

$$= \theta \cdot \frac{1}{\theta^3} \int_0^\infty u^2 e^{-u}\, du = \frac{2}{\theta^2}.$$

In Example 2.37, the expected value computed was the *second moment* of the random variable X. The definition of the moments of a random variable is given in Definition 2.20.

Definition 2.20 Let X be a random variable. Then, for $n \in \mathbb{N}$,

i) the nth moment of the random variable X is $\mu'_n = E(X^n)$.
ii) the nth central moment of the random variable X is $\mu_n = E[(X - \mu)^n]$.

Note that the first central moment, $\mu_1 = E(X - \mu)$, is 0 for every random variable since $E(X - \mu) = E(X) - \mu = \mu - \mu = 0$. The second central moment, μ_2, is called the *variance* and is denoted by σ^2. The variance measures the spread of a random variable and is discussed in detail in Section 2.5.2.

Theorem 2.12 shows that when the nth moment $E(X^n)$ exists, then so does $E(X^k)$ for $1 \leq k \leq n$.

Theorem 2.12 *If X is a random variable and $E(|X|^n) < \infty$ for some $n \in \mathbb{N}$, then $E(|X|^k) < \infty$ for $1 \leq k \leq n$.*

Proof. Because the proofs are similar for discrete and continuous random variables, only the proof for continuous random variables will be given; the proof when X is a discrete random variable follows by replacing integration with summation.

Let X be a continuous random variable with $E(|X|^n) < \infty$ for some $n \in \mathbb{N}$. Let $k \in \mathbb{N}$ be arbitrary but fixed with $k \leq n$. Then,

$$E(|X|^k) = \int_{\mathcal{S}_X} |x|^k f(x)\, dx = \int_{\mathcal{S}_X \cap \{|X| \leq 1\}} |x|^k f(x)\, dx + \int_{\mathcal{S}_X \cap \{|X| > 1\}} |x|^k f(x)\, dx$$

$$\leq \int_{\mathcal{S}_X \cap \{|X| \leq 1\}} |x|^k f(x)\, dx + \underbrace{\int_{\mathcal{S}_X \cap \{|X| > 1\}} |x|^n f(x)\, dx}_{|x|^k \leq |x|^n \text{ since } |X| > 1}$$

$$\leq \underbrace{\int_{\mathcal{S}_X \cap \{|X| \leq 1\}} (1 + |x|^n) f(x)\, dx}_{|x|^k \leq 1 + |x|^n \text{ since } |X| \leq 1} + \int_{\mathcal{S}_X \cap \{|X| > 1\}} |x|^n f(x)\, dx$$

$$= \int_{\mathcal{S}_X \cap \{|X| \leq 1\}} f(x)\, dx + \int_{\mathcal{S}_X \cap \{|X| \leq 1\}} |x|^n f(x)\, dx + \int_{\mathcal{S}_X \cap \{|X| > 1\}} |x|^n f(x)\, dx$$

$$= P(|X| \leq 1) + E(|X|^n) < \infty. \qquad \blacksquare$$

Theorem 2.13 (Properties of Expected Values) *Let X be a random variable and $g(x)$ a real-valued function. If $E(X)$ and $E[g(X)]$ exist, then*

i) $E(c) = c$, $\forall c \in \mathbb{R}$.

ii) $E(aX + b) = aE(X) + b$, $\forall a, b \in \mathbb{R}$.

iii) $E(ag(X) + b) = aE[g(X)] + b$, $\forall a, b \in \mathbb{R}$.

Proof. The proof will only be given for the discrete case since the proof for continuous case is similar with summation replaced by integration.

Let X be a discrete random variable, and let $g(x)$ be a real-valued function with $E(X) < \infty$ and $E[|g(X)|] < \infty$.

i) Let $c \in \mathbb{R}$, then $E(c) = \sum_{x \in S} cf(x) = c \underbrace{\sum_{x \in S} f(x)}_{=1} = c$.

ii) Let $a, b \in \mathbb{R}$, then

$$E(aX + b) = \sum_{x \in S_X} (ax + b)f(x) = a \underbrace{\sum_{x \in S_X} xf(x)}_{=E(X)} + b \underbrace{\sum_{x \in S_X} f(x)}_{=1}$$

$$= aE(X) + b.$$

iii) The proof of part (iii) is left an exercise. ∎

Theorem 2.14 *Let X be a random variable, and let $g_1(x), g_2(x), \ldots, g_n(x)$ be functions defined on S_X with $E[g_i(X)] < \infty$ for $i = 1, 2, \ldots, n$.*

i) *If $g_1(x) \le g_2(x)$ on S_x, then $E[g_1(X)] \le E[g_2(X)]$.*

ii) *$E[ag_1(X) + bg_2(X)] = aE[g_1(X)] + bE[g_2(X)]$, $\forall a, b \in \mathbb{R}$.*

iii) *$E\left[\sum_{i=1}^{n} g_i(X)\right] = \sum_{i=1}^{n} E[g_i(X)]$.*

Proof. The proof will only be given for the discrete case since the proof for the continuous case is similar with summation replaced by integration.

Let X be a discrete random variable, and let $g_1(x), g_2(x), \ldots, g_n(x)$ be functions defined on S_X with $E[g_i(X)] < \infty$ for $i = 1, 2, \ldots, n$.

i) Suppose that $g_1(x) \le g_2(x)$ on S_X and consider $E[g_1(x)]$.

$$E[g_1(x)] = \sum_{x \in S_X} g_1(x)f(x) \le \underbrace{\sum_{x \in S_X} g_2(x)f(x)}_{g_1(x) \le g_2(x)} = E[g_2(X)].$$

ii) The proof of part (ii) is left as an exercise.

iii) The proof of part (iii) is left as an exercise. ∎

2.5.2 The Variance of a Random Variable

The second central moment is called the variance and measures the spread associated with the random variable X. Moreover, the mean and the variance are often used to summarize the distribution of a random variable. The *standard deviation* of a random variable X, denoted by $SD(X)$, is the squareroot of the variance and is often used in place of the variance when summarizing the distribution of a random variable because the units of the standard deviation are the same as the units of X.

Definition 2.21 Let X be a random variable with $E(X^2) < \infty$. The variance of X, denoted by $\text{Var}(X)$ or σ_X^2, is defined to be

$$\text{Var}(X) = E[(X - \mu)^2],$$

where $\mu = E(X)$. The standard deviation of X, denoted by $SD(X)$ or σ_X, is $SD(X) = \sqrt{\text{Var}(X)}$.

Note that $\text{Var}(X) \geq 0$, and moreover, $\text{Var}(X) = 0$ only when there exists $c \in \mathbb{R}$ such that $P(X = c) = 1$. Three alternate ways for computing the variance are given in Theorem 2.15.

Theorem 2.15 *If X is a random variable with mean $E(X) = \mu$ and finite variance $\text{Var}(X)$, then*

i) $\text{Var}(X) = E(X^2) - \mu^2$.
ii) $\text{Var}(X) = E[X(X - 1)] + E(X) - \mu^2$.
iii) $\text{Var}(X) = E[X(X - \mu)]$.

Proof. Let X be a random variable with mean $E(X) = \mu$ and finite variance $\text{Var}(X) = \sigma^2$.

i) Consider $\text{Var}(X) = E[(X - \mu)^2]$.

$$\text{Var}(X) = E[(X - \mu)^2] = E(X^2 - 2\mu X + \mu^2) = E(X^2) - 2\mu E(X) + \mu^2$$
$$= E(X^2) - 2\mu^2 + \mu^2 = E(X^2) - \mu^2.$$

ii) The proof of part (ii) is left as an exercise.
iii) The proof of part (iii) is left as an exercise. ∎

For most random variables, the variance of a random variable is the easiest to compute using either $E(X^2) - \mu^2$ or $E[X(X - 1)] + E(X) - \mu^2$. In particular, the variance of a discrete random variable is often easiest to compute using $E[X(X - 1)] + E(X) - \mu^2$, and the variance of a continuous random variable is often the easiest to compute using $E(X^2) - \mu^2$; however, exceptions to these general rules do exist.

Example 2.38 Let X be a discrete random variable with pdf $f(x) = \frac{1}{2^x}$ for $x \in \mathbb{N}$. The mean of X was found in Example 2.34 with $E(X) = 2$, and because X is discrete, $\text{Var}(X)$ will be computed using $\text{Var}(X) = E[X(X-1)] + E(X) - E(X)^2$. Consider $E[X(X-1)]$.

$$E[X(X-1)] = \sum_{x=1}^{\infty} x(x-1)\frac{1}{2^x} = \underbrace{\sum_{x=2} x(x-1)\frac{1}{2^x}}_{\text{since } x(x-1)=0 \text{ for } x=1}$$

$$= \frac{1}{2^2}\sum_{x=2} x(x-1)\frac{1}{2^{x-2}} = \frac{1}{2^2} \cdot \frac{2}{(1-\frac{1}{2})^3} = 4$$

since $\sum_{x=2} x(x-1)p^{x-2} = \dfrac{d^2}{dp^2}\left[\dfrac{1}{1-p}\right] = \dfrac{2}{(1-p)^3}$ for $|p| < 1$.

Thus, $\text{Var}(X) = E[X(X-1)] + E(X) - E(X)^2 = 4 + 2 - 2^2 = 2$, and the standard deviation is $SD(X) = \sqrt{2}$.

Example 2.39 Let X be a continuous random variable with pdf $f(x) = 3e^{-3x}$ for $x > 0$. The mean of X was found in Example 2.35 with $E(X) = \frac{1}{3}$, and because X is continuous, $\text{Var}(X)$ will be computed using $\text{Var}(X) = E(X^2) - \mu^2$. Consider $E(X^2)$.

$$E(X^2) = \int_0^{\infty} x^2 \cdot 3e^{-3x}\, dx = \underbrace{3\int_0^{\infty} \left(\frac{u}{3}\right)^2 e^{-u}\, \frac{du}{3}}_{u=3x}$$

$$= 3 \cdot \frac{1}{3^3}\underbrace{\int_0^{\infty} u^2 e^{-u}\, du = \frac{1}{9}(-u^2 - 2u - 2)e^{-u}\Big|_0^{\infty}}_{\text{integrating by parts twice}}$$

$$= \frac{2}{9}.$$

Thus, $\text{Var}(X) = E(X^2) - \mu^2 = \frac{2}{9} - \left(\frac{1}{3}\right)^2 = \frac{1}{9}$, and the standard deviation is $SD(X) = \frac{1}{3}$.

Theorem 2.16 *If X be a random variable with $\text{Var}(X) < \infty$, then $\forall a, b \in \mathbb{R}$*

i) $\text{Var}(aX + b) = a^2\text{Var}(X)$.

ii) $SD(aX + b) = |a|SD(X)$.

Proof. Let X be a random variable with mean μ and finite $Var(X)$, and let a, $b \in \mathbb{R}$ be arbitrary but fixed.

i) First, $E(aX + b) = a\mu + b$. Consider $Var(aX + b)$.

$$\begin{aligned}
Var(aX + b) &= E[(aX + b - E(aX + b))^2] = E[(aX + b - (a\mu + b))^2] \\
&= E[(aX - a\mu)^2] = E[a^2(X - \mu)^2] = a^2 E[(X - \mu)^2] \\
&= a^2 Var(X).
\end{aligned}$$

ii) Part (ii) follows directly from part (i), since $SD(aX + b) = \sqrt{Var(aX + b)}$. ∎

From Theorem 2.16, it follows that $Var(X) = Var(X + b)$, and thus, variability does not change when a constant is added to X; however, multiplying X by a constant $a \neq \pm 1$ does change the variability. In particular, for $|a| < 1$, $Var(aX) < Var(X)$ and $|a| > 1$, $Var(aX) > Var(X)$.

Example 2.40 Let X be a random variable with mean μ and variance σ^2. Let $Z = \frac{X-\mu}{\sigma}$, then Z is called a *standardized random variable* because

$$E(Z) = E\left(\frac{X - \mu}{\sigma}\right) = \frac{1}{\sigma}E(X - \mu) = 0$$

and

$$Var(Z) = Var\left(\frac{X - \mu}{\sigma}\right) = \frac{1}{\sigma^2}Var(X - \mu) = \frac{1}{\sigma^2}Var(X) = 1.$$

Theorem 2.17 and Corollaries 2.1 and 2.2 provide important upper bounds on specific probability statements.

Theorem 2.17 *If X is a random variable, $g(x)$ is a nonnegative real-valued function on $\mathcal{S}_X$, and $c > 0$, then*

$$P(g(X) \geq c) \leq \frac{E[g(X)]}{c}.$$

Proof. The proof will only be given for the continuous case since the proof for the discrete case is similar with integration replaced by summation.

Let X be a random variable, $g(x)$ a nonnegative real-valued function on $\mathcal{S}_X$, and let $c > 0$. Consider $E[g(X)]$.

$$\begin{aligned}
E[g(X)] &= \int_{\mathcal{S}_X} g(x)f(x)\,dx \\
&= \int_{\{g(X) \geq c\} \cap \mathcal{S}_x} g(x)f(x)\,dx + \int_{\{g(X) < c\} \cap \mathcal{S}_x} g(x)f(x)\,dx
\end{aligned}$$

$$\geq \int_{\{g(X)\geq c\}\cap S_X} cf(x)\, dx + \int_{\{g(X)<c\}\cap S_X} g(x)f(x)\, dx$$

$$\geq \int_{\{g(X)\geq c\}\cap S_X} cf(x)\, dx$$

since $\int_{\{g(X)<c\}\cap S_X} g(x)f(x)\, dx > 0$.

Thus,

$$E[g(X)] \geq \int_{\{g(X)\geq c\}} cf(x)\, dx = cP(g(X) \geq c)$$

and hence,

$$P(g(X) \geq c) \leq \frac{E[g(X)]}{c}. \qquad \blacksquare$$

Corollaries 2.1 and 2.2 are special cases of Theorem 2.17.

Corollary 2.1   (Markov's Inequality) *If X is a nonnegative random variable with $E(X) < \infty$, then for $k > 0$,*

$$P(X \geq k) \leq \frac{E(X)}{k}.$$

Proof. Follows directly from Theorem 2.17 with $g(x) = x$. $\qquad \blacksquare$

Corollary 2.2   (Chebyshev's Inequality) *If X is a random variable with finite mean μ and variance σ^2, then for $k \geq 1$,*

$$P(|X - \mu| \geq k\sigma) \leq \frac{1}{k^2}$$

or equivalently,

$$P(|X - \mu| \leq k\sigma) \geq 1 - \frac{1}{k^2}.$$

Proof. Follows directly from Theorem 2.17 since

$$\{x : |X - \mu| \geq k\sigma\} = \{x : (X - \mu)^2 \geq k^2\sigma^2\}. \qquad \blacksquare$$

Example 2.41   Let X be a random variable with mean μ and variance σ^2, then by Chebyshev's Inequality:

- $P(|X - \mu| \leq 2\sigma) \geq 1 - \frac{1}{2^2} = 0.75$. Thus, there is a 75% chance of observing a value of X that is within two standard deviations of the mean.
- $P(|X - \mu| \leq 3\sigma) \geq 1 - \frac{1}{3^2} = 0.\overline{8}$. Thus, there is a 89% chance of observing a value of X that is within three standard deviations of the mean.

- $P(|X - \mu| \leq 4\sigma) \geq 1 - \frac{1}{4^2} = 0.9375$. Thus, there is a 93.75% chance of observing a value of X that is within four standard deviations of the mean.

Another important inequality is Jensen's Inequality, which is given in Theorem 2.18.

Theorem 2.18 (Jensen's Inequality) *If X is a random variable with finite mean $E(X)$, $g(x)$ is a function defined on $\mathcal{S}_X$ with $E[g(X)] < \infty$, then*

i) for a convex function $g(x)$, $E[g(X)] \geq g(E(X))$.
ii) for a concave function $g(x)$, $E\left[g(X)\right] \leq g\left(E(X)\right)$.

Example 2.42 Let X be a nonnegative random variable with $E(X^2) < \infty$ and $E\left[\sqrt{X}\right] < \infty$. Since x^2 is a convex function and $\sqrt{x}$ is a concave function, by Jensen's Inequality, it follows that

$$E(X^2) \geq E(X)^2 \quad \text{and} \quad E\left[\sqrt{X}\right] \leq \sqrt{E(X)}.$$

2.5.3 Moment Generating Functions

Recall that the nth moment of a random variable X is $\mu'_n = E(X^n)$ for $n \in \mathbb{N}$, and when $E\left[e^{tX}\right]$ exists for every t in some interval centered at 0, all of the moments of X can be determined from $E\left[e^{tX}\right]$, which is called the *moment generating function* (MGF).

Definition 2.22 Let X be a random variable with pdf $f(x)$. If there exists $h > 0$ such that $E(e^{tX})$ exists for every $t \in (-h, h)$, then the MGF associated with X is defined to be $M_X(t) = E(e^{tX})$.

Thus, when X is a discrete random variable with pdf $f(x)$ defined on $\mathcal{S}_X$ and the MGF exists, the MGF of X is

$$E(e^{tX}) = \sum_{x \in S} e^{tx} f(x).$$

Similarly, when X is a continuous random variable with pdf $f(x)$ defined on $\mathcal{S}_X$ and the MGF exists, the MGF of X is

$$E(e^{tX}) = \int_{x \in S} e^{tx} f(x)\, dx.$$

Examples of the MGF for a discrete and a continuous random variable are given in Examples 2.43 and 2.44, respectively.

Example 2.43 Let X be a discrete random variable with pdf $f(x) = \frac{e^{-\lambda}\lambda^x}{x!}$ for $x \in \mathbb{W}$. Consider $E[e^{tX}]$.

$$
\begin{aligned}
E(e^{tX}) &= \sum_{x=0}^{\infty} e^{tx} \frac{e^{-\lambda}\lambda^x}{x!} \\
&= e^{-\lambda} \sum_{x=0}^{\infty} \frac{(e^t\lambda)^x}{x!} \\
&= e^{-\lambda} e^{e^t\lambda} = e^{\lambda(e^t-1)},
\end{aligned}
$$

which exists $\forall t \in \mathbb{R}$. Thus, $M_X(t) = e^{\lambda(e^t-1)}$ for $t \in \mathbb{R}$.

Example 2.44 Let X be a continuous random variable with pdf $f(x) = 5e^{-5(x-1)}$ for $x > 1$. Consider $E[e^{tX}]$.

$$
\begin{aligned}
M(t) = E(e^{tX}) &= \int_1^{\infty} e^{tx} \cdot 5e^{-5(x-1)} \, dx \\
&= \int_1^{\infty} 5e^{-5x+5+tx} \, dx \\
&= 5e^5 \int_0^{\infty} 5e^{-x(5-t)} \, du \\
&= 5e^5 \left[\frac{-e^{-x(5-t)}}{5-t} \right] \Bigg|_1^{\infty} \\
&= \frac{5e^t}{5-t},
\end{aligned}
$$

which exists $\forall t \in (-5, 5)$. Thus, the MGF of X is $M_X(t) = \frac{5e^t}{5-t}$ for $t \in (-5, 5)$.

It is important to note that there are discrete and continuous random variables that do not have MGFs. For example, if X does not have a finite first moment $E(X)$, such as in Example 2.36, then the MGF of X does not exist. The relationship between a MGF and its moments is given in Theorem 2.19.

Theorem 2.19 *If X is a random variable and $E(X^k) < \infty$, $\forall k \in \mathbb{N}$, then the MGF of X is $M_X(t) = \sum_{k=0}^{\infty} \frac{E(X^k)t^k}{k!}$.*

Proof. Let X be a random variable and $E(X^k) < \infty$, $\forall k \in \mathbb{N}$. Let $M_X(t)$ exist for $t \in (-h, h)$ for some $h > 0$. Since $e^{tx} = \sum_{k=0}^{\infty} \frac{(tx)^k}{k!}$ when $t \in (-h, h)$, it follows that

$$
M_X(t) = E[e^{tX}] = E\left[\sum_{k=0}^{\infty} \frac{(tx)^k}{k!} \right] = \sum_{k=0}^{\infty} \frac{E(X^k)t^k}{k!}. \qquad \blacksquare
$$

Several important properties of MGFs are given in Theorem 2.20.

Theorem 2.20 *If X is a random variable with MGF $M_X(t)$, then*

  i) $M_X(0) = 1$.
  ii) $M_{aX+b}(t) = e^{tb}M_X(at)$, $\forall a, b \in \mathbb{R}$.
  iii) $\frac{d}{dt}[M_X(t)]_{t=0} = M'_X(0) = E(X)$.
  iv) $\frac{d}{dt^n}[M_X(t)]_{t=0} = E(X^n)$.
  v) $\text{Var}(X) = M''_X(0) - [M'_X(0)]^2$.

Proof. Let X be a random variable with MGF $M_X(t)$.

  i) Consider $M(0)$.

$$M(0) = E(e^{0 \cdot X}) = E(e^0) = E(1) = 1.$$

  ii) Consider $M_{aX+b}(t)$.

$$M_{aX+b}(t) = E(e^{(aX+b)t}) = E(e^{atX}e^{tb}) = e^{tb}E(e^{(at)X}) = e^{tb}M(at).$$

  iii) Consider $M'_X(0)$.

$$\begin{aligned} M'_X(0) &= \frac{d}{dt}[E^{tX}]_{t=0} = E\left[\frac{d}{dt}[E^{tX}]\right]_{t=0} \\ &= E[Xe^{tX}]_{t=0} = E[Xe^{0 \cdot X}] \\ &= E(X). \end{aligned}$$

  iv) The proof of part (iv) is left as an exercise.
  v) Part (v) follows directly from part (iii) and part (iv) since $\text{Var}(X) = E(X^2) - E(X)^2$. ∎

Example 2.45 Let X be a discrete random variable with pdf $\binom{n}{x}p^x(1-p)^{n-x}$ for $x = 0, 1, 2, \ldots, n$. The MGF of X is

$$\begin{aligned} E[e^{tX}] &= \sum_{x=0}^{n} e^{tx}\binom{n}{x}p^x(1-p)^{n-x} = \sum_{x=0}^{n}\binom{n}{x}(p\,e^t)^x(1-p)^{n-x} \\ &= \underbrace{\sum_{x=0}^{n}\binom{n}{x}a^x b^{n-x}}_{a=p\ e^t,\ b=1-p,\ \text{and binomial expansion}} = (a+b)^n \end{aligned}$$

$$= (p\,e^t + (1-p))^n.$$

Now, the mean and variance of X can be found using the MGF of X. In particular,

$$\begin{aligned} E(X) &= M'_X(0) = \frac{d}{dt}[(p\,e^t + (1-p))^n]_{t=0} \\ &= n(p\,e^t + (1-p))^{n-1}p\,e^t\Big|_{t=0} = np. \end{aligned}$$

Using $\text{Var}(X) = M''_X(0) - M'_X(0)^2$, the variance can be shown to be $np(1-p)$.

Example 2.46 Let X be a random variable with MGF $M_X(t)$ and let $Y = aX + b$. The MGF of X can be used to prove that $E(Y) = a\mu + b$ and $Var(X) = a^2\sigma^2$. First, consider $M_Y(t)$.

$$M_Y(t) = M_{aX+b}(t) = e^{tb}M_X(at).$$

Now, $E(Y) = M'_Y(0)$, and

$$M'_Y(t) = \frac{d}{dt}[e^{tb}M_X(at)] = be^{tb}M_X(at) + e^{tb}M'_X(at) \cdot a.$$

Hence, $E(Y) = M'_Y(0) = be^0 M_X(0) + ae^0 M'_X(0) = b + aE(X)$.
The variance of Y can be verified by using $Var(Y) = M''_Y(0) - [M'_Y(0)]^2$, which is left as an exercise.

Theorem 2.21 shows that when the MGF of a random variable exists, it can be used to determine the distribution of a random variable.

Theorem 2.21 *Let X and Y be random variables with MGFs $M_X(t)$ and $M_Y(t)$ existing on a common interval $(-h, h)$. If $M_X(t) = M_Y(t)$ for all $t \in (-h, h)$, then $F_X(u) = F_Y(u)$, $\forall u \in \mathbb{R}$.*

Example 2.47 Let X be a discrete random variable with $M_X(t) = (0.25e^t + 0.75)^{12}$. Then, $M_X(t)$ is of the form $(p\,e^t + (1-p))^n$, which is the MGF of the random variable in Example 2.45, which has pdf $\binom{n}{x} p^x (1-p)^{n-x}$ for $x = 0, 1, 2, \ldots, n$. Thus, X has pdf $f(x) = \binom{12}{x} 0.25^x 0.75^{12-x}$ for $x = 0, 1, 2, \ldots, 12$.

Problems

2.5.1 Let X be a discrete random variable with pdf $f(x) = \frac{1}{4} \cdot \left(\frac{3}{4}\right)^x$ for $x \in \mathbb{W}$. Determine

a) $E(X)$.

b) $E[X(X-1)]$.

c) $Var(X)$.

d) $E(2X + 4)$.

2.5.2 Let X be a discrete random variable with pdf $f(x) = \binom{2}{x} \cdot \frac{1}{4}$ for $x = 0, 1, 2$. Determine

a) $E(X)$.

b) $E(X^2)$.

c) $Var(X)$.

d) $E[X(X+1)]$.

2.5.3 Let X be a discrete random variable with pdf $f(x) = \binom{2}{x} p^x (1-p)^{2-x}$ for $x = 0, 1, 2$ and $p \in (0, 1)$. Determine

a) $E(X)$.

b) $E(X^2)$.

c) $Var(X)$.

d) $E[4X^2 + 2]$.

2.5.4 Let X be a discrete random variable with pdf $f(x) = \frac{2x}{n(n+1)}$ for $x = 1, 2, \ldots, n$. Determine

a) $E(X)$.

b) $E\left(\frac{1}{X}\right)$.

2.5.5 Let X be a discrete random variable with pdf $f(x) = \frac{10-x}{55}$ for $x = 0, 2, \ldots, 9$. Determine $E(X)$.

2.5.6 Let X be a discrete random variable with pdf $f(x) = \frac{e^{-2}2^x}{x!}$ for $x \in \mathbb{W}$. Determine

a) $E(X)$.

b) $E(2^X)$.

2.5.7 Let X be a discrete random variable with $E(X) = 2.5, E(X^2) = 6.7$, and pdf $f(x) = p_1 I_{\{1\}}(x) + p_2 I_{\{2\}}(x) + p_3 I_{\{3\}}(x)$ for $x = 1, 2, 3$. Determine the values of p_1, p_2, and p_3.

2.5.8 Let X be a random variable with pdf $f(x) = \frac{1}{3}e^{-\frac{x}{3}}$ for $x > 0$. Determine

a) $E(X)$.

b) $E(X^2)$.

c) $E(X^n)$.

d) $E(X^{10})$.

2.5.9 Let X be a random variable with pdf $f(x) = \begin{cases} \frac{x}{3} & 0 < x < 2 \\ 2(1 - \frac{x}{3}) & 2 \leq x < 3 \end{cases}$.

Determine

a) $E(X)$.

b) $E(X - 2)$.

c) $E(X^2)$.

d) $E(|X - 2|)$.

2.5.10 Let X be a random variable with pdf $f(x) = \theta x^{\theta-1}$ for $x \in (0, 1)$ and $\theta > 0$. Determine

a) $E(X)$.

b) $E(X^n)$ for $n \in \mathbb{N}$.

2.5.11 Let X be a continuous random variable with pdf $f(x) = 5x^4$ for $x \in (0, 1)$. Determine

a) $E(X)$.

b) $E(X^2)$.

c) $E(X^n)$.

d) $E[-\ln(X)]$.

2.5.12 Let X be a continuous random variable with pdf $f(x) = \frac{2}{x^3}$ for $x \in (1, \infty)$. Determine

a) $E(X)$.

b) $E\left(\frac{1}{X}\right)$.

c) $E\left(\frac{1}{X^n}\right)$.

d) $E\left(1 - \frac{1}{X^2}\right)$.

2.5.13 Let X be a continuous random variable with pdf $f(x) = \theta x^{\theta-1}$ for $x \in (0, 1)$ and $\theta > 0$. Determine

a) $E[-\ln(X)]$. b) $E[X(1 - X)]$.

2.5.14 Let X be a continuous random variable with pdf $f(x) = 3(1 - x)^2$ for $x \in (0, 1)$. Determine

a) $E(X)$. b) $E[(1 - X)^n)]$.
c) $E[-\ln(1 - X)]$. d) $E[-\ln[X(1 - X)]]$.

2.5.15 Let X be a continuous random variable with pdf $f(x) = \frac{1}{\sqrt{2\pi}} e^{-\frac{1}{2}(x-5)^2}$ for $x \in \mathbb{R}$. Determine

a) $E(X)$. b) $E[(X - 5)^3]$.

2.5.16 Suppose that X is a random variable representing cost per unit sold, ignoring initial costs, and suppose that $E(X) = 2.5$. If there is an initial cost of $500 for a 1000 unit order, what is the
a) expected total cost for a 1000 unit order?
b) the variance of the total cost for a 1000 unit order when $\mathrm{Var}(X) = 0.01$?
c) the standard deviation of the total cost for a 1000 unit order when $\mathrm{Var}(X) = 0.01$?

2.5.17 Let X be a discrete random variable with pdf $f(x) = \frac{2}{3} \cdot \left(\frac{1}{3}\right)^{x-1}$ for $x \in \mathbb{N}$. Determine

a) $M_X(t)$. b) $E(X)$.
c) $E(X^2)$. d) $\mathrm{Var}(X)$.

2.5.18 Let X be a discrete random variable with pdf $f(x) = \frac{e^{-3}3^x}{x!}$ for $x \in \mathbb{W}$. Determine

a) $M_X(t)$. b) $E(X)$.
c) $E(X^2)$. d) $\mathrm{Var}(X)$.

2.5.19 Let X be a continuous random variable with pdf $f(x) = xe^{-x}$ for $x > 0$. Determine

a) $M_X(t)$. b) $E(X)$.
c) $E(X^2)$. d) $\mathrm{Var}(X)$.

2.5.20 Let X be a discrete random variable with $M_X(t) = \frac{e^{3t}}{6} + \frac{e^{4t}}{2} + \frac{e^{5t}}{3}$. Determine

a) $E(X)$. b) $E(X^2)$.
c) $\mathrm{Var}(X)$. d) $SD(X)$.

2.5.21 Let X be a continuous random variable with pdf $f(x) = \frac{1}{\sqrt{2\pi}}e^{-\frac{x^2}{2}}$ for $x \in \mathbb{R}$. Determine the MGF of $Y = X^2$ (i.e. $E(e^{tX^2})$).

2.5.22 Let X be a random variable with MGF $M_X(t) = \frac{1}{(1-\theta t)}$ for $t \in \left(-\frac{1}{\theta}, \frac{1}{\theta}\right)$ and $Y = \frac{2}{\theta}X$. Determine

a) $M_Y(t)$.   b) $E(Y)$.

c) $E(Y^2)$.   d) $\text{Var}(Y)$.

2.5.23 Let X be a random variable with MGF $M_X(t) = \left(\frac{e^t}{4} + \frac{3}{4}\right)^{10}$ for $t \in \mathbb{R}$ and $Y = 10 - X$. Determine

a) $M_Y(t)$.   b) $E(Y)$.

c) $E(Y^2)$.   d) $\text{Var}(Y)$.

2.5.24 Let X be a random variable with MGF $M_X(t) = e^{\frac{t^2}{2}}$ for $t \in \mathbb{R}$ and $Y = 5X - 3$. Determine

a) $M_Y(t)$.   b) $E(Y)$.

c) $E(Y^2)$.   d) $\text{Var}(Y)$.

2.5.25 Let X be a random variable with MGF $M_X(t) = e^{2t^2}$ for $t \in \mathbb{R}$ and $Y = 2X + 5$. Determine

a) $M_Y(t)$.   b) $E(Y)$.

c) $E(Y^2)$.   d) $\text{Var}(Y)$.

2.5.26 Let X be a random variable with MGF $M_X(t)$ and $Y = \frac{X-\mu}{\sigma}$. Use properties of the MGF to show that $E(Y) = 0$ and $\text{Var}(Y) = 1$.

2.5.27 Let X be a random variable with $E(X^n) = 3^n$, $\forall n \in \mathbb{N}$. Determine $M_X(t)$ and the values of t for which it exists.

2.5.28 Let X be a random variable with $E(X^n) = \frac{n!}{\lambda^n}$, $\forall n \in \mathbb{N}$ and $\lambda > 0$. Determine $M_X(t)$ and the values of t for which it exists.

2.5.29 Let X be a continuous random variable with pdf $f(x) = \frac{1}{\pi} \cdot \frac{1}{1+x^2}$ for $x \in \mathbb{R}$. Show that $E(X)$ does not exist.

2.5.30 Let X be a discrete random variable with pdf $f(1) = 0.20, f(2) = p_2$, and $f(3) = p_3$. If $E(X) = 2.1$, determine the values of p_1 and p_2.

2.5.31 Let X be a continuous random variable with pdf $f(x) = \frac{1}{\theta}e^{-\frac{x}{\theta}}$ for $x > 0$ and $\theta > 0$. Show that
   a) $E(X) = \theta$.   b) $E(X^n) = n!\theta^n$.

2.5.32 Let X be a discrete random variable with support $S = \{0, 1, 2, 3, \dots \}$. If $E(X)$ exists, show that $E(X) = \sum_{i=0}^{\infty} P(X > i)$.

2.5.33 Let X be a continuous random variable with support $S = (0, \infty)$. If $E(X)$ exists, show that $E(X) = \int_0^{\infty}[1 - F(x)]\, dx$.

2.5.34 Let X be a continuous random variable, and suppose that $E[g(X)] < \infty$. Show that $E[ag(X) + b] = aE[g(X)] + b, \forall a, b \in \mathbb{R}$.

2.5.35 Let X be a continuous random variable, and suppose that $E[g_i(X)] < \infty$ for $i = 1, 2$. Show that $E[g_1(X) + g_2(X)] = E[g_1(X)] + E[g_2(X)]$.

2.5.36 Let X be a continuous random variable with pdf $f(x)$ on $\mathbb{R}$. If $\forall x \in \mathbb{R}$ $f(-x) = f(x)$, show $E(X^{2k+1}) = 0, \forall k \in \mathbb{W}$.

2.5.37 Let X be a random variable with $E(X^2) < \infty$. Show
   a) $\text{Var}(X) = E[X(X - 1)] + E(X) - E(X)^2$.
   b) $\text{Var}(X) = E[X(X - \mu)]$.

2.5.38 Show that $E[(X - \alpha)^2]$ is minimized for $\alpha = \mu$.

2.5.39 Show that $P(|X - \mu| \geq k) \leq \frac{E[|X-\mu|^r]}{k^r}$, for $k, r > 0$.

2.5.40 Show $E(\sqrt{X}) \leq \sqrt{E(X)}$.

2.5.41 Let X be a nonnegative random variable with MGF $M_X(t)$. If $c, t > 0$, show that $P(X \geq c) \leq \frac{M_X(t)}{e^{tc}}$.

2.5.42 Let X be a random variable with MGF $M_X(t)$. Show that $\frac{d^n}{dt^n}[M_X(t)]_{t=0} = E(X^n)$.

2.5.43 Let the functions x^+ and x^- be defined by $x^+ = \begin{cases} 0 & x \leq 0 \\ x & x > 0 \end{cases}$ and
$x^- = \begin{cases} -x & x < 0 \\ x & x \geq 0 \end{cases}$. If X is a continuous random variable, show
   a) $E(|X|) = E(X^+) - E(X^-)$.
   b) $E(X) = E(X^+) + E(X^-)$.
   c) $E(X) < \infty$ if $E(X^+)$ and $E(X^-)$ are finite.

2.5.44 If Z has MGF $M_Z(t) = e^{\frac{t^2}{2}}$ for $t \in \mathbb{R}$, determine
a) the MGF of $X = \sigma Z + \mu$.
b) $E(X)$ using the MGF of X.
c) $\text{Var}(X)$ using the MGF of X.

2.5.45 Let X be a random variable with MGF $M_X(t)$ and $Y = aX + b$. Using the MGF of X, show that $\text{Var}(Y) = a^2 \text{Var}(X)$.

2.6 Expected Values for Random Vectors

Expectations are also defined for random vectors; however, with a random vector $\vec{X}$, the expectation is only defined for a function $h(\vec{X})$ where the function $h(x_1, \ldots, x_k)$ maps $\mathcal{S}_{\vec{X}}$ to $\mathbb{R}$. In this case, $h(\vec{X})$ is a random variable, and the definition of $E[h(\vec{X})]$ is given in Definition 2.23.

Definition 2.23 Let $\vec{X}$ be a k-dimensional random vector with joint pdf $f(\vec{x})$ on $\mathcal{S}_{\vec{X}}$, and let $h(x_1, \ldots, x_k)$ be a function mapping $\mathbb{R}^k$ to $\mathbb{R}$. The expected value of $h(X_1, \ldots, X_k)$ is

$$E[h(X_1, \ldots, X_n)] = \underbrace{\sum_{x_k} \sum_{x_{k-1}} \cdots \sum_{x_1}}_{(x_1, \ldots, x_k) \in \mathcal{S}_{\vec{X}}} h(x_1, \ldots, x_k) f(\vec{x})$$

when $\vec{X}$ is discrete and the expected value exists, and

$$E[h(X_1, \ldots, X_n)] = \underbrace{\int \int \cdots \int}_{(x_1, \ldots, x_k) \in \mathcal{S}_{\vec{X}}} h(x_1, \ldots, x_k) f(\vec{x}) \, dx_1 \cdots dx_k$$

when $\vec{X}$ is continuous and the expected value exists.

Example 2.48 Let $\vec{X} = (X_1, X_2)$ be a continuous random vector with joint pdf $f(x_1, x_2) = 15x_1^2 x_2$ for $0 < x_1 < x_2 < 1$. Let $h_1(x_1, x_2) = x_1 x_2$ and $h_2(x_1, x_2) = x_2 - x_1$. Then,

$$E[h_1(X_1, X_2)] = E(X_1 X_2) = \int_0^1 \int_0^y x_1 x_2 \cdot 15x_1^2 x_2 \, dx_1 \, dx_2$$

$$= \int_0^1 \int_0^y 15x_1^3 x_2^2 \, dx_1 \, dx_2 = \frac{15}{28}$$

and

$$E[h_2(X_1, X_2)] = E(X_2 - X_1) = \int_0^1 \int_0^{x_2} (x_2 - x_1) \cdot 15x_1^2 x_2 \, dx_1 \, dx_2$$

$$= \int_0^1 \int_0^{x_2} 15x_1^2 x_2^2 \, dx_1 \, dx_2 - \int_0^1 \int_0^y 15x_1^3 x_2 \, dx_1 \, dx_2$$

$$= \frac{5}{24}.$$

Note that the moments of each of the component random variables making up $\vec{X}$ can be found by taking $h_i(x_1, x_2, \ldots, x_k) = x_i^n$ so that $E[h_i(\vec{X})] = E(X_i^n)$. In particular, the mean and variance of the random variable X_i can be found either using the joint pdf of $\vec{X}$ or from the marginal pdf of X_i. For example, when $\vec{X}$ is a continuous bivariate random vector, the expected value of X_1 using the joint pdf is

$$E(X_1) = \int_{x_2} \int_{x_1} x_1 f(x_1, x_2) \, dx_1 \, dx_2$$

$$= \int_{x_1} \int_{x_2} x_1 f(x_1, x_2) \, dx_2 \, dx_1$$

$$= \int_{x_1} x_1 \underbrace{\left[\int_{x_2} f(x_1, x_2) \, dx_2 \right]}_{f_1(x_1)} dx_1$$

$$= \int_{S_{X_1}} x_1 f_1(x_1) \, dx_1.$$

Similarly, the component random variable variances can also be found using either the joint pdf of $\vec{X}$ or the marginal pdf of X_1.

Example 2.49 Let $\vec{X}$ be a continuous random vector with joint pdf $f(x_1, x_2) = e^{-x_2}$ for $0 < x_1 < x_2 < \infty$. The marginal pdfs are $f_1(x_1) = e^{-x_1}$ for $0 < x_1 < \infty$ and $f_2(x_2) = x_2 e^{-x_2}$ for $0 < x_2 < \infty$. Using the marginal pdfs, the expected values of X_1 and X_2 are

$$E(X_1) = \int_0^\infty x_1 e^{-x_1} \, dx_1 = 1$$

and

$$E(X_2) = \int_0^\infty x_2 \cdot x_2 e^{-x_2} \, dx_2 = 2.$$

Using the joint pdf,

$$E(X_1) = \int_0^\infty \int_0^{x_2} x_1 \, e^{-x_2} \, dx_1 \, dx_2 \int_0^\infty \frac{x_2^2}{2} e^{-x_2} \, dx_2 = 1$$

and

$$E(X_2) = \int_0^\infty \int_0^{x_2} x_2 \cdot x_2 \, e^{-x_2} \, dx_1 \, dx_2 = \int x_2^2 \, e^{-x_2} \, dx_2 = 2.$$

Also,

$$E(X_1^2) = \int_0^\infty \int_0^{x_2} x_1^2 \, e^{-x_2} \, dx_1 \, dx_2 = \int_0^\infty \frac{x_2^3}{3} e^{-x_2} \, dx_2 = 2$$

and therefore, $\mathrm{Var}(X_1) = E(X_1^2) - E(X_1)^2 = 2 - 1^2 = 1$.

2.6.1 Properties of Expectation with Random Vectors

Theorem 2.22 shows that the expectation operator is a linear operator, which follows from the linearity of integrals and summations.

Theorem 2.22 *If $\vec{X}$ is a random vector with support $\mathcal{S}_{\vec{X}} \subset \mathbb{R}^k$ and $h_1(x_1, \dots, x_k)$ and $h_2(x_1, \dots, x_k)$ are functions mapping $\mathcal{S}_{\vec{X}}$ to $\mathbb{R}$, then $\forall \alpha, \beta \in \mathbb{R}$*

$$E[\alpha h_1(\vec{X}) + \beta h_2(\vec{X})] = \alpha E[h_1(\vec{X})] + \beta E[h_2(\vec{X})].$$

Proof. This theorem follows from the linearity of integrals and summations. ∎

Note that as a result of Theorem 2.22, $E(X_1 + X_2) = E(X_1) + E(X_2)$. Theorem 2.22 can be extended to handle more than two functions of $\vec{X}$ yielding

$$E\left[\sum_{i=1}^n \alpha_i h_i(\vec{X})\right] = \sum_{i=1}^n \alpha_i E[h_i(\vec{X})]$$

and

$$E\left[\sum_{i=1}^n \alpha_i X_i\right] = \sum_{i=1}^n \alpha_i E(X_i).$$

Sums and linear combinations of two or more random variables are discussed in detail in Section 2.7.

Recall that when $X_1, X_2, \dots, X_k$ are independent random variables, the joint pdf is the product of the marginal pdfs and the joint support is the Cartesian product of the supports of the individual random variables. Theorem 2.23 shows that the expected value of a product of independent random variables is the product of their expectations.

Theorem 2.23 *If $\vec{X}$ is a random vector of independent random variables and $h_i(x_i)$ are functions for $i = 1, 2, \ldots, k$, then $E\left[\prod_{i=1}^{k} h_i(x_i)\right] = \prod_{i=1}^{k} E[h_i(x_i)]$.*

Proof. The proof will only be given for continuous random vectors since the proof for discrete random vectors is similar with integration replaced by summation.

Let $\vec{X}$ be a random vector of independent random variables, and let $h_i(x_i)$ be a function for $i = 1, 2, \ldots, k$. Then, since $X_1, \ldots, X_k$ are independent random variables, $f(x_1, \ldots, x_k) = \prod_{i=1}^{k} f_i(x_i)$, $\mathcal{S}_{\vec{X}} = \mathcal{S}_{X_1} \times \mathcal{S}_{X_2} \cdots \times \mathcal{S}_{X_k}$, and

$$E\left[\prod_{i=1}^{k} h_i(x_i)\right] = \underbrace{\int \int \cdots \int}_{(x_1, x_2, \ldots, x_k) \in \mathcal{S}_{\vec{X}}} \prod_{i=1}^{k} h_i(x_i) f(x_1, \ldots, x_k) \, dx_1 \cdots dx_k$$

$$= \int_{\mathcal{S}_{X_k}} \int_{\mathcal{S}_{X_{k-1}}} \cdots \int_{\mathcal{S}_{X_1}} \prod_{i=1}^{k} h_i(x_i) f_i(x_i) \, dx_1 \cdots dx_k$$

$$= \prod_{i=1}^{k} \int_{\mathcal{S}_{X_i}} h_i(x_i) f_i(x_i) \, dx_i$$

$$= \prod_{i=1}^{k} E[h_i(X_i)]. \qquad \blacksquare$$

Corollary 2.3 *Let $X_1, \ldots, X_n$ be independent random variables. If $E(X_i) < \infty$ for $i = 1, \ldots, n$, then $E\left[\prod_{i=1}^{n} X_i\right] = \prod_{i=1}^{n} E(X_i)$.*

Proof. Corollary 2.3 follows directly from Theorem 2.23 with $h_i(x_i) = x_i$. $\qquad \blacksquare$

Thus, when X_1 and X_2 are independent random variables, it follows from Corollary 2.3 that $E(X_1 X_2) = E(X_1)E(X_2)$.

Example 2.50 Let $X_1, X_2, \ldots, X_n$ be iid random variables with $E(X_i) = \mu$ for $i = 1, \ldots, n$. Then,

$$E\left[\prod_{i=1}^{n} X_i\right] = \prod_{i=1}^{n} E(X_i) = \mu^n$$

and

$$E\left[\left(\prod_{i=1}^{n} X_i\right)_n\right] = \prod_{i=1}^{n} E\left(X_i^{\frac{1}{n}}\right) = E\left[X_1^{\frac{1}{n}}\right]^n.$$

Independence also plays an important role when computing the variance of the sum of two or more random variables. Theorem 2.24 provides formulas for the variance of the sum of two independent random variables.

Theorem 2.24 *If X_1 and X_2 are independent random variables with finite variances, then*

i) $\operatorname{Var}(X_1 + X_2) = \operatorname{Var}(X_1) + \operatorname{Var}(X_2)$.
ii) $\operatorname{Var}(aX_1 + bX_2) = a^2\operatorname{Var}(X_1) + b^2\operatorname{Var}(X_2)$, $\forall a, b \in \mathbb{R}$.

Proof. Let X_1 and X_2 be independent random variables with finite variances, and let $\mu_1 = E(X_1)$ and $\mu_2 = E(X_2)$.

i) Consider $\operatorname{Var}(X_1 + X_2)$.

$$\operatorname{Var}(X_1 + X_2) = E[(X_1 + X_2)^2] - (\mu_1 + \mu_2)^2$$

$$= E[X_1^2 + 2X_1X_2 + X_2^2] - (\mu_1^2 + 2\mu_1\mu_2 + \mu_2^2)$$

$$= E(X_1^2) - \mu_1^2 + E(X_2^2) - \mu_2^2 + 2E(X_1X_2) - 2\mu_1\mu_2$$

$$= \operatorname{Var}(X_1) + \operatorname{Var}(X_2) + \underbrace{2E(X_1)E(X_2) - 2\mu_1\mu_2}_{0}$$

$$= \operatorname{Var}(X_1) + \operatorname{Var}(X_2).$$

ii) The proof of part (ii) is left as an exercise. ∎

Note that when X_1 and X_2 are independent random variables, Theorem 2.24 shows that $X_1 + X_2$ and $X_1 - X_2$ have the same variance. In particular,

$$\operatorname{Var}(X_1 + X_2) = \operatorname{Var}(X_1 - X_2) = \operatorname{Var}(X_1) + \operatorname{Var}(X_2).$$

Theorem 2.25 *If X_1 and X_2 are independent random variables with MGFs $M_1(t)$ and $M_2(t)$, then $M_{X_1+X_2}(t) = M_1(t)M_2(t)$.*

Proof. Let X_1 and X_2 be independent random variables with MGFs $M_1(t)$ and $M_2(t)$. Consider $M_{X_1+X_2}(t)$.

$$M_{X_1+X_2}(t) = E[e^{t(X_1+X_2)}] = E[e^{tX_1}e^{tX_2}]$$

$$= \underbrace{E[e^{tX_1}]E[e^{tX_2}]}_{\text{by Theorem 2.23}}$$

$$= M_1(t)M_2(t). \qquad \blacksquare$$

Example 2.51 Let X_1 and X_2 be independent random variables, and let $M_1(t) = (0.3e^t + 0.7)^5$ and $M_2(t) = (0.3e^t + 0.7)^{10}$. Then, by Theorem 2.25, the MGF of $X_1 + X_2$ is

$$M_{X_1 + X_2} = M_1(t)M_2(t) = (0.3e^t + 0.7)^5 \times (0.3e^t + 0.7)^{10}$$
$$= (0.3e^t + 0.7)^{15}.$$

By Example 2.45, the MGF of $X_1 + X_2$ is the MGF of a random variable with pdf $f(x) = \binom{15}{x} 0.3^x (1 - 0.3)^{15-x}$ for $x = 0, 1, 2, \ldots, 15$.

Theorems 2.24 and 2.25 are generalized in Section 2.7 for linear combinations of more than two independent random variables.

2.6.2 Covariance and Correlation

The variance of a random variable measures the variability in the values the random variable takes on. For a pair of random variables, the *covariance* measures the joint variation in the variables. In particular, the covariance measures the strength of the linear relationship between two variables.

Definition 2.24 Let X_1 and X_2 be random variables with finite variances and means $E(X_1) = \mu_1$ and $E(X_2) = \mu_2$. The covariance of X_1 and X_2 is defined to be

$$\text{Cov}(X_1, X_2) = E[(X_1 - \mu_1)(X_2 - \mu_2)]$$

and is denoted by σ_{12}.

The covariance between two random variables, say X_1 and X_2, exists when their variances are finite. A positive covariance means that large values of X_1 tend to be more likely to occur with large values of X_2, and smaller values of X_1 more likely to occur with smaller values of X_2. On the other hand, a negative covariance means that large values of X_1 tend to be more likely to occur with smaller values of X_2, and smaller values of X_1 more likely to occur with larger values of X_2.

Theorem 2.26 *If X_1 and X_2 are random variables with finite variances and means $E(X_1) = \mu_1$ and $E(X_2) = \mu_2$, then $\text{Cov}(X_1, X_2) = E(X_1 X_2) - \mu_1 \mu_2$.*

Proof. Let X_1 and X_2 be random variables with finite variances and means. Let $E(X_1) = \mu_1$ and $E(X_2) = \mu_2$, and consider $\text{Cov}(X_1, X_2)$.

$$\text{Cov}(X_1, X_2) = E[(X_1 - \mu_1)(X_2 - \mu_2)]$$
$$= E[X_1 X_2 - \mu_2 X_1 - \mu_1 X_2 + \mu_1 \mu_2]$$
$$= E(X_1 X_2) - \mu_2 E(X_1) - \mu_1 E(X_2) + \mu_1 \mu_2$$
$$= E(X_1 X_2) - \mu_2 \mu_1 - \mu_1 \mu_2 + \mu_1 \mu_2$$
$$= E(X_1 X_2) - \mu_1 \mu_2. \qquad \blacksquare$$

Example 2.52 Let $\vec{X} = (X_1, X_2)$ be a continuous random vector with joint pdf $f(x_1, x_2) = 15x_1^2 x_2$ for $0 < x_1 < x_2 < 1$. Then,

$$E(X_1) = \int_0^1 \int_0^y x_1 \cdot 15x_1^2 x_2 \, dx_1 \, dx_2 = \int_0^1 \int_0^y 15x_1^3 x_2 \, dx_1 \, dx_2 = \frac{5}{8}$$

and

$$E(X_2) = \int_0^1 \int_0^y x_2 \cdot 15x_1^2 x_2 \, dx_1 \, dx_2 = \int_0^1 \int_0^y 15x_1^2 x_2^2 \, dx_1 \, dx_2 = \frac{5}{6}.$$

From Example 2.48, $E(X_1 X_2) = \frac{15}{28}$. Thus,

$$\text{Cov}(X_1, X_2) = \frac{15}{28} - \frac{5}{8}\left(\frac{5}{6}\right) = \frac{5}{336}.$$

Several important properties of the covariance are given in Theorem 2.27.

Theorem 2.27 (Properties of Covariances) *If X_1 and X_2 are random variables with finite variances, then*

i) $\text{Cov}(X_1, X_2) = \text{Cov}(X_2, X_1)$.
ii) $\text{Cov}(X_1, a) = 0, \ \forall a \in \mathbb{R}$.
iii) $\text{Cov}(X_1 + b, X_2 + c) = \text{Cov}(X_1, X_2) \ \forall b, c \in \mathbb{R}$.
iv) $\text{Cov}(cX_1, dX_2) = cd\text{Cov}(X_1, X_2) \ \forall c, d \in \mathbb{R}$.
v) $\text{Cov}(aX_1 + b, cX_2 + d) = ac\text{Cov}(X_1, X_2) \ \forall a, b, c, d \in \mathbb{R}$.
vi) $\text{Cov}(X_1, X_1) = \text{Var}(X_1)$.

Proof. Let X_1 and X_2 be random variables with finite variances.

i) $\text{Cov}(X_1, X_2) = E(X_1 X_2) - \mu_1 \mu_2 = E(X_2 X_1) - \mu_2 \mu_1 = \text{Cov}(X_2, X_1)$.
ii) Let $a \in \mathbb{R}$ be arbitrary but fixed. Then,

$$\text{Cov}(X_1, a) = E(aX_1) - a\mu_1 = aE(X_1) - a\mu_1 = 0.$$

iii) Let $b, c \in \mathbb{R}$ be arbitrary but fixed. Then,

$$\text{Cov}(X_1 + b, X_2 + c) = E[(X_1 + b - \mu_1 - b)(X_2 + c - \mu_2 - c)]$$
$$= E[(X_1 - \mu_1)(X_2 - \mu_2)] = \text{Cov}(X_1, X_2).$$

iv) The proof of part (iv) is left as an exercise.
v) The proof of part (v) is left as an exercise.
vi) The proof of part (vi) is left as an exercise. ∎

Note that the covariance behaves somewhat similar to the variance. In particular, adding a constant to either of the random variables does not affect the covariance; however, multiplying either or both random variables by a constant does. Theorem 2.28 shows that the covariance is a linear function of the first argument when the second argument is fixed.

Theorem 2.28 *If X_1, X_2, and X_3 are random variables with finite variances, then*

$$\text{Cov}(X_1 + X_2, X_3) = \text{Cov}(X_1, X_3) + \text{Cov}(X_2, X_3).$$

Proof. The proof of Theorem 2.28 is left as an exercise. ∎

Example 2.53 Let X_1, X_2, and X_3 be random variables with $\text{Var}(X_1) = 2$, $\text{Var}(X_2) = 5$, $\text{Var}(X_3) = 4$, $\text{Cov}(X_1, X_2) = -1$, $\text{Cov}(X_1, X_3) = -2$, and $\text{Cov}(X_2, X_3) = 1.5$. Then, by Theorem 2.27,

$$\text{Cov}(X_1, X_3) = \text{Cov}(X_3, X_1) = -2.$$
$$\text{Cov}(2X_1, 4X_2 - 3) = 2 \cdot 4\text{Cov}(X_1, X_2) = -8.$$
$$\text{Cov}(-X_2, 3X_3) = -3\text{Cov}(X_2, X_3) = -4.5.$$
$$\text{Cov}(X_1, X_1) = \text{Var}(X_1) = 2.$$

Theorem 2.29 *If X_1 and X_2 are independent random variables with finite variances, then $\text{Cov}(X_1, X_2) = 0$.*

Proof. Let X_1 and X_2 be independent random variables with finite variances. Consider $\text{Cov}(X_1, X_2)$.

$$\text{Cov}(X_1, X_2) = E(X_1 X_2) - \mu_1 \mu_2 = \underbrace{E(X_1)E(X_2)}_{X_1, \ X_2 \text{ ind.}} - \mu_1 \mu_2 = 0.$$

 ∎

While Theorem 2.29 states that the covariance is 0 for independent random variables, the converse is not true. That is, $\text{Cov}(X_1, X_2) = 0$ does not imply that X_1 and X_2 are independent. Example 2.54 provides an example of dependent variables whose covariance is 0.

Example 2.54 Let $\vec{X} = (X_1, X_2)$ be a discrete random vector with joint pdf given in Table 2.7.
Then, X_1 and X_2 are dependent random variables since

$$P(X_1 = 0, X_2 = -1) = \frac{1}{8} \neq P(X_1 = 0)P(X_2 = -1) = \frac{9}{64}.$$

Furthermore, it is easy to verify that $E(X_1) = \frac{5}{8}$, $E(X_2) = 0$, and $E(X_1 X_2) = 0$. Thus, $\text{Cov}(X_1, X_2) = E(X_1 X_2) - E(X_1)E(X_2) = 0 - \frac{1}{2} \cdot 0 = 0$. Therefore, X_1 and X_2 are dependent random variables with $\text{Cov}(X_1, \vec{X}_2) = 0$.

Table 2.7 The joint pdf of $\vec{X}$ for Example 2.54.

X_1	X_2		
	-1	0	1
0	$\frac{1}{8}$	$\frac{1}{8}$	$\frac{1}{8}$
1	$\frac{1}{4}$	$\frac{1}{8}$	$\frac{1}{4}$

Recall that when X_1 and X_2 are independent random variables, then the variance of $X_1 + X_2$ and the variance of $X_1 - X_2$ are both equal to $\mathrm{Var}(X_1) + \mathrm{Var}(X_2)$. Theorem 2.30 provides the general form of the variance for the sum and difference of two random variables.

Theorem 2.30 *If X_1 and X_2 are random variables with finite variances, then*

i) $\mathrm{Var}(X_1 + X_2) = \mathrm{Var}(X_1) + \mathrm{Var}(X_2) + 2\mathrm{Cov}(X_1, X_2)$.
ii) $\mathrm{Var}(X_1 - X_2) = \mathrm{Var}(X_1) + \mathrm{Var}(X_2) - 2\mathrm{Cov}(X_1, X_2)$.

Proof. Let X_1 and X_2 be random variables with finite variances.

i) Consider $\mathrm{Var}(X_1 + X_2)$.

$$
\begin{aligned}
\mathrm{Var}(X_1 + X_2) &= E[(X_1 + X_2)^2] - (\mu_1 + \mu_2)^2 \\
&= E(X_1^2) + 2E(X_1 X_2) + E(X_2^2) - \mu_1^2 - 2\mu_1\mu_2 - \mu_2^2 \\
&= E(X_1^2) - \mu_1^2 + E(X_2^2) - \mu_2^2 + 2E(X_1 X_2) - 2\mu_1\mu_2 \\
&= \mathrm{Var}(X_1) + \mathrm{Var}(X_2) + 2\mathrm{Cov}(X_1, X_2).
\end{aligned}
$$

ii) The proof of part (ii) is left as an exercise. ∎

Example 2.55 Let X_1 and X_2 be random variables with $\mathrm{Var}(X_1) = 10$, $\mathrm{Var}(X_2) = 25$ and $\mathrm{Cov}(X_1, X_2) = -5$. Then, by Theorem 2.30,

$$
\begin{aligned}
\mathrm{Var}(X_1 + X_2) &= \mathrm{Var}(X_1) + \mathrm{Var}(X_2) + 2\mathrm{Cov}(X_1, X_2) \\
&= 10 + 25 + 2(-5) = 25.
\end{aligned}
$$

$$
\begin{aligned}
\mathrm{Var}(X_1 - X_2) &= \mathrm{Var}(X_1) + \mathrm{Var}(X_2) - 2\mathrm{Cov}(X_1, X_2) \\
&= 10 + 25 - 2(-5) = 45.
\end{aligned}
$$

Theorem 2.31 (Cauchy–Schwarz Inequality) *If X_1 and X_2 are random variables with $\mathrm{Var}(X_1) = \sigma_1^2 < \infty$ and $\mathrm{Var}(X_2) = \sigma_2^2 < \infty$, then*

$$
|\mathrm{Cov}(X_1, X_2)| \le \sigma_1 \sigma_2
$$

with equality if and only if $P(X_2 = aX + b) = 1$ for some $a, b \in \mathbb{R}$.

Proof. Let X_1 and X_2 be random variables with finite variances σ_1^2 and σ_2^2. Let $U = \frac{X_1}{\sigma_1}$ and $V = \frac{X_2}{\sigma_2}$. Then, $\text{Var}(U) = \text{Var}(V) = 1$, and by Theorem 2.27, $\text{Cov}(U, V) = \frac{\text{Cov}(X_1, X_2)}{\sigma_1 \sigma_2}$.

Now, consider $\text{Var}(U - V)$.

$$\text{Var}(U - V) = \text{Var}(U) + \text{Var}(V) - 2\text{Cov}(U, V)$$
$$= 2 - 2\text{Cov}(U, V) = 2 - 2\frac{\text{Cov}(X_1, X_2)}{\sigma_1 \sigma_2}.$$

Thus, $0 \leq \text{Var}(U - V) = 2 - 2\frac{\text{Cov}(X_1, X_2)}{\sigma_1 \sigma_2}$, and therefore, $\text{Cov}(X_1, X_2) \leq \sigma_1 \sigma_2$.

Now consider $\text{Var}(U + V)$.

$$\text{Var}(U + V) = \text{Var}(U) + \text{Var}(V) + 2\text{Cov}(U, V)$$
$$= 2 + 2\text{Cov}(U, V) = 2 + 2\frac{\text{Cov}(X_1, X_2)}{\sigma_1 \sigma_2}.$$

Thus, $0 \leq \text{Var}(U + V) = 2 + 2\frac{\text{Cov}(X_1, X_2)}{\sigma_1 \sigma_2}$, and therefore, $-\sigma_1 \sigma_2 \leq \text{Cov}(X_1, X_2)$.

Hence, $-\sigma_1 \sigma_2 \leq \text{Cov}(X_1, X_2) \leq \sigma_1 \sigma_2$, and therefore, $|\text{Cov}(X_1, X_2)| \leq \sigma_1 \sigma_2$.

Now, $\text{Cov}(X_1, X_2) = 0$ if and only if $\text{Var}(U - V) = 0$, and $\text{Var}(U - V) = 0$ if and only if $P(U - V = c) = 1$ for some constant $c \in \mathbb{R}$. In this case,

$$1 = P(U - V = c) = P\left(\frac{X_1}{\sigma_1} - \frac{X_2}{\sigma_2} = c\right) = P\left(X_2 = \frac{\sigma_2}{\sigma_1}X_1 - c\sigma_2\right).$$

Similarly, $\text{Cov}(X_1, X_2) = -\sigma_1 \sigma_2$ if and only if $\text{Var}(U + V) = 0$, which means

$$1 = P(U + V = c) = P\left(\frac{X_1}{\sigma_1} + \frac{X_2}{\sigma_2} = c\right) = P\left(X_2 = -\frac{\sigma_2}{\sigma_1}X_1 + c\sigma_2\right). \qquad \blacksquare$$

The Cauchy–Schwarz theorem shows that the covariance is a measure of the strength of the linear relationship between two random variables. An alternative measure of the strength of the linear relationship between two variables is the *correlation coefficient*, which is a unitless measure. The definition of the correlation coefficient is given in Definition 2.25.

Definition 2.25 Let X_1 and X_2 be random variables with finite variances σ_1^2 and σ_2^2, respectively. The correlation coefficient between X_1 and X_2 is defined to be

$$\text{Cor}(X_1, X_2) = \frac{\text{Cov}(X_1, X_2)}{\sigma_1 \sigma_2}$$

and is denoted by ρ_{12}. When $\text{Cor}(X_1, X_2) = 0$, the random variables X_1 and X_2 are said to be uncorrelated.

The correlation between random variables X_1 and X_2 is a unitless measure since it is the ratio of the covariance, which has the units of X_1 times the units of X_2, and the product of the standard deviations, which are also in the units of X_1 and X_2. Because of the Cauchy–Schwarz theorem, $|\text{Cov}(X_1, X_2)| \leq \sigma_1 \sigma_2$, it follows that $|\text{Cor}(X_1, X_2)| \leq 1$.

Example 2.56 Let $\vec{X}$ be a random vector with joint pdf $f(x_1, x_2) = \frac{4x_1}{x_2}$ for $0 < x_1 < x_2 < 1$. Then,

$$E(X_1) = \int_0^1 \int_0^{x_2} x_1 \cdot \frac{4x_1}{x_2}\, dx_1\, dx_2 = \frac{4}{9}.$$

$$E(X_2) = \int_0^1 \int_0^{x_2} x_2 \cdot \frac{4x_1}{x_2}\, dx_1\, dx_2 = \frac{2}{3}.$$

$$E(X_1 X_2) = \int_0^1 \int_0^{x_2} x_1 x_2 \cdot \frac{4x_1}{x_2}\, dx_1\, dx_2 = \frac{1}{3}.$$

$$E(X_1^2) = \int_0^1 \int_0^{x_2} x_1^2 \cdot \frac{4x_1}{x_2}\, dx_1\, dx_2 = \frac{1}{4}.$$

$$E(X_2^2) = \int_0^1 \int_0^{x_2} x_2^2 \cdot \frac{4x_1}{x_2}\, dx_1\, dx_2 = \frac{1}{2}.$$

Thus, $\text{Cov}(X_1, X_2) = \frac{1}{3} - \frac{4}{9}\left(\frac{2}{3}\right) = \frac{1}{27}$, $\text{Var}(X_1) = \frac{1}{4} - \left(\frac{4}{9}\right)^2 = \frac{17}{324}$, $\text{Var}(X_2) = \frac{1}{2} - \left(\frac{2}{3}\right)^2 = \frac{1}{18}$, and therefore,

$$\rho_{12} = \text{Cor}(X_1, X_2) = \frac{\text{Cov}(X_1, X_2)}{\text{SD}(X_1)\text{SD}(X_2)} = \frac{\frac{1}{27}}{\sqrt{\frac{17}{324}}\sqrt{\frac{1}{18}}} = 0.1127.$$

Several properties of the correlation coefficient are given in Theorem 2.32.

Theorem 2.32 (Properties of Correlations) *If X_1 and X_2 are random variables with finite variances σ_1^2 and σ_2^2, then*

i) $\text{Cor}(X_1, X_2) = \text{Cor}(X_2, X_1)$.
ii) $|\text{Cor}(X_1, X_2)| \leq 1$.
iii) $\text{Cor}(aX_1 + b, cX_2 + d) = \text{sign}(ac)\text{Cor}(X_1, X_2)$, $\forall a, b, c, d \in \mathbb{R}$.
iv) $\text{Cor}(X_1, X_2) = \pm 1$ *if and only if* $P(X_2 = aX_1 + b)$ *for some* $a, b \in \mathbb{R}$.

Proof. The proof of Theorem 2.32 is left as an exercise. ∎

Example 2.57 Suppose that X_1 and X_2 are random variables with $\text{Var}(X_1) = 9$, $\text{Var}(X_2) = 4$, and $\text{Cov}(X_1, X_2) = 8$. Then,

$$\text{Cor}(X_1, X_2) = \frac{\text{Cov}(X_1, X_2)}{\sigma_1 \sigma_2} = \frac{8}{3 \cdot 2} = \frac{4}{3}.$$

$$\text{Cor}(-2X_1 + 3, 5X_2 + 5) = \text{sign}(-2 \cdot 5)\text{Cor}(X_1, X_2) = -\frac{4}{3}.$$

$$\text{Cor}(10X_1, 5X_2) = \text{sign}(10 \cdot 5)\text{Cor}(X_1, X_2) = \frac{4}{3}.$$

Theorem 2.33 *If X_1 and X_2 are independent random variables with finite variances, then $\text{Cor}(X_1, X_2) = 0$.*

Proof. Let X_1 and X_2 be independent random variables with finite variances. Then, $\text{Cov}(X_1, X_2) = 0$, and hence, $\text{Cor}(X_1, X_2) = 0$. ∎

Up to this point, a k-dimensional random vector has been taken to be a row vector, say $\vec{X} = (X_1, X_2, \ldots, X_k)$; however, a random vector $\vec{X}$ can also be written

as a column vector, say $\vec{X} = \begin{pmatrix} X_1 \\ X_2 \\ \vdots \\ X_k \end{pmatrix}$. When $\vec{X}$ is a written as column vector, the

expected value of $\vec{X}$ is the *vector of means*

$$E(\vec{X}) = \vec{\mu}_{\vec{X}} = \begin{pmatrix} E(X_1) \\ E(X_2) \\ \vdots \\ E(X_k) \end{pmatrix} = \begin{pmatrix} \mu_1 \\ \mu_2 \\ \vdots \\ \mu_k \end{pmatrix}.$$

The variances and covariances of the component random variables can also be summarized in a $k \times k$ *covariance matrix* denoted by $\Sigma_{\vec{X}} = [\sigma_{ij}]$, where $\sigma_{ij} = \text{Cov}(X_i, X_j)$. Then,

$$\Sigma_{\vec{X}} = \begin{pmatrix} \sigma_1^2 & \sigma_{12} & \cdots & \sigma_{1k} \\ \sigma_{21} & \sigma_2^2 & \cdots & \sigma_{2k} \\ \vdots & \vdots & \vdots & \vdots \\ \sigma_{k1} & \sigma_{k2} & \cdots & \sigma_k^2 \end{pmatrix}.$$

Note that $\Sigma_{\vec{X}}$ is a symmetric matrix since $\text{Cov}(X_i, X_j) = \text{Cov}(X_j, X_i)$.

Example 2.58 For the random vector in Example 2.56, the vector of means and the covariance matrix are

$$\mu_{\vec{X}} = \begin{pmatrix} 4 \\ 9 \\ 2 \\ 3 \end{pmatrix} \quad \text{and} \quad \Sigma_{\vec{X}} = \begin{pmatrix} \frac{17}{324} & \frac{1}{27} \\ \frac{1}{27} & \frac{1}{18} \end{pmatrix}.$$

Example 2.59 Let $X_1, \ldots, X_k$ be independent random variables with $\text{Var}(X_i) = \sigma_i^2$. Then, the covariance matrix associated with $\vec{X}$ is

$$\Sigma_{\vec{X}} = \begin{pmatrix} \sigma_1^2 & 0 & 0 & \cdots & 0 \\ 0 & \sigma_2^2 & 0 & \cdots & 0 \\ 0 & 0 & \sigma_3^2 & \cdots & 0 \\ \vdots & \vdots & \vdots & & \vdots \\ 0 & 0 & 0 & \cdots & \sigma_k^2 \end{pmatrix}.$$

When $X_1, X_2, \ldots, X_k$ are iid random variables, the covariance matrix associated with $\vec{X}$ is

$$\Sigma_{\vec{X}} = \begin{pmatrix} \sigma^2 & 0 & 0 & \cdots & 0 \\ 0 & \sigma^2 & 0 & \cdots & 0 \\ 0 & 0 & \sigma^2 & \cdots & 0 \\ \vdots & \vdots & \vdots & & \vdots \\ 0 & 0 & 0 & \cdots & \sigma^2 \end{pmatrix} = \sigma^2 \cdot I_{k \times k},$$

where $I_{k \times k}$ is the $k \times k$ identity matrix.

2.6.3 Conditional Expectation and Variance

The moments of a conditional distribution, say $X_1 | X_2$, are defined similarly to the moments of a random variable. In particular, the *conditional mean* of $X_1 | X_2$ is defined to be

$$E(X_1 | X_2) = \begin{cases} \sum_{\mathcal{S}'_{X_1}} x_1 f(x_1 | x_2) & \text{when } \vec{X} \text{ is discrete,} \\ \int_{\mathcal{S}'_{X_1}} x_1 f(x_1 | x_2) \, dx_1 & \text{when } \vec{X} \text{ is continuous.} \end{cases}$$

Similarly, for a function $h(X_1)$, the conditional expectation is

$$E[h(X_1) | X_2] = \begin{cases} \sum_{\mathcal{S}'_{X_1}} h(x_1) f(x_1 | x_2) & \text{when } \vec{X} \text{ is discrete,} \\ \int_{\mathcal{S}'_{X_1}} h(x_1) f(x_2 | x_2) \, dx_1 & \text{when } \vec{X} \text{ is continuous.} \end{cases}$$

It is important to note that $E(X_1 | X_2)$ is actually a function of the random variable X_2 and not X_1. For a fixed value of X_2, say $X_2 = x_2$, the conditional mean is written as $E(X_1 | X_2 = x_2)$.

Example 2.60 Let $\vec{X}$ be a continuous random vector with joint pdf $f(x_1, x_2) = \frac{3}{4}(x_2 - x_1)$ for $0 < x_1 < x_2 < 2$. The conditional pdf of $X_1 | X_2$ is $f(x_1 | x_2) = \frac{2(x_2 - x_1)}{x_2^2}$

for $0 < x_1 < x_2$. Thus,

$$E(X_1|X_2) = \int_0^{x_2} x_1 \cdot \frac{2(x_2 - x_1)}{x_2^2} \, dx_1 = \frac{x_2}{3},$$

and for $X_2 = 1$, $E(X_1|X_2 = 1) = \frac{1}{3}$.

Let $h(X_1) = X_1^2$. Then,

$$E(X_1^2|X_2) = \int_0^{x_2} x_1^2 \cdot 2\frac{x_2 - x_1}{x_2^2} \, dx_1 = \frac{x_2^2}{6},$$

and for $X_2 = 1$, $E(X_1^2|X_2 = 1) = \frac{1}{6}$.

Theorem 2.34 *If X_1 and X_2 are random variables, then*

  i) $E[aX_1 + b|X_2] = aE[X_1|X_2] + b, \; \forall a, b \in \mathbb{R}$.
  ii) $E[h_1(X_1) + h_2(X_1)|X_2] = E[h_1(X_1)|X_2] + E[h_2(X_1)|X_2]$.
  iii) $E[g(X_2)h(X_1)|X_2 = x_2] = g(x_2)E(X_1|X_2 = x_2)$.

Proof. The proof will only be given for continuous random variables X_1 and X_2 since the proof for two discrete random variables is similar with integration replaced by summation.

  i) Let $a, b \in \mathbb{R}$ be arbitrary but fixed, and consider $E[aX_1 + b|X_2]$.

$$E[aX_1 + b|X_2] = \int_{S'_{X_1}} (ax_1 + b)f(x_1|x_2) \, dx_1$$

$$= a \int_{S'_{X_1}} x_1 f(x_1|x_2) \, dx_1 + \int_{S'_{X_1}} bf(x_1|x_2) \, dx_1$$

$$= aE(X_1|X_2) + b.$$

  ii) The proof of part (ii) is left as an exercise.
  iii) The proof of part (iii) is left as an exercise. ∎

Example 2.61 Let $\vec{X} = (X_1, X_2)$ be a random vector and suppose the pdf of $X_2|X_1$ is $f(x_2|x_1) = \frac{3x_2^2}{x_1^3}$ for $0 < x_2 < x_1 < 1$. Then,

$$E(2X_2^2 + 1|X_1) = \int_0^{x_1} (2x_2^2 + 1) \cdot \frac{3x_2^2}{x_1^3} \, dx_2 = \frac{6x_1^2}{5} + 1,$$

and for $X_1 = 0.5$, $E(2X_2^2 + 1|X_1 = 0.5) = \frac{6(0.5)^2}{5} + 1 = 1.3$.

Now, consider $E[X_1^2(2X_2^2 + 1)|X_1]$.

$$E[X_1^2(2X_2^2 + 1)|X_1] = \int_0^{x_1} x_1^2(2x_2^2 + 1) \cdot \frac{3x_2^2}{x_1^3} \, dx_2$$

$$= x_1^2 \int_0^{x_1} (2x_2^2 + 1) \cdot \frac{3x_2^2}{x_1^3} \, dx_2$$

$$= x_1^2 E(2X_2^2 + 1 | X_1)$$

$$= x_1^2 \left(\frac{6x_1^2}{5} + 1 \right).$$

The conditional variance of $X_1 | X_2$ is given in Definition 2.26 and is analogous to the definition of an unconditional variance.

Definition 2.26 Let X_1 and X_2 be random variables. The conditional variance of $X_1 | X_2$, denoted by $\mathrm{Var}(X_1 | X_2)$, is defined to be

$$\mathrm{Var}(X_1 | X_2) = E[(X_1 - E(X_1 | X_2))^2 | X_2]$$

provided $E(X_1^2 | X_2) < \infty$.

The properties of the conditional variance are similar to those of the unconditional variance. Theorem 2.35 provides an alternative method for computing the conditional variance along with the variance of $aX_1 + b$ conditioned on X_2.

Theorem 2.35 *If X_1 and X_2 are random variables and $\mathrm{Var}(X_1 | X_2) < \infty$, then*

i) $\mathrm{Var}(X_1 | X_2) = E(X_1^2 | X_2) - E(X_1 | X_2)^2$.
ii) $\mathrm{Var}(aX_1 + b | X_2) = a^2 \mathrm{Var}(X_1 | X_2), \forall a, b \in \mathbb{R}$.

Proof. Let X_1 and X_2 be random variables with $\mathrm{Var}(X_1 | X_2) < \infty$.
i) Consider $\mathrm{Var}(X_1 | X_2)$.

$$\mathrm{Var}(X_1 | X_2) = E[(X_1 - E(X_1 | X_2))^2 | X_2]$$

$$= E[X_1^2 - 2X_1 E(X_1 | X_2) + E(X_1 | X_2)^2 | X_2]$$

$$= E(X_1^2 | X_2) - 2E \left[\underbrace{X_1 E(X_1 | X_2) | X_2}_{g(X_2)} \right] + E \left[\underbrace{E(X_1 | X_2)^2 | X_2}_{g(X_2)} \right]$$

$$= E(X_1^2 | X_2) - 2E(X_1 | X_2)E(X_1 | X_2) + E(X_1 | X_2)^2$$

$$= E(X_1^2 | X_2) - E(X_1 | X_2)^2.$$

ii) The proof of part (ii) is left as an exercise. ∎

Example 2.62 Let X_1 and X_2 be random variables, and suppose that $f(x_1 | x_2) = x_2 e^{-x_1 x_2}$ for $0 < x_1 < \infty$ and $x_2 > 0$. Then,

$$E(X_1 | X_2) = \int_0^\infty x_1 \cdot x_2 e^{-x_1 x_2} \, dx_1 = \frac{1}{x_2}$$

and

$$E(X_1^2|X_2) = \int_0^\infty x_1^2 \cdot x_2 e^{-x_1 x_2} \, dx_1 = \frac{2}{x_2^2}.$$

Thus,

$$\text{Var}(X_1|X_2) = E(X_1^2|X_2) - E(X_1|X_2)^2 = \frac{2}{x_2^2} - \left(\frac{1}{x_2}\right)^2 = \frac{1}{x_2^2}.$$

Theorem 2.36 relates the expected value and the variance of a random variable X_1 to the conditional expectation and the conditional variance of $X_1|X_2$.

Theorem 2.36 *If X_1 and X_2 are random variables and $\text{Var}(X_1)$ is finite, then*

i) $E[E(X_1|X_2)] = E(X_1)$ *(Law of Iterated Expectation).*
ii) $E[E(h(X_1)|X_2)] = E[h(X_1)]$.
iii) $\text{Var}(X_1) = E[\text{Var}(X_1|X_2)] + \text{Var}[E(X_1|X_2)]$.

Proof. The proofs will only be given for discrete random variables since the proof for continuous random variables is similar with summation replaced by integration.

Let X_1 and X_2 be random variables with $\text{Var}(X_1) < \infty$.

i) Consider $E[E(X_1|X_2)]$.

$$E[E(X_1|X_2)] = \sum_{x_2} E(X_1|X_2) f_2(x_2)$$

$$= \sum_{x_2} \left[\sum_{x_1} x_1 f(x_1|x_2) \right] f_2(x_2)$$

$$= \sum_{x_2} \sum_{x_1} x_1 f(x_1|x_2) f_2(x_2)$$

$$= \sum_{x_2} \sum_{x_1} x_1 f(x_1, x_2)$$

$$= E(X_1).$$

ii) The proof of part (ii) is left as an exercise.
iii) Consider $E[\text{Var}(X_1|X_2)]$.

$$E[\text{Var}(X_1|X_2)] = E[E(X_1^2|X_2) - E(X_1|X_2)^2]$$
$$= \underbrace{E(X_1^2)}_{\text{by part (ii)}} - E[E(X_1|X_2)^2]$$

$$= E(X_1^2) - \underbrace{E(X_1)^2 + E(X_1)^2}_{0} - E[E(X_1|X_2)^2]$$

$$= \text{Var}(X_1) + E(X_1)^2 - E[E(X_1|X_2)^2]$$

$$= \text{Var}(X_1) + E[E(X_1|X_2)]^2 - E[E(X_1|X_2)^2]$$

$$= \text{Var}(X_1) - [E[E(X_1|X_2)^2] - E[E(X_1|X_2)]^2]$$

$$= \text{Var}(X_1) - \text{Var}[E(X_1|X_2)].$$

Thus, $\text{Var}(X_1) = E[\text{Var}(X_1|X_2)] + \text{Var}[E(X_1|X_2)]$. ∎

Example 2.63 Let X_1 and X_2 be random variables with $E(X_1|X_2) = \frac{3}{4}X_2$, $E(X_2) = 10$, $\text{Var}(X_1|X_2) = \frac{3}{16}X_2$, and $\text{Var}(X_2) = 10$. Then,

$$E(X_1) = E[E(X_1|X_2)] = E\left(\frac{3}{4}X_2\right) = \frac{3}{4} \cdot 10 = 7.5,$$

$$\text{Var}[E(X_1|X_2)] = \text{Var}\left(\frac{3}{4}X_2\right) = \frac{9}{16}\text{Var}(X_2) = \frac{90}{16},$$

$$E[\text{Var}(X_1|X_2)] = E\left[\frac{3}{16}X_2\right] = \frac{30}{16}.$$

Thus, $\text{Var}(X_1) = E[\text{Var}(X_1|X_2)] + \text{Var}[E(X_1|X-2)] = \frac{90}{16} + \frac{30}{16} = 7.5$.

Problems

2.6.1 Let $\vec{X} = (X_1, X_2)$ be a random vector with pdf $f(x_1, x_2) = \frac{x_1 x_2}{60}$, for $x_1 = 1, 2, 3$ and $x_2 = 1, 2, 3, 4$. Determine

a) $E(X_1)$.

b) $E(X_2)$.

c) $E(X_1^2)$.

d) $E(X_1 X_2)$.

e) $\text{Var}(X_1)$.

f) $\text{Cov}(X_1, X_2)$.

2.6.2 Let $\vec{X} = (X_1, X_2)$ be a random vector with pdf $f(x_1, x_2) = \frac{x_1 x_2}{65}$, for $x_1 = 1, \ldots, x_2$ and $x_2 = 1, 2, 3, 4$. Determine

a) $E(X_1)$.

b) $E(X_2)$.

c) $E(X_1^2)$.

d) $E(X_1 X_2)$.

e) $\text{Var}(X_1)$.

f) $\text{Cov}(X_1, X_2)$.

2.6.3 Let $\vec{X} = (X_1, X_2)$ be a random vector with joint pdf $f(x_1, x_2) = 24x_1 x_2$, $0 < x_1 + x_2 < 1, x_1 > 0, x_2 > 0$. Determine

a) $E(X_1)$.

b) $E(X_2)$.

c) $E(X_1^2)$.

d) $E(X_1 X_2)$.

e) $\text{Var}(X_1)$.

f) $\text{Cov}(X_1, X_2)$.

2.6.4 Let $\vec{X} = (X_1, X_2)$ be a random vector with pdf $f(x_1, x_2) = 24x_1x_2$ for $0 < x_1 + x_2 < 1, x_1 > 0, x_2 > 0$. Determine

a) $f_1(x_1)$.   b) $f_2(x_2)$.
c) $E(X_1 | X_2)$.   d) $E(X_1^2 | X_2)$.
e) $\text{Var}(X_1 | X_2)$.   f) $E[\text{Var}(X_1 | X_2)]$.

2.6.5 Let $\vec{X} = (X_1, X_2)$ be a random vector with pdf $f(x_1, x_2) = \frac{3}{2}(x_1^2 + x_2^2)$, for $0 < x_1 < 1, 0 < x_2 < 1$. Determine

a) $E(X_1)$.   b) $E(X_2)$.
c) $E(X_1^2)$.   d) $E(X_1X_2)$.
e) $\text{Var}(X_1)$.   f) $\text{Cov}(X_1, X_2)$.

2.6.6 $\vec{X} = (X_1, X_2)$ be a random vector with pdf $f(x_1, x_2) = \frac{3}{2}(x_1^2 + x_2^2)$, for $0 < x_1 < 1, 0 < x_2 < 1$. Determine

a) $f_2(x_2)$.
b) $E(X_1 | X_2)$.
c) $E(X_1 | X_2 = 0.75)$.

2.6.7 Let $\vec{X} = (X_1, X_2)$ be a random vector with pdf $f(x_1, x_2) = 2e^{-(x_1 + x_2)}$, for $0 < x_1 < x_2 < \infty$. Determine

a) $E(X_1)$.   b) $E(X_2)$.
c) $E(X_1^2)$.   d) $E(X_2^2)$.
e) $\text{Var}(X_1)$.   f) $\text{Var}(X_2)$.
g) $E(X_1X_2)$.   h) $\text{Cov}(X_1, X_2)$.
i) the covariance matrix Σ associated with $\vec{X}$.

2.6.8 Let $\vec{X} = (X_1, X_2)$ be a random vector with pdf $f(x_1, x_2) = 4x_1^2$, for $0 < x_2 < x_1 < 1$. Determine

a) $E(X_1)$.   b) $E(X_2)$.
c) $E(X_1^2)$.   d) $E(X_2^2)$.
e) $\text{Var}(X_1)$.   f) $\text{Var}(X_2)$.
g) $E(X_1X_2)$.   h) $\text{Cov}(X_1, X_2)$.
i) the covariance matrix Σ associated with $\vec{X}$.

2.6.9 Let X_1 and X_2 be random variables with variances $\sigma_1^2 = 4$ and $\sigma_2^2 = 9$ and $\text{Cov}(X_1, X_2) = -3$. Determine

a) $\text{Cov}(5 - 2X_1, X_2 + 4)$.   b) $\text{Cov}(3X_1 + 5, 7 - 3X_2)$.
c) $\text{Cov}(2X_1, X_1 + X_2)$.   d) $\text{Cov}(X_1 - 2, X_1 + 2)$.

2.6.10 Let X_1 and X_2 be independent random variables with variances $\sigma_1^2 = 5$ and $\sigma_2^2 = 2$. Determine

a) $\mathrm{Var}(X_1 - X_2)$.

b) $\mathrm{Var}(3X_1 + 4X_2)$.

c) $\mathrm{Var}(5X_1 - 4X_2)$.

d) $\mathrm{Var}(3X_1 + X_2 + 5)$.

2.6.11 Let X_1 and X_2 be random variables with variances $\sigma_1^2 = 4$ and $\sigma_2^2 = 5$ and $\mathrm{Cov}(X_1, X_2) = 2$. Determine

a) $\mathrm{Var}(X_1 - X_2)$.

b) $\mathrm{Var}\left(\frac{X_1 + X_2}{2}\right)$.

2.6.12 Let X_1 and X_2 be independent random variables with variances $\sigma_1^2 = 10$ and $\sigma_2^2 = 15$. Determine

a) $\mathrm{Var}(X_1 - X_2)$.

b) $\mathrm{Var}\left(\frac{X_1 + X_2}{2}\right)$.

2.6.13 If X_1 and X_2 are independent random variables with finite variances, show that $\mathrm{Var}(X_1 - X_2) = \mathrm{Var}(X_1) + \mathrm{Var}(X_2)$.

2.6.14 Let X_1 and X_2 be independent random variables with finite variances. Let $a, b \in \mathbb{R}$, prove that $\mathrm{Var}(aX_1 + bX_2) = a^2\mathrm{Var}(X_1) + b^2\mathrm{Var}(X_2)$.

2.6.15 Let X and X_2 be independent random variables with finite variances. Prove that $\mathrm{Var}(X_1X_2) = E(X_1)^2\mathrm{Var}(X_2) + E(X_2^2)\mathrm{Var}(X_1)$.

2.6.16 Let X and X_2 be random variables with finite variances, and suppose that $\mathrm{Var}(X_1) = \mathrm{Var}(X_2)$. Prove that $\mathrm{Cov}(X_1 + X_2, X_1 - X_2) = 0$.

2.6.17 Let X_1 and X_2 be random variables with finite variances, and let $a, b, c, d \in \mathbb{R}$. Prove that

a) $\mathrm{Cov}(aX_1, bX_2) = ab\mathrm{Cov}(X_1, X_2)$.

b) $\mathrm{Cov}(aX_1 + b, cX_2 + d) = ac\mathrm{Cov}(X_1, X_2)$.

c) $\mathrm{Cov}(X_1, X_1) = \mathrm{Var}(X_1)$.

2.6.18 Let X_1 and X_2 be random variables with finite variances. Prove

a) $\mathrm{Cov}(-X_1, X_2) = -\mathrm{Cov}(X_1, X_2)$.

b) $\mathrm{Cov}(X_1, -X_1) = -\mathrm{Var}(X_1)$.

2.6.19 Let X_1, X_2, and X_3 be random variables with finite variances. Prove that

a) $\mathrm{Cov}(X_1 + X_2, X_3) = \mathrm{Cov}(X_1, X_3) + \mathrm{Cov}(X_2, X_3)$.

b) $\mathrm{Cov}(aX_1 + bX_2, cX_3) = ac\mathrm{Cov}(X_1, X_3) + bc\mathrm{Cov}(X_2, X_3)$, $\forall a, b, c \in \mathbb{R}$.

2.6.20 Let X_1 and X_2 be random variables with finite variances. Prove
a) $\text{Cor}(X_1, X_2) = \text{Cor}(X_2, X_1)$.
b) $|\text{Cor}(X_1, X_2)| \leq 1$.
c) $\text{Cor}(aX_1, bX_2) = \text{sign}(ab)\text{Cov}(X_1, X_2)$.
d) $\text{Cor}(aX_1 + b, cX_2 + d) = \text{sign}(ac)\text{Cov}(X_1, X_2)$.
e) $\text{Cor}(X_1, X_2) = \pm 1$ if and only if $P(X_2 = aX_1 + b)$ for some $a, b \in \mathbb{R}$.

2.6.21 Let X_1 and X_2 be random variables with variances σ_1^2 and σ_2^2. Prove that $\text{Var}(X_1 - X_2) = \text{Var}(X_1) + \text{Var}(X_2) - 2\text{Cov}(X_1, X_2)$.

2.6.22 Let X_1 and X_2 be random variables with variances σ_1^2 and σ_2^2 with $\text{Cov}(X_1, X_2) < 0$. Show that
a) $\text{Var}(X_1 + X_2) \leq \text{Var}(X_1) + \text{Var}(X_2)$.
b) $\text{Var}(X_1 - X_2) \geq \text{Var}(X_1) + \text{Var}(X_2)$.

2.6.23 Let X_1 and X_2 be random variables with finite means μ_1 and μ_2 and variances σ_1^2 and σ_2^2. Determine
a) $\text{Cov}\left(\frac{X_1 - \mu_1}{\sigma_1}, \frac{X_2 - \mu_2}{\sigma_2}\right)$.
b) $\text{Cov}(X_1 - X_2, X_1 + X_2)$.

2.6.24 Let X_1 and X_2 be iid random variables with finite mean μ and variance σ^2. Determine the values of a_1 and a_2 minimizing $\text{Var}(a_1 X_1 + a_2 X_2)$ subject to $E(a_1 X_1 + a_2 X_2) = \mu$.

2.6.25 Let X_1 and X_2 be random variables with $\text{Var}(X_1 | X_2) < \infty$. Prove that $\text{Var}(aX_1 + b | X_2) = a^2 \text{Var}(X_1 | X_2)$, $\forall a, b \in \mathbb{R}$.

2.6.26 Let X_1, X_2, and X_3 be random variables with finite variances $\text{Var}(X_1) = \sigma_1^2, \text{Var}(X_2) = \sigma_2^2$, and $\text{Var}(X_3) = \sigma_3^2$. Show that $\text{Cor}(X_1, X_2 + X_3) \neq \text{Cor}(X_1, X_2) + \text{Cor}(X_1, X_3)$.

2.6.27 Let X_1 and X_2 be random variables, and suppose that $E(X_2) = 12$, $\text{Var}(X_2) = 5, E(X_1 | X_2) = 5X_2 + 3$, and $\text{Var}(X_1 | X_2) = 2X_2$. Determine
a) $E(X_1)$.   b) $\text{Var}(X_1)$.

2.6.28 Let X_1 and X_2 be random variables, and suppose that $f(x_1 | x_2) = \binom{x_2}{x_1}\left(\frac{1}{2}\right)^2$ for $x_1 = 0, 1, \ldots, x_2$ and $x_2 = 10, 20$. If $f_2(10) = \frac{2}{3}$ and $f_2(20) = \frac{1}{3}$, determine
a) $E(X_1 | X_2)$.   b) $E(X_1)$.

2.6.29 Let X_1 and X_2 be random variables, and suppose that $f(x_1|x_2) = \frac{1}{x_2}$ for $0 < x_1 < x_2$ and $x_2 > 0$. If $f_2(x_2) = e^{-x_2}$, determine

a) $E(X_1|X_2)$.
b) $E(X_1)$.
c) $E(X_1^2|X_2)$.
d) $E(X_1^2)$.
e) $V(X_1)$.
f) $E(X_1|X_2 = 5)$.

2.6.30 Let X_1 and X_2 be continuous random variables. If $h(x)$ and $g(x)$ are functions, prove

a) $E[h(X_1) + g(X_1)|X_2] = E[h(X_1)|X_2] + E[g(X_1)|X_2]$.
b) $E[h(X_1)g(X_2)|X_2 = x_2] = g(x_2)E[h(X_1)|X_2]$.
c) $E[E[h(X_1)|X_2]] = E[h(X_1)]$.

2.6.31 Let X_1 and X_2 be iid random variables with common MGF $M(t) = e^{\frac{t^2}{2}}$ for $t \in \mathbb{R}$. Determine the MGF of

a) $X_1 + X_2$.
b) $X_1 - X_2$.

2.6.32 Let X_1 and X_2 be independent random variables with MGFs $M_1(t) = e^{3(e^t-1)}$ and $M_2(t) = e^{5(e^t-1)}$ for $t \in \mathbb{R}$. Determine

a) the MGF of $X_1 + X_2$.
b) $E(X_1 + X_2)$.

2.6.33 Let X_1, X_2, and X_3 be iid random variables with common MGF $M(t) = 0.2e^t + 0.8$ for $t \in \mathbb{R}$. Determine the MGF of

a) $X_1 + X_2 + X_3$.
b) $E(X_1 + X_2 + X_3)$.

2.6.34 Let X_1 and X_2 be independent continuous random variables. Prove that $E(X_1|X_2) = E(X_1)$.

2.6.35 Let X_1 and X_2 be random variables with $\text{Var}(X_1|X_2) < \infty$. Prove that $\text{Var}(X_1) \geq E[\text{Var}(X_1|X_2)]$.

2.7 Sums of Random Variables

In the theory of statistics, it is common for an estimator to be based on a sum or linear combination of a collection of random variables $X_1, \ldots, X_n$. The pdf of a linear combination of random variables, say $T = T(X_1, \ldots, X_n)$, can be difficult to determine; however, the expected value and the variance of the linear transformation T are only based on the first two moments of $X_1, \ldots, X_n$. In this section, the expected value, variance, and MGFs of a linear transformation of the random variables $X_1, X_2, \ldots, X_n$ are discussed.

Theorem 2.37 *If* $X_1, \ldots, X_n$ *are random variables with* $E(X_i) = \mu_i$ *for* $i = 1, \ldots, n$, *then*

i) $E\left(\sum_{i=1}^{n} X_i\right) = \sum_{i=1}^{n} \mu_i.$

ii) $E\left(\sum_{i=1}^{n} a_i X_i\right) = \sum_{i=1}^{n} a_i \mu_i,$ *for* $a_1, a_2, \ldots, a_n \in \mathbb{R}.$

Proof. The proof will only be given for continuous random variables $X_1, \ldots, X_n$ since the proof for a collection of discrete random variables is similar with integration replaced by summation.

Let $X_1, \ldots, X_n$ be continuous random variables with joint pdf $f(x_1, \ldots, x_n)$ on $\mathcal{S}_{\bar{X}}$, and let $E(X_i) = \mu_i$ for $i = 1, 2, \ldots, n$.

i)

$$E\left[\sum_{i=1}^{n} X_i\right] = \int \cdots \int_{\mathcal{S}_{\bar{X}}} \sum_{i=1}^{n} x_i f(x_1, \ldots, x_n) \, dx_1 \cdots dx_n$$

$$= \sum_{i=1}^{n} \int \cdots \int_{\mathcal{S}_{\bar{X}}} x_i f(x_1, \ldots, x_n) \, dx_1 \cdots dx_n$$

$$= \sum_{i=1}^{n} E(X_i).$$

ii) The proof of part (ii) is left as an exercise. ∎

The general formula for the variance of a sum and a linear combination of random variables $X_1, \ldots, X_n$ is given in Theorem 2.38.

Theorem 2.38 *If* $X_1, \ldots, X_n$ *are random variables with* $\mathrm{Var}(X_i) = \sigma_i^2$ *and* $\mathrm{Cov}(X_i, X_j) = \sigma_{ij}$ *for* $i \neq j$, *then*

i) $\mathrm{Var}\left(\sum_{i=1}^{n} X_i\right) = \sum_{i=1}^{n} \sigma_i^2 + 2 \sum_{i<j} \sigma_{ij}.$

ii) $\mathrm{Var}\left(\sum_{i=1}^{n} a_i X_i\right) = \sum_{i=1}^{n} a_i^2 \sigma_i^2 + 2 \sum_{j=1}^{n} \sum_{i<j} a_i a_j \sigma_{ij}$ *for* $a_1, a_2, \ldots, a_n \in \mathbb{R}.$

Proof. Let $X_1, \ldots, X_n$ are random variables with $\mathrm{Var}(X_i) = \sigma_i^2$ and $\mathrm{Cov}(X_i, X_j) = \sigma_{ij}$ for $i \neq j$.

i)

$$\text{Var}\left[\sum_{i=1}^{n} X_i\right] = E\left[\left(\sum_{i=1}^{n} X_i\right)^2\right] - \left[E\left(\sum_{i=1}^{n} X_i\right)\right]^2$$

$$= E\left[\sum_{j=1}^{n}\sum_{i=1}^{n} X_i X_j\right] - \left(\sum_{i=1}^{n} \mu_i\right)^2$$

$$= \sum_{j=1}^{n}\sum_{i=1}^{n} E(X_i X_j) - \sum_{j=1}^{n}\sum_{i=1}^{n} \mu_i \mu_j$$

$$= \sum_{j=1}^{n}\sum_{i=1}^{n} \text{Cov}(X_i, X_j)$$

$$= \sum_{i=1}^{n} \text{Cov}(X_i, X_i) + \sum_{j=1}^{n}\sum_{i\neq j}^{n} \text{Cov}(X_i, X_j)$$

$$= \sum_{i=1}^{n} \text{Var}(X_i) + 2\sum_{j=1}^{n}\sum_{i<j}^{n} \text{Cov}(X_i, X_j).$$

ii) The proof of part (ii) is left as an exercise. ∎

Example 2.64 Let X_1, X_2, X_3 be random variables with means $\mu_1 = 3$, $\mu_2 = 5$, $\mu_3 = -1$, variances $\sigma_1 = 2, \sigma_2 = 1$, $\sigma_3 = 5$, and covariances $\sigma_{12} = -0.5$, $\sigma_{13} = 0.75$, $\sigma_{23} = -1$. Let $T = 3X_1 - X_2 + 2X_3$. Then,

$E(T) = E(3X_1 - X_2 + 2X_3) = 3\mu_1 - \mu_2 + 2\mu_3 = 2.$

$\text{Var}(T) = \text{Var}(3X_1 - X_2 + 2X_3) = 9\sigma_1^2 + \sigma_2^2 + 4\sigma_3^2 - 6\sigma_{12} + 12\sigma_{13} - 4\sigma_{23} = 153.$

Corollary 2.4 provides the variance for a linear combination of independent random variables, and Corollary 2.5 provides the variance for a linear combination of iid random variables.

Corollary 2.4 *If $X_1, \dots, X_n$ are independent random variables with* $\text{Var}(X_i) = \sigma_i^2$, *then*

i) $\text{Var}\left(\sum_{i=1}^{n} X_i\right) = \sum_{i=1}^{n} \sigma_i^2.$

ii) $\text{Var}\left(\sum_{i=1}^{n} a_i X_i\right) = \sum_{i=1}^{n} a_i^2 \sigma_i^2.$

Proof. Since $X_1, \dots, X_n$ are independent random variables, $\text{Cov}(X_i, X_j) = 0$ for $i \neq j$, and thus, Corollary 2.4 follows directly from Theorem 2.38. ∎

Corollary 2.5 *If $X_1, \ldots, X_n$ are iid with common mean $E(X) = \mu$ and common variance $\mathrm{Var}(X) = \sigma^2$, then*

i) $E\left(\displaystyle\sum_{i=1}^{n} X_i \right) = n\mu.$

ii) $E\left(\displaystyle\sum_{i=1}^{n} a_i X_i \right) = \mu \displaystyle\sum_{i=1}^{n} a_i.$

iii) $\mathrm{Var}\left(\displaystyle\sum_{i=1}^{n} X_i \right) = n\sigma^2.$

iv) $\mathrm{Var}\left(\displaystyle\sum_{i=1}^{n} a_i X_i \right) = \sigma^2 \displaystyle\sum_{i=1}^{n} a_i^2.$

Proof. Corollary 2.4 follows directly from Theorem 2.37 and Corollary 2.4 since $X_1, \ldots, X_n$ are iid random variables. ∎

A common and important linear transformation in statistics is the average of a collection of iid random variables $X_1, \ldots, X_n$. For a collection of random variables $X_1, \ldots, X_n$, the average is $\frac{1}{n} \sum_{i=1}^{n} X_i$, which is denoted by $\overline{X}$. The mean and variance of $\overline{X}$ are given in Corollary 2.6 for a collection of iid random variables.

Corollary 2.6 *If $X_1, \ldots, X_n$ are iid random variables with common mean μ, common variance σ^2, then*

i) $E(\overline{X}) = \mu.$

ii) $\mathrm{Var}(\overline{X}) = \dfrac{\sigma^2}{n}.$

Proof. Let $X_1, \ldots, X_n$ be iid random variables with common mean μ, common variance σ^2.

i)

$$E(\overline{X}) = E\left(\frac{1}{n} \sum_{i=1}^{n} X_i \right) = \frac{1}{n} \sum_{i=1}^{n} E(X_i) = \frac{1}{n} \sum_{i=1}^{n} \mu = \frac{1}{n}(n\mu) = \mu.$$

ii)

$$\mathrm{Var}(\overline{X}) = \mathrm{Var}\left(\frac{1}{n} \sum_{i=1}^{n} X_i \right) = \frac{1}{n^2} \sum_{i=1}^{n} \mathrm{Var}(X_i) = \frac{1}{n^2} \sum_{i=1}^{n} \sigma^2 = \frac{1}{n^2}(n\sigma^2)$$
$$= \frac{\sigma^2}{n}.$$

∎

As will be seen in later chapters, $\overline{X}$ is commonly used as an estimator of the mean of a distribution, and the properties of $\overline{X}$ as an estimator are discussed in Chapter 4.

Let $\vec{X}$ be a k-dimensional column random vector with vector of means $\vec{\mu}$ and covariance matrix Σ. If $a \in \mathbb{R}^k$ is a column vector of scalars, then $a^{\mathsf{T}}\vec{X}$ is a linear transformation of $\vec{X}$ since $a^{\mathsf{T}}\vec{X} = \sum_{i=1}^{k} a_i X_i$. Theorem 2.39 provides a matrix-algebra approach for finding the mean and variance of a linear transformation of a random vector.

Theorem 2.39 *Let $\vec{X}$ be a k-dimensional column random vector with mean vector $\vec{\mu}$ and covariance matrix Σ. If $a \in \mathbb{R}^k$, then*

i) $E(a^{\mathsf{T}}\vec{x}) = a^{\mathsf{T}}\vec{\mu}$.
ii) $\mathrm{Var}(a^{\mathsf{T}}\vec{X}) = a^{\mathsf{T}}\mathrm{Cov}(\vec{X})a$.

Proof. Let $\vec{X}$ be a k-dimensional column random vector with mean vector $\vec{\mu}$ and covariance matrix Σ, let $a \in \mathbb{R}^k$ be arbitrary but fixed, and let $T = a^{\mathsf{T}}\vec{X}$.

i) Since $T = \sum_{i=1}^{k} a_i X_i$, it follows that

$$E(a^{\mathsf{T}}\vec{x}) = E\left(\sum_{i=1}^{k} a_i X_i \right) = \sum_{i=1}^{k} a_i \mu_i = a^{\mathsf{T}}\vec{\mu}.$$

ii) Since $T = \sum_{i=1}^{k} a_i X_i$, it follows that

$$\mathrm{Var}(a^{\mathsf{T}}\vec{X}) = \mathrm{Cov}(a^{\mathsf{T}}\vec{X}, a^{\mathsf{T}}\vec{X}) = \mathrm{Cov}\left(\sum_{i=1}^{k} a_i X_i, \sum_{i=1}^{k} a_i X_i \right)$$

$$= \sum_{j=1}^{k} \sum_{i=1}^{k} a_i a_j \mathrm{Cov}(X_i, X_j) = a^{\mathsf{T}}\Sigma a. \qquad \blacksquare$$

Example 2.65 Let $\vec{X} = \begin{pmatrix} X_1 \\ X_2 \\ X_3 \end{pmatrix}$ be a random vector with $\vec{\mu} = \begin{pmatrix} 5 \\ 8 \\ 4 \end{pmatrix}$, and

$\Sigma = \begin{pmatrix} 3 & 1 & 1 \\ 1 & 2 & 0 \\ 1 & 0 & 5 \end{pmatrix}$. Let $a = \begin{pmatrix} 1 \\ -1 \\ 1 \end{pmatrix}$. Then, $a^{\mathsf{T}}\vec{X} = X_1 - X_2 + X_3$ and

$$E(a^{\mathsf{T}}\vec{X}) = E(X_1 - X_2 + X_3) = a^{\mathsf{T}}\vec{\mu} = (1, -1, 1)\begin{pmatrix} 5 \\ 8 \\ 4 \end{pmatrix} = 1.$$

$$\mathrm{Var}(a^{\mathsf{T}}\vec{X}) = a^{\mathsf{T}}\Sigma a = (1 \quad -1 \quad 1)\begin{pmatrix} 3 & 1 & 1 \\ 1 & 2 & 0 \\ 1 & 0 & 5 \end{pmatrix}\begin{pmatrix} 1 \\ -1 \\ 1 \end{pmatrix} = 10.$$

Example 2.66 Let $X_1, X_2, \ldots, X_n$ be iid random variables with common mean μ and common variance σ^2. Then, $\vec{\mu} = (\mu, \mu, \ldots, \mu)^{\mathrm{T}}$ and $\Sigma = \sigma^2 I_k$. Let $a^{\mathrm{T}} = \left(\frac{1}{n}, \frac{1}{n}, \ldots, \frac{1}{n}\right)$. Then, $a^{\mathrm{T}} \vec{X} = \overline{X}$ and by Corollary 2.6,

$$E(a^{\mathrm{T}} \vec{X}) = a^{\mathrm{T}} \vec{\mu} = \mu.$$

$$\mathrm{Var}(a^{\mathrm{T}} \vec{X}) = a^{\mathrm{T}} \Sigma a = a^{\mathrm{T}} \sigma^2 I_k a = \frac{\sigma^2}{n}.$$

When $X_1, X_2, \ldots, X_n$ are independent random variables, it may be possible to determine the distribution of the sum of the random variables from the moment generating function of the sum. In particular, the MGF of the sum of independent random variables is given in Theorem 2.40, and Corollary 2.7 provides the MGF for the sum of iid random variables.

Theorem 2.40 Let $X_1, \ldots, X_n$ be independent random variables with MGFs $M_1(t), \ldots, M_n(t)$. If $S = \sum_{i=1}^{n} X_i$, then $M_S(t) = \prod_{i=1}^{n} M_i(t)$.

Proof. Let $X_1, X_2, \ldots, X_n$ be independent random variables with MGFs $M_1(t), \ldots, M_n(t)$, and let $S = \sum_{i=1}^{n} X_i$.

$$M_S(t) = E[e^{tS}] = E\left[e^{\sum_{i=1}^{n} tX_i}\right]$$

$$= \prod_{i=1}^{n} \underbrace{E[e^{tX_i}]}_{X_1,\ldots,X_n \text{ ind.}}$$

$$= \prod_{i=1}^{n} M_i(t).$$

∎

Corollary 2.7 *If* $X_1, X_2, \ldots, X_n$ *are iid random variables with common MGF* $M(t)$ *and* $S = \sum_{i=1}^{n} X_i$, *then* $M_S(t) = M(t)^n$.

Proof. The proof of Corollary 2.7 follows directly from Theorem 2.40 with $M_i(t) = M(t)$. ∎

When the MGF of a sum of independent random variables $X_1, X_2, \ldots, X_n$ is equal to the MGF of a random variable Y with known pdf, it follows from Theorem 2.21 that $\sum_{i=1}^{n} X_i$ and Y have the same distribution. Thus, the distribution of the sum of independent random variables can often be determined by find the MGF of the sum and then matching it with a known distribution.

Example 2.67 Let $X_1, X_2, \ldots, X_n$ be independent random variables with MGFs $M_i(t) = (p\, e^t + (1 - p))^{k_i}$ for $k_i \in \mathbb{N}$; a random variable with MGF $M_i(t)$

has pdf $f(x) = \binom{k_i}{x} p^x (1-p)^{k_i - x}$ for $x = 0, 1, 2, \ldots, k_i$ (see Example 2.45). Let $S = \sum_{i=1}^{n} X_i$. Then, by Theorem 2.40,

$$M_S(t) = \prod_{i=1}^{n} M_i(t) = \prod_{i=1}^{n} (p\, e^t + (1-p))^{k_i} = (p\, e^t + (1-p))^{\sum_{i=1}^{n} k_i},$$

which is the MGF of a random variable Y having pdf $f(y) = \binom{s}{y} p^y (1-p)^{s-y}$ for $y = 0, 1, \ldots, s$, where $s = \sum_{i=1}^{n} k_i$.

Example 2.68 Let X be a continuous random variable with pdf $f(x) = \frac{1}{\sqrt{2\pi\sigma^2}} e^{-\frac{1}{2\sigma^2}(x-\mu)^2}$ for $x \in \mathbb{R}$ and MGF $M_X(t) = e^{\frac{t^2\sigma^2}{2} + \mu t}$ for $t \in \mathbb{R}$. The random variable X is called a *normal random variable* with mean μ and variance σ^2. Let $X_1, X_2, \ldots, X_n$ be iid normal random variables with mean μ and variance σ^2, and let $S = \sum_{i=1}^{n} X_i$. Then, by Corollary 2.7,

$$M_S(t) = \prod_{i=1}^{n} M(t) = \left(e^{\frac{t^2\sigma^2}{2} + \mu t} \right)^n = e^{\frac{nt^2\sigma^2}{2} + n\mu t},$$

which is the MGF of a normal random variable with mean $n\mu$ and variance $n\sigma^2$.

Problems

2.7.1 Let X_1, X_2, and X_3 be independent random variables with $\mu_1 = 10$, $\mu_2 = 8$, $\mu_3 = 5$, $\sigma_1^2 = 4$, $\sigma_2^2 = 3$, $\sigma_3^2 = 5$. If $Y_1 = 2X_1 + 3X_2 - 4X_3$ and $Y_2 = X_1 - 2X_2 + X_3$, determine

a) $E(Y_1)$.   b) $\text{Var}(Y_1)$.
c) $E(Y_2)$.   d) $\text{Var}(Y_2)$.

2.7.2 Let X_1, X_2, X_3, X_4 be iid random variables with common mean μ and common variance σ^2 and $Y = X_1 - X_2 - X_3 + 2X_4$. Determine

a) $E(Y)$.   b) $\text{Var}(Y)$.

2.7.3 Let X_1, X_2, X_3, X_4, X_5 be iid random variables with common mean μ and common variance σ^2 and $Y = 5X_1 - X_2 - X_3 - X_4 - X_5$. Determine

a) $E(Y)$.   b) $\text{Var}(Y)$.

2.7.4 Let X_1, X_2, X_3, X_4, X_5 be iid random variables with common mean μ and common variance σ^2 and $Y = \frac{1}{4}(X_1 + X_2) + \frac{1}{10}X_3 + \frac{3}{10}X_4 + \frac{1}{10}X_5$. Determine

a) $E(Y)$.   b) $\mathrm{Var}(Y)$.

2.7.5 Let X_1 and X_2 be iid random variables with common mean μ and variance σ^2. Let $Y = \frac{X_1 - X_2}{\sqrt{2}\sigma}$. Determine

a) $E(Y)$.   b) $\mathrm{Var}(Y)$.

2.7.6 Let X_1 and X_2 be iid random variables with common mean $\mu = 5$ and common variance $\sigma^2 = 4$. Determine the values of a_1 and a_2 such that $E(a_1 X_1 + a_2 X_2) = 15$ and $\mathrm{Var}(a_1 X_1 + a_2 X_2) = 20$.

2.7.7 Let $X_1, X_2, \ldots, X_n$ be iid random variables with common mean μ and variance σ^2. Determine $E\left[\frac{1}{n+1}\left(2X_1 + \sum_{i=2}^{n} X_i\right)\right]$.

2.7.8 Let $X_1, \ldots, X_n$ be iid random variables with common mean μ and variance σ^2. Determine $E(\overline{X}^2)$.

2.7.9 Let $X_1, \ldots, X_n$ be iid random variables with common mean μ and variance σ^2, and let $S = \frac{1}{n}\sum_{i=1}^{n} \frac{X_i - \mu}{\sigma}$. Determine

a) $E(S)$.   b) $\mathrm{Var}(S)$.

2.7.10 Let $X_1, \ldots, X_n$ be iid random variables with common mean μ and common variance σ^2, and let $\overline{X} = \sum_{i=1}^{n} X_i$ and $S_n^2 = \frac{1}{n}\sum_{i=1}^{n} X_i^2 - \overline{X}^2$. Determine

a) $E\left[\frac{1}{n}\sum_{i=1}^{n} X_i^2\right]$.

b) $E(S_n^2)$.

2.7.11 Let $X_1, \ldots, X_n$ be iid random variables with common mean μ and common variance σ^2, and let $T = \sum_{i=1}^{n} (-1)^{i+1} X_i$. Determine

a) $E(T)$.   b) $\mathrm{Var}(T)$.

2.7.12 Let $\vec{X}$ be a k-dimensional discrete random vector and let $a_1, \ldots, a_k \in \mathbb{R}$. If $E(X_i^2) < \infty$ for $i = 1, 2, \ldots, k$, prove that $E\left[\sum_{i=1}^{k} a_i X_i\right] = \sum_{i=1}^{k} a_i \mu_i$.

2.7.13 Let $X_1, \ldots, X_n$ be iid random variables with common mean μ and common variance $\mathrm{Var}(X_i) = \sigma^2$. Determine the linear combination $\sum_{i=1}^n a_i X_i$ with the smallest variance subject to $E\left[\sum_{i=1}^n a_i X_i\right] = \mu$.

2.7.14 Let $X_1, \ldots, X_n$ be independent random variables with common mean μ and $\mathrm{Var}(X_i) = \sigma_i^2$. Determine the linear combination $\sum_{i=1}^n a_i X_i$ with smallest variance subject to $E\left[\sum_{i=1}^n a_i X_i\right] = \mu$.

2.7.15 Let $X_1, \ldots, X_n$ be random variables with $\mathrm{Var}(X_i) = \sigma_i^2 < \infty$ for $i = 1, \ldots, n$. Show that

a) $\mathrm{Cov}\left(\sum_{i=1}^n a_i X_i, \sum_{j=1}^n b_j X_j\right) = \sum_{j=1}^n \sum_{i=1}^n a_i b_j \mathrm{Cov}(X_i, X_j).$

b) $\mathrm{Var}\left(\sum_{i=1}^n a_i X_i\right) = \sum_{j=1}^n \sum_{i=1}^n a_i a_j \mathrm{Cov}(X_i, X_j).$

c) $\mathrm{Var}\left(\sum_{i=1}^n a_i X_i\right) = \sum_{i=1}^n a_i^2 \mathrm{Var}(X_i) + 2 \sum_{j=1}^n \sum_{i<j} a_i a_j \mathrm{Cov}(X_i, X_j).$

2.7.16 Let $X_1, \ldots, X_n$ be independent random variables with MGFs $M_i(t) = (0.5e^t + 0.5)^{k_i}$ for $k_i \in \mathbb{N}$, and let $T = \sum_{i=1}^n X_i$. Determine the MGF of T and the pdf of T.

2.7.17 Let $X_1, \ldots, X_n$ be iid random variables with common MGF $M(t) = 0.6e^t + 0.4$, and let $T = \sum_{i=1}^n X_i$. Determine the MGF of T and the pdf of T.

2.7.18 Let X be a random variable with pdf $f(x) = \frac{e^{-\lambda}\lambda^x}{x!}$ for $x \in \mathbb{W}$ and MGF $M_X(t) = e^{\lambda(e^t - 1)}$ for $t \in \mathbb{R}$. Let $X_1, \ldots, X_n$ be iid random variables distributed as X, and let $T = \sum_{i=1}^n X_i$. Determine the MGF of T and the pdf of T.

2.7.19 Let $X_1, \ldots, X_n$ be independent normal random variables with $E(X_i) = \mu_i$ and $\mathrm{Var}(X_i) = \sigma_i^2$ (see Example 2.68), and let $T = \sum_{i=1}^n X_i$. Determine the

a) MGF of T.
c) $\mathrm{Var}(T)$.

b) $E(T)$.
d) $E(T^2)$.

2.7.20 Let $X_1, \dots, X_n$ be iid random variables with common MGF $M(t)$. Determine the MGF of $\overline{X}$.

2.7.21 Let $X_1, \dots, X_n$ be iid random variables with common MGF $M(t) = \frac{1}{1-\theta t}$, and let $m'_k = \frac{1}{n} \sum_{i=1}^{n} X_i^k$ be the kth sample moment. Determine

a) $E(m'_k)$.

b) $E(m'_2)$.

c) $E(m'_4)$.

d) $E\left(\frac{1}{24} m'_4\right)$.

2.7.22 Let $\vec{X} = (X_1, X_2, X_3)^{\mathsf{T}}$ be a random vector with

$$\vec{\mu} = \begin{pmatrix} 5 \\ 10 \\ 6 \end{pmatrix} \quad \text{and} \quad \Sigma = \begin{pmatrix} 9 & -2 & -1 \\ -2 & 12 & 2 \\ -1 & 2 & 16 \end{pmatrix},$$

and let $a^{\mathsf{T}} = (2, -4, 5)$. Determine

a) $E(a^{\mathsf{T}} \vec{X})$.

b) $\mathrm{Var}(a^{\mathsf{T}} \vec{X})$.

2.7.23 Let $\vec{X} = (X_1, X_2, X_3, X_4)^{\mathsf{T}}$ be a random vector with

$$\vec{\mu} = \begin{pmatrix} 3 \\ 0 \\ 5 \\ 2 \end{pmatrix} \quad \text{and} \quad \Sigma = \begin{pmatrix} 1 & 0 & -1 & 1 \\ 0 & 2 & 2 & -1 \\ -1 & 2 & 3 & 0 \\ 1 & -1 & 0 & 2 \end{pmatrix},$$

and let $a^{\mathsf{T}} = (1, -1, 2, -2)$. Determine

a) $E(a^{\mathsf{T}} \vec{X})$.

b) $\mathrm{Var}(a^{\mathsf{T}} \vec{X})$.

2.7.24 Let $\vec{X}$ be a k-dimensional random vector with covariance matrix Σ. Show that Σ is a positive semidefinite matrix (i.e. $a^T \Sigma a \geq 0$, $\forall a \in \mathbb{R}^k$).

2.8 Case Study – How Many Times was the Coin Tossed?

The distribution of the number of heads in a sequence of identical and independent coin tosses depends on two parameters, namely, the probability of flipping a head on a single toss, say p, and the number of times the coin will be tossed, say N. Generally, in a coin toss problem, the value of N is fixed, the coin is tossed $N = n$ times with the number of heads X in the n tosses being the random variable of interest. An alternative problem arises when coin is tossed N times and x heads are observed, but the number of tosses N is unknown. In this case, the question is how many times was the coin tossed given that x heads were tossed. For example, if $x = 11$ heads are observed, how many times was the coin tossed?

Since N is a random variable, this question cannot be explicitly answered; however, the most likely value of N given x heads were tossed and the expected number of tosses can be determined from the conditional distribution $N|X$. The determination of the conditional distribution $N|X$ will require a probability model for N and will utilize the Law of Total Probability for discrete random variables.

2.8.1 The Probability Model

Suppose that a fair coin (i.e. $p = 0.5$) is to be flipped N times, where N is a random variable, and let X be the number of heads flipped. Because the coin is fair, it is easy to show that the probability that $X = x$ heads are flipped given that the coin is tossed $N = n$ times is

$$P(X = x|N = n) = f(x|n) = \binom{n}{x}\left(\frac{1}{2}\right)^n$$

for $x = 0, 1, \ldots, n$.

Suppose that the probability model governing the number of times the coin will be tossed (i.e. N) is assumed to have pdf

$$f_N(n) = \frac{e^{-20}20^n}{n!}$$

for $n \in \mathbb{W}$; note that other probability models for N can also be used.

Two questions that will be addressed under this probability model are:

- If $X = x$, what is the most likely value of N, the number of times the coin was flipped?
- If $X = x$, what is the expected value of N, the number of times the coin was flipped?

The answer to each of these questions is based on the conditional distribution of $N|X = x$. In particular, the most likely value of N given that $X = x$ is the largest value of $f(n|x)$, and the expected value of N given that $X = x$ is $E(N|X = x)$.

Now, the conditional distribution of N given $X = x$ is $f(n|x) = \dfrac{f(n,x)}{f_X(x)}$ and the support of $N|X = x$ is $S_{N'} = \{x, x + 1, x + 2, \ldots, \}$. Under this probability model, the joint distribution of (N, X) is

$$f(n, x) = f(x|n)f_N(n) = \binom{n}{x}\left(\frac{1}{2}\right)^n \times \frac{e^{-20}20^n}{n!}$$

$$= \frac{n!}{x!(n-x)!}\left(\frac{1}{2}\right)^n \times \frac{e^{-20}20^n}{n!}$$

$$= \frac{e^{-20}10^n}{x!(n-x)!}$$

for $n \in \mathbb{W}$ and $x = 0, 1, \ldots, n$.

Thus, the marginal pdf of X is

$$f_X(x) = \sum_{n=x}^{\infty} f(n, x) = \sum_{n=x}^{\infty} \frac{e^{-20}10^n}{x!(n-x)!}$$

$$= \frac{e^{-20}10^x}{x!} \sum_{n=x}^{\infty} \frac{10^{n-x}}{(n-x)!}$$

$$= \frac{e^{-20}10^x}{x!} \sum_{n=x}^{\infty} \frac{10^{n-x}}{(n-x)!}$$

$$= \frac{e^{-20}10^x}{x!} \underbrace{\sum_{u=0}^{\infty} \frac{10^u}{u!}}_{u=n-x}$$

$$= \frac{e^{-20}10^x}{x!} \times e^{10} = \frac{e^{-10}10^x}{x!}$$

for $x \in \mathbb{W}$.

The conditional distribution of $N|X$ is

$$f(n|x) = \frac{\frac{e^{-20}10^n}{x!(n-x)!}}{\frac{e^{-10}10^x}{x!}} = \frac{e^{-10}10^{n-x}}{(n-x)!}$$

for $n = x, x+1, \ldots$

Now, suppose that a fair coin has been tossed N times and $X = 11$ heads have been observed. Then, the distribution of $N|X = 11$ is

$$f(n|11) = \frac{e^{-10}10^{n-11}}{(n-11)!},$$

which is plotted in Figure 2.6.

Thus, the most likely values of N given $X = 11$ are $N = 20$ and $N = 21$, which is not too far from the intuitive guess of $N \approx 2X = 22$ tosses.

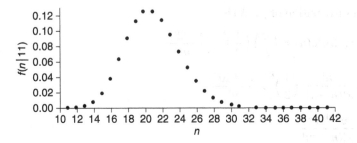

Figure 2.6 A plot of $f(n|x = 11)$ for $x = 11, 12, \ldots, 41$.

Now, the expected number of tosses given $X = 11$ is

$$E(N|X = 11) = \sum_{n=11}^{\infty} n \cdot f(n|x = 11) = \sum_{n=11}^{\infty} n \cdot \frac{e^{-10}10^{n-11}}{(n-11)!}$$

$$= \underbrace{\sum_{u=0}^{\infty} (u + 11) \frac{e^{-10}10^{u}}{u!}}_{u=n-11}$$

$$= \sum_{u=0}^{\infty} u \frac{e^{-10}10^{u}}{u!} + \sum_{u=0}^{\infty} 11 \frac{e^{-10}10^{u}}{u!}$$

$$= 10e^{-10} \underbrace{\sum_{u=1}^{\infty} \frac{10^{u-1}}{(u-1)!}}_{e^{10}} + 11 \underbrace{\sum_{u=0}^{\infty} \frac{e^{-10}10^{u}}{u!}}_{1}$$

$$= 10 + 11 = 21.$$

Thus, $E(N|X = 11) = 21$, which is again close to the intuitive guess of $2x = 22$.

In the formulation of this problem, the distribution of N given by $f_N(n)$ is referred to as the *prior distribution* and contains information on how many times the coin is usually tossed. The distribution of $N|X$ is called the *posterior distribution* and incorporates both the prior information on N and the observed information X into the conditional distribution. The distribution of N given X clearly depends on the prior distribution, and therefore, different answers are obtained when a different prior distribution is used.

Problems

2.8.1 Determine the most likely value of $N|X$ when

    a) $X = 0$.   b) $X = 5$.

2.8.2 Determine $E(N|X)$ when

    a) $X = 0$.   b) $X = 5$.

2.8.3 Suppose that the probability of flipping a head on each toss is $p = 0.4$, determine

    a) $f(n, x)$.
    b) $f_X(x)$.
    c) $f(n|x)$.
    d) the most likely value of N given $X = 10$.
    e) $E(N|X = 10)$.

2.8.4 Suppose that $f_N(n) = \dfrac{e^{-\lambda}\lambda^x}{x!}$ for $x \in \mathbb{W}$ and $\lambda \in \mathbb{N}$. Determine

    a) the most likely value of N given $X = 10$.
    b) $E(N|X = 10)$.

3

Probability Models

Probability models that are commonly used in statistical modeling are discussed in this chapter. In particular, each probability model is given with the pdf and support, the mean, the variance, the moment-generating function (MGF), and the R commands that are used for computing values of the probability density function (pdf) and cumulative distribution function (CDF). For the discrete probability models, the setting for which the model is appropriate is also provided. The key information on each probability model is summarized in Tables A.1 and A.2.

3.1 Discrete Probability Models

Recall that when X is a discrete random variable, the support S is at the most a countable set, the pdf is $f(x) = P(X = x)$, the mean is $E(X) = \sum_{x \in S} x f(x)$, the variance is $\text{Var}(X) = E[(X - \mu)^2]$, and the MGF is $M_X(t) = \sum_{x \in S} e^{tx} f(x)$. Also, the variance of a discrete random variables is often best computed using the alternative formula $\text{Var}(X) = E[(X(X - 1)] + E(X) - E(X)^2$.

3.1.1 The Binomial Model

The binomial probability model is often used to model chance experiments involving repeated trials of a dichotomous experiment or when sampling a population with replacement. For example, the binomial is often used to model the number of heads in n coin tosses, the number of hits in n at bats, the number of wins on a poker machine in n plays, and the number of guilty votes in the initial jury vote in a jury of n individuals.

The binomial probability model can be used to model the number of successes in a chance experiment consisting of n trials under the following conditions.

Mathematical Statistics: An Introduction to Likelihood Based Inference, First Edition. Richard J. Rossi.
© 2018 John Wiley & Sons, Inc. Published 2018 by John Wiley & Sons, Inc.

3.1.1.1 Binomial Setting

- The chance experiment consists of n independent trials.
- Each trial results in one of two outcomes, say success (S) and failure (F).
- The probability of a success on each of the n trials is constant with $P(S) = p$ and $P(F) = 1 - p = q$.
- The random variable X is the number of successes in the n trials.

A random variable X having a binomial distribution is denoted by $X \sim \text{Bin}(n, p)$, where $n \in \mathbb{N}$ is the number of trials, and p the probability of success. The pdf of $X \sim \text{Bin}(n, p)$ is

$$f(x) = \binom{n}{x} p^x q^{n-x}, \quad x = 0, 1, 2, \ldots, n; \quad p \in [0, 1], \quad p + q = 1.$$

When $n = 1$ (i.e. $X \sim \text{Bin}(1, p)$), a binomial random variable X is called a *Bernoulli random variable*, and the pdf of a Bernoulli random variable is

$$f(x) = p^x q^{1-x}, \quad x = 0, 1; \quad p \in [0, 1], \quad p + q = 1.$$

The R command used for computing the pdf of a binomial random variable $X \sim \text{Bin}(n, p)$ at $x \in \mathcal{S}$ is dbinom(x,n,p), and the R command pbinom(x,n,p) is used to compute the value of the CDF at $x \in \mathcal{S}$. That is, dbinom(x,n,p) $= f(x) = P(X = x)$ and pbinom(x,n,p) $= F(x) = \sum_{i=0}^{x} \binom{n}{i} p^i q^{n-i}$, for $x \in \mathcal{S}$.

Theorem 3.1 *If $X \sim \text{Bin}(n, p)$, then*

i) $E(X) = np$.
ii) $\text{Var}(X) = npq$.
iii) $M_X(t) = (p\,e^t + q)^n$.

Proof. Let $X \sim \text{Bin}(n, p)$.

i)

$$E(X) = \sum_{x=0}^{n} x \binom{n}{x} p^x (1-p)^{n-x}$$

$$= \sum_{x=1}^{n} \frac{n!}{(x-1)!(n-x)!} p^x (1-p)^{n-x}$$

$$= np \sum_{x=1}^{n} \frac{(n-1)!}{(x-1)!(n-x)!} p^{x-1} (1-p)^{n-x}$$

$$= np \sum_{u=0}^{n-1} \underbrace{\frac{(n-1)!}{u!(n-1-u)!} p^u (1-p)^{n-1-u}}$$

$$u = x-1, \quad n-x = n-1-u$$

$$= np \sum_{u=0}^{n-1} \binom{n-1}{u} p^u (1-p)^{n-1-u}$$

$$\underbrace{\phantom{np \sum_{u=0}^{n-1} \binom{n-1}{u} p^u (1-p)^{n-1-u}}}_{1}$$

$$= np.$$

ii) $\text{Var}(X) = E[X(X-1)] + E(X) - E(X)^2$ and

$$E[X(X-1)] = \sum_{x=0}^{n} x(x-1) \binom{n}{x} p^x (1-p)^{n-x}$$

$$= \sum_{x=2}^{n} \frac{n!}{(x-2)!(n-x)!} p^x (1-p)^{n-x}$$

$$= n(n-1)p^2 \sum_{x=2}^{n} \frac{(n-2)!}{(x-2)!(n-x)!} p^{x-2} (1-p)^{n-x}$$

$$= n(n-1)p^2 \underbrace{\sum_{u=0}^{n-2} \frac{(n-2)!}{u!(n-2-u)!} p^u (1-p)^{n-2-u}}$$

$$u = x-2, \quad n-x = n-2-u$$

$$= n(n-1)p^2 \underbrace{\sum_{u=0}^{n-2} \binom{n-2}{u} p^u (1-p)^{n-2-u}}_{1}$$

$$= n(n-1)p^2.$$

Thus,

$$\text{Var}(X) = n(n-1)p^2 + np - (np)^2 = np(1-p).$$

iii) The proof of part (iii) is left as an exercise.   ■

Example 3.1   Suppose that a basketball player makes 40% of his three-point shot attempts. If the player shoots 10 three-point shots in a game, and X is the number of three-point shots made, then assuming that the shots are independent and the probability of making each shot is 0.40, $X \sim \text{Bin}(10, 0.4)$. The probability the player makes five of his 10 shots is

$$f(5) = \binom{10}{5} 0.4^5 (1-0.4)^{10-5} = 0.2001.$$

The mean and variance of the random variable X are $\mu = 10(0.4) = 4$ and $\sigma^2 = 10(0.4)(0.6) = 2.4$.

The R command for computing $f(5)$ is `dbinom(5,10,0.4)`, where $x = 5, n = 10$, and $p = 0.4$. The command `dbinom(c(0:10),10,0.4)`

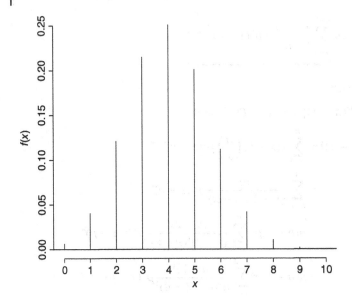

Figure 3.1 A plot of the pdf of $X \sim \text{Bin}(10, 0.4)$.

computes the value of $f(x)$ for each $x \in \{0, 1, 2, \ldots, 10\}$, and the R commands used to create the plot of the pdf of $X \sim \text{Bin}(10, 0.4)$ given in Figure 3.1 are

```
> s=c(0:10) # s is the support of X
> plot(s,dbinom(s,10,0.4),type="h",ylab="f(x)")
> abline(0,0)# puts in a reference line at f(x)=0
```

3.1.2 The Hypergeometric Model

The *hypergeometric* probability model can be used to model a chance experiment involving sampling without replacement from a collection of two types of distinct objects. In particular, the hypergeometric probability model can be used when the following conditions are satisfied.

3.1.2.1 Hypergeometric Setting

- There are M successes and N failures in a collection of $M + N$ objects.
- n objects will be drawn at random and without replacement from $M + N$ objects.
- The random variable X is the number of successes selected in the n draws.

A random variable X having a hypergeometric distribution is denoted by $X \sim \text{Hyper}(M, N, n)$, where M is the number of successes, N is the number of failures, and $n \in \mathbb{N}$ is the number of objects selected. The pdf of a hypergeometric random variable $X \sim \text{Hyper}(M, N, n)$ is

$$f(x) = \frac{\binom{M}{x}\binom{N}{n-x}}{\binom{M+N}{n}}, \quad x = 0, 1, 2, \ldots, \min(M, n); \quad M, N, n \in \mathbb{N}.$$

The R command for computing values of the pdf of $X \sim \text{Hyper}(M, N, n)$ at a value $x \in \mathcal{S}$ is dhyper(x, M, N, n), and phyper(x, M, N, n) is used to compute the value of the CDF for $x \in \mathcal{S}$.

Theorem 3.2 *If $X \sim \text{Hyper}(M, N, n)$, then*

i) $E(X) = n \dfrac{M}{M+N}.$

ii) $\text{Var}(X) = n \dfrac{MN(M + N - n)}{(M + N)^2(M + N - 1)}.$

Proof. Let $X \sim \text{Hyper}(M, N, n)$.

i) WLOG assume $n < M$, and let $\mathcal{S} = \{0, 1, \ldots, n\}$. Then,

$$E(X) = \sum_{x=0}^{n} x \cdot \frac{\binom{M}{x}\binom{N}{n-x}}{\binom{M+N}{n}} = \sum_{x=1}^{n} \frac{\frac{M!}{(x-1)!(n-x)!}\binom{N}{n-x}}{\binom{M+N}{n}}$$

$$= \frac{M}{\frac{M+N}{n}} \sum_{x=1}^{n} \frac{\binom{M-1}{x-1}\binom{N}{n-x}}{\binom{M+N-1}{n-1}} = \frac{nM}{M+N} \underbrace{\sum_{u=0}^{n-1} \frac{\binom{M-1}{x-1}\binom{N}{n-x}}{\binom{M+N-1}{n-1}}}_{u=x-1 \text{ sum}=1}$$

$$= \frac{nM}{M+N}.$$

ii) The proof of part (ii) is left as an exercise. ∎

The MGF of a hypergeometric random variable does exist; however, it is not useful.

Example 3.2 In a particular lottery game, a player chooses six numbers without replacement from the pool of numbers 1 to 51; the order in which the numbers are selected is unimportant. The proprietor of the lottery then chooses six numbers without replacement and at random from the numbers 1 to 51, these are the winning numbers; the order in which the proprietor's numbers are selected is also unimportant. Let X be the number of the proprietor's numbers matched by a player. Then, $X \sim \text{Hyper}(6, 45, 6)$ and

$$P(X = x) = \frac{\binom{6}{x}\binom{45}{6-x}}{\binom{51}{6}}, \quad x = 0, 1, 2, 3, 4, 5, 6.$$

For example, the probability that a player matches all 6 numbers is

$$P(X = 6) = \frac{\binom{6}{6}\binom{45}{0}}{\binom{51}{6}} = \text{dhyper}(6,6,45,6) = 0.000000056,$$

and the probability that a player matches three or fewer numbers is

$$P(X \leq 3) = \text{phyper}(3,6,45,6) = 0.9992.$$

The expected number of matches is $E(X) = 6 \times \frac{6}{51} = 0.706$, and the variance is $\text{Var}(X) = 6 \cdot \frac{6(45)^2}{51^2(50)} = 0.561$.

3.1.3 The Poisson Model

The Poisson probability model can be used to model the number of occurrences of a particular event. For example, the Poisson probability model is often used to model the number of tornadoes occurring in a tornado season, the number of grasshoppers in a field, and the number of hits on a website in a fixed time period. A random variable is a Poisson random variable if it has the following pdf:

$$f(x) = \frac{e^{-\lambda}\lambda^x}{x!}, \quad x \in \mathbb{W}; \quad \lambda \in [0, \infty).$$

A random variable X having a Poisson distribution is denoted by $X \sim \text{Pois}(\lambda)$.

The R command for computing the value of the pdf of $X \sim \text{Pois}(\lambda)$ at $x \in \mathcal{S}$ is dpois(x,lambda), and ppois(x,lambda) computes the value of the CDF for $x \in \mathcal{S}$.

Theorem 3.3 *If $X \sim \text{Pois}(\lambda)$, then*

i) $E(X) = \lambda$.
ii) $\text{Var}(x) = \lambda$.
iii) $M_X(t) = e^{\lambda(e^t - 1)}$.

Proof. Let $X \sim \text{Pois}(\lambda)$

i)

$$E(X) = \sum_{x=0}^{\infty} x \cdot \frac{e^{-\lambda}\lambda^x}{x!} = e^{-\lambda} \sum_{x=1}^{\infty} \frac{\lambda^x}{(x-1)!}$$

$$= \lambda\, e^{-\lambda} \sum_{x=1}^{\infty} \frac{\lambda^{x-1}}{(x-1)!} = \lambda\, e^{-\lambda} \underbrace{\sum_{u=0}^{\infty} \frac{\lambda^u}{u!}}_{u=x-1}$$

$$= \lambda\, e^{-\lambda}\, e^{\lambda} = \lambda.$$

ii) The proof of part (ii) is left as an exercise.

iii) The proof of part (iii) is left as an exercise.

■

Example 3.3 Suppose that the number of grasshoppers in a one-square yard plot in a field is known to follow a Poisson distribution with $\lambda = 5$. Then, the probability that there are four grasshoppers in one-square yard plot in this field is

$$P(X = 4) = \frac{e^{-5} 5^4}{4!} = \text{dpois}(4,5) = 0.1755,$$

and the probability that there are more than 10 grasshoppers in one-square yard is

$$P(X > 10) = 1 - P(X \le 10) = 1 - \text{ppois}(10,5) = 0.0028.$$

The expected number of grasshoppers in a one-square yard plot in this field is $E(X) = 5$, and the variance is $\text{Var}(X) = 5$.

The R commands used to create the plot of the CDF of $X \sim \text{Pois}(5)$ shown in Figure 3.2 are

```
> x=c(0:25) effective support of X
> y=ppois(x,5)#CDF over effective support
> plot(x,y,type="s",ylab="CDF")
> abline(0,0)#adds an x-axis at height 0
```

3.1.4 The Negative Binomial Model

The negative binomial probability model can be used to model chance experiments where trials of an experiment are repeated until the rth success occurs. A random variable X will follow a negative binomial distribution under the following conditions.

3.1.4.1 Negative Binomial Setting

- A chance experiment that results in either a success or a failure will be repeated independently until the rth success occurs.
- The probability of a success on each of the trials is constant with $P(S) = p$.
- The random variable X is the number of failures preceding the rth success.

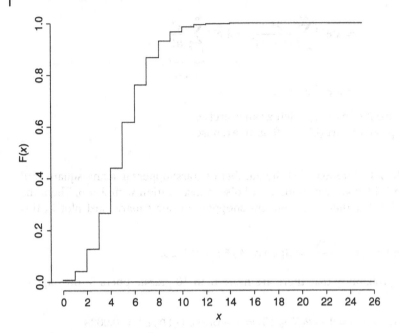

Figure 3.2 The CDF for $X \sim \text{Pois}(5)$.

A random variable X having a negative binomial distribution is denoted by $X \sim \text{NegBin}(r, p)$, where r is the number of successes required to terminate the chance experiment, and p is the probability of a success on each trial. The pdf of a negative binomial random variable $X \sim \text{NegBin}(r, p)$ is

$$f(x) = \binom{r + x - 1}{x} p^r q^x, \quad x \in \mathbb{W}; \quad p \in [0, 1], p + q = 1.$$

The R command for computing values of the pdf of $X \sim \text{NegBin}(r, p)$ at $x \in \mathcal{S}$ is dnbinom(x,r,p), and pnbinom(x,r,p) computes the value of the CDF for $x \in \mathcal{S}$.

When $r = 1$, a negative binomial random variable is called a *geometric random variable*, denoted by $X \sim \text{Geo}(p)$. The pdf of a $X \sim \text{Geo}(p)$ is

$$f(x) = pq^x, \quad x \in \mathbb{W}; \quad p \in [0, 1], p + q = 1,$$

and for $x \in \mathbb{W}$, the CDF is $F(x) = 1 - q^{x+1}$. The R command for computing the value of the pdf of $X \sim \text{Geo}(p)$ is dgeom(x,p), and pgeom(x,p) computes the value of the CDF at $x \in \mathcal{S}$.

Theorem 3.4 *Let* $X \sim \text{NegBin}(p, r)$, *then*

i) $E(X) = \dfrac{rq}{p}$.

ii) $\mathrm{Var}(X) = \dfrac{rq}{p^2}$.

iii) $M_X(t) = \dfrac{p^r}{(1 - qe^t)^r}$.

Proof. Let $X \sim \mathrm{NegBin}(p, r)$.

i)

$$E(X) = \sum_{x=0}^{\infty} x \cdot \binom{r + x - 1}{x} p^r (1 - p)^x = p^r \sum_{x=1}^{\infty} \frac{(r + x - 1)!}{(x - 1)!(r - 1)!} q^x$$

$$= p^r \underbrace{\sum_{u=0}^{\infty} \frac{(r + u)!}{u!(r - 1)!} q^{u+1}}_{u = x - 1} = rp^r q \sum_{u=0}^{\infty} \binom{r + u}{u} q^u$$

$$= rp^r q \times \frac{1}{p^{r+1}} = \frac{rq}{p}.$$

ii) The proof of part (ii) is left as an exercise.
iii) The proof of part (iii) is left as an exercise. ∎

Since a geometric random variable is a negative binomial random variable with $r = 1$, it follows that the mean of a geometric random variable X is $E(X) = \frac{q}{p}$, the variance is $\mathrm{Var}(X) = \frac{q}{p^2}$, and the MGF is $M_X(t) = \frac{p}{1 - e^t}$.

Example 3.4 A telemarketer will make phone calls until 25 successful contacts are made. Assuming that each phone call is independent and the probability that a contact is made is constant with $p = 0.4$, the pdf of $X =$ the number of phone calls failing to make contact before the 25th contact is

$$f(x) = \binom{25 + x - 1}{x} 0.4^{25} \times 0.6^x, \quad x \in \mathbb{W}.$$

The probability that 15 failed calls are made before the 25th contact is made is

$$P(X = 15) = \binom{39}{15} 0.4^{25} \times 0.6^{15} = \mathtt{dnbinom(15,25,0.4)} = 0.0013,$$

and the probability that at least 15 failed calls will be made before the 25th contact is made is

$$P(X \geq 15) = 1 - P(X \leq 14) = 1 - \mathtt{pnbinom(14,25,0.4)} = 0.9980.$$

The mean number of failed phone calls prior to making the 25th contact is $E(X) = \frac{25(0.6)}{0.4} = 37.5$, and the variance is $\mathrm{Var}(X) = \frac{25(0.6)}{0.4^2} = 93.75$.

Example 3.5 Suppose that a dart player throws a dart until the bullseye is hit. Let X be the number of dart throws prior to the player finally hitting the bullseye for the first time. Assuming that each dart toss is independent, and the probability of hitting the bullseye on each throw is constant with $p = 0.1$, the random variable $X \sim \mathrm{Geo}(0.1)$ and $f(x) = 0.1 \times 0.9^x$ for $x \in \mathbb{W}$. The probability it takes five throws (i.e. four misses) to hit the bullseye for the first time is

$$P(X = 4) = f(4) = 0.1 \times 0.9^4 = \texttt{dgeom(4,0.1)} = 0.0656,$$

and the probability it takes less than 10 tosses to hit the bullseye is

$$P(<10 \text{ tosses}) = P(X \le 8) = F(8) = \texttt{pgeom(8,0.1)} = 0.6126$$

since fewer than 10 tosses means fewer than 9 misses. The mean number of misses before the bullseye is hit for the first time is $E(X) = \frac{0.9}{0.1} = 9$, and the variance is $\mathrm{Var}(X) = \frac{0.9}{0.1^2} = 90$.

Finally, if $X \sim \mathrm{NegBin}(r, p)$ is the random variable associated with the number of failures preceding the rth success, then $Y = X + r$ is the random variable associated with the number of trials required to obtain the rth success. The pdf of Y is

$$f_Y(y) = P(Y = y) = P(X + r = y) = P(X = y - r) = \binom{y-1}{r-1} p^r q^{y-r}$$

for $y = r, r+1, \dots, p \in [0,1], p + q = 1$. The mean, variance, and MGF of Y are

$$E(Y) = E(X + r) = \frac{rq}{p} + r = \frac{r}{p}.$$

$$\mathrm{Var}(Y) = \mathrm{Var}(X + r) = \frac{rq}{p^2}.$$

$$M_Y(t) = M_{X+r}(t) = e^{tr} \frac{p^r}{(1 - q\, e^t)^r} = \frac{(p\, e^t)^r}{(1 - q\, e^t)^r}.$$

Example 3.6 In Example 3.5, X is the number of dart throws missing the bullseye before the first bullseye is hit. In this case, the number of dart throws required to hit the bullseye once is $Y = X + 1$ and

$$f_Y(y) = \binom{y-1}{1-1} 0.1^1 0.9^{y-1} = 0.1(0.9)^{y-1}, \quad y = 1, 2, \dots$$

Thus, the probability that it takes 10 throws to hit the bullseye for the first time is $P(Y = 10) = 0.1(0.9)^9 = 0.0387$, and the mean number of throws taken to hit the bullseye once is $E(Y) = \frac{1}{0.1} = 10$.

3.1.5 The Multinomial Model

The multinomial probability model is a multivariate generalization of the binomial probability model. In particular, the multinomial is a multivariate probability model associated with a k-dimensional random vector used to model chance experiments where n objects are randomly allocated independently to k cells. A random vector $\vec{X} = (X_1, X_2, \ldots, X_k)$ will have a multinomial distribution under the following conditions.

3.1.5.1 Multinomial Setting

1) A chance experiment with k possible outcomes, say $O_1, O_2, \ldots, O_k$, will be repeated independently n times.
2) The probability of outcome O_i on any trial is p_i.
3) $\sum_{i=1}^{k} p_i = 1$.
4) X_i is the number of trials resulting in outcome O_i, and the random vector is $\vec{X} = (X_1, X_2, \ldots, X_k)$.

A k-dimensional random vector $\vec{X}$ having a multinomial distribution is denoted by $\vec{X} \sim \text{MulNom}_k(n, \vec{p})$, where k is the number of cells, n is the number of objects being allocated to the k cells, and $\vec{p} = (p_1, p_2, \ldots, p_k)$ is the vector of cell probabilities. The joint pdf of $\vec{X} \sim \text{MulNom}_k(n, \vec{p})$ is

$$f(x_1, x_2, \ldots, x_k) = \binom{n}{x_1, x_2, \ldots, x_k} p_1^{x_1} p_2^{x_2} \cdots p_k^{x_k}$$

for $x_i \in \mathbb{W}$, $\sum_{i=1}^{k} x_i = n$, $p_i \in [0, 1]$ for $i = 1, \ldots, k$, and $\sum_{i=1}^{k} p_i = 1$.

The R command for computing values of the pdf of $\vec{X} \sim \text{MulNom}_k(r, p)$ at $\vec{x} \in S$ is dmultinom(x,n,p), where $x = (x_1, \ldots, x_k)$ and $p = (p_1, \ldots, p_k)$. R does not have a command for computing the joint CDF for a multinomial random variable.

Theorem 3.5 *If $\vec{X} \sim \text{MulNom}_k(p_1, \ldots, p_k)$, then the marginal distribution of $X_i \sim \text{Bin}(n, p_i)$, for $i = 1, \ldots, k$.*

Proof. Let $\vec{X} \sim \text{MulNom}_k(p_1, \ldots, p_k)$. WLOG consider the marginal pdf of X_1 since the proof for each of the marginal pdfs is similar. Then for $x_1 \in \{0, 1, \ldots, n\}$, the marginal pdf of the random variable X_1 is

$$f_1(x_1) = \underbrace{\sum_{x_k} \cdots \sum_{x_2}}_{\sum_{i=2}^{k} n_i = n - x_1} \binom{n}{x_1, x_2, \ldots, x_k} p_1^{x_1} p_2^{x_2} \cdots p_k^{x_k}$$

$$= \frac{n!}{x_1!} p_1^{x_1} \underbrace{\sum_{x_k} \cdots \sum_{x_2}}_{\sum_{i=2}^{k} x_i = n - x_1} \frac{1}{x_2! \ldots x_k!} p_2^{x_2} \cdots p_k^{x_k}$$

$$= \frac{n!}{x_1!} p_1^{x_1} \underbrace{\sum_{x_k} \cdots \sum_{x_2} \frac{1}{x_2! \ldots x_k!} p_2^{x_2} \cdots p_k^{x_k} \frac{(n-x_1)!}{(n-x_1)!}}_{\sum_{i=2}^{k} x_i = n - x_1}$$

$$= \frac{n!}{x_1!(n-x_1)!} p_1^{x_1} \underbrace{\sum_{x_k} \cdots \sum_{x_2} \frac{(n-x_1)!}{x_2! \ldots x_k!} p_2^{x_2} \cdots p_k^{x_k}}_{\sum_{i=2}^{k} x_i = n - x_1}$$

$$= \binom{n}{x_1} p_1^{x_1} (p_2 + \cdots + p_k)^{n-x_1} = \binom{n}{x_1} p_1^{x_1} (1 - p_1)^{n-x_1}.$$

Therefore, $X_1 \sim \text{Bin}(n, p_i)$. ∎

Example 3.7 In a mayoral election, suppose that 40% of the electorate favors candidate A, 35% favors candidate B, and 25% favors candidate C. Suppose that $n = 10$ voters are selected independently and surveyed, and let X_1 be the number of sampled voters voting for candidate A, X_2 the number voting for candidate B, and X_3 the number voting for candidate C. Then, the random vector $\vec{X} = (X_1, X_2, X_3)$ follows a multinomial distribution with $n = 10, \vec{p} = (0.4, 0.35, 0.25)$. The probability that $\vec{X} = (5, 3, 2)$ is

$$P(\vec{X} = (5, 3, 2)) = \binom{10}{5, 3, 2} 0.4^5 \times 0.35^3 \times 0.25^2$$
$$= \text{dmultinom}(\text{c}(5,3,2),10,\text{c}(0.4,0.35,0.25))$$
$$= 0.0691.$$

The marginal distributions of the random variables X_1, X_2, and X_3 are $X_1 \sim \text{Bin}(10, 0.4), X_2 \sim \text{Bin}(10, 0.35)$, and $X_3 \sim \text{Bin}(10, 0.25)$.

Problems

3.1.1 Suppose that $P(\text{Male Child}) = 0.48$, the sex of each child is an independent event, the probability of a male child is constant for each birth, and X is the number of male children in a family with four children. Determine

a) $E(X)$.
b) $\text{Var}(X)$.
c) $P(X = 3)$.
d) $P(X = 0)$.
e) the probability that a family with four children has at least one male child.
f) the probability that a family with four children has at least one male child given the family has at least one female child.

3.1.2 Suppose a fair coin is tossed 25 times. If X is the number of heads flipped in the 25 tosses, determine

a) $E(X)$.   b) $Var(X)$.

c) $P(X = 12)$.   d) $P(X \geq 20)$.

3.1.3 If a fair six-sided die is rolled independently 10 times, determine
a) the probability of rolling four sixes in 10 rolls.
b) the probability of rolling at least four sixes in 10 rolls.
c) the probability of rolling at least one six in 10 rolls.
d) the expected number of sixes in 10 rolls.
e) the variance of the number of sixes rolled in 10 rolls.

3.1.4 Suppose that 20% of a university student body voted for the renovating the student union building (SUB). Let X be the number of students in a sample of size 10 who voted for renovating the SUB. If 10 students are selected at random and independently, determine
a) the probability that exactly 5 of the sampled students voted for SUB renovation.
b) the probability that at least 7 students voted for SUB renovation.
c) $E(X)$.   d) $Var(X)$.

3.1.5 Suppose that the probability that a new consulting firm is awarded a contract is 0.30 for each job bid on, and consulting jobs bid on are independent. Let X be the number of consulting jobs landed in 15 bids. Determine
a) the probability of landing exactly five consulting contracts.
b) the probability of landing at most three contracts.
c) the probability of landing one, two, or three contracts.

3.1.6 Suppose that 30 wells will be sampled independently and at random in a particular county. If the probability that a well is contaminated by arsenic is 0.25, determine
a) the probability that at least 5 of the 30 randomly selected wells will be found to be contaminated by arsenic.
b) the expected number of sampled wells contaminated by arsenic.
c) the standard deviation of the number of sampled wells contaminated by arsenic.

3.1.7 An 80% accuracy rate is claimed for smart missiles currently being used by the military. If it takes at least 10 missile hits to destroy a particular target and 20 smart missiles are fired independently at the target, determine the probability that the target is destroyed.

3.1.8 Suppose that a mining company is going to have n ore hauling trucks. Based on past records, the probability that each truck is available on a particular day is 0.25. Assuming that the availability of each truck is an independent event, determine the number of trucks n the company should have on hand so that the probability that at least one truck is available on any given day is at least 0.95.

3.1.9 Let $X \sim \text{Bin}(12, 0.25)$. Determine
a) $M_X(t)$.   b) $E(X^2)$.

3.1.10 Let $X \sim \text{Bin}(20, 0.8)$. Determine the most likely value of X.

3.1.11 Let $X \sim \text{Bin}(n, p)$. Show that $\sum_{x=0}^{n} \binom{n}{x} p^x (1 - p)^{n-x} = 1$.

3.1.12 Let $X \sim \text{Bin}(n, p)$. Show that $M_X(t) = (p\, e^t + q)^n$.

3.1.13 Let $X \sim \text{Bin}(n, p)$. Show that $E(X^2) = n(n - 1)p^2 + np$.

3.1.14 Let $X \sim \text{Bin}(n, p)$, and let $Y = n - X$. Show that $Y \sim \text{Bin}(n, q)$.

3.1.15 Let $X_1 \sim \text{Bin}(n_1, p)$ and $X_2 \sim \text{Bin}(n_2, p)$ be independent random variables. Determine
a) $M_{X_1 + X_2}(t)$.   b) the distribution of $X_1 + X_2$.

3.1.16 Let X_1 and X_2 be iid $\text{Bin}(n, p)$ random variables. Determine
a) $M_{X_1 + X_2}(t)$.   b) the distribution of $X_1 + X_2$.

3.1.17 Suppose that a club roster contains 10 women and 15 men, and 5 members will be sampled at random and without replacement to form a club committee. If X is the number of women selected for the committee, determine
a) $P(X = 0)$.   b) $P(X \leq 1)$.
c) $P(X > 2)$.   d) $E(X)$.

3.1.18 Suppose that in a shipment of 100 memory chips, 5 are defective. If 6 chips are sampled at random and without replacement, determine the probability that
a) no defective chips are sampled.
b) at least one defective chip is sampled.
c) at most two defective chips are sampled.

3.1.19 Suppose that an urn contains 20 marbles of which 12 are red, 5 are white, and 3 are black. If 4 marbles will be selected at random and without replacement, determine the probability that
a) no red marbles are selected.
b) at least 1 black marble is selected.
c) at least 2 white marbles are selected.

3.1.20 A company has 12 offices in California and 6 offices in New York. If four offices are selected at random and without replacement for auditing, determine the probability that
a) all four offices are located in New York.
b) all four audited offices are located in California.
c) at least two offices are audited in New York.
d) more offices are audited in New York than in California.

3.1.21 Let $X \sim \text{Hyper}(M, N, n)$. Determine $E[X(X-1)]$.

3.1.22 Let $X \sim \text{Hyper}(M, N, n)$. Show that $\text{Var}(X) = \frac{nMN(M+N-n)}{(M+N)^2(M+N-1)}$.

3.1.23 Suppose that X is the number of tornadoes observed in a particular region of the Midwest during a tornado season. If $X \sim \text{Pois}(\lambda = 8)$, determine
a) the probability that five tornadoes hit this region in a tornado season.
b) the probability that at least one tornado hits this region in a tornado season.
c) the expected number of tornadoes to hit this region in a tornado season.

3.1.24 Suppose X is the number of times the Big Hole River will exceed flood stage in a decade and $X \sim \text{Pois}(\lambda = 4)$. Determine
a) the probability that the Big Hole River does not exceed flood stage in a decade.
b) the probability that the Big Hole River exceeds flood stage seven times in a decade.
c) the probability that the Big Hole River exceeds flood stage at least once in a decade.

3.1.25 Suppose that grasshoppers are distributed at random in a large hay field according to a Poisson distribution with $\lambda = 6$ grasshoppers per one-square yard. Determine

a) the probability that there are at least 10 grasshoppers in a one-square yard plot.
b) the probability that there are fewer than 12 grasshoppers in a one-square yard area.

3.1.26 Suppose that X is the number of hits on a popular website that follows a Poisson distribution with $\lambda = 50$ hits per minute. Determine
a) the probability that there are at least 35 hits on this website in a one-minute period.
b) the probability that there are more than 60 hits on this website in a one-minute period.
c) the probability that there are more than 40 hits but less than 60 hits on this website in a one-minute period.

3.1.27 Let $X \sim \text{Pois}(\lambda)$. Show that $\sum_{x=0}^{\infty} \frac{e^{-\lambda}\lambda^x}{x!} = 1$.

3.1.28 Let $X \sim \text{Pois}(\lambda)$. Show that $M_X(t) = e^{\lambda(e^t-1)}$.

3.1.29 Let $X \sim \text{Pois}(\lambda)$. Show that
a) $E[X(X-1)] = \lambda^2$. 　　　　b) $\text{Var}(X) = \lambda$.
c) $E(X^2) = \lambda^2 + \lambda$.

3.1.30 Let $X \sim \text{Pois}(\lambda)$. Show that $E[X(X-1)(X-2)\cdots(X-n+1)] = \lambda^n$ for $n \in \mathbb{N}$.

3.1.31 Let $X \sim \text{Pois}(\lambda)$, and suppose that $P(X=1) = P(X=2)$. Determine
a) the value of λ. 　　　　b) $P(X \geq 2)$.

3.1.32 Let $X \sim \text{Pois}(\lambda)$, and suppose that $P(X=0) = P(X=3)$. Determine
a) the value of λ. 　　　　b) $P(X < 5)$.

3.1.33 Let $X \sim \text{Pois}(\lambda)$. Determine $E[(-1)^X]$.

3.1.34 Let $X_1 \sim \text{Pois}(\lambda_1)$ and $X_2 \sim \text{Pois}(\lambda_2)$ be independent random variables. Determine
a) $M_{X_1+X_2}(t)$. 　　　　b) the distribution of $X_1 + X_2$.

3.1.35 Let X_1 and X_2 be iid $\text{Pois}(\lambda)$ random variables. Determine
a) $M_{X_1+X_2}(t)$. 　　　　b) the distribution of $X_1 + X_2$.

3.1.36 A golfer on a driving range decides to practice hitting golf balls until a golf ball is driven over 300 yards. Suppose that each golf ball is hit independently, the probability that the golfer hits a golf ball over 300 yards on any attempt is 0.2, and X is the number of attempts prior to hitting a golf ball over 300 yards. Determine

a) $P(X \le 20)$. b) $P(X > 10)$.
c) $E(X)$. d) $\text{Var}(X)$.

3.1.37 A pair of fair dice will be rolled independently until a total of seven is rolled. Let X be the number of nonseven rolls before the first seven is rolled. Determine

a) $P(X \le 2)$. b) $P(X \ge 10)$.
c) $E(X)$. d) $\text{Var}(X)$.

3.1.38 Let $X \sim \text{Geo}(p)$ be the number of failures preceding the first success, and let $Y = X + 1$. Then, Y is the number of trials required to obtain the first success. Determine

a) the pdf of Y and $\mathcal{S}_Y$. b) $F_Y(y)$ for $y \in \mathcal{S}_Y$.
c) $E(Y)$. d) $\text{Var}(Y)$.

3.1.39 Let $X \sim \text{Geo}(p)$. Show that $P(X \ge x + y | X \ge y) = P(X \ge x)$ for $x, y \in \mathbb{W}$.

3.1.40 Let $X \sim \text{Geo}(p)$. Show that

a) $\sum_{x=0}^{\infty} pq^x = 1$. b) $E(e^{tX}) = \frac{p}{1 - q\,e^t}$.

3.1.41 Let X_1 and X_2 be iid $\text{Geo}(p)$ random variables. Determine

a) $M_{X_1 + X_2}(t)$. b) the distribution of $X_1 + X_2$.

3.1.42 Missiles will be fired independently at a target until a fifth missile has hit the target destroying the target. If the probability of a missile hitting the target is 0.6, determine

a) the probability that six missiles must be fired to destroy the target.
b) the probability that at least eight missiles must be fired to destroy the target.

3.1.43 A professional dart thrower will throw darts until the bullseye is hit three times. Assuming that each throw is independent, and the probability of hitting the bullseye on any given toss is 0.40, determine

a) the probability that it takes more than 5 throws to hit the bullseye three times.

b) the probability that it takes more than 10 throws to hit the bullseye three times.

3.1.44 Let $X \sim \text{NegBin}(10, 0.25)$ be the number of failures preceding the 10th success, and let $Y = X + 10$. Then, Y is the number of trials required to obtain the 10th success. Determine

a) the pdf of Y and S_Y.   b) $M_Y(t)$.

c) $E(Y)$.   d) $\text{Var}(Y)$.

3.1.45 Let $X \sim \text{NegBin}(r, p)$. Show that $M_X(t) = \frac{p^r}{(1-q\,e^t)^r}$.

3.1.46 Let $X \sim \text{NegBin}(r, p)$. Show that $\text{Var}(X) = \frac{rq}{p^2}$.

3.1.47 Let $X_1 \sim \text{NegBin}(r_1, p)$ and $X_2 \sim \text{NegBin}(r_2, p)$ be independent random variables. Determine

a) $M_{X_1+X_2}(t)$.   b) the distribution of $X_1 + X_2$.

3.1.48 Let X_1 and X_2 be iid $\text{NegBin}(r, p)$ random variables. Determine

a) $M_{X_1+X_2}(t)$.   b) the distribution of $X_1 + X_2$.

3.1.49 Let $\vec{X} = (X_1, X_2, X_3) \sim \text{MulNom}_3(10, 0.2, 0.5, 0.3)$. Determine

a) $P(\vec{X} = (5, 4, 1))$.   b) $P(\vec{X} = (0, 8, 2))$.

c) $E(X_1)$.   d) $E(X_2)$.

3.1.50 A fair six-sided die is to be rolled independently six times. Let X_i be the number of times outcome i is rolled for $i = 1, 2, \ldots, 6$, and let $\vec{X} = (X_1, \ldots, X_6)$. Determine

a) $P(\vec{X} = (1, 1, 1, 1, 1, 1))$.   b) $P(\vec{X} = (3, 0, 2, 1, 0, 0))$.

c) the distribution of X_1.   d) $P(X_1 = 2)$.

e) $E(X_1)$.   f) $\text{Var}(X_1)$.

3.1.51 The four human blood types determined by the presence or absence of the antigens A and B are A, B, AB, and O. Suppose that the breakdown of human blood types is given as follows: 40% have blood type A, 11% have blood type B, 4% have blood type AB, and 45% have blood type O. If 25 people are independently selected, determine the probability that

a) 10 have blood type A, 5 have blood type B, 2 have blood type AB, and 8 have blood type O.
b) at least 2 have blood type AB.
c) at most 6 have blood type B.

3.1.52 Let $\vec{X} = (X_1, X_2) \sim \text{MulNom}_2(n, p_1, p_2)$. Show that $f(x_1, x_2) = f_1(x_1)$, for $x_1 = 0, 1, \ldots, n$.

3.2 Continuous Probability Models

Unlike discrete probability models, the scenarios under which a continuous probability model is appropriate are not obvious. Often, the appropriate continuous probability model is suggested by prior research or data. There are numerous continuous probability models; however, only the models frequently used in this text will be introduced and discussed in this section.

Recall that a continuous random variable X is a random variable where

- the support, $\mathcal{S}$, is the union of one or more intervals,
- the pdf $f(x)$ is a nonnegative function for which $\int_\mathcal{S} f(x)\, dx = 1$,
- $E(X) = \int_\mathcal{S} x\, f(x)\, dx$,
- $\text{Var}(X) = E[(X - \mu)^2]$,
- and MGF $M_x(t) = \int_\mathcal{S} e^{tx} f(x)\, dx$.

In general, the variance of a continuous random variable is computed using the alternative formula $\text{Var}(X) = E(X^2) - E(X)^2$.

3.2.1 The Uniform Model

The uniform probability model is used when each value in the support $\mathcal{S} = (\alpha, \beta)$ is equally likely and all of the subintervals of $\mathcal{S} = (\alpha, \beta)$ of width h have the same probability. A random variable X is said to have a uniform distribution over the interval (α, β), denoted by $X \sim U(\alpha, \beta)$, when the pdf of X is

$$f(x) = \frac{1}{\beta - \alpha}, \quad x \in (\alpha, \beta); \quad \alpha < \beta, \quad \alpha, \beta \in \mathbb{R}.$$

The CDF of $X \sim U(\alpha, \beta)$ is

$$F(x) = \begin{cases} 0 & x \leq \alpha, \\ \frac{x - \alpha}{\beta - \alpha} & \alpha < x < \beta\ . \\ 1 & x \geq \beta \end{cases}$$

The R command for computing the pdf of $X \sim U(\alpha, \beta)$ at $x \in (\alpha, \beta)$ is `dunif(x, α, β)`, `punif(x, α, β)` computes the value of the CDF at x, and `qunif(p, α, β)` computes the $100 \cdot p$th quantile of X for $p \in (0, 1)$.

Theorem 3.6 *If* $X \sim U(\alpha, \beta)$, *then*

i) $E(X) = \frac{\alpha + \beta}{2}$.

ii) $\text{Var}(X) = \frac{(\beta - \alpha)^2}{12}$.

iii) $M_X(t) = \frac{e^{\beta t} - e^{\alpha t}}{t(\beta - \alpha)}$.

Proof. Let $X \sim U(\alpha, \beta)$.

i)

$$E(X) = \int_\alpha^\beta x \cdot \frac{1}{\beta - \alpha} \, dx = \frac{x^2}{2(\beta - \alpha)} \bigg|_\alpha^\beta$$

$$= \frac{\beta^2 - \alpha^2}{2(\beta - \alpha)} = \frac{\alpha + \beta}{2}.$$

ii) The proof of part (ii) is left as an exercise.

iii) The proof of part (iii) is left as an exercise. ∎

Example 3.8 Let $X_1 \sim U(0, 2)$ and $X_2 \sim U(-1, 3)$. The pdfs of X_1 and X_2 are given in Figure 3.3. The 25th percentiles of X_1 and X_2 are `qunif(0.25, 0,2)` =0.5 and `qunif(0.25,-1,3)` =0, respectively.

Example 3.9 Suppose that $X \sim U(-1, 1)$, then

$$P(0.5 \le X \le 0.8) \int_{0.5}^{0.8} \frac{1}{2} \, dx = \frac{0.8 - 0.5}{2} = 0.15,$$

which can also be computed using the R commands

  `punif(0.8,-1,1) - punif(0.5,-1,1) = 0.15`.

The 78th percentile of X is `qunif(0.78,-1,1) = 0.56`.

Let $Y = |X|$. Then, $\mathcal{S}_Y = (0, 1)$ and

$$P(Y \le y) = P(|X| \le y) = P(-y \le X \le y)$$

$$= F(y) - F(-y) = \frac{y + 1}{2} - \frac{-y + 1}{2} = y.$$

Thus, by the CDF method, $f(y) = 1$ for $\in (0, 1)$, and hence, $Y \sim U(0, 1)$.

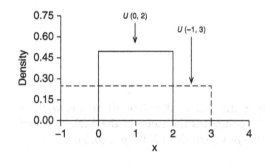

Figure 3.3 The pdfs of $X_1 \sim U(0, 2)$ and $X_2 \sim U(-1, 3)$.

3.2.2 The Gamma Model

The family of gamma distributions are flexible in shape, are long-tail right distributions, and are often used to model waiting times, life data, variables used in climatology, and in finance. A random variable X has a gamma distribution when the pdf is

$$f(x) = \frac{1}{\Gamma(\alpha)\beta^\alpha} x^{\alpha-1} e^{-\frac{x}{\beta}}, \quad x \in (0, \infty); \alpha \in (0, \infty), \beta \in (0, \infty),$$

where $\Gamma(\alpha) = \int_0^\infty x^{\alpha-1} e^{-x}\, dx$ is called the *gamma function*.

Several properties of the gamma function are given in Theorem 3.7.

Theorem 3.7 *Let* $\Gamma(\alpha) = \int_0^\infty x^{\alpha-1} e^{-x}\, dx$ *for* $\alpha > 0$. *Then,*

i) $\Gamma(n) = (n-1)!$ *for* $n \in \mathbb{N}$.
ii) $\Gamma(\alpha + 1) = \alpha\Gamma(\alpha)$ *for* $\alpha > 0$.
iii) $\Gamma\left(\frac{1}{2}\right) = \sqrt{\pi}$.

Proof. The proof of Theorem 3.7 is left as an exercise. ∎

A random variable X having a gamma distribution with parameters α and β is denoted by $X \sim \text{Gamma}(\alpha, \beta)$. When $X \sim \text{Gamma}(\alpha, \beta)$, α is called the *shape parameter* since it controls the shape of a gamma distribution, and β is called the *scale parameter* since larger values of β produce larger variances.

When $X \sim \text{Gamma}(\alpha, \beta)$, the R commands require the identification of the shape and scale parameters. In particular, dgamma (x, shape=α, scale=β) produces the value of the pdf at x, pgamma (x, shape=α, scale=β) produces the value of the CDF at x, and qgamma (p, shape=α, scale=β) produces the $100 \cdot p$th quantile of X.

Theorem 3.8 *If* $X \sim \text{Gamma}(\alpha, \beta)$, *then*

i) $E(X) = \alpha\beta$.
ii) $E(X^n) = \frac{\Gamma(\alpha+n)}{\Gamma(\alpha)}\beta^n$ *for* $n \in \mathbb{N}$.
iii) $\text{Var}(X) = \alpha\beta^2$.
iv) $M_X(t) = (1 - \beta t)^{-\alpha}$.

Proof. Let $X \sim \text{Gamma}(\alpha, \beta)$.

i)

$$E(X) = \int_0^\infty x \cdot \frac{1}{\Gamma(\alpha)\beta^\alpha} x^{\alpha-1} e^{-\frac{x}{\beta}}\, dx = \int_0^\infty \frac{1}{\Gamma(\alpha)\beta^\alpha} x^{\alpha} e^{-\frac{x}{\beta}}\, dx$$

$$= \frac{1}{\Gamma(\alpha)\beta^\alpha} \underbrace{\int_0^\infty (\beta u)^\alpha \, e^{-u} \, \beta \, du}_{u=x/\beta} = \frac{\beta^{\alpha+1}}{\Gamma(\alpha)\beta^\alpha} \int_0^\infty u^\alpha \, e^{-u} \, du$$

$$= \frac{\beta}{\Gamma(\alpha)} \Gamma(\alpha + 1) = \alpha\beta.$$

ii) Let $n \in \mathbb{N}$ be arbitrary but fixed.

$$E(X^n) = \int_0^\infty x^n \cdot \frac{1}{\Gamma(\alpha)\beta^\alpha} x^{\alpha-1} \, e^{-\frac{x}{\beta}} \, dx = \frac{1}{\Gamma(\alpha)\beta^\alpha} \int_0^\infty x^{\alpha+n-1} \, e^{-\frac{x}{\beta}} \, dx$$

$$= \frac{1}{\Gamma(\alpha)\beta^\alpha} \underbrace{\int_0^\infty (\beta u)^{\alpha+n-1} \, e^{-u} \, \beta \, du}_{u=x/\beta} = \frac{\Gamma(\alpha + n)}{\Gamma(\alpha)} \beta^n.$$

iii)

$$\text{Var}(X) = E(X^2) - E(X)^2 = \frac{\Gamma(\alpha + 2)}{\Gamma(\alpha)} \beta^2 - (\alpha\beta)^2$$

$$= \frac{(\alpha + 1)\alpha\Gamma(\alpha)}{\Gamma(\alpha)} \beta^2 - \alpha^2\beta^2$$

$$= \alpha^2\beta^2 + \alpha\beta^2 - \alpha^2\beta^2 = \alpha\beta^2.$$

iv) The proof of part (iv) is left as an exercise. ∎

Example 3.10 Let $X_1 \sim \text{Gamma}(\alpha = 1, \beta = 4), X_2 \sim \text{Gamma}(\alpha = 2, \beta = 4)$, and $X_3 \sim \text{Gamma}(\alpha = 3, \beta = 4)$. The pdfs of X_1, X_2, and X_3 are given in Figure 3.4. Note that X_1, X_2, and X_3 differ only in their values of α and produce three different shapes of the gamma distribution. The means of X_1, X_2, and X_3 are 4, 8, and 12, respectively, and the medians are qgamma(0.5,1,4)=2.77, dgamma(0.5,2,4)=6.71, and qgamma(0.5,3,4)=10.70.

Two commonly used distributions that are special cases of the gamma distribution are the *exponential distribution* and the *Chi-square distribution*.

- An exponential random variable is a gamma random variable with $\alpha = 1$ and is denoted by $X \sim \text{Exp}(\beta)$.
- A Chi-square random variable is a gamma random variable with $\alpha = \frac{r}{2}$ and $\beta = 2$, for $r \in \mathbb{N}$. A Chi-square random variable is denoted by $X \sim \chi_r^2$, and r is called the *degrees of freedom* of the Chi-square distribution.

Table 3.1 gives the pdf, mean, variance, and MGFs of the exponential and Chi-square distributions.

The R commands for the gamma distribution can be used for computing the pdf, CDF, and quantiles of an exponential or Chi-square distribution; however,

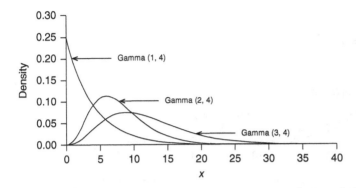

Figure 3.4 A plot of the pdfs of $X_1 \sim$ Gamma(1, 4), $X_2 \sim$ Gamma(2, 4), and $X_3 \sim$ Gamma(3, 4).

Table 3.1 Two special cases of the gamma distribution.

Distribution	pdf	Mean	Variance	MGF
Exponential	$\frac{1}{\beta}e^{-\frac{x}{\beta}}$	β	β^2	$(1 - \beta t)^{-1}$
Chi-square	$\frac{1}{\Gamma(\frac{r}{2})2^{\frac{r}{2}}}x^{\frac{r}{2}-1}e^{-\frac{x}{2}}$	r	$2r$	$(1 - 2t)^{-\frac{r}{2}}$

there are separate commands in R for both the exponential and the Chi-square distributions. In particular, the R commands for the exponential, which are given in terms of the rate which is $\frac{1}{\beta}$, are dexp(x, $\frac{1}{\beta}$), pexp(x, $\frac{1}{\beta}$), and qexp(p, $\frac{1}{\beta}$); for the Chi-square, the R commands are dchisq(x, r), pchisq(x, r), and qchisq(p, r).

Example 3.11 Let $X_1 \sim$ Exp(2), $X_2 \sim$ Exp(4) and $X_3 \sim$ Exp(6.67). The pdfs of X_1, X_2, and X_3 are shown in Figure 3.5.

Example 3.12 Let X be the effective lifetime of a dose of a pain reliever used to treat migraine headaches, and suppose that X follows an exponential distribution with mean $\beta = 4$h. The probability that the pain reliever has a lifetime of more than 6h is

$$P(X \geq 6) = \int_6^\infty \frac{1}{4} e^{-\frac{x}{4}} \, dx = e^{-1.5} = 0.2231 = 1\text{-pexp}(6,1/4).$$

The median lifetime of the pain reliever is qexp(0.5,1/4)=2.77h.

Example 3.13 Let $X_1 \sim \chi_2^2, X_2 \sim \chi_4^2$, and $X_3 \sim \chi_{10}^2$. The pdfs of X_1, X_2, and X_3 are given in Figure 3.6.

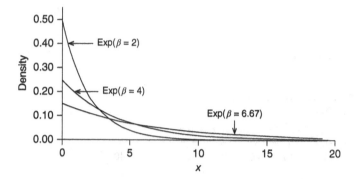

Figure 3.5 A plot of the pdfs of the exponential random variables $X_1, X_2,$ and X_3.

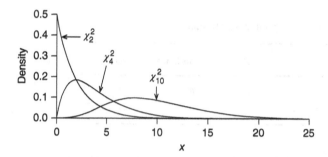

Figure 3.6 A plot of the pdfs of the Chi-square random variables $X_1, X_2,$ and X_3.

3.2.3 The Normal Model

The normal probability model has a bell-shaped distribution that is symmetric about its mean. The normal probability model is used to model many natural phenomena such as weights, heights, temperature, and modeling errors; however, the normal model is only appropriate when a random variable representing the phenomena is distributed symmetrically about its mean.

A random variable X has a normal distribution with parameters μ and σ when the pdf is

$$f(x) = \frac{1}{\sqrt{2\pi\sigma^2}} e^{-\frac{1}{2\sigma^2}(x-\mu)^2}, \quad x \in \mathbb{R}; \ \mu \in \mathbb{R}, \sigma \in \mathbb{R}^+.$$

A normal distribution with $\mu = 0$ and $\sigma = 1$ is called a *standard normal distribution* and is denoted by Z. A random variable X following a normal distribution will be denoted by $X \sim N(\mu, \sigma^2)$.

The R commands for computing the pdf, CDF, and quantiles of $X \sim N(\mu, \sigma^2)$ are dnorm (x, μ, σ), pnorm (x, μ, σ), and qnorm (p, μ, σ), respectively.

Theorem 3.9 *If $X \sim N(\mu, \sigma^2)$, then*

i) $E(X) = \mu$.

ii) $M_X(t) = e^{\frac{t^2 \sigma^2}{2} + \mu t}$.

iii) $\text{Var}(X) = \sigma^2$.

Proof. Let $X \sim N(\mu, \sigma^2)$.

i)

$$E(X) = \int_{-\infty}^{\infty} x \frac{1}{\sqrt{2\pi\sigma^2}} e^{-\frac{1}{2\sigma^2}(x-\mu)^2} \, dx = \underbrace{\int_{-\infty}^{\infty} (y+\mu) \frac{1}{\sqrt{2\pi\sigma^2}} e^{-\frac{1}{2\sigma^2} y^2} \, dy}_{y = x - \mu}$$

$$= \int_{-\infty}^{\infty} y \frac{1}{\sqrt{2\pi\sigma^2}} e^{-\frac{1}{2\sigma^2} y^2} \, dy + \int_{-\infty}^{\infty} \mu \frac{1}{\sqrt{2\pi\sigma^2}} e^{-\frac{1}{2\sigma^2} y^2} \, dy.$$

The first integral is the integral of the odd function $g(y) = y\, e^{-\frac{y^2}{2\sigma^2}}$ over $\mathbb{R}$, and hence, is 0. The second integral is the expected value of the constant μ for $Y \sim N(0, \sigma^2)$, and therefore, $E(X) = 0 + \mu = \mu$.

ii) The proof of part (ii) is left as an exercise.

iii) Using the MGF of X from part (ii), $\text{Var}(X) = E(X^2) - \mu^2 = M_X''(0) - \mu^2$. Now,

$$M_X''(t) = \frac{d^2}{dt^2} M_X(t) = \frac{d}{dt} \left[\frac{d}{dt} M_X(t) \right]$$

$$= \frac{d}{dt} \left[(\sigma^2 t + \mu t) e^{\frac{t^2 \sigma^2}{2} + \mu t} \right]$$

$$= (\sigma^2 t + \mu)^2 e^{\frac{t^2 \sigma^2}{2} + \mu t} + \sigma^2 e^{\frac{t^2 \sigma^2}{2} + \mu t}.$$

Thus, $M_X''(0) = \mu^2 + \sigma^2$, therefore, $\text{Var}(X) = \mu^2 + \sigma^2 - \mu^2 = \sigma^2$. ∎

Example 3.14 Let W be the random variable associated with the weight in pounds of a one-year-old hatchery rainbow trout, and suppose that $W \sim N(0.8, 0.225)$. A plot of the pdf of W is given in Figure 3.7.

The probability that a one-year-old hatchery rainbow weighs between 1 and 1.5 lb is

$$P(1 \le W \le 1.5) = F(1.5) - F(1)$$
$$= \text{pnorm}(1.5, 0.8, 0.15) - \text{pnorm}(1, 0.8, 0.15)$$
$$= 0.999\,998\,5 - 0.908\,788\,8 = 0.091\,209\,69.$$

The 95th percentile of the weights of the one-year-old hatchery rainbow trout is $\text{qnorm}(0.95, 0.8, 0.15) = 1.05$.

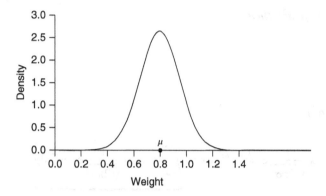

Figure 3.7 The distribution of weights of hatchery rainbow trout in Example 3.14.

Theorem 3.10 *If* $X \sim N(\mu, \sigma^2)$, *then* $Z = \frac{X-\mu}{\sigma} \sim N(0,1)$.

Proof. Let $X \sim N(\mu, \sigma^2)$ and consider $M_Z(t)$.

$$M_Z(t) = M_{\frac{X-\mu}{\sigma}}(t) = M_{\frac{X}{\sigma} - \frac{\mu}{\sigma}}(t) = e^{-\frac{\mu}{\sigma}t} M_X\left(\frac{t}{\sigma}\right)$$

$$= e^{-\frac{\mu}{\sigma}t} \, e^{\frac{(\frac{t}{\sigma})^2\sigma^2}{2} + \mu\frac{t}{\sigma}}$$

$$= e^{\frac{t^2}{2}}.$$

Thus, $M_Z(t)$ is the MGF of an $N(0,1)$ random variable, and therefore, $Z \sim N(0,1)$. ∎

Theorem 3.11 *If* $Z \sim N(0,1)$, *then* $Z^2 \sim \chi_1^2$.

Proof. Let $Z \sim N(0,1)$ and consider the MGF of Z^2.

$$M_{Z^2}(t) = E(e^{tZ^2}) = \int_{-\infty}^{\infty} e^{tz^2} \frac{1}{\sqrt{2\pi}} e^{-\frac{z^2}{2}} \, dz$$

$$= \int_{-\infty}^{\infty} \frac{1}{\sqrt{2\pi}} e^{-\frac{z^2 - 2tz^2}{2}} \, dz = \int_{-\infty}^{\infty} \frac{1}{\sqrt{2\pi}} e^{-\frac{z^2(1-2t)}{2}} \, dz$$

$$= \underbrace{\int_{-\infty}^{\infty} \frac{1}{\sqrt{2\pi}} e^{-\frac{u^2}{2}} \frac{du}{1-2t}}_{u=\sqrt{1-2t}z} = \frac{1}{\sqrt{1-2t}} \underbrace{\int_{-\infty}^{\infty} \frac{1}{\sqrt{2\pi}} e^{-\frac{u^2}{2}} \, du}_{1}$$

$$= \frac{1}{(1-2t)^{\frac{1}{2}}}.$$

Thus, $M_{Z^2}(t) = \frac{1}{(1-2t)^{\frac{1}{2}}}$ that is the MGF of a Chi-square distribution with 1 degree of freedom, and hence, $Z^2 \sim \chi_1^2$. ∎

Corollary 3.1 *If $X \sim N(\mu, \sigma^2)$, then $\left(\frac{X-\mu}{\sigma}\right)^2 \sim \chi_1^2$.*

Proof. The proof of Corollary 3.1 follows directly from Theorems 3.10 and 3.11. ∎

3.2.4 The Log-normal Model

The log-normal probability model is a long-tail right distribution often used to model phenomena where extremely large values occur, but are relatively rare in occurrence. The log-normal distribution is so named because, as Theorem 3.13 shows, the natural logarithm of a log-normal random variable follows a normal distribution. In particular, a random variable X has a log-normal distribution when its pdf is

$$f(x) = \frac{1}{x\sqrt{2\pi\sigma^2}} e^{-\frac{1}{2\sigma^2}(\ln(x)-\mu)^2}, \quad x > 0; \quad \mu \in \mathbb{R}, \quad \sigma > 0.$$

A random variable X having a log-normal distribution with parameters μ and σ is denoted by $X \sim LN(\mu, \sigma^2)$.

The R commands for computing the pdf, CDF, and quantiles of $X \sim LN(\mu, \sigma^2)$ are $\mathtt{dlnorm(x, \mu, \sigma)}$, $\mathtt{plnorm(x, \mu, \sigma)}$, and $\mathtt{qlnorm(p, \mu, \sigma)}$, respectively.

Theorem 3.12 *If $X \sim LN(\mu, \sigma^2)$, then*

i) $E(X) = e^{\mu + \frac{\sigma^2}{2}}$.
ii) $\mathrm{Var}(X) = e^{2\mu + 2\sigma^2} - e^{2\mu + \sigma^2}$.

Proof. The proof of Theorem 3.12 is left as an exercise. ∎

Example 3.15 Let $X_1 \sim LN(1, 2), X_2 \sim LN(1, 0.5)$, and $X_3 \sim LN(1, 1)$ be log-normal random variables. The pdfs of X_1, X_2, and X_3 are given in Figure 3.8.

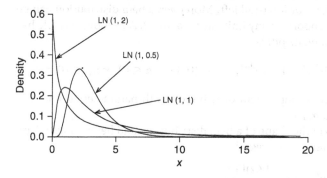

Figure 3.8 The pdfs of the log-normal random variables X_1, X_2, and X_3.

Example 3.16 Let P be the random variable associated with the peak flow in cubic feet per second (cfs) of the Big Hole River at the Melrose gauging station, and suppose that P is known to follow a log-normal distribution with parameters $\mu = 8.5$ and $\sigma = 0.25$. Then, the mean peak flow is

$$E(P) = e^{8.5 + \frac{0.0625}{2}} = 5070.8,$$

and the variance is

$$\text{Var}(P) = e^{2(8.5) + 2(0.0625)} - 5070.8^2 = 1\ 658\ 334.0.$$

The probability that the peak flow exceeds 10 000 cfs is

$$P(P \geq 10\ 000) = 1 - F(10\ 000)$$
$$= 1 - \texttt{plnorm}(10000, 8.5, 0.25) = 0.0022,$$

and the 75th percentile of the peak flows is $\texttt{lnorm}(0.75, 8.5, 0.25)$ $= 5817.5$.

Theorem 3.13 *If $X \sim \text{LN}(\mu, \sigma^2)$, then $\ln(X) \sim N(\mu, \sigma^2)$.*

Proof. Let $X \sim \text{LN}(\mu, \sigma^2)$, and let $Y = \ln(X)$. Then, Y is a one-to-one transformation of X with $S_Y = \mathbb{R}$ and $X = e^Y$. Thus, for $y \in \mathbb{R}$

$$f_Y(y) = f_X(e^y) \left| \frac{d}{dy}[e^y] \right| = \frac{1}{e^y \sqrt{2\pi\sigma^2}} e^{-\frac{1}{2\sigma^2}(\ln(e^y) - \mu)^2} |e^y|$$

$$= \frac{1}{\sqrt{2\pi\sigma^2}} e^{-\frac{1}{2\sigma^2}(y - \mu)^2}.$$

Therefore, $Y = \ln(X) \sim N(\mu, \sigma^2)$. ∎

3.2.5 The Beta Model

The beta distribution provides a family of distributions that are flexible in shape on the interval $(0, 1)$. In particular, a beta distribution can be u-shaped, j-shaped, long-tail right, or long-tail left, Moreover, a beta distribution can be generalized to have support on any finite interval (a, b). A random variable has a beta distribution when its pdf is

$$f(x) = \frac{\Gamma(\alpha + \beta)}{\Gamma(\alpha)\Gamma(\beta)} x^{\alpha-1}(1 - x)^{\beta-1}, \quad x \in (0, 1); \quad \alpha \in (0, \infty), \beta \in (0, \infty).$$

A random variable X having a beta distribution with parameters α and β is denoted by $X \sim \text{Beta}(\alpha, \beta)$.

Both α and β control the shape of a beta distribution, and hence, α and β are called shape parameters. The integral

$$\int_0^1 x^{\alpha-1}(1 - x)^{\beta-1}\, dx = \frac{\Gamma(\alpha)\Gamma(\beta)}{\Gamma(\alpha + \beta)}$$

is called the *beta function* and is denoted by $B(\alpha, \beta)$. The R commands associated with the pdf, CDF, and quantiles of the beta distribution are dbeta(x, α, β), pbeta(x, α, β), and qbeta(p, α, β).

Theorem 3.14 *If* $X \sim \text{Beta}(\alpha, \beta)$, *then*

i) $E(X) = \frac{\alpha}{\alpha+\beta}$.

ii) $\text{Var}(X) = \frac{\alpha\beta}{(\alpha+\beta)^2(\alpha+\beta+1)}$.

iii) for $\alpha = \beta = 1$, $X \sim U(0, 1)$.

Proof. The proof of Theorem 3.14 is left as an exercise. ∎

The MGF of a beta distribution exists; however, it is not useful.

Example 3.17 Let $X_1 \sim (2, 4), X_2 \sim (2, 2.5)$, and $X_3 \sim (2, 2)$. The pdfs of X_1, X_2, and X_3 are given in Figure 3.9.

The means of X_1, X_2, and X_3 are $\mu_1 = 0.33, \mu_2 = 0.\overline{4}$, and $\mu_3 = 0.5$, and their medians are $\tilde{\mu}_1 = 0.314, \tilde{\mu}_2 = 0.436$, and $\tilde{\mu}_3 = 0.5$; the medians were computed using qbeta$(0.5, \alpha, \beta)$.

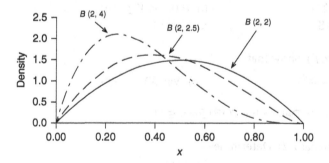

Figure 3.9 A plot of the pdfs of beta random variables X_1, X_2, and X_3.

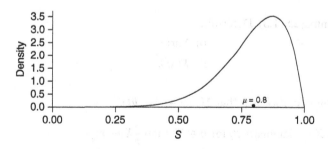

Figure 3.10 The distribution of exam scores in Example 3.18.

Example 3.18 Based on the past experience of teaching an introductory statistics course, midterm exam scores (S) were found to be approximately distributed as a beta distribution with $S \sim \text{Beta}(8, 2)$. A plot of pdf of S is given in Figure 3.10.

The 95th percentile of the midterm exams is qbeta$(0.95, 8, 2) = 0.96$, the median is qbeta$(0.5, 8, 2) = 0.82$, the mean is 0.80, and the probability of a passing score is $P(S \geq 0.70) = 1 - \text{pbeta}(0.70, 8, 20) = 0.804$.

Problems

3.2.1 Let $X \sim U(0, 2)$. Determine

a) $P(X > 0.6)$.

b) $P(0.75 < X \leq 1.6)$.

c) $E(X)$.

d) $\text{Var}(X)$.

3.2.2 Let $X \sim U(-2, 8)$. Determine

a) $P(-1 \leq X \leq 4)$.

b) $P(|X| \geq 1.5)$.

c) $E(X)$.

d) $\text{Var}(X)$.

3.2.3 Let $X \sim U(-1, 1)$, and let $Y = X^2$. Determine

a) $P(Y > 0.25)$.

b) $P(0.5 < Y \leq 0.6)$.

c) $f_Y(y)$ and $\mathcal{S}_Y$.

d) $\text{Var}(Y)$.

3.2.4 Let $X \sim U(\alpha, \beta)$. Show that

a) $E(X^2) = \frac{\beta^2 + \alpha\beta + \alpha^2}{3}$.

b) $\text{Var}(X) = \frac{(\beta - \alpha)^2}{12}$.

3.2.5 Let $X \sim U(\alpha, \beta)$. Determine $E(X^n)$ for $n \in \mathbb{N}$.

3.2.6 Let $X \sim \text{Gamma}(3, 2)$. Determine

a) $P(X \geq 4)$.

b) $E(X)$.

c) $E(X^n)$.

d) $\text{Var}(X)$.

3.2.7 Let $X \sim \text{Gamma}(2.5, 1.5)$. Determine

a) $E(X)$.

b) $\text{Var}(X)$.

c) $E\left(\frac{1}{X}\right)$.

d) $E(\sqrt{X})$.

3.2.8 Let $X \sim \text{Gamma}(\alpha, \beta)$. Show that $M_X(t) = (1 - \beta t)^{-\alpha}$.

3.2.9 Show that if $X \sim \text{Gamma}(n, \beta)$ for $n \in \mathbb{N}$, then $\frac{2}{\beta} X \sim \chi^2_{2n}$.

3.2.10 Let X_1 and X_2 be iid Gamma(α, β) random variables, and let $Y = X_1 + X_2$. Determine

a) $M_Y(t)$. b) the distribution of Y.

3.2.11 Let $X_1 \sim \Gamma(\alpha_1, \beta)$ and $X_2 \sim$ Gamma(α_2, β) be independent random variables, and let $Y = X_1 + X_2$. Determine

a) $M_Y(t)$. b) the distribution of Y.

3.2.12 The value in the support that produces the maximum value of the pdf is called the mode of the random variable. If $X \sim$ Gamma(α, β), determine the mode of X.

3.2.13 Suppose that the time between consecutive 911 calls follows an exponential distribution with $\beta = 10$ min. Determine
a) the mean time between two consecutive 911 calls.
b) the variance of the time between two consecutive 911 calls.
c) the probability of the time between two consecutive 911 calls exceeds 20 min.
d) the probability of the time between two consecutive 911 calls is between 15 and 20 min.
e) the 95th percentile of the times between two consecutive 911 calls.

3.2.14 Suppose that the time to failure of an electronic component follows an exponential distribution with a mean of 90 days. Determine
a) the probability of the time to failure exceeds 120 days.
b) the probability of the time to failure is between 150 and 240 days.
c) the median time to failure.

3.2.15 Let $X \sim$ Exp(β). Determine

a) the median. b) the $100 \cdot p$th percentile.
c) $E(X^2)$. d) $E(X^n)$.

3.2.16 Let $X \sim$ Exp(β). Show that $P(X > \delta + \epsilon | X > \delta) = P(X > \epsilon)$, which is known as the "memoryless property of the exponential."

3.2.17 Let $X \sim$ Exp(β) and $Y = \frac{X}{\beta}$. Determine

a) $f_Y(y)$ and $\mathcal{S}_Y$. b) Var(Y).

3.2.18 Let $X \sim$ Exp(β) and $Y = 2\frac{X}{\beta}$. Show that $Y \sim \chi_2^2$.

3.2.19 Let $X \sim \text{Exp}(\beta)$, and let $Y = X + \tau$ for $\tau \in \mathbb{R}$. Determine
a) $f_Y(y)$ and $\mathcal{S}_Y$.
b) $F_Y(y)$ for $y \in \mathbb{R}$.
c) $E(Y)$.
d) $\text{Var}(Y)$.

3.2.20 Let X_1 and X_2 be iid χ_r^2 random variables, and let $Y = X_1 + X_2$. Determine
a) $M_Y(t)$.
b) the distribution of Y.

3.2.21 Let $X_1 \sim \chi_r^2$ and $X_2 \sim \chi_s^2$ be independent random variables, and let $Y = X_1 + X_2$. Determine
a) $M_Y(t)$.
b) the distribution of Y.

3.2.22 Show that
a) $\Gamma(n) = (n-1)!$ for $n \in \mathbb{N}$.
b) $\Gamma(\alpha + 1) = \alpha\Gamma(\alpha)$ for $\alpha > 0$.
c) $\Gamma\left(\frac{1}{2}\right) = \sqrt{\pi}$.
d) $\Gamma\left(\frac{5}{2}\right) = \frac{3}{4}\sqrt{\pi}$.

3.2.23 Determine
a) $\Gamma(4)$.
b) $\Gamma(8)$.
c) $\Gamma\left(\frac{7}{2}\right)$.
d) $\Gamma\left(\frac{2n+1}{2}\right)$ for $n \in \mathbb{N}$.

3.2.24 Let I be the random variable associated with an individual's score on an IQ exam, and suppose that $I \sim N(100, 225)$. Determine
a) $P(I \geq 135)$.
b) $P(I \leq 70)$.
c) the 95th percentile of the random variable I.

3.2.25 Let G be the random variable associated with an individual's GRE Quantitative score. If $G \sim N(150, 36)$, determine
a) $P(G \geq 165)$.
b) $P(G \leq 145)$.
c) the 90th percentile of the random variable G.

3.2.26 Let W be the random variable associated with the weight of a steelhead returning to the Alsea River hatchery. If $W \sim N(8, 1.2)$, determine
a) the probability that a steelhead returning to the Alsea River hatchery exceeds 9 lb.
b) the probability that a steelhead returning to the Alsea River hatchery weighs between 7 and 11 lb.
c) the weight that only 5% of the steelhead returning to the Alsea River hatchery exceed.

3.2.27 Let $X \sim N(\mu, \sigma^2)$, and suppose that $P(X < 27) = 0.9332$ and $P(X < 28) = 0.9452$. Determine the values of μ and σ^2.

3.2.28 Suppose that $X \sim N(\mu, \sigma^2)$. If the 10th percentile of the distribution of the random variable X is 67.96 and the 95th percentile is 141.125, determine the values of μ and σ.

3.2.29 Let $X \sim N(\mu, \sigma^2)$. Show that $\text{Var}(X) = \sigma^2$.

3.2.30 Let $Z \sim N(0, 1)$. Show that

a) $E(Z^{2n+1}) = 0$.   b) $E(Z^{2n}) = \frac{2^n \Gamma(n+\frac{1}{2})}{\sqrt{\pi}}$.

3.2.31 Let $Z \sim N(0, 1)$. Use the method of CDF to show that $Z^2 \sim \chi_1^2$.

3.2.32 Let X_1 and X_2 be iid $N(\mu, \sigma^2)$ random variables, and let $Y_1 = X_1 + X_2$ and $Y_2 = X_1 - X_2$. Determine

a) $M_{Y_1}(t)$.   b) the distribution of Y_1.
c) $M_{Y_2}(t)$.   d) the distribution of Y_2.

3.2.33 Let $X_1 \sim N(\mu_1, \sigma_1^2)$ and $X_2 \sim N(\mu_2, \sigma_2^2)$ be independent random variables, and let $Y_1 = X_1 + X_2$ and $Y_2 = X_1 - X_2$. Determine

a) $M_{Y_1}(t)$.   b) the distribution of Y_1.
c) $M_{Y_2}(t)$.   d) the distribution of Y_2.

3.2.34 Let $X \sim N(\mu, \sigma^2)$, and let $Y = e^X$. Show that $Y \sim LN(\mu, \sigma^2)$.

3.2.35 Let $X \sim LN(\mu, \sigma^2)$. Show that
a) $E(X) = e^{\mu + \frac{\sigma^2}{2}}$.
b) $\text{Var}(X) = e^{2\mu + 2\sigma^2} - e^{2\mu + \sigma^2}$.

3.2.36 Let $X \sim LN(\mu, \sigma^2)$, and let $Y = \ln(X)$. Show that $Y \sim N(\mu, \sigma^2)$.

3.2.37 Let $X_1 \sim LN(\mu_1, \sigma_1^2)$ and $X_2 \sim LN(\mu_2, \sigma_2^2)$ be independent random variables, and let $Y = \ln(X_1 X_2)$. Show that $Y \sim N(\mu_1 + \mu_2, \sigma_1^2 + \sigma_2^2)$.

3.2.38 Let $X_1 \sim LN(\mu_1, \sigma_1^2)$ and $X_2 \sim LN(\mu_2, \sigma_2^2)$ be independent random variables, and let $Y = \ln\left(\frac{X_1}{X_2}\right)$. Show that $Y \sim N(\mu_1 - \mu_2, \sigma_1^2 + \sigma_2^2)$.

3.2.39 Let $X \sim \text{Beta}(4, 2)$. Determine

a) $F(x), \forall x \in \mathbb{R}$.

b) $P(0.25 < X < 0.5)$.

c) $E(X)$.

d) $\text{Var}(X)$.

e) the 25th percentile.

f) the 75th percentile.

3.2.40 Let $m, n \in \mathbb{N}$. Show that $B(m, n) = \frac{(m-1)!(n-1)!}{(m+n-1)!}$.

3.2.41 If $X \sim \text{Beta}(\alpha, \beta)$ and $\alpha > \beta$, determine the value of x where the pdf attains its maximum (i.e. the mode).

3.2.42 Let $X \sim \text{Beta}(1, 2)$. Determine

a) $F(x)$.

b) $x_{0.5}$.

3.2.43 Let $X \sim \text{Beta}(2, 2)$. Determine

a) $F(x), \forall x \in \mathbb{R}$.

b) $P(0.2 < X < 0.8)$.

c) $E(X)$.

d) $\text{Var}(X)$.

e) the 50th percentile.

f) the 80th percentile.

3.2.44 Let $X \sim \text{Beta}(4, 4)$.

a) $P(X > 0.6)$.

b) $P(0.22 < X < 0.48)$.

c) $E(X)$.

d) $\text{Var}(X)$.

3.2.45 Let $X \sim \text{Beta}(0.5, 2.4)$.

a) $P(X > 0.15)$.

b) $P(0.3 < X < 0.65)$.

c) $E(X)$.

d) $\text{Var}(X)$.

e) $X_{0.25}$.

f) $X_{0.65}$.

3.2.46 Let $X \sim \text{Beta}(\alpha, \beta)$. Determine

a) $E(X^n)$ for $n \in \mathbb{N}$.

b) $E[X^n(1 - X)^m]$ for $n, m \in \mathbb{N}$.

3.2.47 Let $X \sim \text{Beta}(\alpha, \beta)$ and let $Y = 4X$. Determine

a) $f_Y(Y)$ and $\mathcal{S}_Y$.

b) $E(Y)$.

c) $E(Y^2)$.

d) $\text{Var}(Y)$.

3.2.48 Let $X \sim \text{Beta}(\alpha, \beta)$ with $\alpha = \beta$. Show that the distribution of X is symmetric about 0.5.

3.2.49 Let $X \sim \text{Beta}(\alpha, \beta)$. Show that

a) $E(X) = \frac{\alpha}{\alpha+\beta}$.

b) $\text{Var}(X) = \frac{\alpha\beta}{(\alpha+\beta)^2(\alpha+\beta+1)}$.

3.2.50 Let $X \sim \text{Beta}(1,1)$. Show that $X \sim U(0,1)$.

3.2.51 A random variable X is said to follow a Weibull distribution when it has CDF $F(x) = 1 - e^{-\left(\frac{x}{\nu}\right)^{\beta}}$ or $x > 0$, $\nu, \beta > 0$. If X follows a Weibull distribution with parameters ν and β, $X \sim \text{Weib}(\nu, \beta)$, determine

a) $f(x)$ for $x > 0$. b) $E(X)$.
c) $E(X^n)$ for $n \in \mathbb{N}$. d) $\text{Var}(X)$.

3.2.52 Let (X, Y) be a bivariate random variable with pdf

$$f(x,y) = \frac{1}{2\pi\sigma_X\sigma_Y\sqrt{1-\rho^2}} e^{-\frac{1}{2(1-\rho^2)}\left[\left(\frac{x-\mu_X}{\sigma_X}\right)^2 - 2\rho\left(\frac{x-\mu_X}{\sigma_X}\right)\left(\frac{y-\mu_Y}{\sigma_Y}\right) + \left(\frac{y-\mu_Y}{\sigma_Y}\right)^2\right]}$$

for $(x,y) \in \mathbb{R}^2$. Show that
a) $X \sim N(\mu_X, \sigma_X^2)$.
b) $Y \sim N(\mu_Y, \sigma_Y^2)$.
c) $X|Y = y \sim N\left(\mu_X + \rho\frac{\sigma_X}{\sigma_Y}(y - \mu_Y), \sigma_X^2(1 - \rho^2)\right)$.
d) $Y|X = x \sim N\left(\mu_Y + \rho\frac{\sigma_Y}{\sigma_X}(x - \mu_X), \sigma_Y^2(1 - \rho^2)\right)$.
e) that X and Y are independent if and only if $\rho = 0$.

3.3 Important Distributional Relationships

In Section 2.7, the expected value and variance of a linear combination of $X_1, \ldots, X_n$ were discussed. In particular, for independent random variables $X_1, \ldots, X_n$, the mean and variance of a linear combination $T = \sum_{i=1}^{n} a_i X_i$ are

$$E\left[\sum_{i=1}^{n} a_i X_i\right] = \sum_{i=1}^{n} a_i \mu_i.$$

$$\text{Var}\left(\sum_{i=1}^{n} a_i X_i\right) = \sum_{i=1}^{n} a_i^2 \sigma_i^2.$$

In general, determining the exact probability distribution of a function of a collection of random variables is difficult; however, the distribution of a sum of independent random variables can often be determined from the MGF of the sum.

3.3.1 Sums of Random Variables

Many of the results in this section are based on the MGF of the sum of independent random variables. Recall that when $X_1, \ldots, X_n$ be independent random variables and $S = \sum_{i=1}^{n} X_i$, then $M_S(t) = \prod_{i=1}^{n} M_{X_i}(t)$. Theorem 3.15 provides the exact probability distribution of the sum of independent discrete random variables for several cases.

Theorem 3.15 *If $X_1, X_2, \ldots, X_n$ are independent random variables with*

i) $X_i \sim \text{Bin}(1,p)$ *random variables, then* $\sum_{i=1}^{n} X_i \sim \text{Bin}(n,p)$.
ii) $X_i \sim \text{Bin}(k_i,p)$, *then* $\sum_{i=1}^{n} X_i \sim \text{Bin}(\sum_{i=1}^{n} k_i, p)$.
iii) $X_i \sim \text{Pois}(\lambda_i)$, *then* $\sum_{i=1}^{n} X_i \sim \text{Pois}(\sum_{i=1}^{n} \lambda_i)$.
iv) $X_i \sim \text{NegBin}(r_i,p)$, *then* $\sum_{i=1}^{n} X_i \sim \text{NegBin}(\sum_{i=1}^{n} r_i, p)$.
v) $X_i \sim \text{Geo}(p)$, *then* $\sum_{i=1}^{n} X_i \sim \text{NegBin}(n,p)$.

Proof. Let $X_1, \ldots, X_n$ be independent random variables, and let $S = \sum_{i=1}^{n} X_i$.

(i) If $X_i \sim \text{Bin}(1,p)$, then

$$M_S(t) = \prod_{i=1}^{n} M_{X_i(t)} = \prod_{i=1}^{n} (p\, e^t + q) = (p\, e^t + q)^n,$$

which is the MGF of $X \sim \text{Bin}(n,p)$.
(ii)–(v) The proofs of Theorem 3.15 parts (ii)–(v) are left as exercises. ∎

Theorem 3.16 presents several important relationships for the sum of independent gamma random variables.

Theorem 3.16 *If $X_1, X_2, \ldots, X_n$ are independent random variables with*

i) $X_i \sim \text{Exp}(\beta)$, *then* $\sum_{i=1}^{n} X_i \sim \text{Gamma}(n, \beta)$.
ii) $X_i \sim \text{Exp}(\beta)$, *then* $\frac{2}{\beta} \sum_{i=1}^{n} X_i \sim \chi^2_{2n}$.
iii) $X_i \sim \text{Gamma}(\alpha_i, \beta)$, *then* $\sum_{i=1}^{n} X_i \sim \text{Gamma}(\sum_{i=1}^{n} \alpha_i, \beta)$.
iv) $X_i \sim \text{Gamma}(k_i, \beta)$ *where* $k_i \in \mathbb{N}$, *then* $\frac{2}{\beta} \sum_{i=1}^{n} X_i \sim \chi^2_{k^\star}$, *where* $k^\star = 2 \sum_{i=1}^{n} k_i$.
v) $X_i \sim \chi^2_{r_i}$, *then* $\sum_{i=1}^{n} X_i \sim \chi^2_{k^\star}$, *where* $k^\star = \sum_{i=1}^{n} r_i$.

Proof. The proof of Theorem 3.16 is left as an exercise. ∎

Theorem 3.17 and Corollary 3.3 present several important relationships for the sum of independent normal random variables.

Theorem 3.17 *If $X_1, X_2, \ldots, X_n$ are independent normal random variables with $X_i \sim N(\mu_i, \sigma_i^2)$, then*

i) $\sum_{i=1}^{n} X_i \sim N\left(\sum_{i=1}^{n} \mu_i, \sum_{i=1}^{n} \sigma_i^2\right)$.

ii) $\sum_{i=1}^{n} a_i X_i \sim N\left(\sum_{i=1}^{n} a_i \mu_i, \sum_{i=1}^{n} a_i^2 \sigma_i^2\right)$ *for* $a_i \in \mathbb{R}$.

iii) $\sum_{i=1}^{n} \left(\dfrac{X_i - \mu_i}{\sigma_i}\right)^2 \sim \chi_n^2$.

Proof. The proof of Theorem 3.17 is left as an exercise. ∎

Corollary 3.2 *If* $X_1, X_2, \ldots, X_n$ *are iid* $N(\mu, \sigma^2)$ *random variables, then*

i) $\sum_{i=1}^{n} X_i \sim N(n\mu, n\sigma^2)$.

ii) $\overline{X} \sim N\left(\mu, \dfrac{\sigma^2}{n}\right)$.

Proof. Corollary 3.2 follows directly from Theorem 3.17 since $X_1, \ldots, X_n$ are iid. ∎

Corollary 3.3 *If* $Z_1, Z_2, \ldots, Z_n$ *are independent standard normal random variables, then*

i) $\overline{Z} = \dfrac{1}{n} \sum_{i=1}^{n} Z_i \sim N\left(0, \dfrac{1}{n}\right)$.

ii) $\sum_{i=1}^{n} Z_i^2 \sim \chi_n^2$.

Proof. The proof of Corollary 3.3 follows directly from Corollary 3.2 and Theorem 3.17. ∎

When $X_1, X_2, \ldots, X_n$ are iid $N(\mu, \sigma^2)$ random variables with μ and σ^2 are unknown, $\overline{X} = \frac{1}{n} \sum_{i=1}^{n} X_i$ and $S^2 = \frac{1}{n-1} \sum_{i=1}^{n} (X_i - \overline{X})^2$ are commonly used estimators of μ and σ^2. By Theorem 3.17, the distribution of $\overline{X}$ is $N\left(\mu, \frac{\sigma^2}{n}\right)$. Lemma 3.1 and Theorem 3.18 are used to prove Corollary 3.4, which shows that $S^2 \sim \text{Gamma}\left(\frac{n-1}{2}, \frac{2\sigma^2}{n-1}\right)$ when $X_i \overset{iid}{\sim} N(\mu, \sigma^2)$.

Lemma 3.1 *If* $X_1, X_2, \ldots, X_n$ *are iid* $N(\mu, \sigma^2)$ *random variables, then* $\overline{X}$ *and* S^2 *are independent random variables.*

Theorem 3.18 *If* $X_1, X_2, \ldots, X_n$ *are iid* $N(\mu, \sigma^2)$ *random variables, then* $\dfrac{(n-1)S^2}{\sigma^2} \sim \chi_{n-1}^2$.

Proof. Let $X_1, X_2, \ldots, X_n$ be iid $N(\mu, \sigma^2)$ random variables and consider $T = \sum_{i=1}^{n} \left(\dfrac{X_i - \mu}{\sigma}\right)^2$. By Theorem 3.17 part (iv), $T \sim \chi_n^2$, and therefore, $M_T(t) = (1 - 2t)^{-\frac{n}{2}}$.

By Problem 3.3.16

$$T = \sum_{i=1}^{n} \left(\frac{X_i - \mu}{\sigma} \right)^2 = \sum_{i=1}^{n} \left(\frac{X_i - \overline{X}}{\sigma} \right)^2 + n \left(\frac{\overline{X} - \mu}{\sigma} \right)^2$$

$$= \frac{(n-1)S^2}{\sigma^2} + \left(\frac{\overline{X} - \mu}{\frac{\sigma}{\sqrt{n}}} \right)^2.$$

Now, $\left(\frac{\overline{X} - \mu}{\frac{\sigma}{\sqrt{n}}} \right)^2 \sim \chi_1^2$, and since S^2 and $\overline{X}$ are independent by Lemma 3.1, it follows that

$$M_T(t) = (1 - 2t)^{-\frac{n}{2}} = M_{\frac{(n-1)S^2}{\sigma^2}}(t) M_{\left(\frac{\overline{X} - \mu}{\frac{\sigma}{\sqrt{n}}} \right)^2}(t)$$

$$= M_{\frac{(n-1)S^2}{\sigma^2}}(t)(1 - 2t)^{-\frac{1}{2}}.$$

Thus, $M_{\frac{(n-1)S^2}{\sigma^2}}(t) = (1 - 2t)^{-\frac{(n-1)}{2}}$, which is the MGF of a Chi-square random variable with $n - 1$ degrees of freedom. Therefore, $\frac{(n-1)S^2}{\sigma^2} \sim \chi_{n-1}^2$. ∎

Corollary 3.4 *If* $X_1, X_2, \ldots, X_n$ *are iid* $N(\mu, \sigma^2)$ *random variables, then* $S^2 \sim$ *Gamma* $\left(\frac{n-1}{2}, \frac{2\sigma^2}{n-1} \right)$.

Proof. The proof of Corollary 3.4 is left as an exercise. ∎

3.3.2 The *T* and *F* Distributions

The T and F distributions are important probability distributions that are commonly used in statistical analyses involving confidence intervals and hypothesis tests. The T and F distributions are not generally used to model natural phenomena or populations, but do provide the distributions of several important statistics that arise in statistical models where the error structure follows a normal distribution.

The T distribution has support $\mathbb{R}$, is symmetric about zero, and has pdf given by

$$f(t) = \frac{\Gamma\left(\frac{r+1}{2} \right)}{\Gamma\left(\frac{r}{2} \right) \sqrt{\pi r}} \frac{1}{\left(1 + \frac{t^2}{r} \right)^{\frac{r+1}{2}}}, t \in \mathbb{R}; \ r \in \mathbb{N}.$$

A T distribution is denoted by $T \sim t_r$, where r is the *degrees of freedom* of the T distribution.

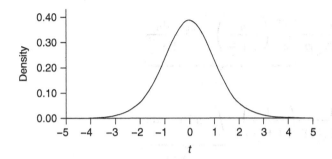

Figure 3.11 The pdf of a T distribution with 10 degrees of freedom.

The R commands for computing the pdf, CDF, and quantiles associated with a T distribution having r degrees of freedom are dt(x, r), pt(x, r), and qt(p, r). A plot of the T distribution with 10 degrees of freedom is given in Figure 3.11.

While a T distribution is symmetric about zero, as is the standard normal distribution, a T distribution has heavier tails than standard normal. Moreover, as the degrees of freedom of a T distribution increase, the T distribution approaches the standard normal distribution, and in fact, as $n \to \infty$ the limiting distribution is the standard normal distribution.

Theorem 3.19 *If $X_1 \sim N(0, 1)$ and $X_2 \sim \chi_r^2$ are independent random variables, then $T = \dfrac{X_1}{\sqrt{\dfrac{X_2}{r}}} \sim t_r$.*

Proof. Let $X_1 \sim N(0, 1)$ and $X_2 \sim \chi_r^2$ be independent random variables. Let $T = \dfrac{X_1}{\sqrt{\dfrac{X_2}{r}}} \sim t_r$ and $S = X_2$. Then, (T, S) is a one-to-one transformation of (X_1, X_2) with $X_1 = T\sqrt{\dfrac{S}{r}}$, $X_2 = S$, and $|J| = \sqrt{\dfrac{S}{r}}$. Thus, since X_1 and X_2 are independent, the joint pdf of T and S is

$$f_{(T,S)}(t, s) = f_1\left(t\sqrt{\frac{s}{r}}\right) f_2(s) \sqrt{\frac{s}{r}} = \frac{1}{\sqrt{2\pi}} e^{-\frac{t^2 s}{2r}} \frac{1}{\Gamma(\frac{r}{2}) 2^{\frac{r}{2}}} s^{\frac{r}{2}-1} e^{-\frac{s}{2}} \sqrt{\frac{s}{r}}$$

$$= \frac{1}{\sqrt{2\pi r} \Gamma(\frac{r}{2}) 2^{\frac{r}{2}}} s^{\frac{r-1}{2}} e^{-\frac{s}{2}\left(1+\frac{t^2}{r}\right)}.$$

Hence,

$$f_T(t) = \int_0^\infty \frac{1}{\sqrt{2\pi r} \Gamma(\frac{r}{2}) 2^{\frac{r}{2}}} s^{\frac{r-1}{2}} e^{-\frac{s}{2}\left(1+\frac{t^2}{r}\right)} \, ds$$

$$= \frac{1}{\sqrt{2\pi r}\Gamma(\frac{r}{2})2^{\frac{r}{2}}} \underbrace{\int_0^\infty \left(\frac{u}{1+\frac{t^2}{r}}\right)^{\frac{r-1}{2}} e^{-\frac{u}{2}} \frac{du}{1+\frac{t^2}{r}}}_{u=s(1+\frac{t^2}{r})}$$

$$= \frac{1}{\sqrt{2\pi r}\Gamma(\frac{r}{2})2^{\frac{r}{2}}} \left(\frac{1}{1+\frac{t^2}{r}}\right)^{\frac{r+1}{2}} \underbrace{\int_0^\infty u^{\frac{r-1}{2}} e^{-\frac{u}{2}}\, du}_{\Gamma(\frac{r+1}{2})2^{\frac{r+1}{2}}}$$

$$= \frac{\Gamma\left(\frac{r+1}{2}\right)}{\sqrt{\pi r}\,\Gamma\left(\frac{r}{2}\right)} \frac{1}{\left(1+\frac{t^2}{2}\right)^{\frac{r+1}{2}}},$$

which is the pdf of a t_r random variable. ∎

Corollary 3.5 *If* $X_1, \dots, X_n$ *are iid* $N(\mu, \sigma^2)$ *random variables, then*
$$T = \frac{\overline{X} - \mu}{\frac{S}{\sqrt{n}}} \sim t_{n-1}.$$

Proof. Let $X_1, \dots, X_n$ be iid $N(\mu, \sigma^2)$ random variables, and let $S = \sqrt{\frac{1}{n-1}\sum_{i=1}^n (X_i - \overline{X})^2}$. Then,

$$T = \frac{\overline{X} - \mu}{\frac{S}{\sqrt{n}}} = \frac{\overline{X} - \mu}{\frac{\sigma}{\sqrt{n}} \frac{S}{\sigma}} = \frac{\frac{\overline{X}-\mu}{\frac{\sigma}{\sqrt{n}}}}{\sqrt{\frac{S^2}{\sigma^2}}}.$$

Now, since $X_1, X_2, \dots, X_n$ are iid $N(\mu, \sigma^2)$ random variables, it follows that $\frac{\overline{X}-\mu}{\frac{\sigma}{\sqrt{n}}} \sim N(0,1)$, $\frac{(n-1)S^2}{\sigma^2} \sim \chi^2_{n-1}$, $\frac{S^2}{\sigma^2} \sim \frac{\chi^2_{n-1}}{n-1}$, and $\overline{X}$ and S^2 are independent random variables. Thus, by Theorem 3.19, $T \sim t_{n-1}$. ∎

Example 3.19 If $X_1, \dots, X_n$ are iid $N(\mu, \sigma^2)$ random variables with μ and σ^2 unknown, then an interval of estimates of μ can be based on the T distribution since $T = \frac{\overline{X}-\mu}{\frac{S}{\sqrt{n}}} \sim t_{n-1}$. To see this, let $t_{n-1, \frac{\alpha}{2}}$ be the value of a T distribution with $n-1$ degrees of freedom such that $P(-t_{n-1,\frac{\alpha}{2}} \le T \le t_{n-1,\frac{\alpha}{2}}) = 1 - \alpha$. Then,

$$1 - \alpha = P(-t_{n-1,\frac{\alpha}{2}} \le T \le t_{n-1,\frac{\alpha}{2}}) = P\left(-t_{n-1,\frac{\alpha}{2}} \le \frac{\overline{X}-\mu}{\frac{S}{\sqrt{n}}} \le t_{n-1,\frac{\alpha}{2}}\right)$$

$$= P\left(\overline{X} - t_{n-1,\frac{\alpha}{2}}\frac{S}{\sqrt{n}} \le \mu \le \overline{X} - t_{n-1,\frac{\alpha}{2}}\frac{S}{\sqrt{n}}\right).$$

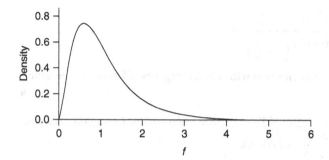

Figure 3.12 The pdf of a F distribution with 6 and 20 degrees of freedom.

Thus, before the sample is collected, there is $(1 - \alpha) \times 100\%$ chance that μ will lie between $\overline{X} - t_{n-1,\frac{\alpha}{2}} \frac{S}{\sqrt{n}}$ and $\overline{X} + t_{n-1,\frac{\alpha}{2}} \frac{S}{\sqrt{n}}$. The interval

$$\left(\overline{X} - t_{n-1,\frac{\alpha}{2}} \frac{S}{\sqrt{n}}, \overline{X} + t_{n-1,\frac{\alpha}{2}} \frac{S}{\sqrt{n}} \right)$$

is called a $(1 - \alpha) \times 100\%$ *confidence interval* for μ.

The F distribution is a long-tail right distribution with support $\mathbb{R}^+$ and has pdf given by

$$f(x) = \frac{\Gamma(\frac{m+n}{2})}{\Gamma(\frac{m}{2})\Gamma(\frac{n}{2})} \frac{m^{\frac{m}{2}} n^{\frac{n}{2}} x^{\frac{m}{2}-1}}{(n+mx)^{\frac{m+n}{2}}}, \quad x > 0; \quad m, n \in \mathbb{N}.$$

A F distribution is denoted by $F \sim F(m, n)$, where m and n are the degrees of freedom of the F distribution.

The R commands for computing the pdf, CDF, and quantiles associated with a F distribution having r degrees of freedom are $df(x, m, n)$, $pf(x, m, n)$, and $qf(p, m, n)$. A plot of a F distribution with 6 and 20 degrees of freedom is given in Figure 3.12.

Theorem 3.20 *If* $T \sim t_n$, *then* $T^2 \sim F(1, n)$. *n degrees of freedom.*

Proof. Let $T \sim t_n$, and let $Y = T^2$. Consider $F_Y(y)$ for $y > 0$.

$$F_Y(y) = P(Y \le y) = P(-\sqrt{y} \le T \le \sqrt{y}) = F_T(\sqrt{y}) - F_T(-\sqrt{y}).$$

Hence,

$$f_Y(y) = f_T(\sqrt{y}) \frac{1}{2\sqrt{y}} - f_T(-\sqrt{y}) \left(-\frac{1}{2\sqrt{y}} \right) \underbrace{= f_Y(\sqrt{y}) \frac{1}{\sqrt{y}}}_{T \text{ is symmetric about } 0}$$

$$= \frac{1}{\sqrt{y}} \frac{\Gamma(\frac{n+1}{2})}{\sqrt{\pi n}\Gamma(\frac{n}{2})} \frac{1}{\left(1 + \frac{y}{n}\right)^{\frac{n+1}{2}}},$$

which is the pdf of an F distribution with 1 and n degrees of freedom. Therefore, $T^2 \sim F(1, n)$. ∎

Theorem 3.21 *If* $X \sim \chi_m^2$ *and* $Y \sim \chi_n^2$ *are independent Chi-square random variables. Then,* $F = \dfrac{X/m}{Y/n} \sim F(m, n)$.

Proof. The proof of Theorem 3.21 is left an exercise. ∎

When $F \sim F(m, n)$ is formed from the ratio of two Chi-square distributions, m is referred to as the *numerator degrees of freedom*, and n is referred to as the *denominator degrees of freedom*.

Theorem 3.22 *Let* $X_1, \ldots, X_n$ *be iid* $N(\mu_X, \sigma^2)$ *random variables and* $Y_1, \ldots, Y_m$ *iid* $N(\mu_Y, \sigma^2)$ *random variables. If* $X_1, \ldots, X_n$ *and* $Y_1, \ldots Y_m$ *are independent collections of random variables, then* $F = \dfrac{\frac{1}{n-1}\sum_{i=1}^{n}(X_i - \overline{X})^2}{\frac{1}{m-1}\sum_{j=1}^{m}(Y_i - \overline{Y})^2} \sim F_{n-1, m-1}$.

Proof. Let $X_1, \ldots, X_n$ be iid $N(\mu_X, \sigma^2)$ random variables, $Y_1, \ldots, Y_m$ be iid $N(\mu_Y, \sigma^2)$ random variables, and suppose that $X_1, \ldots, X_n$ and $Y_1, \ldots Y_m$ are independent collections of random variables. Then, by Theorem 3.18, $\frac{\sum_{i=1}^{n}(X_i - \overline{X})^2}{\sigma^2}$ and $\frac{\sum_{j=1}^{m}(Y_i - \overline{Y})^2}{\sigma^2}$ are distributed as χ_{n-1}^2 and χ_{m-1}^2 random variables. Therefore, $F = \frac{\frac{1}{n-1}\sum_{i=1}^{n}(X_i - \overline{X})^2}{\frac{1}{m-1}\sum_{j=1}^{m}(Y_i - \overline{Y})^2}$ is the ratio of independent Chi-square random variables divided by their degrees of freedom, and hence, $F \sim F_{n-1, m-1}$. ∎

Example 3.20 The F distribution is used to test the utility of the regression model in a regression analysis and the equality of the treatment effects in an analysis of variance. In both cases, the normal distribution is assumed to be the distribution of the error term in the model, and the F statistic is related to the form given in Theorem 3.22.

Problems

3.3.1 Let $X_1, \ldots, X_n$ be iid Bin$(1, p)$ random variables, and let $S = \sum_{i=1}^{n} X_i$. Show that $S \sim$ Bin(n, p).

3.3.2 Let $X_1, \ldots, X_n$ be independent random variables with $X_i \sim$ Bin(k_i, p), and let $S = \sum_{i=1}^{n} X_i$. Show that $S \sim$ Bin$(\sum_{i=1}^{n} k_i, p)$.

3.3.3 Let $X_1, \ldots, X_n$ be iid Pois(λ) random variables, and let $S = \sum_{i=1}^{n} X_i$. Show that $S \sim$ Pois($n\lambda$).

3.3.4 Let $X_1, \ldots, X_n$ be independent random variables with $X_i \sim$ Pois(λ_i), and let $S = \sum_{i=1}^{n} X_i$. Show that $S \sim$ Pois($\sum_{i=1}^{n} \lambda_i$).

3.3.5 Let $X_1, \ldots, X_n$ be iid Geo(p) random variables, and let $S = \sum_{i=1}^{n} X_i$. Show that $S \sim$ NegBin(n, p).

3.3.6 Let $X_1, \ldots, X_n$ be independent NegBin(r_i, p) random variables. Show that $S \sim$ NegBin($\sum_{i=1}^{n} r_i, p$).

3.3.7 Let $X_1, \ldots, X_n$ be iid Exp(β) random variables, and let $S = \sum_{i=1}^{n} X_i$. Show that $S \sim$ Gamma(n, β).

3.3.8 Let $X_1, \ldots, X_n$ be iid Exp(β) random variables, and let $S = \frac{2}{\beta} \sum_{i=1}^{n} X_i$. Show that $S \sim \chi^2_{2n}$.

3.3.9 Let $X_1, \ldots, X_n$ be independent Gamma(α_i, β) random variables, and let $S = \sum_{i=1}^{n} X_i$. Show that $S \sim$ Gamma($\sum_{i=1}^{n}, \alpha_i, \beta$).

3.3.10 Let $X_1, \ldots, X_n$ be independent Gamma(k_i, β) random variables where $k_i \in \mathbb{N}$, and let $S = \frac{2}{\beta} \sum_{i=1}^{n} X_i$. Show that $S = \frac{2}{\beta} \sum_{i=1}^{n} X_i \sim \chi^2_{2k^\star}$ where $k^\star = \sum_{i=1}^{n} k_i$.

3.3.11 Let $X_1, \ldots, X_n$ be independent random variables with $X_i \sim \chi^2_{k_i}$, and let $S = \sum_{i=1}^{n} X_i$. Show that $S \sim \chi^2_{k^\star}$ where $k^\star = \sum_{i=1}^{n} k_i$.

3.3.12 Let $X_1, \ldots, X_n$ be independent random variables with $X_i \sim N(\mu_i, \sigma_i^2)$ and $S = \sum_{i=1}^{n} X_i$. Show that $S \sim N(\sum_{i=1}^{n} \mu_i, \sum_{i=1}^{n} \sigma_i^2)$.

3.3.13 Let $X_1, \ldots, X_n$ be independent random variables with $X_i \sim N(\mu_i, \sigma_i^2)$, and let $S = \sum_{i=1}^{n} a_i X_i$ for $a_i \in \mathbb{R}$. Show that $S \sim N(\sum_{i=1}^{n} a_i \mu_i, \sum_{i=1}^{n} a_i^2 \sigma_i^2)$.

3.3.14 Let $X_1, \ldots, X_n$ be iid $N(\mu, \sigma^2)$ random variables. If $\overline{X} = \frac{1}{n} \sum_{i=1}^{n} X_i$, show that $\overline{X} \sim N(\mu, \frac{\sigma^2}{n})$.

3.3.15 Let $X_1, \ldots, X_n$ be independent random variables with $X_i \sim N(\mu_i, \sigma_i^2)$ and $S = \sum_{i=1}^{n} \left(\frac{X_i - \mu_i}{\sigma_i} \right)^2$. Show that $S \sim \chi^2_n$.

3.3.16   Let $x_1, \ldots, x_n$ be real numbers, and let $S = \sum_{i=1}^{n} (x_i - \mu)^2$. Show that $S = \sum_{i=1}^{n} (x_i - \bar{x})^2 + n(\bar{x} - \mu)^2$.

3.3.17   Let $X_1, \ldots, X_n$ be iid $N(\mu, \sigma^2)$ random variables. If $S = \sum_{i=1}^{n} (X_i - \mu)^2$, show that
a) $S^2 \sim \text{Gamma}\left(\frac{n}{2}, 2\sigma^2\right)$.
b) $E(S^2) = n\sigma^2$.

3.3.18   If $X \sim \chi_m^2$ and $Y \sim \chi_n^2$ are independent χ^2 random variables, show that $F = \frac{\frac{X}{m}}{\frac{Y}{n}} \sim F(m, n)$.

3.4   Case Study – The Central Limit Theorem

One of the most important theorems in statistics is the *Central Limit Theorem*, which is often denoted by CLT. The CLT deals with the asymptotic distribution of $\frac{\bar{X} - \mu}{\frac{\sigma}{\sqrt{n}}}$ for a sequence of iid random variables, say $\{X_i : i \in \mathbb{N}\}$. One reason the CLT is so important is that it provides an easy way to find the approximate distribution of an average or sum of iid random variables, regardless of the underlying distribution of the random variables. Moreover, the CLT is the basis for large sample confidence intervals, large sample hypothesis tests, and plays an important role in building statistical models.

The CLT was first proven by Abraham de Moivre (1733) and Pierre-Simon Laplace (1812) for approximating binomial probabilities. The CLT was later generalized to handle continuous random variables by Siméon Denis Poisson (1829). Since de Moivre's initial work, the CLT has been worked on by many of the best probabilists and mathematicians including Cauchy, Chebyshev, Markov, Lindeberg, Lévy, Lyapunov, and Feller.

3.4.1   Convergence in Distribution

Because the CLT deals with an *asymptotic distribution*, the definition of an asymptotic distribution is first introduced. In particular, the asymptotic distribution of a sequence of random variables is based on a special type of convergence for a sequence of random variables called *convergence in distribution*.

Definition 3.1   Let $\{X_i : i \in \mathbb{N}\}$ a sequence of random variables with $F_i(x)$ the CDF of X_i. The sequence of random variables $\{X_i : i \in \mathbb{N}\}$ is said to converge in distribution to a random variable X, denoted by $X_n \xrightarrow{d} X$, if and only if there exists a random variable X with CDF $F(x)$ such that $\lim_{n \to \infty} F_n(x) = F(x)$ at all points of continuity of $F(x)$. When $X_n \xrightarrow{d} X$, $F(x)$ is called the asymptotic distribution of X_n.

The proof of the CLT given here is based on *Lévy's Continuity Theorem* for MGFs; however, it should be noted that there are many alternate proofs of the CLT.

Theorem 3.23 (Lévy's Continuity Theorem) *Let* $\{X_i : i \in \mathbb{N}\}$ *be a sequence of random variables with* $M_i(t)$ *the MGF of* X_i. *If* X *is a random variable with MGF* $M(t)$ *defined on* $(-h, h)$ *for some* $h > 0$ *and* $\lim_{n \to \infty} M_n(t) = M(t)$, $\forall t \in (-h, h)$, *then* $X_n \xrightarrow{d} X$.

3.4.2 The Central Limit Theorem

The version of the CLT given here requires that the random variables have a MGF; however, there is a stronger version of the CLT requiring only that the random variables have finite variances. The weaker version of the CLT is given here because its proof only relies on previously discussed material.

Theorem 3.24 (Central Limit Theorem) *If* $\{X_i : i \in \mathbb{N}\}$ *is a sequence of iid random variables with MGF* $M(t)$, *mean* μ, *and variance* σ^2, *then*
$$\frac{\overline{X} - \mu}{\frac{\sigma}{\sqrt{n}}} \xrightarrow{d} Z \sim N(0, 1).$$

Proof. Let $\{X_i : i \in \mathbb{N}\}$ be a sequence of iid random variables with MGF $M(t)$, mean μ, and variance σ^2. For $n \in \mathbb{N}$, let $Z_n = \frac{\overline{X}_n - \mu}{\frac{\sigma}{\sqrt{n}}}$, where $\overline{X}_n = \frac{1}{n} \sum_{i=1}^n X_i$.

Consider Z_n.

$$Z_n = \frac{\overline{X}_n - \mu}{\frac{\sigma}{\sqrt{n}}} = \sqrt{n} \frac{\left(\frac{1}{n} \sum_{i=1}^n X_i - \mu \right)}{\sigma} = \frac{1}{\sqrt{n}} \sum_{i=1}^n \left(\frac{X_i - \mu}{\sigma} \right).$$

Now, let $Y_i = \frac{X_i - \mu}{\sigma}$. Then, since $\{X_i : i \in \mathbb{N}\}$ is a sequence of iid random variables, it follows that $\{Y_i : i \in \mathbb{N}\}$ is also a sequence of iid random variables with $E(Y_i) = 0$ and $\text{Var}(Y_i) = 1$.

Consider $M_{Z_n}(t)$.

$$M_{Z_n}(t) = M_{\frac{1}{\sqrt{n}} \sum_{i=1}^n \left(\frac{X_i - \mu}{\sigma} \right)}(t) = M_{\sum_{i=1}^n Y_i} \left(\frac{t}{\sqrt{n}} \right)$$

$$= \prod_{i=1}^n M_{Y_i} \left(\frac{t}{\sqrt{n}} \right) = \left[M_{Y_i} \left(\frac{t}{\sqrt{n}} \right) \right]^n.$$

Expanding $M_{Y_i}\left(\frac{t}{\sqrt{n}}\right)$ in a second-order Taylor's series about $t = 0$,

$$M_{Y_i}\left(\frac{t}{\sqrt{n}}\right) = M_{Y_i}(0) + M'_{Y_i}(0)\frac{t}{\sqrt{n}} + M''_{Y_i}(0)\left(\frac{t}{\sqrt{n}}\right)^2 + o\left(\frac{t^2}{n}\right)$$

$$= 1 + \frac{t^2}{n} + o\left(\frac{t^2}{n}\right)$$

since $M'_{Y_i}(0) = E(Y_i) = 0$ and $M''_{Y_i}(0) = E(Y_i^2) = 1$. Hence,

$$\left[M_{Y_i}\left(\frac{t}{\sqrt{n}}\right)\right]^n = \left[1 + \frac{t^2}{n} + o\left(\frac{t^2}{n}\right)\right]^n$$

and since $\lim_{n\to\infty} o\left(\frac{t^2}{n}\right) = 0$, it follows that

$$\lim_{n\to\infty} M_{Z_n}(t) = \lim_{n\to\infty}\left[M_{Y_i}\left(\frac{t}{\sqrt{n}}\right)\right]^n = \lim_{n\to\infty}\left[1 + \frac{t^2}{n} + o\left(\frac{t^2}{n}\right)\right]^n = e^{\frac{t^2}{2}},$$

which is the MGF of $Z \sim N(0, 1)$. Therefore, by Lévy's Continuity Theorem

$$\frac{\overline{X} - \mu}{\frac{\sigma}{\sqrt{n}}} \xrightarrow{d} Z \sim N(0, 1).$$

∎

The CLT can generally be used to approximate probabilities, construct confidence intervals, and test hypotheses based on an average of iid random variables when $n \geq 30$; however, the accuracy of the approximation provided by the CLT will depend on the underlying distribution. In particular, with long-tailed distributions, such as the log-normal distribution, values of n larger than 30 are generally required.

In practice, the CLT can be used to approximate the distribution of $\overline{X}_n$ rather than Z_n, since for sufficiently large n, the approximate distribution of $\overline{X}_n$ is $N\left(\mu, \frac{\sigma}{\sqrt{n}}\right)$.

Example 3.21 Let $X_1, \ldots, X_{25}$ be iid $\text{Exp}(10)$ random variables. Then, according to the CLT, the approximate distribution of $\overline{X}_{25}$ is $N\left(10, \frac{100}{25}\right)$, and the approximate probability that $\overline{X}_{25}$ is between 8 and 12 is

$$P(8 \leq \overline{X} \leq 12) \approx \text{dnorm}(12,10,2) - \text{dnorm}(8,10,2) = 0.6827.$$

For comparison, the exact distribution of $\overline{X}$ is Gamma(25, 0.4), and the exact probability that $\overline{X}$ is between 8 and 12 is

$$P(8 \leq \overline{X} \leq 12) = \texttt{dgamma(12,25,scale=0.4)}$$
$$- \texttt{dgamma(8,25,scale=0.4)}$$
$$= 0.6860.$$

Thus, in this case, the CLT approximation is off by less than 0.5%.

An alternate version of the CLT is given in Corollary 3.6 for sums of iid random variables.

Corollary 3.6 (Central Limit Theorem for a Sum) *Let $\{X_i : i \in \mathbb{N}\}$ be a sequence of iid random variables with common MGF $M(t)$, mean μ, and variance σ^2. If $S_n = \sum_{i=1}^{n} X_i$, then $\dfrac{S_n - n\mu}{\sigma\sqrt{n}} \xrightarrow{d} N(0, 1)$.*

Proof. The proof of Corollary 3.6 is left as an exercise. ∎

Example 3.22 Let $X_1, \dots, X_n$ be iid random variables with $X_i \sim \text{Bin}(1, p)$. Then, when n is sufficiently large, the CLT for Sums states that the distribution of $\frac{\sum_{i=1}^{n} X_i - np}{\sqrt{npq}}$ is approximately an $N(0, 1)$ distribution. Thus, for $n = 30$ and $p = 0.3$, the probability that $7 < \sum_{i=1}^{30} X_i \leq 10$ can be approximated using $X \approx N(9, 6.3)$. Thus,

$$P\left(7 < \sum_{i=1}^{30} X_i \leq 10\right) \approx \texttt{pnorm(10,9,2.51)} - \texttt{pnorm(7,9,2.51)}$$
$$= 0.4421.$$

By Theorem 3.15, $\sum_{i=1}^{30} X_i \sim \text{Bin}(30, 0.3)$, and therefore, the exact probability of $7 < \sum_{i=1}^{30} X_i \leq 10$ is

$$P\left(7 < \sum_{i=1}^{30} X_i \leq 10\right) = \texttt{pbinom(10,30,0.3)}$$
$$- \texttt{pbinom(7,30,0.3)}$$
$$= 0.4490.$$

The Central Limit for Sums works best for approximating binomial probabilities when $np \geq 5$ and $nq \geq 5$. Also, there are finite population correction factors that can be used to improve the CLT approximation.

Problems

3.4.1 Let $X_1, \ldots, X_{50}$ be iid Pois(4) random variables.
   a) Use CLT to approximate $P(3.8 \le \overline{X} \le 4.2)$.
   b) Determine $P(3.8 \le \overline{X} \le 4.2)$ using the exact distribution of $\sum_{i=1}^{50} X_i$.

3.4.2 Let $X_1, \ldots, X_{36}$ be iid Gamma(1.5, 2) random variables.
   a) Use CLT to approximate $P(2.6 \le \overline{X} \le 3.25)$.
   b) Determine $P(2.6 \le \overline{X} \le 3.25)$ using the exact distribution of $\sum_{i=1}^{36} X_i$.

3.4.3 If $X_1, \ldots, X_{25}$ are iid Exp(10) random variables, show that $\overline{X} \sim$ Gamma(25, 0.4).

3.4.4 Let $X_1, \ldots, X_{100}$ be iid Bin(1, 0.25) random variables.
   a) Use CLT to approximate $P(20 \le \sum_{i=1}^{100} X_i \le 30)$.
   b) Determine $P(20 \le \sum_{i=1}^{100} X_i \le 30)$ using the exact distribution of $\sum_{i=1}^{50} X_i$.

3.4.5 Let $X_1, \ldots, X_{25}$ be iid Geo(0.25) random variables.
   a) Use CLT to approximate $P(60 \le \sum_{i=1}^{25} X_i \le 95)$.
   b) Determine $P(60 \le \sum_{i=1}^{25} X_i \le 95)$ using the exact distribution of $\sum_{i=1}^{25} X_i$.

3.4.6 Prove Corollary 3.6.

4

Parametric Point Estimation

Parametric point estimation is the process of determining and evaluating estimators of the unknown parameters of a specific probability model. In particular, the components of parametric estimation are the following:

i) a probability model $f(x; \theta)$, which is specified up to the values of the unknown parameters;
ii) a set of possible values of θ under consideration, which is called the *parameter space* and is denoted by Θ;
iii) a random sample of n observations from the probability model;
iv) a set of point estimators for the values of the unknown parameters based on the information contained in the random sample;
v) the specific properties of the estimators that are used to evaluate the accuracy and efficiency of the estimator.

For example, suppose that a biased coin is to be flipped n times, with each flip independent of the others where the probability of flipping a head is unknown. In this case, each flip is Bernoulli trial with $X = 1$ when a head is flipped and $X = 0$ when a tail is flipped. The probability model for X is $f(x; \theta) = \theta^x(1-\theta)^{1-x}$ for $x = 0, 1$, where $P(X = 1) = \theta$, the parameter space is $\Theta = [0, 1]$, the n flips constitute a random sample from the probability model, and a point estimator that might be used to estimate θ is $\widehat{\theta} = \overline{X}$. Chapter 4 introduces the properties of parametric point estimators, and in Chapter 5, commonly used methods for determining point estimators are discussed.

4.1 Statistics

The available information for estimating the unknown parameters in the point estimation problem for a specific probability model is contained in a *random sample*.

Mathematical Statistics: An Introduction to Likelihood Based Inference, First Edition. Richard J. Rossi.
© 2018 John Wiley & Sons, Inc. Published 2018 by John Wiley & Sons, Inc.

Definition 4.1 A collection of random variables $X_1, \ldots, X_n$ is called a sample of size n. A sample of n independent random variables $X_1, \ldots, X_n$ is called a random sample.

In practice, a random sample is a collection of random variables collected from a common population or from several different populations. For example, a sample of iid observations might arise when a researcher collects a random sample of n observations from an exponential distribution. A random sample of n observations from different populations is common when building a statistical model. For example, in building a simple linear regression model, a random sample of n independent random variables with $Y_i \sim N(\beta_0 + \beta_1 x_i, \sigma^2)$ for fixed values (i.e. not random) $x_2, x_2, \ldots, x_n$ is collected.

Once a random sample is collected, the information in the sample can be used for estimating the unknown parameters of a population.

Definition 4.2 Given a sample $X_1, \ldots, X_n$, a statistic $T = T(X_1, \ldots, X_n)$ is a function of a sample that does not depend on any unknown parameters or other unknown quantities. An estimator is a statistic that is used for estimating an unknown quantity, and an estimate is the observed value of the estimator.

Example 4.1 In a first course on Statistics, the statistics $\sum_{i=1}^{n} X_i$ and $\sum_{i=1}^{n} X_i^2$ are introduced as two of the basic summary statistics used to summarize a random sample on a random variable X. When the mean and the variance of the random variable X are unknown, the statistics

$$\overline{X} = \frac{1}{n} \sum_{i=1}^{n} X_i$$

and

$$s^2 = \frac{1}{n} \left[\sum_{i=1}^{n} X_i^2 - n\overline{X}^2 \right]$$

are commonly used as estimators of μ and σ^2, respectively.

4.1.1 Sampling Distributions

To evaluate the behavior of an estimator of a parameter θ, the likely values of the estimator, the mean of the estimator, and the variability of the estimator are all important characteristics of the estimator that should be considered. The behavior of an estimator and its effectiveness for estimating a parameter θ are based on the probability distribution of the estimator, which is called the *sampling distribution*.

Definition 4.3 For a sample $X_1, \ldots, X_n$ and a statistic $T = T(X_1, \ldots, X_n)$, the sampling distribution of the statistic T is the probability distribution associated with the random variable T. The pdf of the sampling distribution of T is denoted by $f_T(t; \theta)$.

Example 4.2 Let $X_1, \ldots, X_n$ be a sample of iid $N(\mu, \sigma^2)$ random variables. Then, the sampling distribution of $\overline{X}$ is $N\left(\mu, \frac{\sigma^2}{n}\right)$, and the sampling distribution of S^2 is $\frac{\sigma^2}{n-1}\chi^2_{n-1}$.

Example 4.3 Let $X_1, \ldots, X_n$ be a sample of iid $\text{Bin}(k, p)$ random variables. Then, the sampling distribution of $\sum_{i=1}^{n} X_i$ is $\text{Bin}(nk, p)$.

4.1.2 Unbiased Statistics and Estimators

In many cases, the sampling distribution of statistic T can be determined using the transformation methods for random variables and random vectors given in Chapter 3. However, in many cases, the sampling distribution of a statistic is difficult to determine, and in this case, the mean and variance of the statistic are often used for evaluating the effectiveness of the statistic for estimating θ. Fortunately, many statistics are based on sums of random variables where the mean and standard deviation of the statistic can be found using Theorems 2.37 and 2.38.

In general, when a statistic T is used for estimating a parameter θ, the mean of the sampling distribution of T should be close to θ (i.e. $E(T) \approx \theta$), and when $E(T) = \theta$, the statistic T is called *unbiased* estimator of the parameter θ.

Definition 4.4 A statistic T is said to be an unbiased estimator of a parameter θ when $E(T) = \theta$, $\forall \theta \in \Theta$. A statistic is said to be a biased estimator of θ when $E(T) \neq \theta$, and the bias of a statistic T for estimating a parameter θ is defined to be $Bias(T; \theta) = E(T) - \theta$.

When T is a biased estimator of a parameter θ, and the bias associated with T is less than 0, T tends to underestimate θ on the average, and when the bias of T is greater than 0, T tends to overestimate θ on the average.

Theorem 4.1 *If $X_1, \ldots, X_n$ is a sample of iid random variables with common mean μ, then $\overline{X}$ is an unbiased estimator of μ.*

Proof. The proof of Theorem 4.1 follows directly from Theorem 2.37. ∎

Theorem 4.2 *If $X_1, \ldots, X_n$ is a sample of iid random variables with common variance σ^2, then $S^2 = \frac{1}{n-1} \sum_{i=1}^{n} (X_i - \overline{X})^2$ is an unbiased estimator of σ^2.*

Proof. Let $X_1, \ldots, X_n$ be a sample of iid random variables with common variance σ^2, and let $S^2 = \frac{1}{n-1} \sum_{i=1}^{n} (X_i - \overline{X})^2$.

First, S^2 can be written as

$$S^2 = \frac{1}{n-1} \left(\sum_{i=1}^{n} X_i^2 - n\overline{X}^2 \right).$$

Thus,

$$E(S^2) = E \left[\frac{1}{n-1} \left(\sum_{i=1}^{n} X_i^2 - n\overline{X}^2 \right) \right] = \frac{1}{n-1} \left[\sum_{i=1}^{n} E(X_i)^2 - nE(\overline{X}^2) \right]$$

$$= \frac{1}{n-1} \left[n(\sigma^2 + \mu^2) - n \left(\mu^2 + \frac{\sigma^2}{n} \right) \right] = \frac{(n-1)\sigma^2}{n-1} = \sigma^2.$$

Hence, S^2 is an unbiased estimator of σ^2 when $X_1, \ldots, X_n$ is a sample of iid random variables with common variance σ^2. ■

Example 4.4 Let $X_1, \ldots, X_n$ be a sample of iid random variables with common variance σ^2, and let $S_n^2 = \frac{1}{n} \left(\sum_{i=1}^{n} X_i^2 - n\overline{X}^2 \right) = \frac{n-1}{n} S^2$. Then,

$$E(S_n^2) = E \left(\frac{n-1}{n} S^2 \right) = \frac{n-1}{n} E(S^2) = \frac{n-1}{n} \sigma^2.$$

Hence, S_n^2 is a biased estimator of σ^2, and the bias of S_n^2 for estimating σ^2 is

$$\mathrm{Bias}(S_n^2; \sigma^2) = E(S_n^2) - \sigma^2 = \frac{n-1}{n} \sigma^2 - \sigma^2 = -\frac{\sigma^2}{n}.$$

Thus, S_n^2 underestimates σ^2 on the average.

Example 4.5 Let $X_1, \ldots, X_n$ be a sample of Pois(λ) random variables. Since $E(X_i) = \lambda$, it follows that $E(\overline{X}) = \lambda$, and hence, $\overline{X}$ is an unbiased estimator of λ.

Example 4.6 Let $X_1, \ldots, X_n$ be a sample of Exp $\left(\beta = \frac{1}{\theta} \right)$ random variables with $\Theta = \mathbb{R}^+$. Then, $E(\overline{X}) = \mu = \beta = 1/\theta$, and hence, $\overline{X}$ is an unbiased estimator of $\frac{1}{\theta}$. Thus, it seems reasonable to use $\frac{1}{\overline{X}}$ as an estimator of θ; however,

$$E \left(\frac{1}{\overline{X}} \right) = \frac{n\theta}{n-1} \text{ (see Problem 4.1.5). Hence, } \frac{1}{\overline{X}} \text{ is a biased estimator of } \theta \text{ with}$$

$$\mathrm{Bias} \left(\frac{1}{\overline{X}}; \theta \right) = \frac{n\theta}{n-1} - \theta = \frac{\theta}{n-1}.$$

In Example 4.6, the estimator $T = \frac{1}{\overline{X}}$ is a biased estimator of θ; however, $\frac{n-1}{n} T$ is an unbiased estimator of θ based on T. In general, if T is a biased estimator of θ with $E(T) = c\theta$ for some constant c, then $\frac{1}{c} T$ is an unbiased estimator of θ. While unusual, it is also true that if T is a biased estimator of θ with $E(T) = c\theta + b$ and c and b do not depend on θ, then $\frac{T-b}{c}$ is an unbiased estimator of θ.

Theorem 4.3 *If T_1 and T_2 are unbiased estimators of θ, then $\alpha T_1 + (1 - \alpha)T_2$ is an unbiased estimator of θ, $\forall \alpha \in \mathbb{R}$.*

Proof. The proof of Theorem 4.3 is left as an exercise. ∎

Example 4.7 Let $X_1, \ldots, X_n$ be a sample of random variables with common mean μ, and let $T = \frac{1}{3}X_1 + \frac{2}{3}\overline{X}$. Then, by Theorem 4.3, T is an unbiased estimator of μ since $E(X_1) = \mu = E(\overline{X})$.

Example 4.8 Let $X_1, \ldots, X_n$ be a sample of iid $N(\mu, \sigma_1^2)$ random variables, $Y_1, Y_2, \ldots, Y_m$ be a sample of iid $N(\mu, \sigma_2^2)$ random variables, and suppose that X_i and Y_j are independent random variables for $i = 1, \ldots, n$ and $j = 1, \ldots, m$. Then, by Theorem 4.3, it follows that $\frac{n}{n+m}\overline{X} + \frac{m}{n+m}\overline{Y}$ is an unbiased estimator of μ.

When an estimator is biased, the bias sometimes can be made negligible by taking a sufficiently large sample. An estimator whose bias goes to zero as n goes to infinity is called an *asymptotically unbiased estimator.*

Definition 4.5 A statistic $T_n = T(X_1, \ldots, X_n)$ is said to be an asymptotically unbiased estimator of a parameter θ when $\lim_{n \to \infty} \text{Bias}(T_n; \theta) = 0$.

Example 4.9 In Example 4.4, S_n^2 was shown to be a biased estimator of σ^2 with bias $-\frac{\sigma^2}{n}$. Thus, the asymptotic bias of S_n^2 is

$$\lim_{n \to \infty} \text{Bias}(S_n^2; \sigma^2) = \lim_{n \to \infty} -\frac{\sigma^2}{n} = 0.$$

Hence, S_n^2 is an asymptotically unbiased estimator of σ^2.

Example 4.10 In Example 4.6, $T = \frac{1}{\overline{X}}$ is a biased estimator of θ with $\text{Bias}(1/\overline{X}; \theta) = \frac{\theta}{n-1}$. Since

$$\lim_{n \to \infty} Bias(T; \theta) = \lim_{n \to \infty} \left[\frac{n}{n-1}\theta - \theta\right] = \lim_{n \to \infty} \frac{\theta}{n-1} = 0$$

it follows that $\frac{1}{\overline{X}}$ is an asymptotically unbiased estimator of θ.

4.1.3 Standard Error and Mean Squared Error

While it is important that $E(T) \approx \theta$ for an estimator T of θ, the expected value of T does not measure the precision nor the accuracy of the estimator T. The *accuracy* of a statistic is a function of the bias of the statistic and the variability of the statistic. The standard deviation of the sampling distribution of an estimator is called the *standard error* of the estimator and measures the *precision* of an estimator.

Definition 4.6 The standard deviation of the sampling distribution of a statistic T is called the standard error of the statistic T and is denoted by SE(T).

Note that the accuracy of an estimator is a measure of how close the sampling distribution of an estimator is to θ, while the precision of an estimator measures the spread in the sampling distribution of the estimator. Thus, an estimator can be precise but not accurate. For example, an estimator with large bias might also have high precision, but because of the large bias, the sampling distribution of the estimator will not be close to θ, and therefore, the estimator will not be an accurate estimator.

Theorem 4.4 *If $X_1, \ldots, X_n$ is a sample of iid random variables with common variance σ^2, then $SE(\overline{X}) = \frac{\sigma}{\sqrt{n}}$.*

Proof. Let $X_1, \ldots, X_n$ be a sample of iid random variables with common variance σ^2. Then, by Theorem 2.38

$$\text{Var}(\overline{X}) = \text{Var}\left(\frac{1}{n}\sum_{i=1}^{n} X_i\right) = \frac{1}{n^2}\sum_{i=1}^{n}\text{Var}(X_i) = \frac{n\sigma^2}{n^2} = \frac{\sigma^2}{n}.$$

Therefore, $SE(\overline{X}) = \frac{\sigma}{\sqrt{n}}$. ∎

Example 4.11 Let $X_1, \ldots, X_n$ be a sample of iid Bin($1, p$) random variables. Then, $\overline{X}$ is an unbiased estimator of p, and since $\text{Var}(\overline{X}) = \frac{\sigma^2}{n}$, the standard error of $\overline{X}$ is $SE(\overline{X}) = \sqrt{\text{Var}(\overline{X})} = \sqrt{\frac{\sigma^2}{n}} = \sqrt{\frac{p(1-p)}{n}}$.

Example 4.12 Let $X_1, \ldots, X_n$ be a sample of iid $N(\mu, \sigma^2)$ random variables. Then, S^2 is an unbiased estimator of σ^2, and since $X_1, \ldots, X_n$ are iid $N(\mu, \sigma^2)$, it follows that $S^2 \sim \frac{\sigma^2}{n-1}\chi^2_{n-1}$. Hence,

$$SE(S^2) = \sqrt{\text{Var}(S^2)} = \sqrt{\text{Var}\left(\frac{\sigma^2}{n-1}\chi^2_{n-1}\right)}$$

$$= \sqrt{\frac{\sigma^4}{(n-1)^2}2(n-1)} = \sigma^2\sqrt{\frac{2}{n-1}}.$$

A measure of the accuracy of an estimator T used to estimate a parameter θ is the *Mean Squared Error* (MSE) associated with T.

Definition 4.7 Let $X_1, \ldots, X_n$ be a sample, the MSE of a statistic $T = T(X_1, \ldots, X_n)$ used to estimate a parameter θ is defined to be

$$\text{MSE}(T; \theta) = E[(T - \theta)^2].$$

An estimator with a small MSE is unlikely to be far from the parameter it is estimating. Moreover, when comparing two different estimators of θ, the estimator with the smaller MSE is the more accurate estimator of θ. Theorem 4.5 shows that the MSE associated with an estimator is a function of only the bias and the standard error of the estimator.

Theorem 4.5 *Let $X_1, \ldots, X_n$ be a random sample, and let $T = T(X_1, \ldots, X_n)$ be an estimator of θ. If $\text{Var}(T) < \infty$, then $\text{MSE}(T; \theta) = \text{Bias}(T; \theta)^2 + \text{SE}(T)^2$.*

Proof. Let $X_1, \ldots, X_n$ be a random sample, and let T be an estimator of θ with $\text{Var}(T) < \infty$.

First, note that $(T - \theta)^2 = (T - E(T) + E(T) - \theta)^2$ and

$$(T - E(T) + E(T) - \theta)^2 = (T - E(T))^2 + 2(T - E(T))(E(T) - \theta)$$
$$+ (E(T) - \theta)^2.$$

Thus,

$$\text{MSE}(T; \theta) = E[(T - E(T))^2 + 2(T - E(T))(E(T) - \theta) + (E(T) - \theta)^2]$$
$$= E[(T - E(T))^2] + 2E[(T - E(T))(E(T) - \theta)] + [(E(T) - \theta)^2]$$
$$= \underbrace{E[(T - E(T))^2]}_{\text{Var}(T)} + 2\underbrace{(E(T) - \theta)E[(T - E(T))]}_{0} + \underbrace{[(E(T) - \theta)^2]}_{\text{Bias}(T;\theta)^2}$$
$$= \text{Var}(T) + \text{Bias}(T; \theta)^2.$$

Therefore, $\text{MSE}(T; \theta) = \text{Bias}(T; \theta)^2 + \text{SE}(T)^2$. ∎

When an estimator T is an unbiased estimator of θ, it follows from Theorem 4.5 that $\text{MSE}(T; \theta) = \text{Var}(T) = \text{SE}(T)^2$, and hence, for an unbiased estimator T of θ, the standard error also measures the accuracy of the estimator T.

Example 4.13 Let $X_1, \ldots, X_n$ be a sample of iid $N(\mu, \sigma^2)$. Since S^2 is an unbiased estimator of σ^2 it follows that $\text{MSE}(S^2; \sigma^2) = \text{Var}(S^2)$. Now, $S^2 \sim \frac{\sigma^2}{n-1}\chi^2_{n-1}$, and therefore, $\text{MSE}(S^2; \sigma^2) = \text{Var}(S^2) = \frac{\sigma^4}{(n-1)^2}2(n-1) = \frac{2\sigma^4}{n-1}$.

Example 4.14 Let $X_1, \ldots, X_n$ be a sample of iid $N(\mu, \sigma^2)$. Consider $\text{MSE}(S_n^2; \sigma^2)$.

$$\text{MSE}(S_n^2; \sigma^2) = \text{Var}(S_n^2) + \text{Bias}(S_n^2; \sigma^2)^2$$

$$= \text{Var}\left(\frac{n-1}{n}S^2\right) + \left(\frac{-\sigma^2}{n}\right)^2$$

$$= \frac{(n-1)^2}{n^2}\text{Var}(S^2) + \frac{\sigma^4}{n^2}$$

$$= \frac{\frac{2\sigma^4}{n-1}(n-1)^2 + \sigma^4}{n^2} = \frac{(2n-1)\sigma^4}{n^2}.$$

When the MSE of an estimator T of θ goes to zero as n goes to infinity, the estimator T can be made as accurate as desired by taking a sufficiently large sample, and in this case, the estimator T is said to be an *MSE-consistent* estimator of θ.

Definition 4.8   Let $X_1, \ldots, X_n$ be a random sample, and let $T_n = T(X_1, \ldots, X_n)$ be an estimator of θ. The estimator T_n is said to be an MSE-consistent estimator of θ when $\lim_{n \to \infty} \text{MSE}(T_n; \theta) = 0$.

Theorem 4.6   *If $X_1, \ldots, X_n$ is a sample of random variables, then a statistic T_n is MSE consistent for estimating a parameter θ if and only if $\lim_{n \to \infty} \text{SE}(T_n) = 0$ and $\lim_{n \to \infty} \text{Bias}(T_n; \theta) = 0$.*

Proof. The proof of Theorem 4.6 follows directly from Theorem 4.5.   ∎

Theorem 4.7   *If $X_1, \ldots, X_n$ is a sample of iid random variables with common mean μ and variance σ^2, then $\overline{X}$ is an MSE-consistent estimator of μ.*

Proof. Let $X_1, \ldots, X_n$ be a sample iid random variables with common mean μ and variance σ^2. Then, $\text{Bias}(\overline{X}; \mu) = 0$ and $\text{SE}(\overline{X}) = \frac{\sigma}{\sqrt{n}}$. Thus,

$$\lim_{n \to \infty} \text{MSE}(\overline{X}; \mu) = \lim_{n \to \infty} \frac{\sigma^2}{n} = 0,$$

and therefore, $\overline{X}$ is MSE consistent for μ.   ∎

Theorem 4.8   *If $X_1, \ldots, X_n$ is a sample of iid $N(\mu, \sigma^2)$ random variables, then*

i) S^2 *is MSE consistent for estimating* σ^2.
ii) S_n^2 *is MSE consistent for estimating* σ^2.

Proof. The proof of Theorem 4.8 is left as exercise.   ∎

Thus, both S^2 and S_n^2 are MSE-consistent estimators of σ^2 for a sample of iid normal random variables. Moreover, a more general version of Theorem 4.8 can be proved showing that S^2 and S_n^2 are both MSE consistent for iid random samples from any probability model with $E(X^4) < \infty$.

Example 4.15 Let $X_1, \ldots, X_n$ be a sample of iid $N(0, \sigma^2)$ random variables, and let $T = \frac{1}{n} \sum_{i=1}^{n} X_i^2$. Then, $E(X_i^2) = \sigma^2$ since $E(X_i) = 0$, and therefore, $E(T) = \sigma^2$ and $\text{Bias}(T, \sigma^2) = 0$. Also, since $X_1, \ldots, X_n$ are iid $N(0, \sigma^2)$, $\sum_{i=1}^{n} \frac{X_i^2}{\sigma^2} \sim \chi_n^2$ by Theorem 3.17, and hence, $T \sim \frac{\sigma^2}{n} \chi_n^2$.

Thus, $\text{Var}(T) = \frac{2\sigma^4}{n}$ and $\lim_{n \to \infty} \text{MSE}(T; \sigma^2) = \lim_{n \to \infty} \frac{2\sigma^4}{n} = 0$. Therefore, when $X_1, \ldots, X_n$ are iid $N(0, \sigma^2)$, $T = \frac{1}{n} \sum_{i=1}^{n} X_i^2$ is an MSE-consistent estimator of σ^2.

Let T_1 and T_2 be estimators of a parameter θ, the ratio of the MSEs of T_1 and T_2 is used to compare the accuracy of the two estimators. The ratio $\frac{\text{MSE}(T_1; \theta)}{\text{MSE}(T_2; \theta)}$ is called the *relative efficiency* of the two estimators.

Definition 4.9 Let T_1 and T_2 be estimators of θ. The relative efficiency of T_1 to T_2 is defined to be

$$\text{RE}(T_1, T_2; \theta) = \frac{\text{MSE}(T_1; \theta)}{\text{MSE}(T_2; \theta)},$$

and the asymptotic relative efficiency of T_1 to T_2 is

$$\text{ARE}(T_1, T_2; \theta) = \lim_{n \to \infty} \frac{\text{MSE}(T_1; \theta)}{\text{MSE}(T_2; \theta)}.$$

Note that when $\text{RE}(T_1, T_2; \theta) < 1$, T_1 is a more accurate estimator of θ (i.e. more efficient) than T_2. On the other hand, when $\text{RE}(T_1, T_2; \theta) > 1$, T_2 is a more accurate estimator of θ than T_1; when $RE(T_1, T_2; \theta) = 1$, T_1 and T_2 are equally efficient estimators of θ.

Example 4.16 Let $X_1, \ldots, X_n$ be a sample of iid $N(\mu, \sigma^2)$, and let S_n^2 and S^2 be estimators of σ^2. From Example 4.13, $\text{MSE}(S^2; \sigma^2) = \frac{2\sigma^4}{n-1}$ and from Example 4.14, $\text{MSE}(S_n^2; \sigma^2) = \frac{(2n-1)\sigma^4}{n^2}$. Thus, the relative efficiency of S_n^2 to S^2 is

$$\text{RE}(S_n^2, S^2; \sigma^2) = \frac{\frac{(2n-1)\sigma^4}{n^2}}{\frac{2\sigma^4}{n-1}} = \frac{(2n-1)(n-1)}{2n^2} < 1.$$

Hence, for a fixed sample of n iid normal random variables, S_n^2 is a more efficient estimator of σ^2 than S^2.

The asymptotic relative efficiency of S_n^2 to S^2 is

$$\text{ARE}(S_n^2, S^2; \sigma^2) = \lim_{n \to \infty} \text{RE}(S_n^2, S^2) = \lim_{n \to \infty} \frac{(2n-1)(n-1)}{2n^2} = 1.$$

Thus, for a sufficiently large sample of iid $N(\mu, \sigma^2)$ random variables, S^2 is essentially as efficient as S_n^2.

4.1.4 The Delta Method

In some estimation problems, an estimator T with known standard error is available for estimating θ; however, the goal is to estimate a function of θ, say $g(\theta)$. In this case, $g(T)$ is often used for estimating $g(\theta)$, but when g is a nonlinear function, the standard error of the estimator $g(T)$ is often difficult to determine. In this case, the *Delta method* can be used to approximate the standard error of the estimator $g(T)$.

The Delta Method is based on expanding a function of a random variable in a first-order Taylor's series about its mean. In particular, if X is a random variable with finite mean μ and variance σ^2, and $g(x)$ is a nonlinear function differentiable in an open interval about μ, then the first-order Taylor's series expansion for $g(X)$ is

$$g(X) \approx g(\mu) + (X - \mu)g'(\mu).$$

Based on a first-order Taylor's series for $g(X)$, it follows that $E[g(X)] \approx g(\mu)$ and $\mathrm{Var}[g(X)] \approx \mathrm{Var}(X) \times [g'(\mu)]^2$. Thus, when T is an estimator of θ with $\mathrm{Var}(T) < \infty$, and $g(\theta)$ is a differentiable function of θ, the standard error of $g(T)$ can be approximated using $\mathrm{SE}[g(T)] \approx \mathrm{SE}(T) \times |g'(E(T))|$.

Example 4.17 Let $X_1, \ldots, X_n$ be a sample of iid $\mathrm{Exp}\left(\beta = \frac{1}{\theta}\right)$ random variables. Then, $E(X_i) = \mu = \frac{1}{\theta}$ and $\mathrm{Var}(X_i) = \sigma^2 = \frac{1}{\theta^2}$. Now, since $E(\overline{X}) = \frac{1}{\theta}$, a reasonable estimator of θ is $g(\overline{X}) = \frac{1}{\overline{X}}$. In this case,

$$\mathrm{SE}\left(\frac{1}{\overline{X}}\right) \approx \mathrm{SE}(\overline{X})|g'(\mu)| = \frac{\sigma}{\sqrt{n}}\left|-\frac{1}{\mu^2}\right| = \frac{\sigma}{\sqrt{n}\mu^2} = \frac{\theta}{\sqrt{n}}$$

since $E(X) = \frac{1}{\theta}$ and $\mathrm{Var}(X) = \frac{1}{\theta^2}$.

Example 4.18 Let $X_1, \ldots, X_n$ be a sample of iid $N(\mu, \sigma^2)$ random variables. Then, S^2 is an unbiased estimator of σ^2 and $\mathrm{Var}(S^2) = \frac{2\sigma^4}{n-1}$. Using $S = \sqrt{S^2}$ as an estimator σ seems reasonable, and in this case,

$$\mathrm{SE}(S) \approx \mathrm{SE}(S^2)\left|\frac{1}{2\sqrt{\sigma^2}}\right| = \sqrt{\frac{2\sigma^4}{n-1}} \times \frac{1}{2\sqrt{\sigma^2}} = \sqrt{\frac{\sigma^2}{2(n-1)}}.$$

Problems

4.1.1 Let $X_1, \ldots, X_n$ be a sample of iid $U(0, \theta)$ random variables, and let $T = 2\overline{X}$ be an estimator of θ. Determine
  a) $E(T)$.
  b) $\mathrm{Bias}(T; \theta)$.

c) MSE$(T; \theta)$.

d) whether T is an MSE-consistent estimator of θ.

4.1.2 Let $X_1, \ldots, X_n$ be a sample of iid Gamma$(4, \theta)$, and let $T = \frac{1}{4}\overline{X}$ be an estimator of θ. Determine

a) $E(T)$.

b) Bias$(T; \theta)$.

c) MSE$(T; \theta)$.

d) whether T is an MSE-consistent estimator of θ.

4.1.3 Let $X_1, \ldots, X_n$ be a sample of iid $N(0, \theta)$, and let $T = \frac{1}{n}\sum_{i-1}^{n} X_i^2$ be an estimator of θ. Determine

a) $E(T)$.

b) Bias$(T; \theta)$.

c) the sampling distribution of T.

d) MSE$(T; \theta)$.

e) whether T is an MSE-consistent estimator of θ.

4.1.4 Let $X_1, \ldots, X_n$ be a sample of iid $N(\mu, \theta)$, and let $S = \sqrt{S^2}$ be an estimator of $\sqrt{\theta}$. Using $\frac{(n-1)S^2}{\theta} \sim \chi^2_{n-1}$, determine

a) $E(S)$.

b) Bias$(S; \theta)$.

c) Bias$(S; \theta)$ when $n = 20$.

d) Var(S).

e) whether T is an MSE-consistent estimator of $\sqrt{\theta}$.

f) an unbiased estimator of σ based on S.

4.1.5 Let $X_1, \ldots, X_n$ be a sample of iid Exp$\left(\beta = \frac{1}{\theta}\right)$. Determine

a) the sampling distribution of $\sum_{i=1}^{n} X_i$.

b) $E(X_i)$.

c) $E(\overline{X})$.

d) $E\left(\frac{1}{\overline{X}}\right)$.

e) Bias$\left(\frac{1}{\overline{X}}; \theta\right)$.

f) an unbiased estimator U of θ based on $\frac{1}{\overline{X}}$.

g) whether U is an MSE-consistent estimator of θ.

4.1.6 Let $X_1, \ldots, X_n$ be a sample of iid random variables with common mean μ and variance σ^2, and let $T = \frac{1}{n}\sum_{i=1}^{n}(X_i - \overline{X})X_i$. Determine

a) $E(T)$.

b) Bias$(T; \sigma^2)$.

c) an estimator U based on T that is an unbiased estimator of σ^2.

4.1.7 Show that $s^2 = \frac{1}{n-1} \sum_{i=1}^{n} (x_i - \bar{x}) x_i$.

4.1.8 Let $X_1, \ldots, X_n$ be a sample of iid $N(\mu, \theta)$ random variables, and let $T_c = cS^2$. Determine
a) $E(T_c)$.
b) $MSE(T_c; \theta)$.
c) the estimator $T^\star = c^\star S^2$ that minimizes $MSE(T_c; \theta)$.
d) whether $T^\star$ is an MSE-consistent estimator for θ.
e) $RE(T^\star, S^2; \theta)$.
f) $ARE(T^\star, S^2; \theta)$.

4.1.9 Let $X_1, \ldots, X_n$ be a sample of iid random variables with common mean μ and variance σ^2. Let $\bar{X}^2$ be an estimator of μ^2. Determine
a) $E(\bar{X}^2)$.
b) $Bias(\bar{X}^2; \mu^2)$.
c) $E(\bar{X}^2 - \frac{S^2}{n})$.
d) $Bias(\bar{X}^2 - \frac{S^2}{n}; \mu^2)$.
e) the asymptotic bias of $\bar{X}^2$ for estimating μ^2.

4.1.10 Let $X_1, \ldots, X_n$ be a sample of iid $Bin(1, \theta)$ random variables, and let $T = \bar{X}(1 - \bar{X})$ be an estimator of $Var(X_i) = \theta(1 - \theta)$. Determine
a) $E(T)$.
b) $Bias(T; p(1 - p))$.
c) the asymptotic bias of T for estimating $p(1 - p)$.
d) an unbiased estimator of $p(1 - p)$ based on T.

4.1.11 Let $X_1, \ldots, X_n$ be a sample of iid random variables with common mean μ and variance σ^2. Let $T = \frac{2}{n(n+1)} \sum_{i=1}^{n} i X_i$ be an estimator of μ. Determine
a) $E(T)$.
b) $Var(T)$.
c) $Bias(T; \mu)$.
d) $MSE(T; \mu)$.
e) $RE(T, \bar{X}; \mu)$.
f) $ARE(T, \bar{X}; \mu)$.
g) whether T is an MSE-consistent estimator of μ.

4.1.12 Let $X_1, \ldots, X_n$ be a sample of iid $N(0, \theta)$ random variables, and let $T = \frac{1}{n} \sum_{i=1}^{n} X_i^2$. Determine
a) $E(T)$.
b) $Var(T)$.

c) $RE(T, \theta)$.

d) $ARE(T, S^2; \theta)$.

4.1.13 Prove Theorem 4.3.

4.1.14 Prove Theorem 4.8.

4.1.15 Let $X_1, \ldots, X_n$ be a random sample of iid $N(\mu, \sigma_1^2)$ random variables, $Y_1, \ldots, Y_m$ a random sample of iid $N(\mu, \sigma_2^2)$ random variables, and suppose that X_i and Y_j are independent random variables for $i = 1, \ldots, n$ and $j = 1, \ldots, m$. Let $T = \frac{n}{n+m}\overline{X} + \frac{m}{n+m}\overline{Y}$. Determine

a) $E(T)$.

b) $\text{Var}(T)$.

c) $\text{MSE}(T; \mu)$.

d) $\lim_{n,m \to \infty} \text{MSE}(T; \mu)$.

4.1.16 Let $X_1, \ldots, X_n$ be a random sample of iid $N(\mu, \sigma_1^2)$ random variables, $Y_1, Y_2, \ldots, Y_m$ a random sample of iid $N(\mu, \sigma_2^2)$ random variables, and suppose that X_i and Y_j are independent random variables for $i = 1, \ldots, n$ and $j = 1, \ldots, m$. Let $T_1 = \frac{n}{n+m}\overline{X} + \frac{m}{n+m}\overline{Y}$ and $T_2 = \frac{2}{3}\overline{X} + \frac{1}{3}\overline{Y}$. Determine

a) $\text{Var}(T_1)$. b) $\text{Var}(T_2)$.

c) $RE(T_1, T_2; \mu)$.

4.1.17 Let $X_1, \ldots, X_n$ be a sample of iid $\text{Exp}(\beta)$ random variables. Use the Delta Method to determine the approximate standard error of $\widehat{P}(X > 3) = e^{-\frac{3}{\overline{X}}}$.

4.1.18 Let $X_1, \ldots, X_n$ be a sample of iid $U(0, 2\alpha)$ random variables. Use the Delta Method to determine the approximate standard error of $\widehat{P}(X > 1) = 1 - \frac{1}{2\overline{X}}$.

4.1.19 Let $X_1, \ldots, X_n$ be a sample of iid $\text{Pois}(\lambda)$ random variables. Use the Delta Method to determine the approximate standard error of $\widehat{P}(X = 0) = e^{-\overline{X}}$.

4.1.20 Let $X_1, \ldots, X_n$ be a sample of iid $\text{Exp}(\beta)$ random variables. Use the Delta Method to determine the approximate standard error of $\widehat{\beta^2} = \overline{X}^2$.

4.2 Sufficient Statistics

In parametric estimation, the information about the value of the unknown parameter $\theta \in \Theta$ is contained in a random sample of size n selected from a common pdf $f(x; \theta)$. A statistic that contains all of the relevant information about θ contained in a sample is called a *sufficient statistic*. Fisher introduced the *sufficiency* property of a statistic in (1922) [2].

Throughout this section, the focus will be on a sample of n iid random variables with common probability model $f(x; \theta)$ for $\theta \in \Theta \subset \mathbb{R}^d$.

Definition 4.10 Let $X_1, \ldots, X_n$ be a sample of iid random variables with common pdf $f(x; \theta)$ for $\theta \in \Theta \subset \mathbb{R}^d$. A vector of statistics $\vec{S}(\vec{X}) = (S_1(\vec{X}), \ldots, S_k(\vec{X}))$ is said to be a k-dimensional sufficient statistic for a parameter θ if and only if the conditional distribution of $\vec{X} = (X_1, \ldots, X_n)$ given $S = s$ does not depend on θ for any value of s.

Clearly, the entire sample is always a sufficient statistic for θ since $P(\vec{X}|\vec{X}) = 1$ for all values of θ. Also, it is important to note that sufficient statistics are not unique. For example, if $X_1, \ldots, X_n$ is a random sample from $f(x; \theta)$, then $\bar{X}$ and $2\bar{X}$ are both sufficient for θ. In general, it is preferable to have a sufficient statistic that reduces a sample $X_1, \ldots, X_n$ to a k-dimensional statistic where $k < n$.

Definition 4.11 gives an alternate but equivalent definition of sufficiency.

Definition 4.11 Let $X_1, \ldots, X_n$ be a sample of iid random variables with common pdf $f(x; \theta)$ for $\theta \in \Theta \subset \mathbb{R}^d$. A vector of statistics $\vec{S}(\vec{X}) = (S_1(\vec{X}), \ldots, S_k(\vec{X}))$ is said to be a k-dimensional sufficient statistic for a parameter θ if and only if for any other vector of statistics T the conditional distribution of T given $S = s$ does not depend on θ for every value of s.

The *Principle of Sufficiency* states that when $S(\vec{X})$ is a sufficient statistic for θ, $\vec{x}$ and $\vec{y}$ are observed iid samples of size n from $f(x; \theta)$, and $S(\vec{x}) = S(\vec{y})$, then the information about θ contained in the two samples is the same.

Now, neither Definition 4.10 nor Definition 4.11 is very useful for determining sufficient statistics; however, Theorem 4.9, the *Fisher–Neyman factorization theorem* [2, 3], does provide a useful way of determining sufficient statistics. The Fisher–Neyman factorization theorem is based on the *likelihood function*, which is defined in Definition 4.12.

Definition 4.12 For a sample $X_1, \ldots, X_n$, the likelihood function $L(\theta|\vec{X})$ is the joint pdf of $\vec{X} = (X_1, \ldots, X_n)$. The log-likelihood function $\ell(\theta)$ is the logarithm of the likelihood function.

Note that the likelihood function $L(\theta|\vec{x})$ is a function of θ for a given sample $\vec{X}$. To simplify the notation, the likelihood function is generally denoted by $L(\theta)$; $L(\theta)$ is the notation used for the likelihood function throughout the remainder of this book. Also, the log-likelihood function contains all of the information about θ contained in the likelihood function since the logarithm is a one-to-one function.

Example 4.19 Let $X_1, \ldots, X_n$ be a sample of iid $Exp(\theta)$ random variables with $\Theta = \mathbb{R}^+$. Then, the likelihood function is

$$L(\theta) = \prod_{i=1}^{n} \frac{1}{\theta} e^{-\frac{x_i}{\theta}} = \theta^{-n} e^{-\frac{1}{\theta} \sum_{i=1}^{n} x_i},$$

and the log-likelihood function is

$$\ell(\theta) = \ln[L(\theta)] = -n \ln(\theta) - \frac{\sum_{i=1}^{n} x_i}{\theta}.$$

Example 4.20 Let $X_1, \ldots, X_n$ be a sample of iid $U(0, \theta)$ random variables with $\Theta = \mathbb{R}^+$. Then, the likelihood function is

$$L(\theta) = \prod_{i=1}^{n} \frac{1}{\theta} I_{(0,\theta)}(x_i) = \theta^{-n} \prod_{i=1}^{n} I_{(0,\theta)}(x_i).$$

Note that

$$\prod_{i=1}^{n} I_{(0,\theta)}(x_i) = \begin{cases} 1 & \text{if } x_i \in (0, \theta), \quad \forall i, \\ 0 & \text{otherwise,} \end{cases}$$

and hence,

$$L(\theta) = \begin{cases} \theta^{-n} & \text{if } x_i \in (0, \theta), \quad \forall i, \\ 0 & \text{otherwise.} \end{cases}$$

Thus, $L(\theta) = \theta^{-n} I_{[x_{(n)}, \infty)}(\theta)$.

Theorem 4.9 shows that a sufficient statistic can be found by factoring the likelihood function into two functions, namely, a function of the sufficient statistic and θ and a function of $\vec{x}$ alone.

Theorem 4.9 (Neyman–Fisher Factorization Theorem) *Let* $X_1, \ldots, X_n$ *be a sample of iid random variables with pdf* $f(x; \theta)$ *and parameter space* Θ. *A statistic* $S(\vec{X})$ *is sufficient for* θ *if and only if* $L(\theta)$ *can be factored as*

$$L(\theta) = g(S(\vec{x}); \theta) h(\vec{x}),$$

where $g(S(\vec{x}); \theta)$ *does not depend on* $x_1, \ldots, x_n$ *except through* $S(\vec{x})$, *and* $h(\vec{x})$ *does not involve* θ.

Examples 4.21 and 4.22 use the Neyman–Fisher Factorization Theorem to find sufficient statistics when $\Theta \subset \mathbb{R}$, and Example 4.23 illustrates how to use the Neyman–Fisher factorization theorem to find a vector of statistics that are jointly sufficient for θ when $\Theta \subset \mathbb{R}^2$.

Example 4.21 Let $X_1, \ldots, X_n$ be a sample of iid $\text{Bin}(1, \theta)$ random variables with $\Theta = [0, 1]$. The likelihood function is

$$L(\theta) = f(x_1, \ldots, x_n; \theta) = \prod_{i=1}^{n} \theta^{x_i}(1-\theta)^{1-x_i} = \theta^{\sum_{i=1}^{n} x_i}(1-\theta)^{n-\sum_{i=1}^{n} x_i},$$

which can be factored as

$$L(\theta) = \left(\frac{\theta}{1-\theta} \right)^{\sum_{i=1}^{n} x_i} (1-\theta)^n \times 1 = g(S(\vec{x}); \theta)h(\vec{x}),$$

where $g(S(\vec{x}); \theta) = \left(\frac{\theta}{1-\theta} \right)^{\sum_{i=1}^{n} x_i} (1-\theta)^n$ and $h(\vec{x}) = 1$. Thus, by the Fisher–Neyman factorization theorem, $S(\vec{x}) = \sum_{i=1}^{n} x_i$ is a sufficient statistic for θ.

Example 4.22 Suppose that $X_1, \ldots, X_n$ are iid $\text{Pois}(\theta)$ random variables with $\Theta = \mathbb{R}^+$. The likelihood function is

$$L(\theta) = f(x_1, \ldots, x_n; \theta) = \prod_{i=1}^{n} \frac{e^{-\theta}\theta^{x_i}}{x_i!} = \frac{e^{-n\theta}\theta^{\sum_{i=1}^{n} x_i}}{\prod_{i=1}^{n} x_i!},$$

which can be factored as

$$L(\theta) = \underbrace{e^{-n\theta}\theta^{\sum_{i=1}^{n} x_i}}_{g(S(\vec{x});\theta)} \times \underbrace{\frac{1}{\prod_{i=1}^{n} x_i!}}_{h(\vec{x})},$$

where $S(\vec{x}) = \sum_{i=1}^{n} x_i$. Thus, by the Fisher–Neyman factorization theorem, $S(\vec{X}) = \sum_{i=1}^{n} X_i$ is a sufficient statistic for θ.

Example 4.23 Let $X_1, \ldots, X_n$ be iid $N(\theta_1, \theta_2)$ random variables with $\Theta = \mathbb{R} \times \mathbb{R}^+$. Then, the likelihood function is

$$L(\theta) = f(x_1, \ldots, x_n; \theta_1, \theta_2) = \prod_{i=1}^{n} \frac{1}{\sqrt{2\pi\theta_2}} e^{-\frac{1}{2\theta_2}(x_i - \theta_1)^2}$$

$$= (2\pi\theta_2)^{-\frac{n}{2}} e^{-\frac{1}{2\theta_2}\sum_{i=1}^{n}(x_i - \theta_1)^2} = (2\pi\theta_2)^{-\frac{n}{2}} e^{-\frac{1}{2\theta_2}[\sum_{i=1}^{n} x_i^2 - 2\theta_1 \sum_{i=1}^{n} x_i + \theta_1^2]}$$

$$= \underbrace{(\theta_2)^{-\frac{n}{2}} e^{-\frac{1}{2\theta_2}[\sum_{i=1}^{n} x_i^2 - 2\theta_1 \sum_{i=1}^{n} x_i + n\theta_1^2]}}_{g(S_1(\vec{x}), S_2(\vec{x}); \theta_1, \theta_2)} \times \underbrace{(2\pi)^{-\frac{n}{2}}}_{h(\vec{x})}.$$

Thus, by the Fisher–Neyman factorization theorem, $S = (\sum X_i, \sum X_i^2)$ is a statistic that is jointly sufficient for $\vec{\theta} = (\theta_1, \theta_2)$.

Recall that sufficient statistics are not unique, and Theorem 4.10 shows that a one-to-one transformation of a sufficient statistic is also a sufficient statistic.

Theorem 4.10 *If $X_1, \ldots, X_n$ is a sample of random variables, $S = (S_1, \ldots, S_k)$ a sufficient statistic for θ, and $\phi : \mathbb{R}^d \to \mathbb{R}^d$ is a one-to-one function, then $\phi(S)$ is a sufficient statistic for θ.*

Proof. Let $X_1, \ldots, X_n$ be a sample of random variables, $\phi : \mathbb{R}^d \to \mathbb{R}^d$ be a one-to-one function, and S a sufficient statistic for θ. Then, $S = \phi^{-1}(\phi(S))$, and since S is sufficient for θ, it follows that

$$L(\theta) = g(S; \theta)h(\vec{x}) = g(\phi^{-1}(\phi(S)); \theta)h(\vec{x}) = g^{\star}(\phi(S); \theta)h(\vec{x}),$$

where $g^{\star}(\phi(S); \theta)$ depends on $\vec{x}$ only through S. Therefore, $\phi(S)$ is sufficient for θ by the Fisher–Neyman factorization theorem. ∎

Note that Theorem 4.10 shows that when S is sufficient for θ, then so is cS for $c \neq 0$. In particular, $\overline{X}$ is sufficient for θ whenever $\sum_{i=1}^{n} X_i$ is sufficient for θ.

Example 4.24 Let $X_1, \ldots, X_n$ be a sample of iid $N(\theta_1, \theta_2)$ random variables with $\Theta = \mathbb{R} \times \mathbb{R}^+$. In Example 4.23, $S = (\sum_{i=1}^{n} X_i, \sum_{i=1}^{n} X_i^2)$ was shown to be jointly sufficient for $\vec{\theta} = (\theta_1, \theta_2)$. Let

$$S' = \phi\left(\sum_{i=1}^{n} X_i, \sum_{i=1}^{n} X_i^2 \right) = \left(\frac{1}{n} \sum_{i=1}^{n} X_i, \frac{1}{n} \left[\sum_{i=1}^{n} X_i^2 - n\overline{X}^2 \right] \right) = (\overline{X}, S_n^2).$$

Then, ϕ is a one-to-one function, and by Theorem 4.10, $S' = (\overline{X}, S_n^2)$ is jointly sufficient for $\vec{\theta}$.

Because sufficient statistics are not unique and generally reduce the information in the sample to a statistic of smaller dimension than the sample, it is preferable to use the sufficient statistic that produces the maximal reduction of sample information. The sufficient statistic that reduces the information in the sample to a statistic of the smallest dimension is called a *minimal sufficient statistic*.

Definition 4.13 A sufficient statistic S is a minimal sufficient statistic if and only if S is a sufficient statistic for θ, and when T is any other sufficient statistic for θ, then there exists a function g such that $S = g(T)$.

A method for finding minimal sufficient statistics is due Lehmann and Scheffé [4]. The method proposed by Lehmann and Scheffé is based on the ratio of the likelihood function for two samples of size n, say $X_1, \ldots, X_n$ and $Y_1, \ldots, Y_n$. In particular, Lehmann and Scheffé's method shows that when S is a sufficient statistic for θ and $\frac{L(\theta|\vec{x})}{L(\theta|\vec{y})}$ is independent of θ only when $S(\vec{x}) = S(\vec{y})$, then S is a minimal sufficient statistic for θ.

Example 4.25 Let $X_1, \ldots, X_n$ be a sample of iid Pois(θ) random variables with $\Theta = \mathbb{R}^+$. Then,

$$L(\theta) = \prod_{i=1}^{n} \frac{e^{-\theta}\theta^{x_i}}{x_i!} = \frac{e^{-n\theta}\theta^{\sum x_i}}{\prod x_i!}.$$

Now, let $Y_1, Y_2, \ldots, Y_n$ be another iid sample of Pois(θ) random variables. Then,

$$\frac{L(\theta|\vec{x})}{L(\theta|\vec{y})} = \frac{\frac{e^{-n\theta}\theta^{\sum x_i}}{\prod x_i!}}{\frac{e^{-n\theta}\theta^{\sum y_i}}{\prod y_i!}} = \frac{\prod x_i!}{\prod y_i!}\theta^{\sum x_i - \sum y_i},$$

which is independent of θ if and only if $\sum_{i=1}^{n} x_i = \sum_{i=1}^{n} y_i$. Hence, $\sum_{i=1}^{n} X_i$ is a minimal sufficient statistic for θ.

Finally, Rao [5] and Blackwell [6] proved similar results showing that only estimators based on sufficient statistics should be used for estimating an unknown parameter. The Rao–Blackwell theorem, Theorem 4.11, shows that the variability of any unbiased estimator can be reduced by conditioning on a sufficient statistic, provided the estimator is not already a function of the sufficient statistic.

Theorem 4.11 (Rao–Blackwell Theorem) *Let $X_1, \ldots, X_n$ be iid random variables with pdf $f(x; \theta)$ for $\theta \in \Theta$, S a sufficient statistic for θ, and T an unbiased estimator of $\tau(\theta)$ which is not a function of S alone. If $T^{\star} = E(T|S)$, then*

i) $T^{\star}$ is a function of the sufficient statistic S alone.
ii) $E(T^{\star}) = \tau(\theta)$.
iii) $\mathrm{Var}(T^{\star}) \leq \mathrm{Var}(T), \forall \theta \in \Theta$.

Proof. Let $X_1, \ldots, X_n$ be iid random variables with pdf $f(\cdot; \theta)$ for $\theta \in \Theta$, S a sufficient statistic for θ, and T an unbiased estimator of $\tau(\theta)$ which is not a function of S alone. Let $T^{\star} = E(T|S)$.

i) Since S is a sufficient statistic, the distribution of $T|S$ is free of θ, and therefore, $T^{\star} = E(T|S)$ is a function of S alone.
ii) $E(T^{\star}) = E[E(T|S)] = E(T) = \tau(\theta)$ by Theorem 2.36.
iii) By Theorem 2.36,

$$\mathrm{Var}(T) = \underbrace{E[\mathrm{Var}(T|S)]}_{\geq 0} + \mathrm{Var}[E(T|S)] \geq \mathrm{Var}[E(T|S)] = \mathrm{Var}(T^{\star}).$$

∎

The Rao–Blackwell theorem shows any estimator of θ can be improved by conditioning on a sufficient statistic, and therefore, only estimators of θ based entirely on a sufficient statistic should be used for estimating θ. For example, $\overline{X}$ is an unbiased estimator of μ, and therefore, when S is a sufficient statistic for θ, $T^\star = E(\overline{X}|S)$ is a more accurate estimator of μ than $\overline{X}$; however, if $S = \sum_{i=1}^{n} X_i$, then $\overline{X}$ is already a function of S, and in this case, $T^\star = \overline{X}$.

The estimator $T^\star$ created in the Rao–Blackwell theorem by conditioning an unbiased statistic T on a sufficient statistic S is called the *Rao–Blackwellization* of the estimator T. An example showing the use of the Rao–Blackwell theorem to create a more accurate unbiased estimator is given in Example 4.26.

Example 4.26 Let $X_1, \ldots, X_n$ be iid Pois(θ) random variables. Then, X_1 is an unbiased estimator of θ and from Example 4.22, $S = \sum_{i=1}^{n} X_i$ is sufficient for θ. By Theorem 3.15, the sampling distribution of $S = \sum_{i=1}^{n} X_i \sim$ Pois($n\theta$) and the sampling distribution of $S - X_1 = \sum_{i=2}^{n} X_i \sim$ Pois($(n-1)\theta$).

Let $s \in W$ and consider $P(X_1 = x_1 | S = s)$.

$$P(X_1 = x_1 | S = s) = \frac{P(X_1 = x_1 \text{ and } S = s)}{P(S = s)}$$

$$= \frac{P(X_1 = x_1 \text{ and } \sum_{i=2}^{n} X_i = s - x_1)}{\frac{e^{-n\theta}(n\theta)^s}{s!}}.$$

Now, since $X_1, \ldots, X_n$ are independent random variables it follows that X_1 and $\sum_{i=2}^{n} X_i$ are independent. Thus,

$$P(X_1 = x_1 | S = s) = \frac{P(X_1 = x_1)P(\sum_{i=2}^{n} X_i = s - x_1)}{\frac{e^{-n\theta}(n\theta)^s}{s!}}$$

$$= \frac{\frac{e^{-\theta}\theta^{x_1}}{x_1!} \times \frac{e^{-(n-1)\theta}((n-1)\theta)^{s-x_1}}{(s-x_1)!}}{\frac{e^{-n\theta}(n\theta)^s}{s!}} = \binom{s}{x_1}\left(\frac{1}{n}\right)^{x_1}\left(\frac{n-1}{n}\right)^{s-x_1}.$$

Hence, $X_1|S \sim$ Bin $\left(S, \frac{1}{n}\right)$, and therefore, $T^\star = E(X_1|S) = \frac{S}{n} = \overline{X}$. Therefore, $\overline{X}$ is an unbiased estimator of θ with smaller variance than X_1. In particular, Var(X_1) = θ and Var($\overline{X}$) = $\frac{\theta}{n}$.

4.2.1 Exponential Family Distributions

A flexible family of probability distributions that contains many of the commonly used probability models and is very important in the theory of statistics and statistical modeling is the *exponential family* of distributions. Probability models in the exponential family can be written in the same general form and have special properties that are important in point estimation, hypothesis testing and building statistical models. The exponential family of distributions was introduced independently by Darmois [7], Pitman and Wishart [8], and Koopman [9] in 1935 and 1936.

Definition 4.14 A random variable X with pdf $f(x; \theta)$, support S, and parameter space $\Theta \subset \mathbb{R}^d$ is said to be in the d-parameter exponential family of distributions when $f(x; \theta)$ can be written as

$$f(x; \theta) = c(\theta)h(x)e^{\sum_{j=1}^{k} q_j(\theta)t_j(x)}, \qquad x \in S$$

for some real valued functions $c(\theta)$, $h(x)$, $q_j(\theta)$, and $t_j(x)$ and provided S does not depend on θ. If Θ contains an open rectangle in $\mathbb{R}^d$, then the pdf of X is said to be in the regular exponential family of distributions.

When X is a random variable with exponential family pdf $f(x; \theta)$ with $\Theta \subset \mathbb{R}$, $f(x; \theta)$ is said to be in the *one-parameter exponential family*. Examples of one-parameter exponential family distributions are given in Examples 4.27 and 4.28. An example of a two-parameter exponential family distribution is given in Example 4.29.

Example 4.27 Let $X \sim \text{Bin}(n, \theta)$ with $\Theta = [0, 1]$. Then, X is in the one-parameter exponential family of distributions since $S = \{0, 1, \ldots, n\}$ does not depend on θ and

$$f(x; \theta) = \binom{n}{x} \theta^x (1 - \theta)^{n-x} = \theta^n \binom{n}{x} e^{\ln\left(\frac{\theta}{1-\theta}\right) \cdot x}$$
$$= c(\theta)h(x)e^{q(\theta)t(x)},$$

where $c(\theta) = \theta^n$, $h(x) = \binom{n}{x}$, $q(\theta) = \ln\left(\frac{\theta}{1-\theta}\right)$, and $t(x) = x$. Also, because $\Theta = [0, 1]$ contains an open rectangle in $\mathbb{R}$, it follows that $f(x; \theta)$ is a regular exponential family distribution.

Example 4.28 Let $X \sim \text{Exp}(\theta)$ with $\Theta = \mathbb{R}^+$. Then, X in the one-parameter exponential family of distributions since $S = (0, \infty)$ does not depend on θ and

$$f(x; \theta) = \frac{1}{\theta}e^{-\frac{x}{\theta}} = \frac{1}{\theta} \cdot 1 \cdot e^{-\frac{1}{\theta} \cdot x}$$
$$= c(\theta)h(x)e^{q(\theta)t(x)},$$

where $c(\theta) = \frac{1}{\theta}$, $h(x) = 1$, $q(\theta) = -\frac{1}{\theta}$, and $t(x) = x$. Also, because $\Theta = \mathbb{R}^+$ is an open rectangle in $\mathbb{R}$ it follows that $f(x; \theta)$ is a regular exponential family distribution.

Note that the choice of the functions $c(\theta)$, $h(x)$, $q(\theta)$, and $t(x)$ is not unique. For example, in Example 4.28, $q(\theta)$ was chosen to be $-\frac{1}{\theta}$, which made $t(x) = x$; however, the alternative choice of $q'(\theta) = \frac{1}{\theta}$ and $t(x) = -x$ can also be used to show that $f(x; \theta)$ is an exponential family distribution.

Example 4.29 Let $X \sim N(\theta_1, \theta_2)$ with $\Theta = \mathbb{R} \times \mathbb{R}^+$. Then, X is in the two-parameter exponential family of distribution since S does not depend on θ and

$$f(x; \theta_1, \theta_2) = \frac{1}{\sqrt{2\pi\theta_2}} e^{-\frac{1}{2\theta_2}(x-\theta_1)^2} = \frac{1}{\sqrt{\theta_2}} e^{-\frac{\theta_1^2}{2\theta_2}} \cdot \frac{1}{\sqrt{2\pi}} \cdot e^{\frac{\theta_1}{\theta_2} \cdot x - \frac{1}{2\theta_2} \cdot x^2}$$

$$= c(\theta_1, \theta_2) h(x) e^{q_1(\theta_1, \theta_2) t_1(x) + q_2(\theta_1, \theta_2) t_2(x)},$$

where

$$c(\theta_1, \theta_2) = \frac{1}{\sqrt{\theta_2}} e^{-\frac{\theta_1^2}{2\theta_2}}, \quad h(x) = \frac{1}{\sqrt{2\pi}}, \quad q_1(\theta_1, \theta_2) = \frac{\theta_1}{\theta_2},$$

$$q_2(\theta_1, \theta_2) = -\frac{1}{2\theta_2}$$

and $t_1(x) = x$ and $t_2(x) = x^2$. Since $\Theta = \mathbb{R} \times \mathbb{R}^+$ is an open rectangle in $\mathbb{R}$, $f(x; \theta)$ is a regular exponential family of distribution.

A d-parameter exponential family is said to be *curved* when the dimension of Θ is less than the dimension of the vector $\vec{\theta}$. For example, if $X \sim N(\theta, \theta^2)$, then $f(x; \theta)$ is a curved exponential family distribution since $\dim(\Theta) = 1$ and $\dim(\theta) = 2$. Most of the exponential family probability models used in the parametric estimation are not in the curved exponential family. Moreover, when $f(x; \theta)$ is a regular exponential family distribution, it cannot be a curved exponential family distribution.

Theorem 4.12 *If X is a random variable with pdf $f(x; \theta)$ in the one-parameter exponential family of distributions and $X_1, \ldots, X_n$ is a sample of iid random variables with pdf $f(x; \theta)$, then likelihood function is*

$$L(\theta) = c(\theta)^n \prod_{i=1}^{n} h(x_i) e^{q(\theta) \sum_{i=1}^{n} t(x_i)}.$$

Proof. The proof of Theorem 4.12 is left as an exercise. ∎

Theorem 4.13 *If X is a random with pdf $f(x; \theta)$ in the d-parameter exponential family of distributions, then $\vec{T} = \left(\sum_{i=1}^{n} t_1(X_i), \ldots, \sum_{i=1}^{n} t_k(X_i) \right)$ is a sufficient statistic for θ.*

Proof. The proof of Theorem 4.13 is left as an exercise. ∎

Example 4.30 Let $X_1, \ldots, X_n$ be a sample of iid Bin(n, p) random variables. In Example 4.27, $f(x; \theta)$ was shown to be an exponential family distribution with $t(x) = x$. Thus, by Theorem 4.13, it follows that $T = \sum_{i=1}^{n} X_i$ is a sufficient statistic for θ.

Example 4.31 Let $X_1, \ldots, X_n$ be a sample of iid $N(\theta_1, \theta_2)$ random variables. In Example 4.29, $f(x; \theta_1, \theta_2)$ was shown to be an exponential family distribution with $t_1(x) = x$ and $t_2(x) = x^2$. Thus, by Theorem 4.13, it follows that $\vec{T} = (\sum_{i=1}^{n} X_i, \sum_{i=1}^{n} X_i^2)$ is jointly sufficient for $\vec{\theta} = (\theta_1, \theta_2)$.

Definition 4.15 If $f(x; \theta)$ is a one-parameter exponential family distribution and $q(\theta) = \theta$, then the exponential family is said to be in canonical form.

Theorem 4.14 *If X is a random variable with pdf $f(x; \theta)$ in the one-parameter exponential family of distributions in canonical form, then $E[t(X)] = -\frac{c'(\theta)}{c(\theta)}$.*

Proof. Let X be a random variable with pdf $f(x; \theta)$ in the one-parameter exponential family of distributions in canonical form. First,

$$\int_S c(\theta) h(x) e^{\theta t(x)}\, dx = 1$$

so that

$$\frac{1}{c(\theta)} = \int_S h(x) e^{\theta t(x)}\, dx.$$

Now,

$$\frac{\partial}{\partial \theta} \left[\frac{1}{c(\theta)} \right] = -\frac{c'(\theta)}{c(\theta)^2} = \frac{\partial}{\partial \theta} \int_S h(x) e^{\theta t(x)}\, dx$$

$$= \int_S \frac{\partial}{\partial \theta} [h(x) e^{\theta t(x)}]\, dx$$

$$= \int_S t(x) h(x) e^{\theta t(x)}\, dx = \frac{E[t(X)]}{c(\theta)}.$$

Hence, $E[t(X)] = c(\theta) \times -\frac{c'(\theta)}{c(\theta)^2} = -\frac{c'(\theta)}{c(\theta)}$. ∎

Example 4.32 Let $X \sim \text{Exp}\left(\beta = \frac{1}{\theta}\right)$. Then, the pdf of X is in the exponential family and has canonical form since $f(x; \theta) = \theta e^{-\theta x} = \theta \cdot 1 \cdot e^{\theta(-x)}$. Thus, $c(\theta) = \theta$ and by Theorem 4.14, $E(-X) = -\frac{1}{\theta}$.

Moreover, if $X_1, \ldots, X_n$ is a sample of iid $\text{Exp}\left(\beta = \frac{1}{\theta}\right)$ random variables, then $T = \sum_{i=1}^{n} (-X_i)$ is sufficient for θ and $E(T) = -\frac{n}{\theta}$.

Probability models in the regular exponential family of distributions also have an important property known as *completeness*, an important property in determining the unbiased estimator having the smallest variance.

Definition 4.16 Let $S = S(\vec{X})$ be a statistic with family of sampling distributions $\mathcal{F}_S = \{f_S(s;\theta) : \theta \in \Theta\}$. The family of distributions $\mathcal{F}_S$ is said to be a complete family of distributions when $E[h(S)] = 0$ for all $\theta \in \Theta$ if and only if for all $\theta \in \Theta$, $P(h(S) = 0) = 1$. A statistic whose family of sampling distributions is complete is called a complete statistic.

A sufficient statistic that is complete is called a *complete sufficient statistic*. Also, if S is a complete sufficient statistic then so is $g(S)$ when $g(\cdot)$ is a one-to-one function.

Example 4.33 Let $X_1, \ldots, X_n$ be a sample of iid Pois(θ) random variables with $\Theta = (0, \infty)$. In Example 4.22, $S = \sum_{i=1}^{n} X_i$ was shown to be a sufficient statistic for θ. Let $h(s)$ be any function with $E[h(S)] = 0$ for all $\theta \in (0, \infty)$. Then, $S \sim$ Pois($n\theta$) and

$$0 = E[h(S)] = \sum_{s=0}^{\infty} h(s) \times \frac{e^{-n\theta}(n\theta)^s}{s!} = e^{-n\theta} \sum_{s=0}^{\infty} \frac{h(s)n^s}{s!}\theta^s$$

$$= e^{-n\theta} \sum_{s=0}^{\infty} a(s)\theta^s,$$

where $a(s) = \frac{h(s)n^s}{s!}$. Since $\sum_{s=0}^{\infty} a(s)\theta^s$ is a power series equal to 0 for all $\theta \in (0, \infty)$, it follows that all of its coefficients are equal to zero. Thus, $a(s) = \frac{h(s)n^s}{s!} = 0$ for all $s \in \mathbb{W}$, and hence, $h(s) = 0$ for all $s \in \mathbb{W}$. Therefore, $S = \sum_{i=1}^{n} X_i$ is a complete sufficient statistic for θ.

In general, it can be difficult to show that a statistic is a complete statistic using Definition 4.16; however, when a iid sample is selected from a regular exponential family distribution, Theorem 4.15 shows that the statistic $\vec{T} = \left(\sum_{i=1}^{n} t_1(x_i), \ldots, \sum_{i=1}^{n} t_k(x_i) \right)$ is a complete sufficient statistic.

Theorem 4.15 *If $X_1, \ldots, X_n$ are iid random variables with pdf $f(x;\theta)$ in the regular exponential family of distributions and $f(x;\theta) = c(\theta)h(x)e^{\sum_{q=i}^{k} q_j(\theta)t_j(x)}$ for $x \in S$, then $T(X_1, \ldots, X_n) = \left(\sum_{i=1}^{n} t_1(X_i), \ldots, \sum_{i=1}^{n} t_k(X_i) \right)$ is a complete minimal sufficient statistic for θ.*

Example 4.34 Let $X_1, \ldots, X_n$ be a sample of iid Bin(n, θ) random variables with $\Theta = [0, 1]$. In Example 4.27, $X \sim$ Bin(n, θ) was shown to be in the exponential family of distributions, and since $\Theta = [0, 1]$ contains an open rectangle in $\mathbb{R}$, it follows that $X \sim$ Bin(n, θ) is in the regular exponential family of distributions. The statistic $\sum_{i=1}^{n} X_i$ was shown to be a sufficient statistic for θ in Example 4.30, and therefore, by Theorem 4.15, $\sum_{i=1}^{n} X_i$ is a complete minimal sufficient statistic for θ.

Example 4.35 Let $X_1, \ldots, X_n$ be a sample of iid $\mathrm{Exp}(\theta)$ random variables with $\Theta = (0, \infty)$. In Example 4.28, $X \sim \mathrm{Exp}(\theta)$ was shown to be in the exponential family of distributions with $t(x) = x$. Thus, $\sum_{i=1}^{n} X_i$ is a sufficient statistic for θ, and since $\Theta = (0, \infty)$ is an open rectangle in $\mathbb{R}$, it follows that $\sum_{i=1}^{n} X_i$ is a complete minimal sufficient statistic for θ.

Example 4.36 Let $X_1, \ldots, X_n$ be a sample of iid $N(\theta_1, \theta_2)$ random variables with $\Theta = \mathbb{R} \times \mathbb{R}^+$. In Example 4.29, $X \sim N(\theta_1, \theta_2)$ is shown to be an exponential family distribution with $t_1(x) = x$ and $t_2(x) = x^2$. Since $\Theta = \mathbb{R} \times \mathbb{R}^+$ is an open rectangle in $\mathbb{R}^2$, it follows that $S = (\sum_{i=1}^{n} X_i, \sum_{i=1}^{n} X_i^2)$ is a complete minimal sufficient statistic for $\theta = (\theta_1, \theta_2)$.

Now, $g(S_1, S_2) = \left(\frac{S_1}{n}, \frac{S_2}{n-1} - \frac{n}{n-1} S_1^2 \right)$ is a one-to-one function of S, and therefore, by Theorem 4.10 with it follows $S^\star = (\overline{X}, S^2)$ is also a complete minimal sufficient statistic for (θ_1, θ_2).

Problems

4.2.1 Let $X_1, \ldots, X_n$ be a sample of iid $\mathrm{Geo}(\theta)$ random variables with $\mathbb{S} = \mathbb{W}$ and $\theta \in [0, 1]$.
  a) Determine the likelihood function $L(\theta)$.
  b) Use the Fisher–Neyman factorization theorem to determine a sufficient statistic S for θ.

4.2.2 Let $X_1, \ldots, X_n$ be a sample of iid $\mathrm{Gamma}(\theta, 1)$ random variables with $\theta \in (0, \infty)$.
  a) Determine the likelihood function $L(\theta)$.
  b) Use the Fisher–Neyman factorization theorem to determine a sufficient statistic S for θ.

4.2.3 Let $X_1, \ldots, X_n$ be a sample of iid $\mathrm{Gamma}(4, \theta)$ random variables with $\theta \in (0, \infty)$.
  a) Determine the likelihood function $L(\theta)$.
  b) Use the Fisher–Neyman factorization theorem to determine a sufficient statistic S for θ.

4.2.4 Let $X_1, \ldots, X_n$ be a sample of iid $\mathrm{Beta}(4, \theta)$ random variables with $\theta \in (0, \infty)$.
  a) Determine the likelihood function $L(\theta)$.
  b) Use the Fisher–Neyman factorization theorem to determine a sufficient statistic S for θ.

4.2.5 Let $X_1, \ldots, X_n$ be a sample of iid $N(0, \theta)$ random variables with $\theta \in (0, \infty)$. Determine
  a) a sufficient statistic S for θ.
  b) $E(S)$.
  c) an unbiased estimator of θ that is a function of S.

4.2.6 Let $X_1, \ldots, X_n$ be a sample of iid $N(\theta, 2)$ random variables with $\theta \in \mathbb{R}$. Determine
  a) a sufficient statistic S for θ.
  b) $E(S)$.
  c) an unbiased estimator of θ that is a function of S.

4.2.7 Let $X_1, \ldots, X_n$ be a sample of iid $N(\theta_1, \theta_2)$ random variables with $\theta \in \mathbb{R} \times \mathbb{R}^+$.
  a) Determine the likelihood function $L(\theta_1, \theta_2)$.
  b) Use the Fisher–Neyman factorization theorem to determine a sufficient statistic S for (θ_1, θ_2).

4.2.8 Let $X_1, \ldots, X_n$ be a sample of iid $\text{Beta}(\theta_1, \theta_2)$ random variables with $\theta \in \mathbb{R}^+ \times \mathbb{R}^+$. Use the Fisher–Neyman factorization theorem to determine a sufficient statistic S for $\vec{\theta}$.

4.2.9 Let $X_1, \ldots, X_n$ be a sample of iid $N(\theta, \theta^2)$ random variables with $\theta \in \mathbb{R}^+ \times \mathbb{R}^+$. Use the Fisher–Neyman factorization theorem to determine a sufficient statistic S for $\vec{\theta}$.

4.2.10 Let $X_1, \ldots, X_n$ be a sample of iid random variables with pdf $f(x) = \theta x^{\theta-1}$ for $x \in (0, 1)$ and $\theta \in (0, \infty)$. Use the Fisher–Neyman factorization theorem to determine a statistic for θ.

4.2.11 Let $X_1, \ldots, X_n$ be a sample of iid random variables with $f(x) = \frac{1}{2}\theta e^{-\theta|x|}$ for $x \in \mathbb{R}$ and $\theta \in (0, \infty)$. Use the Fisher–Neyman factorization theorem to determine a sufficient statistic for θ.

4.2.12 Let $X_1, \ldots, X_n$ be a sample of $\text{Bin}(1, \theta)$ random variables with $\theta \in (0, 1)$. Using the Lehmann–Scheffé method, determine a minimal sufficient statistic S for θ.

4.2.13 Let $X_1, \ldots, X_n$ be a sample of $\text{Gamma}(\theta, 1)$ random variables with $\theta \in (0, \infty)$. Using the Lehmann–Scheffé method, determine a minimal sufficient statistic S for θ.

4.2.14 Let $X_1, \ldots, X_n$ be a sample of $N(\theta, 1)$ random variables with $\theta \in \mathbb{R}$. Using the Lehmann–Scheffé method, determine a minimal sufficient statistic S for θ.

4.2.15 Let $X \sim \text{Pois}(\theta)$ with $\Theta = (0, \infty)$.
a) Show that pdf of the random variable X is in the one-parameter regular exponential family of distributions.
b) If $X_1, \ldots, X_n$ is a sample of iid $\text{Pois}(\theta)$ random variables with $\Theta = (0, \infty)$, determine a complete minimal sufficient statistic for θ.

4.2.16 Let $X \sim \text{Geo}(\theta)$ with $\Theta = [0, 1]$.
a) Show that pdf of the random variable X is in the one-parameter regular exponential family of distributions.
b) If $X_1, \ldots, X_n$ is a sample of iid $\text{Geo}(\theta)$ random variables with $\Theta = (0, 1)$, determine a complete minimal sufficient statistic for θ.

4.2.17 Let $X_1, \ldots, X_n$ be a sample of random variables with common pdf $f(x) = \theta x^{\theta-1}$ for $x \in (0, 1)$ and $\Theta = (0, \infty)$.
a) Show that $f(x; \theta)$ is in the regular exponential family of distributions.
b) Let $X_1, \ldots, X_n$ be a sample of iid random variables with pdf $f(x; \theta)$. Determine a complete minimal sufficient statistic for θ.

4.2.18 Let $X \sim N(\theta, 1)$ with $\Theta = (0, \infty)$.
a) Show that pdf of the random variable X is in the one-parameter regular exponential family of distributions.
b) If $X_1, \ldots, X_n$ is a sample of iid $N(0, \theta)$ random variables with $\Theta = (0, \infty)$, determine a complete minimal sufficient statistic for θ.

4.2.19 Let $X_1, \ldots, X_n$ be a sample of $\text{Beta}(\theta_1, \theta_2)$ random variables with $\Theta = \mathbb{R}^+ \times \mathbb{R}^+$.
a) Show that pdf of the random variable X is in the two-parameter regular exponential family of distributions.
b) If $X_1, \ldots, X_n$ is a sample of iid $\text{Beta}(\theta_1, \theta_2)$ random variables with $\Theta = \mathbb{R}^+ \times \mathbb{R}^+$, determine a complete minimal sufficient statistic for $\vec{\theta}$.

4.2.20 Let $X_1, \ldots, X_n$ be a sample of $N(\theta, \theta^2)$ random variables with $\Theta = \mathbb{R} \times \mathbb{R}^+$.

a) Show that pdf of the random variable X is a curved two-parameter regular exponential family of distributions.
b) Determine a sufficient statistic T for θ.
c) Determine a nontrivial statistic $h(T)$ with $E[h(T)] = 0$, $\forall \theta \in \Theta$.

4.2.21 Let $X \sim \text{Gamma}(\theta_1, \theta_2)$ random variables with $\Theta = \mathbb{R}^+ \times \mathbb{R}^+$.
a) Show that the pdf of X with $\Theta = \mathbb{R}^+ \times \mathbb{R}^+$ is in the regular exponential family of distributions.
b) If $X_1, \dots, X_n$ is a sample of iid $\text{Gamma}(\theta_1, \theta_2)$ random variables with $\Theta = \mathbb{R}^+ \times \mathbb{R}^+$, determine a complete minimal sufficient statistic for $\vec{\theta}$.

4.2.22 Let $\vec{X} = (X_1, X_2, X_3, X_4) \sim \text{MulNom}_4(n, \vec{\theta})$ with $\Theta = [0, 1]^4$.
a) Show that $f(x; \theta)$ is in the four-parameter exponential family of distributions.
b) Let $X_1, \dots, X_n$ be a sample of a sample of $\text{MulNom}_4(n, \vec{\theta})$ random variables with $\Theta = (0, 1) \times (0, 1) \times (0, 1) \times (0, 1)$. Determine a complete minimal sufficient statistic for θ.

4.2.23 Let $X \sim U(0, \theta)$ with $\Theta = (0, \infty)$. Show that $f(x; \theta)$ is not in the exponential family of distributions.

4.2.24 Let $X \sim \text{Cauchy}(\theta, 1)$ with $\Theta = \mathbb{R}$ and $f(x; \theta) = \frac{1}{\pi[1+(x-\theta)^2]}$ for $x \in \mathbb{R}$. Show that $f(x; \theta)$ is not in the exponential family of distributions.

4.2.25 Prove Theorem 4.12.

4.2.26 Prove Theorem 4.13.

4.3 Minimum Variance Unbiased Estimators

The goal in parametric point estimation is to find the "best" estimator of the unknown parameter θ. There are many criteria for judging an estimator; however, the most commonly used criterion is that the estimator with smallest MSE is best.

Within the class of all unbiased estimators, the MSE criterion becomes the best unbiased estimator is the estimator with smallest variance. The unbiased estimator of θ having the smallest variance for all $\theta \in \Theta$ is referred to as a *uniformly minimum variance unbiased estimator* or a *UMVUE*.

Definition 4.17 Let $X_1, \ldots, X_n$ be a sample of iid random variables with common pdf $f(x; \theta)$ on parameter space θ. An estimator $T^\star$ of $\tau(\theta)$ is said to be a UMVUE of $\tau(\theta)$ if and only if

i) $E(T^\star) = \tau(\theta), \quad \forall \theta \in \Theta$.

ii) $\text{Var}(T^\star) \leq \text{Var}(T), \quad \forall \theta \in \Theta$ for any other unbiased estimator T of $\tau(\theta)$.

When a complete sufficient statistic S and an unbiased estimator T of θ exist, the Lehmann–Scheffé Theorem, Theorem 4.16, shows that the Rao–Blackwellization of the estimator T, namely $T^\star = E(T|S)$, is a UMVUE of θ.

Theorem 4.16 (Lehmann–Scheffé) *Let $X_1, \ldots, X_n$ be a sample of iid random variables with pdf $f(x; \theta)$ on Θ. If $S = S(\vec{X})$ is a complete sufficient statistic, $T^\star = T^\star(S)$ is a function of S alone, and $T^\star$ is unbiased for $\tau(\theta)$, then $T^\star$ is a UMVUE of $\tau(\theta)$.*

Proof. Let $X_1, \ldots, X_n$ be a sample of iid random variables with pdf $f(x; \theta)$ on Θ, $S = S(\vec{X})$ a complete sufficient statistic, $T^\star = T^\star(S)$ a function of S alone, and suppose that $T^\star$ is unbiased for $\tau(\theta)$. Let T be any other unbiased estimator of $\tau(\theta)$.

Then, by the Rao–Blackwell theorem $T'(S) = E[T|S]$ is also an unbiased estimator $\tau(\theta)$ and T' is a function of S alone. Thus, $E(T^\star - T') = 0, \quad \forall \theta \in \Theta$, and since S is a complete statistic it follows that $P(T^\star = T') = 1, \quad \forall \theta \in \Theta$.

Hence, $T^\star$ is the only unbiased estimator of $\tau(\theta)$ based on S, and by the Rao–Blackwell theorem, $\text{Var}(T^\star) \leq \text{Var}(T)$ for any other unbiased estimator of $\tau(\theta), \forall \theta \in \Theta$. Therefore, $T^\star$ is a UMVUE of $\tau(\theta)$. ∎

Example 4.37 Let $X_1, \ldots, X_n$ be a sample of iid $\text{Bin}(1, \theta)$ random variables with $\Theta = [0, 1]$. Then, by Example 4.27, $\text{Bin}(n, \theta)$ with $\Theta = (0, 1)$ is a one-parameter regular exponential family distribution with $t(x) = x$. Thus, $\sum_{i=1}^n X_i$ is a complete sufficient statistic for θ and $E\left[\sum_{i=1}^n X_i\right] = n\theta$. Since $\overline{X}$ is an unbiased estimator of θ that is a function of the complete sufficient statistic $\sum_{i=1}^n X_i$ alone, by the Lehmann–Scheffé theorem, $\overline{X}$ is a UMVUE of θ.

Example 4.38 Let $X_1, \ldots, X_n$ be a sample of iid $\text{Gamma}(\alpha = 4, \theta)$ random variables with $\Theta = (0, \infty)$. Then, $\text{Gamma}(4, \theta)$ with $\Theta = (0, \infty)$ is in the one-parameter regular exponential family of distributions with $t(x) = x$. Thus, $\sum_{i=1}^n X_i$ is a complete sufficient statistic for θ, and since $E\left[\sum_{i=1}^n X_i\right] = 4n\theta$, it follows that $\frac{1}{4}\overline{X}$ is an unbiased estimator of θ that is a function of the complete sufficient statistic $\sum_{i=1}^n X_i$ alone. Therefore, by the Lehmann–Scheffé theorem, $\frac{1}{4}\overline{X}$ is a UMVUE of θ.

Example 4.39 Let $X_1, \ldots, X_n$ be a sample of iid $N(\theta_1, \theta_2)$ random variables with parameter space $\Theta = \mathbb{R} \times \mathbb{R}^+$. Since $N(\theta_1, \theta_2)$ with $\Theta = \mathbb{R} \times \mathbb{R}^+$ is in the two-parameter regular exponential family of distributions with $t_1(x) = x$ and $t_2(x) = x^2$ it follows that $(\sum_{i=1}^n X_i, \sum_{i=1}^n X_i^2)$ is a complete sufficient statistic for (θ_1, θ_2). Now, $T(S) = (\overline{X}, S^2)$ is a one-to-one function of S, and hence, is also a complete sufficient statistic for (θ_1, θ_2). Moreover, since $\overline{X}$ and S^2 are unbiased estimators of μ and σ^2, respectively, it follows from the Lehmann–Scheffé Theorem that $\overline{X}$ is a UMVUE of θ_1 and S^2 is a UMVUE of θ_2.

4.3.1 Cramér–Rao Lower Bound

When a complete sufficient statistic or an unbiased estimator of a parameter is not readily available, it is difficult to determine whether or not a UMVUE exists; however, under certain *regularity conditions*, a lower bound can be placed on the variance of all unbiased estimators of the parameter, and in this case, any estimator that attains the lower bound is a UMVUE. The regularity conditions required for the lower bound are given in Definition 4.18 for a one-parameter pdf.

Definition 4.18 (Regularity Conditions) Let $X_1, \ldots, X_n$ be iid random variables with pdf $f(x; \theta)$ for $x \in \mathcal{S}$ and $\theta \in \Theta$. The pdf $f(x; \theta)$ is said to satisfy the regularity conditions when

i) $\frac{\partial}{\partial \theta} \ln[f(x; \theta)]$ exists for all $x \in \mathcal{S}$ and $\theta \in \Theta$.

ii) $\frac{\partial}{\partial \theta} \int_{\mathcal{S}} L(\theta) \, dx_1 \cdots dx_n = \int_{\mathcal{S}} \frac{\partial}{\partial \theta} L(\theta) dx_1 \cdots dx_n$, $\forall \theta \in \Theta$ when $f(x; \theta)$ is the pdf of a continuous random variable.

iii) $\frac{\partial}{\partial \theta} \sum_{\mathcal{S}} L(\theta) = \sum_{\mathcal{S}} \frac{\partial}{\partial \theta} L(\theta)$, $\forall \theta \in \Theta$ when $f(x; \theta)$ is the pdf of a discrete random variable.

iv) for any statistic $s(X_1, \ldots, X_n)$, $\frac{\partial}{\partial \theta} \int_{\mathcal{S}} s(x_1, \ldots, x_n) L(\theta) dx_1 \cdots dx_n = \int_{\mathcal{S}} s(x_1, \ldots, x_n) \frac{\partial}{\partial \theta} L(\theta) dx_1 \cdots dx_n \, \forall \theta \in \Theta$ when $f(x; \theta)$ is the pdf of a continuous random variable.

v) for any statistic $s(X_1, \ldots, X_n)$,

$$\frac{\partial}{\partial \theta} \sum_{\mathcal{S}} s(x_1, \ldots, x_n) L(\theta) = \sum_{\mathcal{S}} s(x_1, \ldots, x_n) \frac{\partial}{\partial \theta} L(\theta), \quad \forall \theta \in \Theta$$

when $f(x; \theta)$ is the pdf of a discrete random variable.

vi) $E\left[\left(\frac{\partial}{\partial \theta} \ln[f(x; \theta)]\right)^2\right] < \infty, \quad \forall \theta \in \Theta$.

The regularity conditions are satisfied for all regular exponential family distributions and distributions where the support of $f(x; \theta)$ is bounded and does not depend on θ. Provided the regularity conditions are satisfied, a lower bound can be placed on the variance of any unbiased estimator of θ. The lower bound on the variance of an unbiased estimator is called the *Cramér–Rao lower bound* and was introduced independently in publications by Cramér [10], Rao [5], Darmois [11], and Frechét [12]. The lower bound for estimators of $\tau(\theta)$ is given in Theorem 4.17.

Theorem 4.17 (Cramér–Rao Lower Bound) *Let* $X_1, \ldots, X_n$ *be a sample of iid random variables with pdf* $f(x; \theta)$ *satisfying the regularity conditions. If* $T = T(\vec{X})$ *is any unbiased estimator of* $\tau(\theta)$, *then*

$$\operatorname{Var}(T) \geq \frac{\left[\tau'(\theta)\right]^2}{E\left[\left(\frac{\partial}{\partial \theta} \ln[f(X_1, \ldots, x_n; \theta)]\right)^2\right]} = \operatorname{CRLB}_{\tau(\theta)}.$$

Corollary 4.1 *Let* $X_1, \ldots, X_n$ *be a sample of iid random variables with pdf* $f(x; \theta)$. *If* $T = T(\vec{X})$ *is any unbiased estimator of* θ *then*

$$\operatorname{Var}(T) \geq \frac{1}{E\left[\left(\frac{\partial}{\partial \theta} \ln[f(X_1, \ldots, x_n; \theta)]\right)^2\right]} = \operatorname{CRLB}_{\theta}.$$

When T is an unbiased estimator of $\tau(\theta)$, and $\operatorname{Var}(T)$ attains the Cramér–Rao lower bound, then T is a UMVUE of $\tau(\theta)$, however, it is possible that the variance of a UMVUE of $\tau(\theta)$ does not attain the Cramér–Rao lower bound. That is, the Cramér–Rao lower bound is not a sharp lower bound is not always attainable. An example of a UMVUE that does not attain the Cramér–Rao lower bound is given in Example 4.45.

Note that the Cramér–Rao lower bound for unbiased estimators of $\tau(\theta)$ is

$$\operatorname{CRLB}_{\tau(\theta)} = \left[\frac{\partial \tau(\theta)}{\partial \theta}\right]^2 \operatorname{CRLB}_{\theta},$$

Thus, the lower bound on the variance of the unbiased estimators of $\tau(\theta)$ can easily be computed from the Cramér–Rao lower bound for unbiased estimators of θ.

The denominator in the Cramér–Rao lower bound is called *Fisher's Information number* and is denoted by $I_n(\theta)$. Fisher's Information measures the amount of available information in the sample about the unknown parameter θ. In particular, as Fisher's Information number increases, the Cramér–Rao lower bound decreases, and thus, there is more information about θ in the sample for large values of $I_n(\theta)$.

Definition 4.19 Let X be a random variable with pdf $f(x; \theta)$ with $\theta \in \Theta \subset \mathbb{R}$ satisfying the regularity conditions in Definition 4.18. Fisher's Information for a single observation from a pdf $f(x; \theta)$ is defined to be

$$I(\theta) = E\left[\left(\frac{\partial}{\partial \theta} \ln[f(x; \theta)]\right)^2\right]$$

and for a random sample $X_1, \ldots, X_n$ with joint pdf $f(x_1, \ldots, x_n; \theta)$ Fisher's Information is defined to be

$$I_n(\theta) = E\left[\left(\frac{\partial}{\partial\theta}\ln[f(x_1, \ldots, x_n; \theta)]\right)^2\right].$$

Theorem 4.18 *Let* $X_1, \ldots, X_n$ *be iid random variables with pdf* $f(x; \theta)$ *for* $\theta \in \Theta$. *If* $f(x; \theta)$ *satisfies the regularity conditions in Definition 4.18, then* $I_n(\theta) = nI(\theta)$.

Proof. The proof of Theorem 4.18 is left as an exercise. ∎

Example 4.40 Let $X_1, \ldots, X_n$ be a sample of iid Pois(θ) with $\theta \in (0, \infty)$. Then, $f(x; \theta)$ is a regular exponential family distribution, and therefore, the regularity conditions are satisfied. Now,

$$\frac{\partial}{\partial\theta}\ln[f(x; \theta)] = \frac{\partial}{\partial\theta}[-\theta + x\ln(\theta) - \ln(x!)] = -1 + \frac{x}{\theta}.$$

Hence,

$$I(\theta) = E\left[\left(-1 + \frac{X}{\theta}\right)^2\right] = E\left[1 - 2\frac{X}{\theta} + \frac{X^2}{\theta^2}\right]$$

$$= 1 - 2\frac{E(X)}{\theta} + \frac{E(X^2)}{\theta^2} = 1 - 2\frac{\theta}{\theta} + \frac{\theta + \theta^2}{\theta^2}$$

$$= \frac{1}{\theta}.$$

Hence, $I_n(\theta) = \frac{n}{\theta}$, and the Cramér–Rao lower bound for unbiased estimators of θ is

$$\text{CRLB}_\theta = \frac{1}{I_n(\theta)} = \frac{\theta}{n}.$$

For estimating $\tau(\theta) = P(X = 0) = e^{-\theta}$, the Cramér–Rao lower bound for unbiased estimators of $\tau(\theta)$ is

$$\text{CRLB}_{e^{-\theta}} = \frac{\left[\frac{\partial}{\partial\theta}e^{-\theta}\right]^2}{I_n(\theta)} = \frac{e^{-2\theta}}{\frac{n}{\theta}} = \frac{\theta e^{-2\theta}}{n}.$$

Lemma 4.1 *If* X *be a random variable with pdf* $f(x; \theta)$ *satisfying the regularity conditions in Definition 4.18, then*

i) $E\left[\frac{\partial}{\partial\theta}\ln[f(X; \theta)]\right] = 0.$

ii) $E\left[\left(\frac{\partial}{\partial\theta}\ln[f(X; \theta)]\right)^2\right] = -E\left[\frac{\partial^2}{\partial\theta^2}\ln[f(X; \theta)]\right].$

Proof. The proof of Lemma 4.1 is given for the continuous case only since the proof for a discrete pdf is similar.

Let X be a continuous random variable with pdf $f(x; \theta)$ satisfying the regularity conditions.

i) Consider $E\left[\frac{\partial}{\partial\theta}\ln[f(X;\theta)]\right]$.

$$E\left[\frac{\partial}{\partial\theta}\ln[f(X;\theta)]\right] = \int_S \frac{\partial}{\partial\theta}\ln[f(x;\theta)]f(x;\theta)dx$$

$$= \int_S \frac{\frac{\partial}{\partial\theta}f(x;\theta)}{f(x;\theta)}f(x;\theta)dx = \int_S \frac{\partial}{\partial\theta}f(x;\theta)dx$$

$$= \frac{\partial}{\partial\theta}\int_S f(x;\theta)dx = \frac{\partial}{\partial\theta}(1) = 0.$$

ii) The proof of part (ii) is left as an exercise. ∎

Theorem 4.19 *Let $X_1, \ldots, X_n$ be iid random variables with pdf $f(x; \theta)$ for $\theta \in \Theta$. If $f(x; \theta)$ satisfies the regularity conditions in Definition 4.18, then*

$$I(\theta) = -E\left[\frac{\partial^2}{\partial\theta^2}\ln[f(x;\theta)]\right].$$

Proof. The proof of Theorem 4.19 follows directly from part (ii) of Lemma 4.1 ∎

For probability models satisfying the regularity conditions, it is generally easiest to compute the Cramér–Rao lower bound using $I_n(\theta) = -nE\left[\frac{\partial^2}{\partial\theta^2}\ln(f(x;\theta)\right]$ rather than the form given in the definition of Fisher's Information.

Example 4.41 Let $X_1, \ldots, X_n$ be a sample of iid Geo(θ) random variables with $S = \mathbb{N}$ and $\Theta = [0, 1]$. Since the pdf of a Geo(θ) random variable with $\Theta = [0, 1]$ is a regular exponential family distribution, the regularity conditions are satisfied. Since $\ln[f(x; \theta)] = \ln(\theta) + (x - 1)\ln(1 - \theta)$, it follows that

$$\frac{\partial}{\partial\theta}\ln[f(x;\theta)] = \frac{1}{\theta} - \frac{x-1}{1-\theta}$$

and

$$\frac{\partial^2}{\partial\theta^2}\ln[f(x;\theta)] = -\frac{1}{\theta^2} - \frac{x-1}{(1-\theta)^2}.$$

Since $E(X) = \frac{1}{\theta}$, Fisher's Information number is

$$I_n(\theta) = -nE\left[-\frac{1}{\theta^2} - \frac{X-1}{(1-\theta)^2}\right] = \frac{n}{\theta^2} + n\frac{\frac{1}{\theta}-1}{(1-\theta)^2} = \frac{n}{\theta^2(1-\theta)}.$$

Thus, the Cramér–Rao lower bound on unbiased estimators of θ is

$$\text{CRLB}_\theta = \frac{1}{I_n(\theta)} = \frac{\theta^2(1-\theta)}{n},$$

and the Cramér–Rao lower bound for unbiased estimators of $E(X) = \frac{1}{\theta}$ is

$$\text{CRLB}_{\frac{1}{\theta}} = \left[\frac{\partial}{\partial\theta}\frac{1}{\theta}\right]^2 \text{CRLB}_\theta = \frac{1}{\theta^4} \times \frac{\theta^2(1-\theta)}{n} = \frac{1-\theta}{n\theta^2}.$$

Example 4.42 Let $X_1, \ldots, X_n$ be a sample of iid $\text{Exp}(\theta)$ random variables with $\Theta = (0, \infty)$. Then, the log-likelihood function is $\ln[f(x; \theta)] = -\ln(\theta) - \frac{x}{\theta}$. Thus,

$$\frac{\partial}{\partial\theta}\ln[f(x; \theta)] = -\frac{1}{\theta} + \frac{x}{\theta^2}$$

and

$$\frac{\partial^2}{\partial\theta^2}\ln[f(x; \theta)] = \frac{1}{\theta^2} - \frac{2x}{\theta^3}.$$

Therefore,

$$I_n(\theta) = -nE\left[\frac{1}{\theta^2} - 2\frac{x}{\theta^3}\right] = -\frac{n}{\theta^2} + 2n\frac{E(X)}{\theta^3} = -\frac{n}{\theta^2} + 2n\frac{\theta}{\theta^3} = \frac{n}{\theta^2}.$$

Thus, the Cramér–Rao lower bound for unbiased estimators of θ is

$$\text{CRLB} = \frac{\theta^2}{n}.$$

Now, $E(\overline{X}) = \theta = E(X)$ and $\text{Var}(\overline{X}) = \frac{\sigma^2}{n} = \frac{\theta^2}{n}$. Thus, $\text{Var}(\overline{X})$ attains the Cramér–Rao lower bound, and therefore, $\overline{X}$ is a UMVUE of θ.

Example 4.43 Let $X_1, \ldots, X_n$ be a sample of iid random variables with pdf $f(x; \theta) = \theta x^{\theta-1}$ for $x \in (0, 1)$ and $\theta \in (0, \infty)$. Then, $f(x; \theta)$ is a regular exponential family distribution (see Problem 4.3.17), and therefore, the regularity conditions are satisfied.

Now, $\ln[f(x; \theta)] = \ln(\theta) + (\theta - 1)\ln(x)$, and hence,

$$\frac{\theta}{\partial\theta}\ln[f(x; \theta)] = \frac{1}{\theta} + \ln(x)$$

and

$$\frac{\theta^2}{\partial\theta^2}\ln[f(x; \theta)] = -\frac{1}{\theta}^2.$$

Thus, Fisher's Information number is $I_n(\theta) = \frac{n}{\theta^2}$, and the Cramér–Rao lower bound for unbiased estimators of θ is $\text{CRLB} = \frac{\theta^2}{n}$.

Since $f(x; \theta)$ is a regular exponential family distribution there exists a complete sufficient statistic for θ, and in particular, $\sum_{i=1}^{n} \ln(X_i)$ is a complete sufficient statistic for θ. Now, $E[\ln(X)] = -\frac{1}{\theta}$, and therefore, $-\frac{1}{n} \sum_{i=1}^{n} \ln(X_i)$ is a UMVUE of $\frac{1}{\theta}$.

The Cramér–Rao lower bound is also used to define the *efficiency* of a point estimator.

Definition 4.20 Let $X_1, \ldots, X_n$ be iid random variables with pdf $f(x; \theta)$ on S for $\theta \in \Theta$ satisfying the regularity conditions in Definition 4.18, and let T be an estimator of θ. The efficiency of the estimator T for estimating θ is defined to be

$$e(T; \theta) = \frac{\text{CRLB}_\theta}{\text{Var}(T)}.$$

An estimator T of θ is said to be an efficient estimator of θ when $e(T; \theta) = 1$, $\forall \theta \in \Theta$ and asymptotically efficient when $\lim_{n \to \infty} e(T; \theta) = 1$.

Note that the efficiency of an estimator is always less than or equal to 1 since the Cramér–Rao lower bound is the smallest possible variance for an unbiased estimator of θ, and an estimator with efficiency equal to one for all $\theta \in \Theta$ is a UMVUE since it attains the Cramér-Rao lower bound. Furthermore, since the Cramér–Rao lower bound may not be attainable, there are cases when no efficient estimator of θ exists.

Example 4.44 If $X_1, \ldots, X_n$ are iid random variables with pdf $f(x; \theta) = \theta e^{-\theta x}$ for $\in (0, \infty)$ and $\Theta = (0, \infty)$, then $f(x; \theta)$ is a regular exponential family distribution, and it can be shown that the Cramér–Rao lower bound for unbiased estimators of θ is $\text{CRLB}_\theta = \frac{\theta^2}{n}$. Since $E(X_i) = \frac{1}{\theta}$, it is reasonable to use $\frac{1}{\bar{X}}$ as an estimator of θ.

Now, $\text{Var}(\frac{1}{\bar{X}}) = \frac{n^2 \theta^2}{(n-1)^2(n-2)}$, and therefore, the efficiency of $\frac{1}{\bar{X}}$ for estimating θ is

$$e\left(\frac{1}{\bar{X}}; \theta\right) = \frac{\frac{\theta^2}{n}}{\frac{n^2 \theta^2}{(n-1)^2(n-2)}} = \frac{(n-1)^2(n-2)}{n^3}.$$

For example, when $n = 25$, the efficiency of $\frac{1}{\bar{X}}$ is 0.85, and therefore, $\frac{1}{\bar{X}}$ is 85% as efficient as an estimator of θ that attains the Cramér–Rao lower bound. The estimator $\frac{1}{\bar{X}}$ is an asymptotically efficient estimator of θ since

$$\lim_{n \to \infty} e\left(\frac{1}{\bar{X}}; \theta\right) = \lim_{n \to \infty} \frac{(n-1)^2(n-2)}{n^3} = 1.$$

When $\vec{\theta}$ is a d-dimensional vector of parameters, Cramér–Rao lower bounds on the variance of the unbiased estimators of each component θ_i of $\vec{\theta}$ exist, provided specific regularity conditions on $f(x; \vec{\theta})$ are satisfied. As in the one-parameter case, the regularity conditions on $f(x; \vec{\theta})$ have to do with the differentiability of $\ln[f(x; \vec{\theta})]$ and the ability to interchange the order of integration/summation and partial differentiation. Fortunately, when $f(x; \vec{\theta})$ is a multiparameter regular exponential family distribution, the necessary regularity conditions are satisfied.

Definition 4.21 Let $X_1, \ldots, X_n$ be iid random variables with d-dimensional regular exponential family pdf with $\Theta \subset \mathbb{R}^d$. Fisher's Information matrix is defined to be the $d \times d$ matrix with ijth entry

$$I_n(\vec{\theta})_{ij} = E\left[\frac{\partial}{\partial \theta_i} \ln[f(x_1, \ldots, x_n; \vec{\theta})] \times \frac{\partial}{\partial \theta_j} \ln[f(x_1, \ldots, x_n; \vec{\theta})]\right].$$

Similar to the one-parameter case, when $X_1, \ldots, X_n$ are iid random variable $I_n(\vec{\theta})_{ij} = n I(\vec{\theta})_{ij}$ where

$$I(\vec{\theta})_{ij} = E\left[\frac{\partial}{\partial \theta_i} \ln[f(x; \vec{\theta})] \times \frac{\partial}{\partial \theta_j} \ln[f(x; \vec{\theta})]\right]$$

and alternatively, $I(\vec{\theta})_{ij}$ can be computed as

$$I(\vec{\theta})_{ij} = -E\left[\frac{\partial^2}{\partial \theta_i \, \partial \theta_j} \ln[f(x; \vec{\theta})]\right].$$

Theorem 4.20 *If $X_1, X_2, \ldots, X_n$ are iid random variables with d-dimensional regular exponential family pdf $f(x; \vec{\theta})$ with $\Theta \subset \mathbb{R}^d$, then the Cramér–Rao lower bound for unbiased estimators of $\tau(\theta_i)$ is*

$$\mathrm{CRLB}_{\tau(\theta_i)} = \left[\frac{\partial}{\partial \theta_i} \tau(\theta_i)\right]^2 \times [I_n(\vec{\theta})^{-1}]_{ii},$$

and the Cramér–Rao lower bound for unbiased estimators of θ_i is $[I_n(\vec{\theta})^{-1}]_{ii}$.

Example 4.45 Let $X_1, \ldots, X_n$ be iid $N(\theta_1, \theta_2)$ random variables with $\Theta = \mathbb{R} \times \mathbb{R}^+$. Then, the log-likelihood function is

$$\ln[f(x; \theta_1, \theta_2)] = -\frac{1}{2} \ln(\theta_2) - \frac{1}{2} \ln(2\pi) - \frac{1}{2\theta_2}(x - \theta_1)^2.$$

Since the pdf of a $N(\theta_1, \theta_2)$ random variable with $\Theta = \mathbb{R} \times \mathbb{R}^+$ is a regular exponential family distribution, the regularity conditions are satisfied. The three second order partial derivatives of the log-likelihood function are

$$\frac{\partial^2}{\partial \theta_1^2} \ln[f(x;\theta)] = -\frac{1}{\theta_2},$$

$$\frac{\partial^2}{\partial \theta_2^2} \ln[f(x;\theta)] = \frac{1}{2\theta_2^2} - \frac{(x-\theta_1)^2}{\theta_2^3},$$

$$\frac{\partial^2}{\partial \theta_1 \partial \theta_2} \ln[f(x;\theta)] = -\frac{x-\theta_1}{\theta_2^2}.$$

Thus, Fisher's Information matrix is

$$I_n(\theta_1,\theta_2) = -n \begin{pmatrix} -\frac{1}{\theta_2} & E\left[-\frac{x-\theta_1}{\theta_2^2}\right] \\ E\left[-\frac{x-\theta_1}{\theta_2^2}\right] & E\left[\frac{1}{2\theta_2^2} - \frac{(x-\theta_1)^2}{\theta_2^3}\right] \end{pmatrix} = \begin{pmatrix} \frac{n}{\theta_2} & 0 \\ 0 & \frac{n}{2\theta_2^2} \end{pmatrix}.$$

The Cramér–Rao lower bound for unbiased estimators of θ_1 is

$$\text{CRLB}_{\theta_1} = [I_n(\theta_1,\theta_2)^{-1}]_{11} = \frac{\theta_2}{n},$$

and the Cramér–Rao lower bound for unbiased estimators of θ_2 is

$$\text{CRLB}_{\theta_2} = [I_n(\theta_1,\theta_2)^{-1}]_{22} = \frac{2\theta_2^2}{n}.$$

In Example 4.39, S^2 was shown to be a UMVUE of θ_2. Since $\text{Var}(S^2) = \frac{2\theta_2^2}{n-1} \neq \frac{2\theta_2^2}{n}$, it follows that S^2 is a UMVUE of θ_2 that does not attain the Cramér–Rao lower bound.

Problems

4.3.1 Let $X_1,\ldots,X_n$ be an iid sample of Pois(θ) random variables with $\Theta = (0,\infty)$. Determine
  a) Fisher's Information number for θ.
  b) the Cramér–Rao lower bound for unbiased estimators of θ.
  c) whether or not $\overline{X}$ is UMVUE for θ.
  d) the Cramér–Rao lower bound for unbiased estimators of $P(X = 0)$.

4.3.2 Let $X_1,\ldots,X_n$ be an iid sample of Geo(θ) random variables with $\Theta = [0,1]$. Determine
  a) Fisher's Information number for θ.
  b) the Cramér–Rao lower bound for unbiased estimators of θ.
  c) the Cramér–Rao lower bound for unbiased estimators of $\frac{\theta}{1-\theta}$.

4.3.3 Let $X_1, \dots, X_n$ be an iid sample of Exp(θ) random variables with $\Theta = (0, \infty)$. Determine
a) Fisher's Information number for θ.
b) the Cramér–Rao lower bound for unbiased estimators of θ.
c) whether or not $\overline{X}$ is UMVUE for θ.
d) show that $\frac{n}{n+1}\overline{X}^2$ is UMVUE for θ^2.
e) whether or not $\frac{n}{n+1}\overline{X}^2$ attains the Cramér–Rao lower bound for unbiased estimators of θ^2.

4.3.4 Let $X_1, \dots, X_n$ be an iid sample of Exp $\left(\beta = \frac{1}{\theta}\right)$ random variables with $\Theta = (0, \infty)$. Determine
a) Fisher's Information number for θ.
b) the Cramér–Rao lower bound for unbiased estimators of θ.
c) a UMVUE of θ. *Hint:* Consider $\frac{1}{\overline{X}}$.

4.3.5 Let $X_1, \dots, X_n$ be an iid sample of $N(0, \theta)$ random variables with $\Theta = (0, \infty)$. Determine
a) Fisher's Information number for θ.
b) the Cramér–Rao lower bound for unbiased estimators of θ.
c) a UMVUE for θ.

4.3.6 Let $X_1, \dots, X_n$ be an iid sample of Gamma($4, \theta$) random variables with $\Theta = (0, \infty)$. Determine
a) a complete sufficient statistic for θ.
b) Fisher's Information number for θ.
c) the Cramér–Rao lower bound for unbiased estimators of θ.
d) a UMVUE for 4θ.

4.3.7 Let $X_1, \dots, X_n$ be an iid sample of random variables with pdf $f(x) = \theta x^{\theta-1}$ for $x \in (0, 1)$ and $\Theta = (0, \infty)$. Determine
a) a complete sufficient statistic for θ.
b) Fisher's Information number for θ.
c) the Cramér–Rao lower bound for unbiased estimators of θ.
d) a UMVUE for $\frac{1}{\theta^2}$.

4.3.8 Let $X_1, \dots, X_n$ be an iid sample of Pois($\lambda = e^{-\theta}$) random variables with $\Theta = (0, \infty)$. Determine
a) a complete sufficient statistic for θ.
b) Fisher's Information number for θ.
c) the Cramér–Rao lower bound for unbiased estimators of θ.
d) a UMVUE for Var(X_1).

4.3.9 Let $X_1, \ldots, X_n$ be an iid sample of Beta$(2, \theta_2)$ random variables with $\Theta = \mathbb{R}^+ \times \mathbb{R}^+$. Determine
   a) a complete sufficient statistic for θ.
   b) Fisher's Information number for θ.
   c) the Cramér–Rao lower bound for unbiased estimators of θ.
   d) a UMVUE for $\frac{2\theta+1}{\theta(\theta+1)}$.

4.3.10 Let $X_1, \ldots, X_n$ be an iid sample of LN(θ_1, θ_2) random variables with $\Theta = \mathbb{R} \times \mathbb{R}^+$. Determine
   a) a statistic that is jointly sufficient for (θ_1, θ_2).
   b) the Fisher's Information matrix about $\vec{\theta}$.
   c) the Cramér–Rao lower bound for unbiased estimators of θ_1.
   d) the Cramér–Rao lower bound for unbiased estimators of θ_2.
   e) Show that $\hat{\theta}_1 = \frac{1}{n} \sum_{i=1}^{n} \ln(X_i)$ is UMVUE for θ_1.

4.3.11 Let $X_1, \ldots, X_n$ be iid Gamma(θ_1, θ_2) random variables with $\Theta = \mathbb{R}^+ \times \mathbb{R}^+$. Determine
   a) a statistic that is jointly sufficient for (θ_1, θ_2).
   b) the Fisher's Information matrix about $\vec{\theta}$.
   c) the Cramér–Rao lower bound for unbiased estimators of θ_1.
   d) the Cramér–Rao lower bound for unbiased estimators of θ_2.

4.3.12 Let X be a continuous random variable with pdf $f(x; \theta)$ on $\mathcal{S}$ with $\Theta \subset \mathbb{R}$. If $f(x; \theta)$ satisfies the regularity conditions in Definition 4.18, show $E\left[\left(\frac{\partial \ln[f(x;\theta)]}{\partial \theta}\right)^2\right] = -E\left[\frac{\partial^2 \ln[f(x;\theta)]}{\partial \theta^2}\right]$.

4.3.13 Prove Theorem 4.18.

4.4 Case Study – The Order Statistics

When X is a random variable whose support depends on the value of one or more parameters, the pdf of X is said to be a *range-dependent pdf*. Random variables that have range-dependent pdfs are not exponential family distributions, and hence, sufficient statistics must be found using the definition of a sufficient statistic or the Fisher–Neyman factorization theorem; likewise, a complete statistic must be found using the definition of completeness. Moreover, the regularity conditions needed for the Cramér–Rao lower bound are not satisfied for range-dependent pdfs, and hence, the Cramér–Rao lower bound on the variance of an unbiased estimator of a parameter does not apply.

Examples of range-dependent random variables include the uniform distribution over (θ_1, θ_2) and the two-parameter exponential distribution, which has pdf given by

$$f(x; \theta_1, \theta_2) = \frac{1}{\theta_1} e^{-\left(\frac{x-\theta_2}{\theta_1}\right)}, \quad \text{for} \quad x > \theta_2; \Theta = \mathbb{R}^+ \times \mathbb{R}$$

and is denoted by $\text{Exp}(\theta_1, \theta_2)$.

Also, when X is a random variable with support $\mathcal{S} \neq \mathbb{R}$ that does not depend on any of the parameters of its pdf, $Y = X + \theta^\star$ has a range-dependent pdf for any $\theta^\star \in \mathbb{R}$. For example, if $X \sim \text{Gamma}(\theta_1, \theta_2)$ and $Y = X + \theta^\star$, then Y has a range-dependent pdf for any value of $\theta^\star \in \mathbb{R}$.

When working with a sample of iid random variables having a range-dependent pdf, the sufficient statistics for θ are usually functions of the *order statistic*, which is the vector of ordered sample values. For example, if $X_1, \ldots, X_n$ are iid random variables with range-dependent pdf $f(x; \theta)$ on support $\mathcal{S} = (\theta, \infty)$, then the minimum of the sample values is often a sufficient statistic for θ.

Definition 4.22 If $X_1, \ldots, X_n$ is a sample of iid random variables the order statistic associated with the sample is denoted by $(X_{(1)}, X_{(2)}, \ldots, X_{(n)})$ where $X_{(1)} \leq X_{(2)} \leq \cdots \leq X_{(n)}$. The kth-order statistic is $X_{(k)}$, which is the kth smallest value in the random sample $X_1, \ldots, X_n$.

When $X_1, \ldots, X_n$ are iid random variables with common pdf $f(x; \theta)$, because the order statistic is simply a reordering of the random variables $X_1, \ldots, X_n$ it follows that there are $n!$ possible orderings of the random variables $X_1, \ldots, X_n$. Thus, the joint pdf of the order statistic is

$$f_{(1,\ldots,n)}(x_{(1)}, x_{(2)}, \ldots, x_{(n)}; \theta) = n! \prod_{i=1}^{n} f(x_i; \theta).$$

Theorem 4.21 shows that the order statistic is always a sufficient statistic for a sample of iid random variables.

Theorem 4.21 *If $X_1, \ldots, X_n$ is a sample of iid random variables with common pdf $f(x; \theta)$ on $\mathcal{S}$ with $\Theta \subset \mathbb{R}^d$, then $\vec{S} = (X_{(1)}, X_{(2)}, \ldots, X_{(n)})$ is a sufficient statistic for $\vec{\theta}$.*

Proof. Let $X_1, \ldots, X_n$ be a sample of iid random variables with common pdf $f(x; \vec{\theta})$ on $\mathcal{S}$ with $\vec{\theta} \in \Theta \subset \mathbb{R}^d$, and let $S = (X_{(1)}, X_{(2)}, \ldots, X_{(n)})$. Then, the conditional pdf of $\vec{X}|S$ is

$$f(\vec{x}|\vec{s}) = \frac{f(x_1, \ldots, x_n)}{f(s_1, \ldots, s_n)}$$

$$= \frac{\prod_{i=1}^{n} f(x_i; \vec{\theta})}{n! \prod_{i=1}^{n} f(x_i; \vec{\theta})} = \frac{1}{n!}$$

since $\vec{s}$ is simply a reordering of $\vec{x}$.

Thus, since $f(\vec{x}|x_{(1)}, \ldots, x_{(n)})$ does not depend on $\vec{\theta}$, $S = (X_{(1)}, \ldots, X_{(n)})$ is a sufficient statistic for $\vec{\theta}$. ∎

The first-order statistic is $X_{(1)} = \text{Min}\{X_i\}$, and the nth order statistic is $X_{(n)} = \text{Max}\{X_i\}$, which are the most important order statistics when the support is of the form $S = (\theta_1, \theta_2)$. The order statistics are often used in estimating the median, quantiles, midrange, range, and interquartile range of a distribution. Theorem 4.22 and its two corollaries show that when sampling a range-dependent pdf, under certain conditions $X_{(1)}$ and $X_{(n)}$ will be sufficient statistics for θ.

Theorem 4.22 *If $X_1, \ldots, X_n$ is a sample of iid random variables with pdf $f(x; \theta) = c(\vec{\theta})h(x)$ on $S = (\theta_1, \theta_2)$ with $\Theta = \mathbb{R} \times (\theta_1, \infty)$, then $S = (X_{(1)}, X_{(n)})$ is a sufficient statistic for $\vec{\theta}$.*

Proof. Let $X_1, \ldots, X_n$ be a sample of iid random variables with common pdf $f(x; \theta) = c(\vec{\theta})h(x)$ on $S = (\theta_1, \theta_2)$ with $\Theta = \mathbb{R} \times (\theta_1, \infty)$. Then,

$$L(\vec{\theta}) = \prod_{i=1}^{n} c(\theta)h(x_i)I_{(\theta_1, \theta_2)}(x_i) = c(\vec{\theta})^n \prod_{i=1}^{n} h(x_i)I_{(\theta_1, \theta_2)}(x_i).$$

Now,

$$\prod_{i=1}^{n} h(x_i)I_{(\theta_1, \theta_2)}(x_i) = \begin{cases} \prod_{i=1}^{n} h(x_i) & \text{only when } \theta_1 \leq x_i \leq \theta_2 \text{ for } i = 1, \ldots, n \\ 0 & \text{otherwise} \end{cases}$$

and since $\theta_1 \leq x_i \leq \theta_2$ for $i = 1, \ldots, n$ if and only if $x_{(1)} \geq \theta_1$ and $x_{(n)} \leq \theta_2$, it follows that

$$L(\vec{\theta}) = c(\vec{\theta})^n I_{(x_{(1)}, \infty)}(\theta_1) I_{(-\infty, x_{(n)})}(\theta_2) \prod_{i=1}^{n} h(x_i) = g(x_{(1)}, x_{(n)}; \vec{\theta})h(\vec{x})$$

with $g(x_{(1)}, x_{(n)}; \vec{\theta}) = c(\vec{\theta})^n I_{(x_{(1)}, \infty)}(\theta_1) I_{(-\infty, x_{(n)})}(\theta_2)$ and $h(\vec{x}) = \prod_{i=1}^{n} h(x_i)$. Thus, by the Fisher–Neyman factorization theorem, $S = (X_{(1)}, X_{(n)})$ is a sufficient statistic for $\vec{\theta}$. ∎

Corollary 4.2 *If $X_1, \ldots, X_n$ is a sample of iid random variables with pdf $f(x; \theta) = c(\theta)h(x)$ on (θ, α) where $\alpha \in \mathbb{R}$ or $\alpha = \infty$ and $\Theta = (-\infty, \alpha)$, then $X_{(1)}$ is a sufficient statistic for θ.*

Proof. The proof of Corollary 4.2 is left as an exercise. ∎

Corollary 4.3 *If $X_1, \ldots, X_n$ is a sample of iid random variables with pdf $f(x; \theta) = c(\theta)h(x)$ on (α, θ) where $\alpha \in \mathbb{R}$ or $\alpha = -\infty$ and $\Theta = (\alpha, \infty)$, then $X_{(n)}$ is a sufficient statistic for θ.*

Proof. The proof of Corollary 4.3 is left as an exercise. ∎

For example, when $X_1, \ldots, X_n$ is a sample of iid $U(\theta_1, \theta_2)$ random variable, $(X_{(1)}, X_{(n)})$ is jointly sufficient for $\vec{\theta} = (\theta_1, \theta_2)$, and when $X_1, \ldots, X_n$ is a sample of iid $U(0, \theta)$ random variables, $X_{(n)}$ is a sufficient statistic for θ.

Example 4.46 Let $X_1, \ldots, X_n$ be a sample of iid random variables with pdf $f(x; \theta) = e^{-(x-\theta)}$ on $S = (\theta, \infty)$ for $\theta \in \mathbb{R}$. Then, $f(x; \theta)$ is of the form $c(\theta)h(x)$ on $S = (\theta, \infty)$, and therefore, by Corollary 4.2 it follows that $X_{(1)}$ is sufficient for θ.

For a sample $X_1, \ldots, X_n$ of iid random variables, the CDFs of the sampling distributions of $X_{(1)}$ and $X_{(n)}$ are based entirely on the common CDF of $X_1, \ldots, X_n$. The CDFs of $X_{(1)}$ and $X_{(n)}$ are given in Theorem 4.23.

Theorem 4.23 *If $X_1, \ldots, X_n$ is a sample of iid random variables with common CDF $F(x; \theta)$ and support S, then the CDF of the sampling distribution of*

i) $X_{(1)}$ is $F_{(1)}(x; \theta) = [1 - F(x; \theta)]^n$ for $x \in S$.
ii) $X_{(n)}$ is $F_{(n)}(x; \theta) = [F(x; \theta)]^n$ for $x \in S$.

Proof. Let $X_1, \ldots, X_n$ be iid random variables with common CDF $F(x; \theta)$ and support S.

i) For $x \in S$, $F_{(1)}(x) = P(X_{(1)} \leq x) = 1 - P(X_{(1)} > x)$, however, $X_{(1)} > x$ if and only if $X_i > x$ for $i = 1, 2, \ldots, n$. Thus, since $X_1, \ldots, X_n$ is a sample of iid random variables

$$F_{(1)}(x) = 1 - P(X_1 > x, X_2 > x, \ldots, X_n > x) = 1 - \prod_{i=1}^{n} P(X_i > x)$$

$$= [1 - F(x)]^n.$$

ii) The proof of part (ii) is left as an exercise. ∎

The pdfs associated with $X_{(1)}$ and $X_{(n)}$ can be determined from their CDFs. In particular, Corollary 4.4 provides the pdfs of $X_{(1)}$ and $X_{(n)}$ when $X_1, \ldots, X_n$ are iid continuous random variables.

Corollary 4.4 *If $X_1, \ldots, X_n$ is a sample of iid continuous random variables with common pdf $f(x; \theta)$, support S, and CDF $F(x; \theta)$, then the pdf of the sampling distribution of*

i) $X_{(1)}$ is $f_{(1)}(x; \theta) = nf(x; \theta)[1 - F(x; \theta)]^{n-1}$ for $x \in S$.
ii) $X_{(n)}$ is $f_{(n)}(x; \theta) = nf(x; \theta)[F(x; \theta)]^{n-1}$ for $x \in S$.

Proof. The proof of Corollary 4.4 is left as an exercise. ∎

Note that the pdf of $X_{(1)}$ and $X_{(n)}$ for a sample of iid discrete random variables can also be found from their CDFs using $f(x_k) = F(x_k) - F(x_{k-1})$ for $x_k \in \mathcal{S}$, but in general, this approach does not yield a simple form for the pdf when working with discrete random variables.

Regardless of whether the sample is from a discrete or a continuous probability model, the sampling distributions of $X_{(1)}$ and $X_{(n)}$ are used to determine their means, variances, biases, MSEs, and the overall effectiveness of $X_{(1)}$ and $X_{(n)}$ as estimators.

Example 4.47 Let $X_1, \ldots, X_n$ be a sample of iid $U(0, \theta)$ random variables with CDF $F(x) = 1 - \frac{x}{\theta}$ for $x \in (0, \theta)$. Then, by Corollary 4.3, $X_{(n)}$ is sufficient for θ, and by Corollary 4.4, the pdf of $X_{(n)}$ is

$$f_{(n)}(x) = \frac{n}{\theta}\left[1 - \left(1 - \frac{x}{\theta}\right)\right]^{n-1} = \frac{nx^{n-1}}{\theta^n}, \quad x \in (0, \theta).$$

Thus,

$$E[X_{(n)}] = \int_0^\theta x \cdot \frac{nx^{n-1}}{\theta^n} \, dx = \frac{n}{n+1}\theta$$

and

$$E\left[X_{(n)}^2\right] = \int_0^\theta x^2 \cdot \frac{n}{\theta^n} x^{n-1} \, dx = \frac{n}{n+2}\theta^2.$$

Hence, $E[X_{(n)}] = \frac{n}{n+1}\theta$ and $\mathrm{Var}(X_{(n)}) = \frac{n\theta^2}{(n+2)(n+1)^2}$.

Moreover, $X_{(n)}$ is a biased estimator of θ with $\mathrm{Bias}(X_{(n)}; \theta) = -\frac{\theta}{n+1}$, but since $\lim_{n\to\infty}\mathrm{Bias}(X_{(n)}; \theta) = 0$ and $\lim_{n\to\infty}\mathrm{Var}(X_{(n)}) = 0$ it follows that $X_{(n)}$ is an MSE-consistent estimator of θ.

Now, let $h(x)$ be any function with $E[h(X_{(n)})] = 0$, $\forall \theta \in (0, \infty)$. Then, $\forall \theta \in (0, \infty)$

$$0 = \int_0^\theta h(x) \cdot \frac{nx^{n-1}}{\theta^n} \, dx = \int_0^\theta \frac{n}{\theta^n} h(x) x^{n-1} \, dx.$$

Thus, by the Fundamental Theorem of Calculus, it follows that

$$0 = \frac{\partial}{\partial\theta}(0) = \frac{\partial}{\partial\theta} \int_0^\theta \frac{n}{\theta^n} h(x) x^{n-1} \, dx$$

$$= \frac{n}{\theta^n} h(\theta)\theta^{n-1} = \frac{n}{\theta} h(\theta).$$

Hence, $\frac{n}{\theta}h(\theta) = 0$, $\forall \theta \in (0, \infty)$, and hence, $h(x) = 0$ with probability 1. Therefore, $X_{(n)}$ is also a complete sufficient statistic for θ.

Finally, since $T = \frac{n+1}{n}X_{(n)}$ is unbiased estimator of θ based on a complete sufficient statistic, it follows by the Lehmann–Scheffé theorem that T is a UMVUE of θ.

Example 4.48 Let $X_1, \dots, X_n$ be a sample of iid $\text{Exp}(\beta, \theta)$ random variables with $\beta > 0$ known and $\theta \in \mathbb{R}$. The common pdf of the random variables $X_1, \dots, X_n$ is $f(x; \beta, \theta) = \frac{1}{\beta}e^{-\frac{(x-\theta)}{\beta}}$ for $x > \theta$.

By Corollary 4.2, $X_{(1)}$ is sufficient for θ, and by Corollary 4.4, the pdf of $X_{(1)}$ is

$$f_{(1)}(x) = n\frac{1}{\beta}e^{-\frac{(x-\theta)}{\beta}}\left[1 - (1 - e^{-\frac{(x-\theta)}{\beta}})\right]^{n-1} = \frac{n}{\beta}e^{-\frac{n(x-\theta)}{\beta}}, \quad x > \theta.$$

Thus, $X_{(1)} \sim \text{Exp}(\frac{\beta}{n}, \theta)$, and $E[X_{(1)}] = \theta + \frac{\beta}{n}$ and $\text{Var}[X_{(1)}] = \frac{\beta^2}{n^2}$.

Hence, $X_{(1)}$ is a biased estimator of θ with $\text{Bias}(X_{(1)}; \theta) = \frac{\beta}{n}$. However, since $\lim_{n\to\infty}\text{Bias}(X_{(1)}; \theta) = 0$ and $\lim_{n\to\infty}\text{Var}[X_{(1)}] = 0$, $X_{(1)}$ is an MSE-consistent estimator of θ.

Suppose that $h(x)$ is any function with $E[h(X_{(1)}] = 0$. Then, $\forall \theta \in \Theta = \mathbb{R}$

$$0 = \int_{\theta}^{\infty} h(x) \cdot \frac{n}{\beta}e^{-\frac{n(x-\theta)}{\beta}}\,dx$$

and by the Fundamental Theorem of Calculus,

$$0 = \frac{\partial}{\partial\theta}(0) = \frac{\partial}{\partial\theta}\int_{\theta}^{\infty} h(x) \cdot \frac{n}{\beta}e^{-\frac{n(x-\theta)}{\beta}}\,dx$$
$$= -h(\theta)\frac{n}{\beta}, \quad \forall \theta \in \mathbb{R}.$$

Thus, $h(\theta) = 0$ with probability 1, and $X_{(1)}$ is a complete statistic.

Now, since $X_{(1)}$ is a complete sufficient statistic for θ and $T = X_{(1)} - \frac{\beta}{n}$ is an unbiased estimator of θ, by the Lehmann–Scheffé theorem, T is a UMVUE of θ.

Problems

4.4.1 Let $X_1, \dots, X_n$ be a sample of iid $\text{Exp}(\theta)$ random variables with $\theta \in (0, \infty)$.

   a) Determine the pdf of $X_{(1)}$.
   b) Determine $E[X_{(1)}]$.
   c) Determine $\text{Var}[X_{(1)}]$.

d) Show that $T = nX_{(1)} \sim \text{Exp}(\theta)$.
e) Show that $T = nX_{(1)}$ is not an MSE-consistent estimator of θ.

4.4.2 Let $X_1, \ldots, X_n$ be iid random variables with pdf $f(x; \theta) = e^{-(x-\theta)}$ on $\mathcal{S} = (\theta, \infty)$ with $\Theta = \mathbb{R}$.
a) Show $X_{(1)}$ is a sufficient statistic for θ.
b) Determine $f_{(1)}(x)$.
c) Determine $E[X_{(1)}]$.
d) Show $X_{(1)}$ is a complete sufficient statistic for θ.
e) Find a UMVUE of θ.

4.4.3 Let $X_1, \ldots, X_n$ be a sample of iid $U(\theta, 1)$ random variables for $\theta \in (-\infty, 1)$.
a) Show that $X_{(1)}$ is sufficient for θ.
b) Determine the pdf of $X_{(1)}$.
c) Determine $E[X_{(1)}]$.
d) Determine $\text{Var}[X_{(1)}]$.
e) Determine $\text{MSE}(X_{(1)}; \theta)$.
f) Is $X_{(1)}$ an MSE-consistent estimator of θ?
g) Show that $X_{(1)}$ is a complete sufficient statistic.
h) Determine a UMVUE of θ.

4.4.4 Let $X_1, \ldots, X_n$ be a sample of iid $\text{Exp}(\theta_1, \theta_2)$ random variables with common pdf $f(x; \theta_1, \theta_2) = \frac{1}{\theta_1} e^{-\frac{x-\theta_2}{\theta_1}}$ for $x > \theta_2$ and $\Theta = \mathbb{R} \times \mathbb{R}^+$.
a) Show that $S = (X_{(1)}, \sum_{i=1}^{n} X_i)$ is jointly sufficient for (θ_1, θ_2).
b) Determine the pdf of $X_{(1)}$.
c) Determine $E[X_{(1)}]$.
d) Determine $E[X_{(1)}^2]$.
e) Determine $\text{Var}[X_{(1)}]$.
f) Is $X_{(1)}$ an MSE-consistent estimator of θ_2?
g) Given $S = (X_{(1)}, \sum_{i=1}^{n} X_i)$ is a complete sufficient statistic for (θ_1, θ_2), determine the UMVUEs of θ_1 and θ_2.

4.4.5 Let $X_1, \ldots, X_n$ be a sample of iid random variables with pdf $f(x; \theta) = \frac{6}{\theta^3} x(\theta - x)$ on $\mathcal{S} = (0, \theta)$ with $\Theta = \mathbb{R}^+$. Determine
a) $F(x)$.
b) $f_{(n)}(x)$.

4.4.6 Let $X_1, \ldots, X_n$ be a sample of iid random variables with pdf $f(x; \theta) = \frac{3x^2}{\theta^3}$ on $\mathcal{S} = (0, \theta)$ with $\Theta = \mathbb{R}^+$. Determine
a) a sufficient statistic for θ.
b) $F(x)$.

c) $f_{(n)}(x)$.

d) $E[X_{(n)}]$.

e) $\mathrm{Var}[X_{(n)}]$.

f) an unbiased estimator of θ based on $X_{(n)}$.

4.4.7 Let $X_1, \ldots, X_n$ be a sample of iid random variables with pdf $f(x; \theta) = e^{x-\theta}$ on $\mathcal{S} = (-\infty, \theta)$ with $\Theta = \mathbb{R}$. Determine

a) a sufficient statistic for θ.

b) $F(x)$.

c) $f_{(n)}(x)$.

d) $E[X_{(n)}]$.

e) an unbiased estimator of θ based on $X_{(n)}$.

4.4.8 Let $X_1, \ldots, X_n$ be a sample of iid random variable with pdf $f(x; \theta) = \frac{1}{2^{x-\theta+1}}$ on $\mathcal{S} = \{\theta, \theta + 1, \theta + 2, \ldots\}$ with $\Theta = \mathbb{N}$. Determine

a) a sufficient statistic for θ.

b) $F_{(1)}(x)$.

c) $f_{(1)}(x)$.

d) $E[X_{(1)}]$.

4.4.9 Prove Corollary 4.2.

4.4.10 Prove Corollary 4.3.

4.4.11 Prove Theorem 4.23 part (ii).

4.4.12 Prove Corollary 4.4.

5

Likelihood-based Estimation

Given a probability model $f(x; \vec{\theta})$ and a parameter space Θ, there are several different approaches that can be used to determine an estimator of $\vec{\theta}$. In particular, the Method of Moments, Bayesian, and Maximum Likelihood are three approaches that can be used to find estimators. In this chapter, only the Maximum Likelihood and Bayesian methods of estimation, which are based on the likelihood function, are discussed.

Recall that the likelihood function for a sample $X_1, \ldots, X_n$ is the joint pdf of the random vector $\vec{X} = (X_1, \ldots, X_n)$, that is, $L(\theta) = f(X_1, \ldots, X_n; \vec{\theta})$, and the log-likelihood function is $\ell(\vec{\theta}) = \ln[L(\vec{\theta})]$. When $X_1, \ldots, X_n$ is a sample of iid random variables, $L(\vec{\theta}) = \prod_{i=1}^{n} f(x_i; \vec{\theta})$, and when $X_1, \ldots, X_n$ is a sample of independent random variables, $L(\vec{\theta}) = \prod_{i=1}^{n} f_i(x_i; \vec{\theta})$.

In a parametric model $f(x; \theta)$, the likelihood function links the observed data and the probability model so that statistical inferences can be made about θ. The importance of the likelihood function is illustrated by the *Law of Likelihood*.

The Law of Likelihood: *Let $X_1, \ldots, X_n$ be a sample of iid random variables with common pdf $f(x; \vec{\theta})$ and parameter space Θ. For $\vec{\theta} \in \Theta$, the larger the value of $L(\vec{\theta})$ is, the more $\vec{\theta}$ agrees with the observed data. Thus, the degree to which the information in the sample supports a parameter value $\vec{\theta}_0 \in \Theta$, in comparison to another value $\vec{\theta}_1 \in \Theta$, is equal to the ratio of their likelihoods $\Lambda(\vec{\theta}_0, \vec{\theta}_1) = \frac{L(\vec{\theta}_0)}{L(\vec{\theta}_1)}$. In particular, the information in the sample is in better agreement with $\vec{\theta}_1$ than $\vec{\theta}_0$ when $\Lambda < 1$, and the information in the sample agrees better with $\vec{\theta}_0$ than $\vec{\theta}_1$ when $\Lambda > 1$.*

Theorem 5.1 shows that the likelihood function depends on the information in a sample only through a sufficient statistic.

Mathematical Statistics: An Introduction to Likelihood Based Inference, First Edition. Richard J. Rossi.
© 2018 John Wiley & Sons, Inc. Published 2018 by John Wiley & Sons, Inc.

Theorem 5.1 *Let $X_1, \ldots, X_n$ be a sample of iid random variables with pdf $f(x; \vec{\theta})$ and parameter space Θ. If S is a sufficient statistic for $\vec{\theta}$, then the likelihood function $L(\vec{\theta})$ depends on $X_1, \ldots, X_n$ only through the sufficient statistic S.*

Proof. Let $X_1, \ldots, X_n$ be a sample iid random variables with pdf $f(x; \vec{\theta})$ and parameter space Θ, and let S be a sufficient statistic. Then, by the Fisher–Neyman factorization theorem,

$$L(\vec{\theta}) = f(x_1, \ldots, x_n; \vec{\theta}) = g(S; \vec{\theta}) h(x_1, \ldots, x_n).$$

Since the likelihood function is a function of $\vec{\theta}$, it follows that $L(\vec{\theta})$ only depends on $X_1, \ldots, X_n$ through $g(S; \vec{\theta})$. Therefore, $L(\vec{\theta})$ depends on $X_1, \ldots, X_n$ only through the sufficient statistic S. ∎

Note that since the natural logarithm is a one-to-one function on $\mathbb{R}^+$, no information about $\vec{\theta}$ is lost when the log-likelihood function $\ell(\vec{\theta}) = \ln[L(\vec{\theta})]$ is used in place of $L(\vec{\theta})$. Moreover, for many probability models, it is easier to work with the log-likelihood function than with the likelihood function when $X_1, \ldots, X_n$ is a sample of iid random variables since

$$\ell(\vec{\theta}) = \ln\left[\prod_{i=1}^{n} f(x_i; \vec{\theta})\right] = \sum_{i=1}^{n} \ln f(x_i; \vec{\theta}).$$

When $\Theta \subset \mathbb{R}$, and the log-likelihood function is differentiable with respect to θ, the first derivative of $\ell(\theta)$ is called the *Score function*; when $\Theta \subset \mathbb{R}^d$, and $\frac{\partial \ell(\vec{\theta})}{\partial \theta_i}$ exists for $i = 1, \ldots, d$, the Score function is the vector of first partial derivatives of $\ell(\vec{\theta})$ with respect to $\vec{\theta}$.

Definition 5.1 Let $X_1, \ldots, X_n$ be a sample of random variables with likelihood function $L(\vec{\theta})$ for $\vec{\theta} \in \Theta$. If the log-likelihood function $l(\vec{\theta})$ is differentiable with respect to $\vec{\theta}$, then the Score function is defined to be

i) $\dot{\ell}(\theta) = \frac{\partial}{\partial \theta} l(\theta)$, when $\Theta \subset \mathbb{R}$.

ii) the vector $\dot{\ell}(\vec{\theta}) = \left(\frac{\partial}{\partial \theta_1} l(\vec{\theta}), \ldots, \frac{\partial}{\partial \theta_d} l(\vec{\theta})\right)$, when $\Theta \subset \mathbb{R}^d$.

When $X_1, \ldots, X_n$ is a sample of iid random variables with pdf $f(x; \theta)$ with $\Theta \subset \mathbb{R}$, and $f(x; \theta)$ satisfies the regularity conditions in Definition 4.18, the Score function $\dot{\ell}(\theta)$ exists and $\dot{\ell}(\theta) = \sum_{i=1}^{n} \frac{\partial}{\partial \theta} \ln[f(x_i; \theta)]$. Similarly, when $\Theta \subset \mathbb{R}^d$,

$$\dot{\ell}(\vec{\theta}) = \begin{pmatrix} \sum_{i=1}^{n} \frac{\partial}{\partial \theta_1} \ln\left[f(x_i; \vec{\theta})\right] \\ \vdots \\ \sum_{i=1}^{n} \frac{\partial}{\partial \theta_d} \ln\left[f(x_i; \vec{\theta})\right] \end{pmatrix}.$$

For a sample of iid random variables, Theorem 5.2 shows that the Score function is directly related to Fisher's information.

Theorem 5.2 *If* $X_1, \ldots, X_n$ *is a sample of iid random variables with pdf* $f(x; \theta)$ *and* $\Theta \subset \mathbb{R}$ *satisfying the regularity conditions in Definition 4.18, then*

i) $E[\dot{\ell}(\theta)] = 0, \forall \theta \in \Theta$.

ii) $\mathrm{Var}[\dot{\ell}(\theta)] = nI(\theta), \forall \theta \in \Theta$.

Proof. Let $X_1, \ldots, X_n$ be a sample of iid random variables with pdf $f(x; \theta)$ and $\theta \in \Theta \subset \mathbb{R}$ satisfying the regularity conditions in Definition 4.18. For $\theta \in \Theta$, $\frac{\partial}{\partial \theta} \ln[f(X_1; \theta)], \ldots, \frac{\partial}{\partial \theta} \ln[f(X_n; \theta)]$ is a collection of iid random variables, and by Lemma 4.1, $E\left[\frac{\partial}{\partial \theta} \ln[f(X_i; \theta)]\right] = 0$ and $E\left[\left(\frac{\partial}{\partial \theta} \ln[f(X_i; \theta)]\right)^2\right] = I(\theta)$.

i) $$E[\dot{\ell}(\theta)] = E\left[\sum_{i=1}^n \frac{\partial}{\partial \theta} \ln[f(x_i; \theta)]\right] = \sum_{i=1}^n E\left[\sum_{i=1}^n \frac{\partial}{\partial \theta} \ln[f(x_i; \theta)]\right] = 0.$$

ii) Since $E[\dot{\ell}(\theta)] = 0$, $\mathrm{Var}[\dot{\ell}(\theta)] = E[\dot{\ell}(\theta)^2]$. Now,

$$\dot{\ell}(\theta)^2 = \sum_{i=1}^n \sum_{j=1}^n \frac{\partial}{\partial \theta} \ln[f(x_i; \theta)] \frac{\partial}{\partial \theta} \ln[f(x_j; \theta)]$$

$$= \sum_{i=1}^n \left(\frac{\partial}{\partial \theta} \ln[f(x_i; \theta)]\right)^2 + \sum_{i \neq j} \frac{\partial}{\partial \theta} \ln[f(x_i; \theta)] \frac{\partial}{\partial \theta} \ln[f(x_j; \theta)].$$

Let $Y_i = \frac{\partial}{\partial \theta} \ln[f(X_i; \theta)]$. Then, $Y_1, \ldots, Y_n$ are iid random variables, and since $Y_1, \ldots, Y_n$ are independent random variables,

$$\mathrm{Cov}(Y_i, Y_j) = E\left[\frac{\partial}{\partial \theta} \ln[f(x_i; \theta)] \frac{\partial}{\partial \theta} \ln[f(x_j; \theta)]\right] = 0$$

for $i \neq j$. Thus,

$$\mathrm{Var}[\dot{\ell}(\theta)] = E[\dot{\ell}(\theta)^2] = E\left[\sum_{i=1}^n \left(\frac{\partial}{\partial \theta} \ln[f(x_i; \theta)]\right)^2\right]$$

$$= E\left[\sum_{i=1}^n \left(\frac{\partial}{\partial \theta} \ln[f(x_i; \theta)]\right)^2\right] = \sum_{i=1}^n E\left[\left(\frac{\partial}{\partial \theta} \ln[f(x_i; \theta)]\right)^2\right]$$

$$= \sum_{i=1}^n I(\theta) = nI(\theta). \qquad \blacksquare$$

Theorem 5.3 *If* $X_1, \ldots, X_n$ *is a sample of iid random variables with pdf* $f(x; \theta)$ *and* $\theta \in \Theta \subset \mathbb{R}$ *satisfying the regularity conditions in Definition 4.18, then* $\forall \theta \in \Theta$

$$\frac{1}{\sqrt{n}} \dot{\ell}(\theta) \xrightarrow{d} N(0, I(\theta)).$$

Proof. Since $\frac{\partial}{\partial\theta}\ln[f(x_1;\theta)], \ldots, \frac{\partial}{\partial\theta}\ln[f(x_n;\theta)]$ are iid random variables with mean 0 and variance $I(\theta)$, Theorem 5.3 follows directly from the Central Limit Theorem for Sums given in Corollary 3.6. ∎

In Section 5.1, the Score function is used in finding an estimator of θ, and in Chapter 6, the Score function is used for testing claims about θ.

5.1 Maximum Likelihood Estimation

One of the most widely used methods for estimating the unknown parameters of a parametric probability model $f(x;\theta)$ is *maximum likelihood estimation*, which was widely popularized and used extensively by Fisher [2]. The method of maximum likelihood estimation (MLE) is based on the behavior of the likelihood function, and in particular, the *maximum likelihood principle*.

The Maximum Likelihood Principle: *Given a random sample $X_1, \ldots, X_n$ and a parametric model $f(x_1, \ldots, x_n; \theta)$, choose as the estimator of θ, say $\widehat{\theta}(\vec{X})$, the value of $\theta \in \Theta$ that maximizes the likelihood function.*

The maximum likelihood principle states that the estimator of a parameter θ should be the value of $\theta \in \Theta$ that maximizes the likelihood function, which according to the Law of Likelihood is the value of θ that best agrees with the observed data. The formal definition of the *maximum likelihood estimator* (MLE) of θ is given in Definition 5.2.

Definition 5.2 The MLE of θ, denoted by $\widehat{\theta}$, is a solution to the maximization problem $\max_{\theta\in\Theta} L(\theta)$.

It is important to note that a maximum likelihood estimate must reside in the parameter space Θ; however, an MLE is not necessarily unique. In some cases, the value of θ maximizing the likelihood function can occur at one of the boundaries of Θ, and therefore, in these cases, the parameter space will be taken to be a closed interval rather than an open interval so that the boundary values are included in Θ. For example, when $X \sim \text{Bin}(n, \theta)$, the parameter space $\Theta = [0, 1]$ will be used rather than $(0, 1)$.

5.1.1 Properties of MLEs

There are several reasons for using the method of maximum likelihood to estimate the parameters in a parametric probability model. In particular, MLEs are usually functions of a sufficient statistic, they are invariant under parameter transformations, and for large samples, they are often asymptotically unbiased and MSE consistent. Theorems 5.4 and 5.5 show that a unique MLE is a function of a sufficient statistic, and an MLE is invariant under transformations of θ.

Theorem 5.4 *If* $X_1, \ldots, X_n$ *is a sample of iid random variables with pdf* $f(x; \theta)$ *for* $\theta \in \Theta$, S *is a sufficient statistic for* θ, *and* $\hat{\theta}$ *is the unique MLE of* θ, *then* $\hat{\theta}$ *is a function of the sufficient statistic* S.

Proof. Let $X_1, \ldots, X_n$ be a sample of iid random variables with pdf $f(x; \theta)$ for $\theta \in \Theta$, S a sufficient statistic for θ, and $\hat{\theta}$ the unique MLE of θ. Since S is sufficient for θ, by the Fisher–Neyman factorization theorem, $L(\theta) = g(S(\vec{x}; \theta))h(\vec{x})$ for some functions g and h. Thus, the likelihood function is

$$L(\theta) = \prod_{i=1}^{n} f(x_i; \theta) = g(S(\vec{x}); \theta)h(\vec{x}),$$

and therefore,

$$\max_{\theta \in \Theta} L(\theta) = \max_{\theta \in \Theta} g(S(\vec{x}); \theta)h(\vec{x}) = h(\vec{x})\max_{\theta \in \Theta} g(S(\vec{x}); \theta).$$

Thus, maximizing $L(\theta)$ is equivalent to maximizing $g(S(\vec{x}); \theta)$, and since the maximum likelihood estimate is unique, $\hat{\theta}$ can only depend on $X_1, \ldots, X_n$ through a function of the sufficient statistic S. ∎

As a result of Theorem 5.4 and the Lehmann–Scheffé theorem, when $\hat{\theta}$ is the unique MLE of θ based on a complete sufficient statistic and $\hat{\theta}$ is an unbiased estimator of θ, then $\hat{\theta}$ is a UMVUE of θ.

Theorem 5.5 (**The Invariance Property of the MLE**) *If* $\hat{\theta}$ *is the MLE of* θ, *then for any function* $\tau(\theta)$, *the MLE of* $\tau(\theta)$ *is* $\tau(\hat{\theta})$.

Proof. The proof of Theorem 5.5 depends upon the induced likelihood function and can be found in Casella and Berger [13]. ∎

The invariance property of MLEs is sometimes referred to as the "plug-in rule" since the MLE of θ can be plugged into any function of $\tau(\theta)$ to create the MLE of $\tau(\theta)$. For example, when $\hat{\theta}$ is the MLE of $\text{Var}(X)$, by the invariance property of the MLE, $\sqrt{\hat{\theta}}$ is the MLE of the standard deviation of X.

Example 5.1 If $X_1, \ldots, X_n$ are iid Pois(θ) random variables and $\hat{\theta}$ is the MLE of θ, then the MLE of $P(X = 1) = \theta e^{-\theta}$ is $\hat{\theta} e^{-\hat{\theta}}$.

While there is no guarantee that an MLE will be an unbiased estimator, Theorem 5.6 shows that under certain regularity conditions, not only is the

MLE asymptotically unbiased but it is also MSE consistent, asymptotically normally distributed, and asymptotically efficient.

Theorem 5.6 (Large Sample Properties of an MLE) *Let $X_1, \ldots, X_n$ be a sample of iid random variables with pdf $f(x; \theta)$ for $\Theta \subset \mathbb{R}$ satisfying the regularity conditions in Definition 4.18 and let $\hat{\theta}$ be the MLE of θ. Then,*

i) $\hat{\theta}$ *is an asymptotically unbiased estimator of θ.*

ii) $\hat{\theta}$ *is MSE consistent for θ.*

iii) $\sqrt{n}(\hat{\theta} - \theta) \xrightarrow{d} N(0, I^{-1}(\theta))$.

iv) $\hat{\theta}$ *is asymptotically efficient estimator of θ.*

The exact variance of a maximum likelihood estimator can be difficult to compute, but for a sample of iid random variables from a pdf satisfying the regularity conditions in Definition 4.18 and for a sufficiently large sample size, the *asymptotic variance* can be used to approximate the variance of an MLE. In particular, Theorem 5.6 shows that the asymptotic variance, denoted by $\mathrm{AsyVar}(\hat{\theta})$, is the Cramér–Rao lower bound on the variance of unbiased estimators of θ; the *asymptotic standard error* of the MLE is the square root of the asymptotic variance.

In practice, there are two ways to estimate the asymptotic variance of an MLE. The first method uses the plug-in rule with $\hat{\theta}$ and $I(\theta)$, and in this case, the asymptotic variance of the MLE is $\mathrm{AsyVar}(\hat{\theta}) = \frac{1}{nI(\hat{\theta})}$. The second method is also based on the plug-in rule with $\hat{\theta}$ and $\ddot{\ell}(\theta)$, and in this case, $\mathrm{AsyVar}(\hat{\theta}) = -\frac{1}{\ddot{\ell}(\hat{\theta})}$. Both methods are asymptotically equivalent and converge to the true asymptotic variance as $n \to \infty$.

An analogous result holds for multiparameter distributions with $\Theta \subset \mathbb{R}^d$, provided the pdf satisfies the necessary regularity conditions, that is, when $X_1, \ldots, X_n$ is a sample of iid random variables with common pdf $f(x; \vec{\theta})$ for $\Theta \subset \mathbb{R}^d$, $f(x; \vec{\theta})$ satisfies the regularity conditions given in Definition 4.18, and $\hat{\theta} = (\hat{\theta}_1, \ldots, \hat{\theta}_d)$ is the MLE of $\vec{\theta}$, then $\hat{\theta}_i$ is asymptotically unbiased, MSE consistent, asymptotically efficient, and

$$\sqrt{n}(\hat{\theta}_i - \theta_i) \xrightarrow{d} N(0, I^{-1}(\vec{\theta})_{ii}),$$

where $I(\theta)$ is the $d \times d$ Fisher's information matrix. Also, in the multiparameter case, the asymptotic variance of $\hat{\theta}_i$ is $\mathrm{AsyVar}(\hat{\theta}_i) = \frac{1}{n} I_{ii}(\theta)^{-1}$.

5.1.2 One-parameter Probability Models

Determining the MLE for a parametric probability model $f(x; \theta)$ when $\Theta \subset \mathbb{R}$ involves determining the value or values of $\theta \in \Theta$ that maximize the likelihood function $L(\theta)$. Since the solution to an optimization problem is unchanged by a strictly monotone increasing transformation of the objective function, the MLE can also be found by maximizing the log-likelihood function.

When $X_1, \ldots, X_n$ is a sample of iid random variables with pdf $f(x; \theta)$ for $\Theta \subset \mathbb{R}$ and the log-likelihood function is differentiable with respect to θ, the MLE of θ can be found by finding the critical points of $l(\theta)$ and determining whether or not a critical point maximizes $l(\theta)$. In particular, when $l(\theta)$ is twice differentiable, any critical point in Θ satisfying $\dot{\ell}(\theta) = 0$ and $\ddot{\ell}(\theta) < 0$ is an MLE of θ. Thus, when the log-likelihood is differentiable, the MLE is found by setting the Score function equal to zero and solving for θ. Note that the log-likelihood function is differentiable with respect to θ when the parametric model $f(x; \theta)$ is a regular exponential family distribution, and in this case, the MLE is a solution to $\dot{\ell}(\theta) = 0$.

Examples 5.2–5.4 illustrate the process of determining the MLE when the log-likelihood function is differentiable.

Example 5.2 Let $X_1, \ldots, X_n$ be a sample of iid Bernoulli random variables with $\Theta = [0, 1]$. The likelihood function associated with $X_1, \ldots, X_n$ is

$$L(\theta) = \prod_{i=1}^{n} \theta^{x_i}(1-\theta)^{1-x_i} = \theta^{\sum_{i=1}^{n} x_i}(1-\theta)^{n-\sum_{i=1}^{n} x_i},$$

and hence, $l(\theta) = \sum_{i=1}^{n} x_i \ln(\theta) + \left(n - \sum_{i=1}^{n} x_i\right) \ln(1-\theta)$. Since the log-likelihood function is differentiable with respect to θ, the Score function is

$$\dot{\ell}(\theta) = \frac{\sum_{i=1}^{n} x_i}{\theta} - \frac{n - \sum_{i=1}^{n} x_i}{1-\theta}.$$

Setting $\dot{\ell}(\theta) = 0$ and solving for θ yields

$$0 = \dot{\ell}(\theta) = \frac{\sum_{i=1}^{n} x_i}{\theta} - \frac{n - \sum_{i=1}^{n} x_i}{1-\theta} \Longleftrightarrow \frac{\sum_{i=1}^{n} x_i}{\theta} = \frac{n - \sum_{i=1}^{n} x_i}{1-\theta}$$

$$\Longleftrightarrow (1-\theta) \sum_{i=1}^{n} x_i = \theta \left(n - \sum_{i=1}^{n} x_i\right) \Longleftrightarrow \sum_{i=1}^{n} x_i = \theta n$$

$$\Longleftrightarrow \theta = \frac{\sum_{i=1}^{n} x_i}{n} = \bar{x}.$$

The second derivative of $l(\theta)$ evaluated at $\theta = \bar{x}$ is

$$\ddot{\ell}(\bar{x}) = -\frac{\sum_{i=1}^{n} x_i}{\bar{x}^2} - \frac{n - \sum_{i=1}^{n} x_i}{(1-\bar{x})^2} < 0$$

and since $\bar{x} \in [0, 1]$, it follows that the MLE of θ is $\hat{\theta} = \bar{x}$.

Moreover, since $f(x; \theta)$ is a regular exponential family distribution with complete sufficient statistic $\sum_{i=1}^{n} X_i$ and $E(\bar{X}) = \theta$, it follows that $\hat{\theta} = \bar{X}$ is a UMVUE of θ. The standard error of $\hat{\theta}$ is

$$SE(\hat{\theta}) = \sqrt{Var(\bar{X})} = \sqrt{\frac{\sigma^2}{n}} = \sqrt{\frac{\theta(1-\theta)}{n}}.$$

Example 5.3   Let $X_1, \ldots, X_n$ be a sample of iid Geo(θ) random variables with $S = \mathbb{W}$ and $\Theta = [0, 1]$. The likelihood function associated with $X_1, \ldots, X_n$ is

$$L(\theta) = \prod_{i=1}^{n} \theta(1 - \theta)^{x_i} = \theta^n (1 - \theta)^{\sum_{i=1}^{n} x_i},$$

and the log-likelihood function is $l(\theta) = n \ln(\theta) + \ln(1 - \theta) \sum_{i=1}^{n} x_i$. The log-likelihood function is differentiable with respect to θ, and

$$\dot{\ell}(\theta) = \frac{n}{\theta} - \frac{\sum x_i}{1 - \theta}.$$

Setting $\dot{\ell}(\theta) = 0$ and solving for θ yields

$$0 = \dot{\ell}(\theta) = \frac{n}{\theta} - \frac{\sum_{i=1}^{n} x_i}{1 - \theta} \iff \frac{n}{\theta} = \frac{\sum_{i=1}^{n} x_i}{1 - \theta}$$

$$\iff n(1 - \theta) = \theta \sum_{i=1}^{n} x_i \iff n = \theta \left(n + \sum_{i=1}^{n} x_i \right)$$

$$\iff \theta = \frac{n}{n + \sum_{i=1}^{n} x_i} = \frac{1}{1 + \bar{x}}.$$

Since $\frac{1}{1+\bar{x}}$ is in $\Theta = [0, 1]$, and

$$\ddot{\ell}(\theta) = -\frac{n}{\theta^2} - \frac{\sum_{i=1}^{n} x_i}{(1 - \theta)^2} < 0, \quad \forall \theta \in \Theta$$

the MLE of θ is $\widehat{\theta} = \frac{1}{1+\bar{x}}$.

In this case, there is no explicit formula for the standard error of $\widehat{\theta}$; however, for sufficiently large samples, the asymptotic standard error can be used. Since Fisher's information for $X \sim$ Geo(θ) with $S = \mathbb{W}$ is

$$I(\theta) = E\left[-\frac{\partial^2}{\partial \theta^2} \ln[\theta(1 - \theta)^X] \right] = E\left[\frac{1}{\theta^2} + \frac{X}{(1 - \theta)^2} \right] = \frac{1}{\theta^2(1 - \theta)},$$

the asymptotic variance of $\widehat{\theta}$ is AsyVar($\widehat{\theta}$) = $\frac{\theta^2(1-\theta)}{n}$. Thus, for sufficiently large samples SE($\widehat{\theta}$) $\approx \theta \sqrt{\frac{1-\theta}{n}}$.

Example 5.4   Let $X_1, \ldots, X_n$ be a sample of iid random variables with pdf $f(x; \theta) = \theta \, e^{-\theta x}$ for $x \in (0, \infty)$ and $\Theta \in (0, \infty)$. The likelihood function associated with $X_1, \ldots, X_n$ is

$$L(\theta) = \prod_{i=1}^{n} \theta \, e^{-\theta x_i} = \theta^n \, e^{-\theta \sum_{i=1}^{n} x_i},$$

and the log-likelihood function is $l(\theta) = n \ln(\theta) - \theta \sum_{i=1}^{n} x_i$.

Hence, the Score function is $\dot{\ell}(\theta) = \frac{n}{\theta} - \sum_{i=1}^{n} x_i$, and setting $\dot{\ell}(\theta) = 0$ and solving for θ yields

$$0 = \dot{\ell}(\theta) = \frac{n}{\theta} - \sum_{i=1}^{n} x_i \iff \frac{n}{\theta} = \sum_{i=1}^{n} x_i$$

$$\iff n = \theta \sum_{i=1}^{n} x_i \iff \theta = \frac{n}{\sum_{i=1}^{n} x_i} = \frac{1}{\bar{x}}.$$

Since $\frac{1}{\bar{x}} \in \Theta$ and $\ddot{\ell}(\theta) = -\frac{n}{\theta^2} < 0$, $\forall \theta \in \Theta$, the MLE of θ is $\hat{\theta} = \frac{1}{\bar{x}}$.

Furthermore, since $nI(\theta) = E[-\ddot{\ell}(\theta)] = \frac{n}{\theta^2}$, for sufficiently large sample sizes $\text{SE}(\hat{\theta}) \approx \frac{\theta}{\sqrt{n}}$. However, in this case, the exact variance of $\hat{\theta}$ can be determined using the fact that $\sum_{i=1}^{n} X_i \sim \text{Gamma}\left(n, \frac{1}{\theta}\right)$. The exact variance of $\hat{\theta} = \frac{1}{\bar{X}}$ is $\frac{n\theta^2}{(n-1)^2(n-2)}$, and thus, $\text{SE}(\hat{\theta}) = \frac{\theta}{n-1}\sqrt{\frac{n}{n-2}}$.

Example 5.5 Let $X_1, \ldots, X_n$ be a sample of iid Gamma(θ, 1) random variables. Then,

$$L(\theta) = \Gamma(\theta)^{-n} e^{-\sum_{i=1}^{n} x_i} \prod_{i}^{n} x_i^{\theta-1}$$

and $l(\theta) = -n \ln[\Gamma(\theta)] - \sum_{i=1}^{n} x_i + (\theta - 1) \sum_{i=1}^{n} \ln(x_i)$. Thus, the Score function is $\dot{\ell}(\theta) = -n\psi(\theta) - \sum_{i=1}^{n} \ln(x_i)$, where $\psi(\theta) = \frac{\partial}{\partial \theta} \ln[\Gamma(\theta)]$ is called the *digamma function*.

Thus, the MLE of θ is the solution to $\dot{\ell}(\theta) = -n\psi(\theta) - \sum_{i=1}^{n} \ln(x_i) = 0$, however in this case, there is no analytic solution to this equation. Therefore, the value of the MLE must be found numerically and requires the observed values of the random variables $X_1, \ldots, X_n$.

For some probability models with a differentiable log-likelihood, there will be no analytic solution to $\dot{\ell}(\theta) = 0$, and in this case, the MLE must be found using numerical methods. When there is no analytic solution to $\dot{\ell}(\theta) = 0$, a statistical package such as R can be used to find the MLE numerically. In particular, the R package *fitdistrplus* can be used with an observed sample to find the MLE for most of the commonly used probability models including the normal, log-normal, exponential, Poisson, Cauchy, gamma, logistic, negative binomial, geometric, beta, and Weibull probability models.

The R commands used to find the MLE with the R package `fitdistrplus` are

```
> install.packages("fitdistrplus")
> library("fitdistrplus")
> fitdist(x,dist,...)
```

where x is a vector containing the sampled values, dist is the R name of the distribution being fit. For more information and examples on using the fitdist command type ?fitdist in R. The output of the fitdist command includes the MLE and its asymptotic standard error. The MLE for a sample of iid Gamma random variables is found using fitdistrplus in Example 5.6.

Example 5.6 Suppose that a sample of $n = 40$ iid Gamma($\theta, 4$) random variables produces the data in Table 5.1.

The MLE of θ can be found using the fitdist command in R, assuming that the observed data have been read into the object x. The fitdist command requires the specification of the object containing the data, the distribution, and initial guess of the value of θ using the start option, and any fixed (i.e. known) arguments using fix.arg. The R command and its output for the data in Table 5.1 are

```
> fitdist(x,"gamma",start=list(shape=2),
  + fix.arg=list(scale=4))
  Fitting of the distribution 'gamma' by maximum likelihood
  Parameters:
        estimate  Std. Error
  shape 10.07611  0.4895095
```

Thus, the MLE of θ is $\widehat{\theta} = 10.076$, and the asymptotic standard error of the MLE is 0.490.

Among the probability distributions where the MLE must be found numerically are the beta, gamma, Weibull, logistic, and Gumbel distributions; more complicated probability models such as finite mixture models and generalized linear models may also require that the MLEs be found numerically.

For probability models where the log-likelihood function is not differentiable with respect the θ, the MLE of θ must be found by carefully analyzing the behavior of the likelihood function over the parameter space Θ. The most common reason for the log-likelihood function to be nondifferentiable is that the support of $f(x; \theta)$ depends on θ. Thus, when the support of $f(x; \theta)$ depends on θ, the likelihood function must take into account the support of the underlying probability model.

Table 5.1 A sample of $n = 40$ iid observations from a Gamma($\theta, 4$) random variable.

44.52	61.50	31.56	34.36	30.17	33.64	32.02	44.91	45.60	54.92
58.14	44.01	26.36	30.46	25.68	57.76	37.35	27.22	30.17	39.78
29.33	22.39	28.47	39.87	48.68	34.67	53.28	63.46	31.22	28.56
46.10	30.57	51.00	66.42	29.75	86.94	45.49	22.51	25.68	43.81

When the support S of a probability model depends on θ, say S_θ, the likelihood function is

$$L(\theta) = \prod_{i=1}^{n} f(x_i; \theta) I_{S_\theta}(x_i) = \prod_{i=1}^{n} f(x_i; \theta) \prod_{i=1}^{n} I_{S_\theta}(x_i),$$

where $I_{S_\theta}(x_i) = \begin{cases} 1 & \text{when } x_i \in S_\theta \\ 0 & \text{otherwise} \end{cases}$.

Then, since $I_{S_\theta}(x_i)$ is 0 or 1 for each observation x_i, it follows that $\prod_{i=1}^{n} I_{S_\theta}(x_i)$ is also 0 or 1. Thus, the likelihood function is

$$L(\theta) = \begin{cases} \displaystyle\prod_{i=1}^{n} f(x_i; \theta) & \text{when } x_i \in S_\theta, \forall i, \\ 0 & \text{otherwise,} \end{cases}$$

and the maximum value of the likelihood function occurs only for a θ value where $\prod_{i=1}^{n} I_{S_\theta}(x_i) = 1$.

Thus, the first step in analyzing the likelihood function is to determine the subset of Θ, say Θ^+, where $\prod_{i=1}^{n} I_{S_\theta}(x_i) = 1$. The second step in finding the MLE is to determine the value of $\theta \in \Theta^+$ maximizing $\prod_{i=1}^{n} f(x_i; \theta)$ on Θ^+.

Examples 5.7 and 5.8 illustrate the process used for finding the MLE when the support of the probability model depends on θ.

Example 5.7 Let $X_1, \ldots, X_n$ be a sample of iid $U(0, \theta]$ random variables with $\Theta = \mathbb{R}^+$. Since $S = (0, \theta]$ depends on θ, the pdf can be written as $f(x; \theta) = \frac{1}{\theta} I_{(0,\theta]}(x)$. Then, the likelihood function is

$$L(\theta) = \prod_{i=1}^{n} \frac{1}{\theta} I_{(0,\theta]}(x_i) = \theta^{-n} \prod_{i=1}^{n} I_{(0,\theta]}(x_i)$$

$$= \begin{cases} \theta^{-n} & \text{if } x_i \in (0, \theta], \quad \forall i, \\ 0 & \text{otherwise,} \end{cases}$$

and the set of values of θ for which $\prod_{i=1}^{n} I_{(0,\theta]}(x_i) = 1$ is

$$\Theta^+ = \{\theta \in \mathbb{R} : 0 < x_i \le \theta, \, i = 1, \ldots, n\} = \{\theta \in \mathbb{R}^+ : \theta \ge x_{(n)}\}.$$

Thus, the maximum value of $L(\theta)$ occurs only when $\theta \ge x_{(n)}$, and since $L(\theta) = \theta^{-n}$ is a decreasing function on Θ^+, it follows that $L(\theta)$ attains its maximum value at $\theta = x_{(n)}$. Therefore, the MLE of θ is $\hat{\theta} = X_{(n)}$.

Example 5.8 Let $X_1, \ldots, X_n$ be a sample of iid $\text{Exp}(1, \theta)$ random variables with pdf $f(x; \theta) = e^{-(x-\theta)}$ for $x \in [\theta, \infty)$ and $\Theta = \mathbb{R}$. Then, the likelihood function is

$$L(\theta) = \prod_{i=1}^{n} e^{-(x-\theta)} I_{[\theta,\infty)}(x_i) = e^{-\sum_{i=1}^{n}(x_i-\theta)} \prod_{i=1}^{n} I_{[\theta,\infty)}(x_i),$$

and the set of values of θ for which $\prod_{i=1}^{n} I_{[\theta,\infty]}(x_i) = 1$ is

$$\Theta^+ = \{\theta \in \mathbb{R} : x_i \geq \theta, \ i = 1, \ldots, n\} = \{\theta \in \mathbb{R}^+ : \theta \leq x_{(1)}\}.$$

Thus, the maximum value of $L(\theta)$ occurs only when $\theta \leq x_{(1)}$, and since $L(\theta) = e^{n\theta - \sum_{i=1}^{n} x_i}$ is increasing on Θ^+, it follows that $L(\theta)$ attains its maximum value at $\theta = x_{(1)}$. Therefore, the MLE of θ is $\hat{\theta} = X_{(1)}$.

For some probability models, only a subset of the parameter space will be considered, that is, the MLE must be an element of a *restricted parameter space*. For example, the unrestricted parameter space for the binomial probability model is $\Theta = [0, 1]$, but when the only possible values of θ under consideration are $\{0.5, 0.8\}$, the parameter space for the probability model is a restricted parameter space. In some cases, the log-likelihood on a restricted parameter space will not be differentiable, and in other cases, the boundaries of the restricted parameter space must be dealt with when determining the MLE. In either case, the MLE must be a value of the restricted parameter space.

When the restricted parameter space is at the most a countable subset of the general parameter space, the log-likelihood function is not differentiable; in this case, the *likelihood ratio* can often be used to analyze the behavior of the likelihood function on the restricted parameter space. For $\theta_1 < \theta_2$, the likelihood ratio is $\Lambda(\theta_1, \theta_2) = \frac{L(\theta_1)}{L(\theta_2)}$, and the likelihood ratio can be used to determine where $L(\theta)$ is increasing and decreasing. In most cases, the MLE will occur at the change point where $\Lambda(\theta_1, \theta_2)$ goes from being less than one to greater than one or vice versa.

Example 5.9 Let $X_1, \ldots, X_n$ be a sample of iid $\text{Bin}(k, \theta)$ random variables on the restricted parameter space $\Theta = \{0.3, 0.5\}$ and with k known. Then, the likelihood function is

$$L(\theta) = \prod_{i=1}^{n} \binom{k}{x_i} \theta^{x_i}(1-\theta)^{k-x_i} = \theta^{\sum_{i=1}^{n} x_i}(1-\theta)^{nk-\sum_{i=1}^{n} x_i} \prod_{i=1}^{n} \binom{k}{x_i},$$

and the likelihood ratio is

$$\Lambda(0.3, 0.5) = \frac{L(0.3)}{L(0.5)} = \frac{0.3^{\sum_{i=1}^{n} x_i}(0.7)^{kn-\sum_{i=1}^{n} x_i}}{0.5^{kn}}$$

$$= \left(\frac{0.3}{0.7}\right)^{\sum_{i=1}^{n} x_i}\left(\frac{0.7}{0.5}\right)^{kn}.$$

Thus, the MLE of θ is

$$\hat{\theta} = \begin{cases} 0.3 & \text{when } \Lambda(0.3, 0.5) > 1 \\ 0.5 & \text{when } \Lambda(0.3, 0.5) < 1 \end{cases}$$

and when $\Lambda(0.3, 0.5) = 1$, both values of θ agree with the observed data equally well, and in this case, the MLE $\hat{\theta}$ is not unique.

The determination of the MLE θ actually requires the values of k, n, and $\sum_{i=1}^{n} X_i$. For example, suppose that $k = 5, n = 10$, and $\sum_{i=1}^{10} = 35$. Then, $\Lambda(0.3, 0.5) = 0.348$, and thus, $\hat{\theta} = 0.5$. On the other hand, if $k = 5, n = 10$, and $\sum_{i=1}^{10} = 21$, $\Lambda(0.3, 0.5) = 444.199$ and $\hat{\theta} = 0.3$.

When the restricted parameter space is a subinterval of the unrestricted parameter space, the MLE can often be found by considering the MLE on the unrestricted parameter space and then adjusting appropriately when the value of θ maximizing the likelihood function is not in the restricted parameter space.

Example 5.10 Let $X_1, \ldots, X_n$ be a sample of iid $N(\theta, 1)$ random variables with $\Theta = [0, \infty)$. Then, the log-likelihood function is $l(\theta) = -\frac{n}{2}\ln[2\pi] - \frac{1}{2}\sum_{i=1}^{n}(x_i - \theta)^2$, and the Score function is $\dot{\ell}(\theta) = \sum_{i=1}^{n}(x_i - \theta) = n(\bar{x} - \theta)$.

Since the unrestricted parameter space is $\mathbb{R}$, the MLE on the unrestricted parameter space is $\hat{\theta} = \bar{x}$; however, $\bar{x}$ may not be in the restricted parameter space $\Theta = [0, \infty)$. Now, since the Score function indicates that $l(\theta)$ is increasing for $\theta \leq \bar{x}$ and decreasing for $\theta > \bar{x}$ it follows that when $\bar{x} \geq 0$, the MLE of θ is $\bar{x}$, and when $\bar{x} < 0$, the MLE is the smallest value in $[0, \infty)$ which is zero. Therefore, the MLE of θ for the restricted parameter space $\Theta = [0, \infty)$ is

$$\hat{\theta} = \begin{cases} 0 & \text{when } \bar{x} < 0 \\ \bar{x} & \text{when } \bar{x} \geq 0 \end{cases}.$$

5.1.3 Multiparameter Probability Models

When the parametric probability model is a multiparameter model with $\Theta \subset \mathbb{R}^d$, the MLE is the vector of estimates $\hat{\theta} = (\hat{\theta}_1, \ldots, \hat{\theta}_d) \in \Theta$ maximizing the likelihood function. Maximizing $L(\theta)$ over a parameter space Θ is more complicated in a d-parameter probability model than it is in a one-parameter model; however, when the log-likelihood is twice differentiable with respect to each parameter θ_i, the MLE can still be found by maximizing the log-likelihood function.

In particular, when $\Theta \subset \mathbb{R}^2$ and the log-likelihood is twice differentiable with respect to θ_1 and θ_2, the MLE of $\vec{\theta}$ is a solution to the simultaneous system of equations

$$0 = \dot{\ell}_1(\vec{\theta}) = \frac{\partial \ell(\theta)}{\partial \theta_1}$$

$$0 = \dot{\ell}_2(\vec{\theta}) = \frac{\partial \ell(\theta)}{\partial \theta_2}$$

provided $\ddot{\ell}_{11}(\hat{\theta})\ddot{\ell}_{22}(\hat{\theta}) - [\ddot{\ell}_{12}(\hat{\theta})]^2 > 0$ and $\ddot{\ell}_{11}(\hat{\theta}) < 0$, where $\ddot{\ell}_{ij}(\vec{\theta}) = \frac{\partial^2 \ell(\theta)}{\partial \theta_i \, \partial \theta_j}$.

In the multiparameter case, the MLE is still invariant under transformations, and under the regularity conditions in Definition 4.18, the component MLEs are asymptotically unbiased, MSE consistent, and asymptotically normally distributed.

Example 5.11 Let $X_1, \ldots, X_n$ be a sample of iid $N(\theta_1, \theta_2)$ random variables with $\Theta = \mathbb{R} \times \mathbb{R}^+$. Then, the likelihood function is

$$L(\theta_1, \theta_2) = \prod_{i=1}^{n} \frac{1}{\sqrt{2\pi\theta_2}} e^{-\frac{1}{2\theta_2}(x_i - \theta_1)^2} = (2\pi\theta_2)^{-n/2} \, e^{-\frac{1}{2\theta_2} \sum (x_i - \theta_1)^2}$$

and the log-likelihood function is

$$\ell(\theta_1, \theta_2) = -\frac{n}{2} \ln(2\pi\theta_2) - \frac{1}{2\theta_2} \sum (x_i - \theta_1)^2.$$

Since the log-likelihood function is twice differentiable with respect to both θ_1 and θ_2, the MLE of $\vec{\theta}$ is a solution to the simultaneous system of equations

$$0 = \dot{\ell}_1(\theta_1, \theta_2) = \frac{2}{2\theta_2} \sum (x_i - \theta_1), \tag{5.1}$$

$$0 = \dot{\ell}_2(\theta_1, \theta_2) = -\frac{n}{2\theta_2} + \frac{1}{2\theta_2^2} \sum (x_i - \theta_1)^2. \tag{5.2}$$

First, solving $\dot{\ell}_1(\theta_1, \theta_2) = 0$ for θ_1 yields

$$0 = \dot{\ell}_1(\theta_1, \theta_2) = \frac{1}{\theta_2} \sum (x_i - \theta_1) \Longleftrightarrow 0 = \sum x_i - n\theta_1$$

$$\Longleftrightarrow 0 = n\bar{x} - n\theta_1 \Longleftrightarrow \theta_1 = \bar{x}.$$

Then, plugging $\theta_1 = \bar{x}$ into the equation $0 = \dot{\ell}_2(\theta_1, \theta_2)$ and solving for θ_2 yields

$$0 = -\frac{n}{2\theta_2} + \frac{1}{2\theta^2} \sum (x_i - \bar{x})^2 \Longleftrightarrow \frac{n}{2\theta_2} = \frac{1}{2\theta^2} \sum (x_i - \bar{x})^2$$

$$\Longleftrightarrow \theta_2 = \frac{1}{n} \sum (x_i - \bar{x})^2.$$

A quick check of the second partial derivatives properties of $\ell(\theta_1, \theta_2)$ shows that the log-likelihood function attains its maximum at $\theta_1 = \bar{x}$ and $\theta_2 = \frac{1}{n} \sum (x_i - \bar{x})^2$, and hence, the MLEs of θ_1 and θ_2 are $\hat{\theta}_1 = \bar{x}$ and $\hat{\theta}_2 = \frac{1}{n} \sum (x_i - \bar{x})^2$.

Moreover, since $\hat{\theta}_1 = \bar{X}$ is an unbiased estimator of θ_1 based on a complete sufficient statistic, it follows that $\hat{\theta}_1$ is a UMVUE of θ_1. On the other hand, $\hat{\theta}_2 = \frac{1}{n} \sum_{i=1}^{n} (X_i - \bar{X})^2$ is a biased estimator of θ_2 and is not a UMVUE of θ_2.

Finally, by the invariance property of MLEs, the MLE of $SD(X) = \sqrt{\theta_2}$ is $\sqrt{\widehat{\theta}_2} = \sqrt{\frac{1}{n}\sum_{i=1}^{n}(X_i - \overline{X})^2}$, and the MLE of the coefficient of variation, $CV = \frac{\sqrt{\theta_2}}{|\theta_1|}$, is

$$\widehat{CV} = \frac{\sqrt{\widehat{\theta}_2}}{|\widehat{\theta}_1|} = \frac{\sqrt{\frac{1}{n}\sum_{i=1}^{n}(x_i - \overline{x})^2}}{|\overline{x}|}.$$

As is the case with some of the one-parameter probability models, there are multiparameter probability models with differentiable log-likelihood functions where the MLE must be found numerically. In particular, the gamma, beta, Weibull, and Gumbel probability models all require that the MLE be found numerically.

Example 5.12 Let $X_1, \ldots, X_n$ be a sample of iid Gamma(θ_1, θ_2) random variables. Then, the likelihood function is

$$L(\theta_1, \theta_2) = \prod_{i=1}^{n} \frac{1}{\Gamma(\theta_1)\theta_2^{\theta_1}} x_i^{\theta_1 - 1} e^{-\frac{x_i}{\theta_2}} = \frac{e^{-\frac{1}{\theta_2}\sum_{i=1}^{n} x_i} \prod_{i=1}^{n} x_i^{\theta_1 - 1}}{[\Gamma(\theta_1)\theta_2^{\theta_1}]^n},$$

and the log-likelihood function is

$$\ell(\theta_1, \theta_2) = -n\ln[\Gamma(\theta_1)] - n\theta_1\ln(\theta_2) - \frac{1}{\theta_2}\sum_{i=1}^{n} x_i + (\theta_1 - 1)\sum_{i=1}^{n}\ln(x_i).$$

Thus, the partial derivatives of $\ell(\theta_1, \theta_2)$ with respect to θ_1 and Rθ_2 are

$$\dot{\ell}_1(\theta_1, \theta_2) = -n\psi(\theta_1) - n\ln(\theta_2) + \sum_{i=1}^{n}\ln(x_i),$$

$$\dot{\ell}_2(\theta_1, \theta_2) = -\frac{n\theta_1}{\theta_2} + \frac{1}{\theta_2^2}\sum_{i=1}^{n} x_i.$$

However, there are no analytic solutions to $\dot{\ell}_1(\theta_1, \theta_1) = \dot{\ell}_2(\theta_1, \theta_2) = 0$, and thus, the MLEs of θ_1 and θ_2 must be found numerically.

For example, suppose that the sample of $n = 50$ iid observations shown in Table 5.2 comes from a Gamma(θ_1, θ_2) distribution. Again, the R package *fitdistrplus* can be used to determine the MLEs of θ_1 and θ_2 using the command `fitdist(xg,"gamma",start=list(shape=1,scale=2))`, where the data in Table 5.2 is stored in the object xg.

Table 5.2 An observed sample of $n = 50$ observations from $X \sim \text{Gamma}(\theta_1, \theta_2)$.

40.59	26.61	11.03	14.38	10.50	40.21	20.41	11.71	14.14	22.63
13.43	8.00	12.73	22.72	31.12	18.01	35.67	46.09	15.02	12.80
28.60	14.47	28.39	50.34	13.79	71.45	28.02	8.08	10.49	26.41
11.86	9.19	28.06	20.75	34.96	25.39	15.68	38.72	17.92	30.39
36.47	26.03	23.60	22.71	31.79	31.18	31.01	33.48	16.77	14.15

The output of the command `fitdist(xg, "gamma")` is

```
> fitdist(xg,"gamma",start=list(shape=1,scale=2))
    Fitting of the distribution 'gamma' by maximum likelihood
    Parameters:
            estimate Std. Error
    shape 4.114569   0.7903262
    scale 5.870123   1.1989513
```

The standard errors reported by `fitdist` are the asymptotic standard errors computed from the Hessian matrix

$$H(\vec{x}) = \begin{pmatrix} \ddot{\ell}_{11}(\hat{\theta}_1, \hat{\theta}_2) & \ddot{\ell}_{12}(\hat{\theta}_1, \hat{\theta}_2) \\ \ddot{\ell}_{21}(\hat{\theta}_1, \hat{\theta}_2) & \ddot{\ell}_{22}(\hat{\theta}_1, \hat{\theta}_2) \end{pmatrix},$$

which is an estimator of $I_n(\theta_1, \theta_2)$. In particular, the estimate of the asymptotic variance of $\hat{\theta}_i$ based on the Hessian matrix is $[H(\vec{x})^{-1}]_{ii}$.

A visual assessment of the fit of the observed sample to the gamma probability model is produced by the R commands

```
> ff1=fitdist(xg,"gamma",start=list(shape=1,scale=2))
> plot(ff1)
```

The visual assessment of the model fit to the data in Table 5.2 is shown in Figure 5.1.

When the support of a multiparameter probability model depends on one or more of the parameters, the log-likelihood function is not differentiable over Θ. As is the case with one-parameter distributions, the likelihood function for a multiparameter range dependent probability model must be expressed using the indicator function of the support. Again, maximizing the likelihood function requires identifying the subset of Θ, say, Θ^+, where $L(\vec{\theta}) > 0$, and the MLE of θ is the value of $\theta \in \Theta^+$ that maximizes the likelihood function.

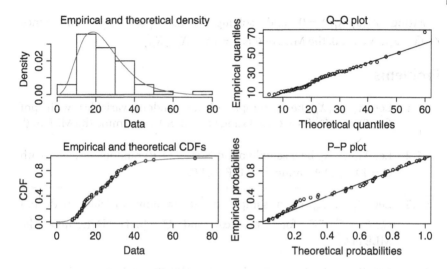

Figure 5.1 The R visual assessments of the gamma model fit to the data in Table 5.2.

Example 5.13 Let $X_1, \ldots, X_n$ be a sample of iid $\mathrm{Exp}(\theta_1, \theta_2)$ random variables with pdf $f(x; \theta_1, \theta_2) = \frac{1}{\theta_1} e^{-\frac{1}{\theta_1}(x-\theta_2)}$ for $x \in [\theta_2, \infty)$ and $\Theta = \mathbb{R}^+ \times \mathbb{R}$. Then, $f(x; \theta_1, \theta_2)$ is a range dependent probability model, and the likelihood function is

$$L(\theta_1, \theta_2) = \prod_{i=1}^{n} \frac{1}{\theta_1} e^{-\frac{1}{\theta_1}(x_i - \theta_2)} I_{[\theta_2, \infty)}(x_i)$$

$$= \theta_1^{-n} e^{-\frac{1}{\theta_1} \sum_{i=1}^{n}(x_i - \theta_2)} \prod_{i=1}^{n} I_{[\theta_2, \infty)}(x_i)$$

$$= \theta_1^{-n} e^{-\frac{1}{\theta_1} \sum_{i=1}^{n}(x_i - \theta_2)} \prod_{i=1}^{n} I_{(-\infty, x_{(1)}]}(\theta_2)$$

since

$$\prod_{i=1}^{n} I_{[\theta_2, \infty)}(x_i) = \begin{cases} 1 & \text{if } x_i \in [\theta_2, \infty), \ \forall i, \\ 0 & \text{otherwise.} \end{cases} = \begin{cases} 1 & \text{if } x_{(1)} \geq \theta_2, \\ 0 & \text{otherwise.} \end{cases}$$

Thus, $L(\theta_1, \theta_2) > 0$ only when $x_{(1)} \geq \theta_2$, $\forall \theta_1 \in \mathbb{R}^+$. Furthermore, for a fixed value of $\theta_1 \in \mathbb{R}^+$ and $x_{(1)} \geq \theta_2$, $L(\theta_1, \theta_2) = \theta_1^{-n} e^{-\frac{1}{\theta_1} \sum_{i=1}^{b}(x_i - \theta_2)}$ is increasing in θ_2. Therefore, for a fixed value of θ_1, $L(\theta_1, \theta_2)$ attains its maximum only when $\theta_2 = x_{(1)}$.

Now, for $\theta_2 = x_{(1)}$, $\ell(\theta_1, x_{(1)}) = -n \ln(\theta_1) - \frac{1}{\theta_1} \sum (x_i - x_{(1)})$ is differentiable with respect to θ_1, and $\ell_1(\theta_1, x_{(1)}) = -\frac{n}{\theta_1} + \frac{1}{\theta_1^2} \sum (x_i - x_{(1)})$.

Setting $\dot{\ell}_1(\theta_1, X_{(1)}) = 0$ and solving yields $\theta_1 = \bar{x} - x_{(1)}$, and since $\ddot{\ell}_{11}(\bar{x} - x_{(1)}, x_{(1)}) < 0$, the MLE of θ is $\hat{\theta} = (\bar{X} - X_{(1)}, X_{(1)})$.

Problems

5.1.1 Let $X_1, \ldots, X_n$ be a sample of iid random variables with pdf $f(x; \theta) = \theta x^{\theta-1}$ for $x \in (0, 1)$ and $\Theta = (0, \infty)$. Determine the MLE of θ.

5.1.2 Let $X_1, \ldots, X_n$ be a sample of iid NegBin$(4, \theta)$ random variables with $\Theta = [0, 1]$. Determine the MLE $\hat{\theta}$ of θ.

5.1.3 Let $X_1, \ldots, X_n$ be a sample of iid random variables with pdf $f(x; \theta) = \frac{3x^2}{\theta^3} e^{-\frac{x^3}{\theta^3}}$ for $x \in (0, \infty)$ and $\Theta = [0, 1]$. Determine the MLE of θ.

5.1.4 Let $X_1, \ldots, X_n$ be a sample of iid $N(0, \theta)$ random variables with $\Theta = (0, \infty)$. Determine
a) the MLE $\hat{\theta}$ of θ.
b) $E(\hat{\theta})$.
c) the asymptotic variance of the MLE of θ.
d) the MLE of $SD(X_i) = \sqrt{\theta}$.

5.1.5 Let $X_1, \ldots, X_n$ be a sample of iid $N(0, \theta^2)$ random variables with $\Theta = (0, \infty)$. Determine
a) the MLE of θ.
b) the asymptotic variance of the MLE of θ.
c) the MLE of $Var(X_i) = \theta^2$.

5.1.6 Let $X_1, \ldots, X_n$ be a sample of iid LN$(\theta, 1)$ random variables with $\Theta = \mathbb{R}$. Determine the
a) MLE $\hat{\theta}$ of θ.
b) $E(\hat{\theta})$.
c) $Var(\hat{\theta})$.
d) MLE of Median$(X_i) = e^\theta$.

5.1.7 Let $X_1, \ldots, X_n$ be a sample of iid Bin(m, θ) random variables with $\Theta = [0, 1]$. Determine
a) the MLE $\hat{\theta}$ of θ.
b) $E(\hat{\theta})$.
c) $Var(\hat{\theta})$.
d) the Cramér–Rao lower bound for unbiased estimators of θ.
e) whether or not $\hat{\theta}$ is a UMVUE of θ.

5.1.8 Let $X_1, \ldots, X_n$ be a sample of iid Gamma(α, θ) random variables with α known and $\Theta = (0, \infty)$. Determine
a) the MLE $\widehat{\theta}$ of θ.
b) $E(\widehat{\theta})$.
c) $\text{Var}(\widehat{\theta})$.
d) the Cramér–Rao lower bound for unbiased estimators of θ.
e) whether or not $\widehat{\theta}$ is a UMVUE of θ.

5.1.9 Let $X_1, \ldots, X_n$ be a sample of iid random variables with pdf $f(x : \theta) = \frac{\theta}{x^{\theta+1}}$ for $x \in (1, \infty)$ and $\Theta = (0, \infty)$.
a) Determine the MLE $\widehat{\theta}$ of θ.
b) Using the fact that $\sum_{i=1}^{n} \ln(X_i) \sim \text{Gamma}\left(n, \frac{1}{\theta}\right)$, determine $E(\widehat{\theta})$ and $\text{Var}(\widehat{\theta})$.
c) Determine the asymptotic variance of the MLE of θ.

5.1.10 Let $X_1, \ldots, X_n$ be a sample of iid Pois(θ) random variables with $\Theta = \mathbb{R}^+$. Determine
a) the Score function associated with θ.
b) the variance of the Score function.

5.1.11 Let $X_1, \ldots, X_n$ be a sample of iid $N(\theta, \sigma^2)$ random variables with $\Theta = \mathbb{R}$ and σ^2 known. Determine
a) the Score function associated with θ.
b) the variance of the Score function.

5.1.12 Let $X_1, \ldots, X_n$ be a sample of iid $N(\mu, \theta)$ random variables with $\Theta = \mathbb{R}^+$ and μ known. Determine
a) the Score function associated with θ.
b) the variance of the Score function.

5.1.13 The data given in Table 5.3 represent a sample of $n = 50$ iid observations from a Gamma($\theta, \beta = 2.5$) distribution.

Table 5.3 A random sample of $n = 50$ observations from a Gamma($\theta, 2.5$) distribution.

7.572	19.503	5.475	11.189	15.734	5.490	4.775	11.674	3.10	14.521
15.493	6.736	15.669	14.782	3.817	7.353	9.790	8.920	5.414	5.527
9.259	4.167	11.399	5.011	12.072	21.469	18.784	9.121	12.554	10.918
7.150	6.506	10.373	5.556	5.260	17.426	8.551	1.926	26.855	10.438
8.424	19.585	8.328	9.077	22.866	4.658	9.545	9.044	3.771	8.648

a) Use the observed data and R to determine the MLE of θ.
b) Use the observed data and R to determine the approximate standard error of $\widehat{\theta}$.
c) Determine the MLE of $E(X)$.
d) Determine the MLE of $\text{Var}(X)$.
e) Produce the R visual assessment of the models fit to the data.

5.1.14 The data given in Table 5.4 represent a sample of $n = 50$ iid observations from a Beta($\theta, \beta = 4$) distribution.
a) Use the observed data and R to determine the MLE of θ.
b) Use the observed data and R to determine the approximate standard error of $\widehat{\theta}$.
c) Determine the MLE of $E(X)$.
d) Determine the MLE of $\text{Var}(X)$.
e) Produce the R visual assessment of the models fit to the data.

5.1.15 The data given in Table 5.5 represent a sample of $n = 50$ iid observations from a Beta($\alpha = 1.2, \theta$) distribution.
a) Use the observed data and R to determine the MLE of θ.
b) Use the observed data and R to determine the approximate standard error of $\widehat{\theta}$.
c) Determine the MLE of $E(X)$.
d) Determine the MLE of $\text{Var}(X)$.

5.1.16 Let $X_1, \ldots, X_n$ be a sample of iid random variables with pdf $f(x : \theta) = \frac{1}{\theta}$ for $x \in \{1, 2, \ldots, \theta\}$ and $\Theta = \mathbb{N}$. Determine the MLE of θ.

Table 5.4 A random sample of $n = 50$ observations from a Beta($\theta, 4$) distribution.

0.191	0.176	0.174	0.179	0.223	0.315	0.425	0.292	0.218	0.051
0.227	0.164	0.048	0.208	0.009	0.340	0.137	0.344	0.161	0.492
0.396	0.233	0.262	0.216	0.202	0.203	0.211	0.040	0.279	0.453
0.140	0.231	0.237	0.021	0.279	0.558	0.152	0.045	0.111	0.034
0.328	0.182	0.105	0.412	0.166	0.033	0.146	0.283	0.096	0.085

Table 5.5 A random sample of $n = 50$ observations from a $Beta(1.2, \theta)$ distribution.

0.518	0.058	0.422	0.327	0.159	0.111	0.121	0.133	0.091	0.380
0.579	0.388	0.237	0.178	0.326	0.158	0.235	0.105	0.119	0.133
0.236	0.319	0.063	0.160	0.312	0.351	0.671	0.256	0.416	0.354
0.254	0.040	0.126	0.504	0.421	0.361	0.393	0.152	0.057	0.042
0.258	0.223	0.240	0.029	0.234	0.041	0.161	0.536	0.242	0.718

5.1.17 Let X be a single observation on a Bin$(\theta, 0.5)$ random variables with $\Theta = \mathbb{N}$.
a) If $X = 5$, determine the MLE of θ.
b) If $X = x$, determine the MLE of θ.

5.1.18 Let $X_1, \ldots, X_{20}$ be a sample of iid Bin$(1, \theta)$ random variables with $\Theta = \{0.5, 0.8\}$. Determine
a) the likelihood ratio function $\Lambda(0.5, 0.8)$.
b) the MLE $\widehat{\theta}$ of θ when $\sum_{i=1}^{20} x_i = 11$.
c) the MLE $\widehat{\theta}$ of θ when $\sum_{i=1}^{20} x_i = 15$.

5.1.19 Let $X_1, \ldots, X_{25}$ be a sample of iid Pois(θ) random variables with $\Theta = \{2, 5\}$. Determine
a) the likelihood ratio function $\Lambda(2, 5)$.
b) the MLE $\widehat{\theta}$ of θ when $\sum_{i=1}^{25} x_i = 68$.
c) the MLE $\widehat{\theta}$ of θ when $\sum_{i=1}^{25} x_i = 102$.

5.1.20 Let $X_1, \ldots, X_{50}$ be a sample of iid Pois(θ) random variables with $\Theta = \{2.5, 4, 5\}$. Determine
a) the MLE $\widehat{\theta}$ of θ when $\sum_{i=1}^{50} x_i = 168$.
b) the MLE $\widehat{\theta}$ of θ when $\sum_{i=1}^{50} x_i = 198$.
c) the MLE $\widehat{\theta}$ of θ when $\sum_{i=1}^{50} x_i = 236$.

5.1.21 Let $X_1, \ldots, X_n$ be a sample of iid $U(0, \theta]$ random variables. Determine
a) the MLE $\widehat{\theta}$ of θ.
b) $E(\widehat{\theta})$.
c) $\text{Var}(\widehat{\theta})$.

5.1.22 Let $X_1, \ldots, X_n$ be a sample of iid $U[\theta, 1)$ random variables. Determine
a) the MLE $\widehat{\theta}$ of θ.
b) $E(\widehat{\theta})$.
c) $\text{Var}(\widehat{\theta})$.

5.1.23 Let $X_1, \ldots, X_n$ be a sample of iid random variables with pdf $f(x; \theta) = \left(\frac{e^{-\theta}}{1+e^{-\theta}}\right)^x \left(\frac{1}{1+e^{-\theta}}\right)^{1-x}$, $x = 0, 1$, and $\theta \in [0, \infty)$. Determine
a) a sufficient statistic for θ.
b) the MLE of θ.
c) the MLE of $P(X = 1)$.
d) the Cramér–Rao lower bound for estimating θ.

5.1.24 Let $\vec{X} = (X_1, X_2, X_3)$ be an observation from MulNom$_3(n, \theta_1, \theta_2, \theta_3)$ with $\Theta = [0, 1] \times [0.1] \times [0, 1] = [0, 1]^3$. Determine the MLE of $\vec{\theta} = (\theta_1, \theta_2, \theta_3)$.

5.1.25 Let $\vec{X} = (X_1, \dots, X_k)$ be an observation from $\mathrm{MulNom}_k(n, \theta_1, \dots, \theta_k)$ with $\Theta = [0, 1]^k$. Determine the MLE of $\vec{\theta} = (\theta_1, \theta_2, \dots, \theta_k)$.

5.1.26 Let $X_1, \dots, X_n$ be a sample of iid $\mathrm{LN}(\theta_1, \theta_2)$ random variables with $\Theta = \mathbb{R} \times \mathbb{R}^+$.
   a) Determine the MLE of $\vec{\theta} = (\theta_1, \theta_2)$.
   b) Determine the MLE of $E(X_i) = e^{\theta_1 + \frac{\theta_2}{2}}$.
   c) Using the fact that $\ln(X_i) \sim N(\theta_1, \theta_2)$, determine $E(\widehat{\theta}_1)$.
   d) Determine whether or not $\widehat{\theta}_1$ is a UMVUE of θ_1.
   e) Determine $E(\widehat{\theta}_2)$.
   f) Determine whether or not $\widehat{\theta}_2$ is a UMVUE of θ_2.

5.1.27 Let $X_1, \dots, X_n$ be a sample of iid $N(\theta, \theta^2)$ random variables with $\Theta = \mathbb{R}^+$. Determine the MLE of θ.

5.1.28 Let $X_1, \dots, X_n$ be a sample of iid $U(\theta_1, \theta_2)$ random variables with $\Theta = \{(\theta_1, \theta_2) \in \mathbb{R}^2 : \theta_1 < \theta_2\}$. Determine the MLE of $\vec{\theta} = (\theta_1, \theta_2)$.

5.1.29 In Example 5.13, $X_1, \dots, X_n$ are iid $\mathrm{Exp}(\theta_1, \theta_2)$ random variables, and the MLEs of θ_1 and θ_2 are $\widehat{\theta}_1 = \overline{X} - X_{(1)}$ and $\widehat{\theta}_2 = X_{(1)}$, respectively. Determine
   a) $E(\widehat{\theta}_1)$.
   b) $E(\widehat{\theta}_2)$.
   c) an unbiased estimator of θ_1 based on $\widehat{\theta}_1$.
   d) an unbiased estimator of θ_2 based on $\widehat{\theta}_2$.

5.1.30 Let $X_1, \dots, X_n$ be a sample of iid random variables with pdf $f(x; \theta_1, \theta_2) = \theta_1 \, e^{-\theta_1(x - \theta_2)}$ with $S = [\theta_2, \infty)$ and $\Theta = \mathbb{R}^+ \times \mathbb{R}$. Determine
   a) $L(\theta_1, \theta_2)$.
   b) the MLE of $\vec{\theta} = (\theta_1, \theta_2)$.
   c) $E(\widehat{\theta}_2)$.

5.1.31 A random sample of $n = 25$ observations from $X \sim \mathrm{Gamma}(\theta_1, \theta_2)$ is given in Table 5.6.
   a) Use the observed data and R to determine the MLEs of θ_1 and θ_2.
   b) Use the observed data and R to determine the approximate standard errors of $\widehat{\theta}_1$ and $\widehat{\theta}_2$.
   c) Determine the MLE of $E(X)$.
   d) Determine the MLE of $\mathrm{Var}(X)$.
   e) Perform a visual assessment of the models fit to the data.

Table 5.6 A random sample of $n = 25$ observations from a Gamma(θ_1, θ_2) distribution.

3.54	10.30	9.49	3.97	14.14	5.72	13.49	15.87	8.81
9.74	7.33	10.44	6.93	13.71	17.13	23.50	12.65	3.70
16.39	6.98	6.78	26.06	12.03	5.97	11.44		

5.1.32 A random sample of $n = 30$ observations from $X \sim$ Gamma(θ_1, θ_2) is given in Table 5.7.
   a) Use the observed data and R to determine the MLEs of θ_1 and θ_2.
   b) Use the observed data and R to determine the approximate standard errors of $\widehat{\theta}_1$ and $\widehat{\theta}_2$.
   c) Determine the MLE of $E(X)$.
   d) Determine the MLE of Var(X).

5.1.33 A random sample of $n = 50$ observations from Beta(θ_1, θ_2) is given in Table 5.8.
   a) Use the observed data and R to determine the MLEs of θ_1 and θ_2.
   b) Use the observed data and R to determine the approximate standard errors of $\widehat{\theta}_1$ and $\widehat{\theta}_2$.
   c) Determine the MLE of $E(X)$.
   d) Determine the MLE of Var(X).
   e) Perform a visual assessment of the models fit to the data.

Table 5.7 A random sample of $n = 30$ observations from a Gamma(θ_1, θ_2) distribution.

2.45	2.74	3.20	1.72	2.45	1.97	3.61	2.48	1.77	2.98
2.06	3.31	2.40	5.62	0.94	3.18	1.38	1.69	3.92	3.65
2.87	5.42	1.67	2.07	1.18	3.04	2.11	2.92	2.33	1.83

Table 5.8 A random sample of $n = 50$ observations from a Beta(θ_1, θ_2) distribution.

0.809	0.494	0.820	0.551	0.831	0.844	0.471	0.716	0.655	0.461
0.753	0.556	0.708	0.730	0.639	0.512	0.516	0.513	0.294	0.869
0.450	0.393	0.465	0.485	0.630	0.567	0.610	0.647	0.574	0.695
0.499	0.633	0.627	0.309	0.882	0.553	0.714	0.784	0.431	0.500
0.668	0.638	0.610	0.316	0.160	0.709	0.316	0.741	0.774	0.781

5.1.34 A Weibull random variable has pdf $f(x; \theta_1, \theta_2) = \frac{\theta_1}{\theta_2} \left(\frac{x}{\theta_2} \right)^{\theta_1 - 1} e^{-\left(\frac{x}{\theta_2} \right)^{\theta_1}}$ for $x \in \mathbb{R}^+$ and $\Theta = \mathbb{R}^+ \times \mathbb{R}^+$. A random sample of $n = 50$ observations from $X \sim$ Weibull(θ_1, θ_2) is given in Table 5.9.
  a) Use the observed data and R to determine the MLEs of θ_1 and θ_2.
  b) Use the observed data and R to determine the approximate standard errors of $\hat{\theta}_1$ and $\hat{\theta}_2$.
  c) Perform a visual assessment of the models fit to the data.

5.1.35 A logistic random variable has pdf $f(x; \theta_1, \theta_2) = \dfrac{e^{-\frac{(x-\theta_1)}{\theta_2}}}{\theta_2 \left(1 + e^{-\frac{(x-\theta_1)}{\theta_2}} \right)^2}$ for $x \in \mathbb{R}$ and $\Theta = \mathbb{R} \times \mathbb{R}^+$. A random sample of $n = 50$ observations from $X \sim$ Logis(θ_1, θ_2) is given in Table 5.10.
  a) Use the observed data and R to determine the MLEs of θ_1 and θ_2.
  b) Use the observed data and R to determine the approximate standard errors of $\hat{\theta}_1$ and $\hat{\theta}_2$.
  c) Perform a visual assessment of the models fit to the data.

5.1.36 Let $X_1, \ldots, X_n$ be a sample of iid Exp(θ) random variables with $\Theta = [5, \infty)$. Determine the MLE of θ.

5.1.37 Let $X_1, \ldots, X_n$ be a sample of iid Bin$(1, \theta)$ random variables with $\Theta = [0.5, 1]$. Determine the MLE of θ.

Table 5.9 A random sample of $n = 50$ observations from a Weibull(θ_1, θ_2) distribution.

1.446	2.153	1.599	2.588	1.740	1.167	1.473	1.685	1.938	2.622
1.224	1.012	1.687	0.993	1.664	0.875	0.410	1.147	0.712	1.664
2.946	1.689	1.398	0.692	2.020	1.490	2.353	1.995	0.365	1.912
1.027	2.726	1.228	1.648	1.823	0.923	1.707	1.928	1.601	1.436
0.727	2.928	0.381	2.689	1.997	1.713	0.727	1.442	2.143	0.463

Table 5.10 A random sample of $n = 50$ observations from a Logis(θ_1, θ_2) distribution.

64.288	47.139	51.879	44.758	45.054	50.031	52.197	46.307	50.652	53.595
51.465	53.045	46.914	47.459	46.302	45.666	53.043	54.923	51.892	45.752
49.929	55.875	43.328	53.652	46.903	51.196	48.063	48.357	52.378	49.381
54.620	53.367	49.086	52.343	56.840	52.486	52.366	54.040	50.640	51.365
52.308	50.718	50.534	50.477	48.657	52.844	46.573	52.687	43.633	52.645

5.1.38 Let $X_1, \ldots, X_n$ be a sample of iid $N(0, \theta)$ random variables with $\Theta = [1, \infty)$. Determine the MLE of θ.

5.1.39 Let $Y_1, Y_2, \ldots, Y_n$ be a sample of independent $N(\theta_0 + \theta_1 x_i, \theta_2)$ random variables where $x_1, \ldots, x_n$ are fixed constants and $\Theta = \mathbb{R} \times \mathbb{R} \times \mathbb{R}^+$. Determine the MLE of
a) $\vec{\theta} = (\theta_0, \theta_1, \theta_2)$.
b) $E(Y^*) = \theta_0 + \theta_1 x^*$ where x^* is a constant.

5.1.40 If $\widehat{\theta}$ is the MLE of θ and $\tau(\theta)$ is a one-to-one function, prove that the MLE of $\tau(\theta)$ is $\tau(\widehat{\theta})$.

5.2 Bayesian Estimation

An alternative to MLE, which also utilizes the information in the likelihood function, is *Bayesian point estimation*. Bayesian estimation differs from classical methods of point estimation, including MLE, by treating the parameter θ in a probability model as a random variable; in the classical or frequentist approach to point estimation, θ is treated as a fixed unknown real number, not a random variable.

5.2.1 The Bayesian Setting

In Bayesian point estimation, the value of θ in a parametric model $f(x; \theta)$ is a random variable having its own pdf $\pi(\theta; \lambda)$; $\pi(\theta; \lambda)$ is called the *prior distribution* and λ is called a *hyperparameter* of the prior. The prior distribution is based on the possible values of θ in the parametric model (i.e. Θ) and reflects the uncertainty about θ before the data have been collected. The choice of the prior can be made subjectively or objectively and can be based on prior information about θ, mathematical convenience, or the total lack of prior information on θ. Subjective priors are believed to contain important information about θ, while objective priors or noninformative priors tend to emphasize the information that is contained in the likelihood function.

In the Bayesian approach, the information about θ contained in the prior distribution and the information about θ contained in a random sample from the probability model $f(x; \theta)$ are used to estimate θ. When θ is a random variable, the parametric model $f(x; \theta)$ that generates a random sample is actually the conditional distribution of X given θ, and hence, the pdf of X will denoted by $f(x|\theta)$ rather than $f(x; \theta)$.

Inferences about θ in the Bayesian approach are based on the distribution of θ given the observed values of a random sample, say $x_1, \ldots, x_n$, which is called the *posterior distribution* and is denoted by $f(\theta|\vec{x})$. Bayes theorem, Theorem 1.14, is used to determine the posterior distribution $f(\theta|\vec{x})$ from the distribution of $\vec{X}|\theta$ and the prior distribution of θ.

In particular, when θ is a continuous random variable, Bayes theorem states that the pdf of the posterior distribution is

$$f(\theta|\vec{x}) = \frac{f(\vec{x}, \theta; \lambda)}{f_{\vec{x}}(\vec{x})} = \frac{f(\vec{x}|\theta)\pi(\theta; \lambda)}{\int_{\mathcal{S}_\theta} f(\vec{x}|\theta)\pi(\theta; \lambda) \, d\theta}.$$

Similarly, when θ is a discrete random variable, Bayes theorem states that the pdf of the posterior distribution is

$$f(\theta|\vec{x}) = \frac{f(\vec{x}, \theta; \lambda)}{f_{\vec{x}}(\vec{x})} = \frac{f(\vec{x}|\theta)\pi(\theta; \lambda)}{\sum_{\theta \in \mathcal{S}_\theta} f(\vec{x}|\theta)\pi(\theta; \lambda) \, d\theta}.$$

Note that in both the cases, the posterior distribution of θ given $\vec{x}$ is proportional to the product of the likelihood function associated with the sample and the prior distribution of θ, that is, $f(\theta|\vec{x}) \propto L(\theta)\pi(\theta; \lambda)$. Thus, the posterior distribution combines the information about θ in the prior distribution and the likelihood function to produce an updated distribution containing all of the available information on θ. Moreover, in the Bayesian estimation, all inferences about θ are based on the posterior distribution.

Theorem 5.7 shows that the posterior distribution depends on the sample $\vec{x}$ only through a sufficient statistic for θ.

Theorem 5.7 *If $X_1, \ldots, X_n$ is a sample of iid random variables with common pdf $f(x|\theta)$, S a sufficient statistic for θ, and $\pi(\theta; \lambda)$ a prior distribution for θ, then the posterior distribution of θ given $\vec{X}$ depends on the sample only through the sufficient statistic S.*

Proof. The proof will only be given for a continuous prior distribution since the proof for a discrete prior is similar with the integration replaced by a summation.

Let $X_1, \ldots, X_n$ be a sample of iid random variables with common pdf $f(x|\theta)$, S a sufficient statistic for θ, and $\pi(\theta; \lambda)$ the prior distribution of θ. Then, since S is sufficient for θ, by the Fisher–Neyman factorization theorem, the joint distribution of $X_1, \ldots, X_n$ can be factored as $f(\vec{x}|\theta) = g(S; \theta)h(\vec{x})$ for some functions g and h.

Thus, the pdf of the posterior distribution is

$$\begin{aligned} f(\theta|\vec{x}) &= \frac{f(\vec{x}|\theta)\pi(\theta)}{\int_{\mathcal{S}_\theta} f(\vec{x}|\theta)\pi(\theta; \lambda) \, d\theta} = \frac{g(S; \theta)h(\vec{x})\pi(\theta; \lambda)}{\int_{\mathcal{S}_\theta} g(S; \theta)h(\vec{x})\pi(\theta; \lambda) \, d\theta} \\ &= \frac{g(S; \theta)\pi(\theta; \lambda)}{\int_{\mathcal{S}_\theta} g(S; \theta)\pi(\theta; \lambda) \, d\theta}, \end{aligned}$$

which is a function of S and θ alone. Thus, $f(\theta|\vec{x})$ depends on $\vec{X}$ only through the sufficient statistic S. ∎

Examples 5.14 and 5.15 illustrate the use of Bayes theorem in deriving the posterior distribution of $\theta|\vec{x}$.

Example 5.14 Let $X_1, \ldots, X_n$ be a sample of iid $\text{Bin}(1, \theta)$ random variables with $\Theta = [0, 1]$, and suppose that the prior distribution of θ is the noninformative prior $U[0, 1]$. The likelihood function for the observed sample is

$$L(\theta) = \prod_{i=1}^{n} \theta^{x_i}(1-\theta)^{1-x_i} = \theta^{\sum_{i=1}^{n} x_i}(1-\theta)^{n-\sum_{i=1}^{n} x_i}.$$

Now, since the prior distribution is continuous, the posterior distribution is

$$f(\theta|\vec{x}) = \frac{L(\theta)\pi(\theta; \lambda)}{\int_{S_\theta} L(\theta)\pi(\theta; \lambda)\, d\theta} = \frac{\theta^{\sum_{i=1}^{n} x_i}(1-\theta)^{n-\sum_{i=1}^{n} x_i}}{\int_0^1 \theta^{\sum_{i=1}^{n} x_i}(1-\theta)^{n-\sum_{i=1}^{n} x_i}\, d\theta}$$

$$= \frac{\theta^{\sum_{i=1}^{n} x_i}(1-\theta)^{n-\sum_{i=1}^{n} x_i}}{\dfrac{\Gamma\left(\sum_{i=1}^{n} x_i + 1\right)\Gamma\left(n - \sum_{i=1}^{n} x_i + 1\right)}{\Gamma(n+2)}}$$

$$= \frac{\Gamma(n+2)}{\Gamma\left(\sum_{i=1}^{n} x_i + 1\right)\Gamma\left(n - \sum_{i=1}^{n} x_i + 1\right)}\theta^{\sum_{i=1}^{n} x_i}(1-\theta)^{n-\sum_{i=1}^{n} x_i},$$

which is the pdf of a $\text{Beta}\left(\sum_{i=1}^{n} x_i + 1, n - \sum_{i=1}^{n} x_i + 1\right)$ random variable. Note that $f(\theta|\vec{x})$ only depends on the observed sample through the sufficient statistic $\sum_{i=1}^{n} X_i$.

Thus, without any sample data, the mean value of θ based on the prior distribution is $E(\theta) = 0.5$; however, given the information in the observed sample, the posterior mean is $E(\theta|\vec{x}) = \frac{\sum_{i=1}^{n} x_i + 1}{n+2} = \frac{n\bar{x}+1}{n+2}$. For example, for a sample of $n = 25$ with $\sum_{i=1}^{25} x_i = 21$, the posterior mean is $E(\theta|\vec{x}) = \frac{21+1}{25+2} = 0.815$.

Example 5.15 Let $X_1, \ldots, X_n$ be a sample of iid $\text{Pois}(\theta)$ with $\Theta = \mathbb{R}^+$ and suppose that the prior distribution of θ is $\text{Exp}(\lambda)$. Then, the likelihood function is

$$L(\theta) = \prod_{i=1}^{n} \frac{e^{-\theta}\theta^{x_i}}{x_i!} = \frac{e^{-n\theta}\theta^{\sum_{i=1}^{n} x_i}}{\prod_{i=1}^{n} x_i!}$$

and since the prior is continuous, the posterior distribution is

$$f(\theta|\vec{x}) = \frac{L(\theta)\pi(\theta; \lambda)}{\int_{S_\theta} L(\theta)\pi(\theta; \lambda)\, d\theta} = \frac{\dfrac{e^{-n\theta}\theta^{\sum_{i=1}^{n} x_i}}{\prod_{i=1}^{n} x_i!} \times \dfrac{1}{\lambda}e^{-\frac{\theta}{\lambda}}}{\int_0^\infty \dfrac{e^{-n\theta}\theta^{\sum_{i=1}^{n} x_i}}{\prod_{i=1}^{n} x_i!} \times \dfrac{1}{\lambda}e^{-\frac{\theta}{\lambda}}\, d\theta}$$

$$= \frac{\dfrac{1}{\lambda \prod_{i=1}^{n} x_i!} e^{-n\theta} \theta^{\sum_{i=1}^{n} x_i} \times e^{-\frac{\theta}{\lambda}}}{\dfrac{1}{\lambda \prod_{i=1}^{n} x_i!} \displaystyle\int_0^\infty e^{-n\theta} \theta^{\sum_{i=1}^{n} x_i} \times e^{-\frac{\theta}{\lambda}} \, d\theta}$$

$$= \frac{e^{-\theta(n+\frac{1}{\lambda})} \theta^{\sum_{i=1}^{n} x_i}}{\underbrace{\displaystyle\int_0^\infty \theta^{\sum_{i=1}^{n} x_i} e^{-\theta(n+\frac{1}{\lambda})} \, d\theta}_{\text{gamma integral}}}$$

$$= \frac{e^{-\theta(n+\frac{1}{\lambda})} \theta^{\sum_{i=1}^{n} x_i}}{\left(n+\frac{1}{\lambda}\right)^{\sum_{i=1}^{n} x_i + 1} \Gamma\left(\sum_{i=1}^{n} x_i + 1\right)}$$

$$= \frac{e^{-\theta(n+\frac{1}{\lambda})} \theta^{\sum_{i=1}^{n} x_i}}{\left(n+\frac{1}{\lambda}\right)^{(\sum_{i=1}^{n} x_i + 1)} \Gamma\left(\sum_{i=1}^{n} x_i + 1\right)}.$$

Thus, $\theta|\vec{x} \sim \text{Gamma}\left(\sum_{i=1}^{n} x_i + 1, \frac{1}{n+\frac{1}{\lambda}}\right)$. Note that without the information in the random sample, the mean value of θ based on the prior distribution is $E(\theta) = \lambda$; however, given the information in the observed sample, the posterior mean of θ is $E(\theta|\vec{x}) = \frac{\sum_{i=1}^{n} x_i + 1}{n+\frac{1}{\lambda}} = \frac{\lambda n}{\lambda n+1}\overline{x} + \frac{\lambda}{\lambda n+1}$. Thus, the posterior mean is a weighted average of the sample mean $\overline{X}$ and the mean of the prior distribution λ.

5.2.2 Bayesian Estimators

After the posterior distribution has been determined, all Bayesian inferences about θ are based on the posterior distribution. A Bayesian point estimator is determined from the posterior distribution for a particular *loss function*.

Definition 5.3 For an estimator T of θ, a loss function $\mathcal{L}(T; \theta)$ is a nonnegative real-valued function with $\mathcal{L}(\theta; \theta) = 0$.

Commonly used loss functions in Bayesian estimation include the *squared error loss function*, $\mathcal{L}_2(T; \theta) = (T - \theta)^2$, and the *absolute deviation loss function*, $\mathcal{L}_1(T; \theta) = |T - \theta|$. Note that both the $\mathcal{L}_2$ and $\mathcal{L}_1$ loss functions are specific cases of the more general $\mathcal{L}_p$ loss function, which is $\mathcal{L}_p(T; \theta) = |T - \theta|^p$ for $p > 0$. The expected value of the loss function, taken as a function of θ, is called the *risk function*.

Definition 5.4 Let $X_1, \ldots, X_n$ be a sample. For an estimator $T(\vec{X})$ of θ and a loss function $\mathcal{L}(T; \theta)$, the risk function associated with T and $\mathcal{L}$ is the expected value of the loss function with respect to $f(x_1, \ldots, x_n|\theta)$. The risk function for an estimator T and a loss function $\mathcal{L}$ is denoted by $R(T;\theta) = E[\mathcal{L}(T;\theta)|\theta]$.

When the loss function is the squared error loss function, the risk function is the mean squared error (MSE) of the estimator. When the loss function is the absolute loss function, then the risk function is referred to as the *mean absolute deviation* (MAD) of the estimator.

In the frequentist approach, an estimator is often chosen to minimize the expected value of the loss function, which is called the *risk* associated with the estimator. On the other hand, in the Bayesian approach, estimators are chosen to minimize the expected value of the risk function, which is called the *Bayes risk*.

Definition 5.5 If $X_1, \ldots, X_n$ is a sample of iid random variables with common pdf $f(x|\theta)$, the Bayes risk of an estimator T associated with a loss function $\mathcal{L}$ and a prior distribution $\pi(\theta; \lambda)$ is $E_\theta[R(T;\theta)]$.

Bayesian estimators are judged on their Bayes risk with respect to the prior distribution, and an estimator that has minimum Bayes risk for a particular loss function and prior distribution is called a *Bayesian estimator* of θ.

Definition 5.6 Let $\mathcal{L}$ be a loss function and $\pi(\theta; \lambda)$ a prior distribution on θ. An estimator $T^\star$ of θ with minimum Bayes risk is called a Bayesian estimator of θ.

Thus, $T^\star$ is the Bayesian estimator of θ when $E_\theta[R(T^\star;\theta)] \le E_\theta[R(T;\theta)]$ for every other estimator T of θ. Theorem 5.8 shows that a Bayesian estimator is the estimator that minimizes $\int_\Theta \mathcal{L}(T;\theta)f(\theta|\vec{x})\pi(\theta; \lambda)\,d\theta$, which is called the *Bayes posterior risk*.

Theorem 5.8 *Let $X_1, \ldots, X_n$ be a random sample with joint pdf $f(\vec{x}|\theta)$, $\pi(\theta; \lambda)$ a prior distribution for θ, and $\mathcal{L}$ a loss function. The Bayes estimator of θ is the estimator $T^\star$ that minimizes $\int_\Theta \mathcal{L}(T;\theta)f(\theta|\vec{x})\pi(\theta; \lambda)\,d\theta$.*

Proof. The proof is only given for a continuous prior since the proof for a discrete prior is similar with the integration changed to summation.

Let $X_1, \ldots, X_n$ be a random sample with joint pdf $f(\vec{x}|\theta)$, $\pi(\theta; \lambda)$ a prior distribution for θ, and $\mathcal{L}$ a loss function and consider $E_\theta[R(T;\theta)]$.

$$E_\theta[R(T;\theta)] = \int_\Theta \int_{\mathcal{S}_{\vec{x}}} \mathcal{L}(T;\theta)f(\vec{x}|\theta)\pi(\theta; \lambda)\,d\vec{x}\,d\theta$$

$$= \int_{\mathcal{S}_{\vec{x}}} \left[\int_\Theta \mathcal{L}(T;\theta)f(\vec{x}|\theta)\pi(\theta; \lambda)\,d\theta \right] d\vec{x}$$

$$= \int_{\mathcal{S}_{\vec{x}}} \left[\int_{\Theta} \mathcal{L}(T; \theta) f(\vec{x}|\theta) \pi(\theta; \lambda) \frac{f(\vec{x})}{f(\vec{x})} \, d\theta \right] d\vec{x}$$

$$= \int_{\mathcal{S}_{\vec{x}}} \left[\int_{\Theta} \mathcal{L}(T; \theta) f(\theta|\vec{x}) \, d\theta \right] f(\vec{x}) d\vec{x}.$$

Since $\mathcal{L}$ is nonnegative, it follows that $E_{\theta}[R(T; \theta)]$ is minimized by minimizing $\int_{\Theta} \mathcal{L}(T; \theta) f(\theta|\vec{x}) \, d\theta$ for each $\vec{x}$. ∎

The key difference between the frequentist and Bayesian approaches is that the goal of the frequentist approach is to determine the estimator or θ that minimizes the risk function, while the goal of the Bayesian approach is to determine the estimator that minimizes the Bayes risk.

Theorem 5.9 shows that the mean of the posterior distribution is the Bayes estimator under the squared error loss function.

Theorem 5.9 *If $X_1, \ldots, X_n$ is a sample of iid random variables with common pdf $f(x|\theta)$, $\pi(\theta; \lambda)$ is a prior distribution for θ, and $\mathcal{L}(\hat{\theta}; \theta)$ is the squared error loss function, then the mean of posterior distribution, $E[\theta|\vec{x}]$, is the Bayesian estimator of θ.*

Proof. The proof is only given for a continuous prior since the proof for a discrete prior is similar with the integration changed to summation.

Let $X_1, \ldots, X_n$ be a sample of iid random variables with common pdf $f(x|\theta)$, $\pi(\theta; \lambda)$ a prior distribution for θ, and let $\mathcal{L}(\hat{\theta}; \theta)$ be the squared error loss function.

Then, by Theorem 5.8, the Bayes estimator of θ is the estimator $T^{\star}$ that minimizes the posterior risk $h(T) = \int_{\Theta} (T - \theta)^2 f(\theta|\vec{x}) \, d\theta$. Now, $h'(T) = 0$ if and only if

$$\int_{\Theta} 2(T - \theta) f(\theta|\vec{x}) \, d\theta \quad = \quad 0$$

$$\Longleftrightarrow \int_{\Theta} T f(\theta|\vec{x}) \, d\theta = \int_{\Theta} \theta f(\theta|\vec{x}) \, d\theta$$

$$\Longleftrightarrow T = \int_{\Theta} \theta f(\theta|\vec{x}) \, d\theta.$$

Thus, the Bayes estimator of θ under the squared error loss function is the mean of the posterior distribution. ∎

Example 5.16 Let $X_1, \ldots, X_n$ be a sample of iid Geo(θ) random variables with $\mathcal{S} = \mathbb{W}$ and $\Theta = [0, 1]$, and let $\theta \sim \text{Beta}(\alpha, \beta)$. By Theorem 5.9, the Bayesian estimator of θ under the squared error loss function is the mean of the posterior distribution.

Now, the likelihood function for the sample is

$$L(\theta) = \prod_{i=1}^{n} \theta(1-\theta)^{x_i} = \theta^n(1-\theta)^{\sum_{i=1}^{n} x_i},$$

and therefore, the posterior distribution is

$$f(\theta|\vec{x}) = \frac{\theta^n(1-\theta)^{\sum_{i=1}^{n} x_i} \times \dfrac{\Gamma(\alpha+\beta)}{\Gamma(\alpha)\Gamma(\beta)} \theta^{\alpha-1}(1-\theta)^{\beta-1}}{\int_0^1 \theta^n(1-\theta)^{\sum_{i=1}^{n} x_i} \times \dfrac{\Gamma(\alpha+\beta)}{\Gamma(\alpha)\Gamma(\beta)} \theta^{\alpha-1}(1-\theta)^{\beta-1}\, d\theta}$$

$$= \frac{\theta^{n+\alpha-1}(1-\theta)^{\sum_{i=1}^{n} x_i + \beta - 1}}{\int_0^1 \theta^{n+\alpha-1}(1-\theta)^{\sum_{i=1}^{n} x_i + \beta - 1}\, d\theta}$$

$$= \frac{\Gamma(n + \sum_{i=1}^{n} x_i + \alpha + \beta)}{\Gamma(n+\alpha)\Gamma(\sum_{i=1}^{n} x_i + \beta)} \theta^{n+\alpha-1}(1-\theta)^{\sum_{i=1}^{n} x_i + \beta - 1}.$$

Hence, $\theta|\vec{x} \sim \text{Beta}\left(n + \alpha, \sum_{i=1}^{n} X_i + \beta\right)$, and therefore, the Bayes estimator of θ under the squared error loss function is $E(\theta|\vec{x}) = \frac{n+\alpha}{n + \sum_{i=1}^{n} x_i + \alpha + \beta}$.

Example 5.17 Let X be the number of heads in θ tosses of a fair coin, and let $\theta \sim \text{Pois}(\lambda)$. Then, $X|\theta \sim \text{Bin}(\theta, 0.5)$, and the posterior distribution of $\theta|x$ for $\theta \geq x$ is

$$f(\theta|x) = \frac{\binom{\theta}{x} 0.5^\theta \times \dfrac{e^{-\lambda}\lambda^\theta}{\theta!}}{\sum_{\theta=x}^{\infty} \binom{\theta}{x} 0.5^\theta \times \dfrac{e^{-\lambda}\lambda^\theta}{\theta!}} = \frac{\dfrac{\binom{\theta}{x} e^{-\lambda}(0.5\lambda)^\theta}{\theta!}}{\dfrac{(0.5\lambda)^x e^{-\lambda}}{x!} \sum_{\theta=x}^{\infty} \dfrac{(0.5\lambda)^{\theta-x}}{(\theta-x)!}}$$

$$= \frac{\dfrac{e^{-\lambda}(0.5\lambda)^\theta}{x!(\theta-x)!}}{\dfrac{(0.5\lambda)^x e^{-\lambda}}{x!}} \times e^{0.5\lambda} = \frac{e^{-0.5\lambda}(0.5\lambda)^{\theta-x}}{(\theta-x)!}.$$

The posterior distribution is a shifted Poisson distribution with parameter 0.5λ and support $S_{\theta|x} = \{x, x+1, \dots\}$. The mean of the posterior distribution is $E(\theta|x) = 0.5\lambda + x$, and hence, the Bayesian estimator of θ is $\hat{\theta} = 0.5\lambda + x$ under the squared error loss function.

Thus, the Bayesian estimator of the number of times the coin was tossed is the number of heads observed in the θ flips of the coin plus one-half of the prior mean λ. For example, if the prior distribution of θ is $\text{Pois}(20)$ and $x = 8$ heads were flipped, then the Bayes estimate of the number of times the coin was tossed, minimizing the Bayes risk for the squared error loss, is $\hat{\theta} = 0.5(20) + 8 = 18$.

Theorem 5.10 shows that the median of the posterior distribution is the Bayes estimator under the absolute deviation loss function.

Theorem 5.10 *If $X_1, \ldots, X_n$ is a sample of iid random variables with common pdf $f(x|\theta)$, $\pi(\theta; \lambda)$ is a prior distribution for θ, and $\mathcal{L}(\hat{\theta}; \theta)$ is the absolute deviation loss function, then the median of posterior distribution is the Bayes estimator of θ.*

Proof. The proof of Theorem 5.10 is left as an exercise. ∎

In most cases, there is not a simple formula for the median of the posterior distribution; however, with an observed sample, the median can be found using R as illustrated in Example 5.18.

Example 5.18 In Example 5.15, $X_i|\theta \sim \text{Pois}(\theta)$, the prior distribution is $\text{Exp}(\lambda)$, and the posterior distribution of $\theta|\vec{x}$ is Gamma $\left(\sum_{i=1}^{n} x_i + 1, \frac{1}{n + \frac{1}{\lambda}} \right)$. Thus, the Bayesian estimator for the absolute deviation loss function is the median of the posterior, which can be found using the R command `qgamma(0.5,shape,scale)`. For example, if $X_1, \ldots, X_{25}$ is a random sample with $\sum_{i=1}^{25} x_i = 52$ and hyperparameter $\lambda = 5$, then $\theta|\vec{x} \sim \text{Gamma}(53, 0.04)$. In this case, the median of the posterior distribution is

```
> qgamma(0.5,53,scale=0.04)
    [1]   2.106682
```

For comparison purposes, the MLE of θ for a Poisson distribution is $\overline{X}$, which in this case is $\overline{x} = 2.08$.

Example 5.19 In Example 5.16, $X_i|\theta \sim \text{Geo}(\theta)$, the prior distribution is $\text{Beta}(\alpha, \beta)$, and the posterior distribution of $\theta|\vec{x}$ is $\text{Beta}\left(n + \alpha, \sum_{i=1}^{n} X_i + \beta\right)$. If $\sum_{i=1}^{50} x_i = 149$, $\alpha = 15$, and $\beta = 2$, then posterior distribution is $\theta|\vec{x} \sim \text{Beta}(65,151)$, and the median of the posterior is

```
> qbeta(.5,65,151)
    [1]  0.3003105
```

The MLE of θ for a $X_i \overset{\text{iid}}{\sim} \text{Geo}(\theta)$ is $\hat{\theta} = \frac{1}{\overline{X}+1}$, and for $\sum_{i=1}^{50} x_i = 149$, the value of the MLE is $\hat{\theta} = 0.251$. In this case, the influence of the prior distribution pulls the Bayesian estimate away from the MLE toward the median of the prior. A graph of the prior distribution is given in Figure 5.2, which clearly shows that the prior distribution is concentrated at the upper end of $\Theta = [0, 1]$.

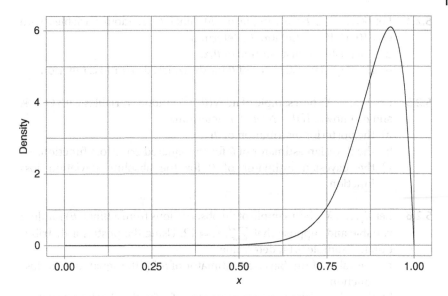

Figure 5.2 The Beta(15, 2) prior for θ in Example 5.19.

Problems

5.2.1 Let $X_1, \ldots, X_n$ be a sample of iid Bin$(1, \theta)$ random variables with $\Theta = \{0.3, 0.5\}$. If θ has prior distribution $\pi(\theta) = \begin{cases} 0.25 & \text{when } \theta = 0.3 \\ 0.75 & \text{when } \theta = 0.5 \end{cases}$.
Determine
a) the posterior distribution of $\theta | \vec{x}$.
b) the posterior distribution of $\theta | \vec{x}$ when $n = 20$ and $\sum x_i = 8$.
c) the Bayesian estimator of θ for the squared error loss function when $n = 20$ and $\sum x_i = 8$.

5.2.2 Let $X_1, \ldots, X_n$ be a sample of iid Bin$(1, \theta)$ random variables with $\Theta = [0, 1]$. If $\theta \sim$ Beta$(1, 2)$, determine
a) the posterior distribution of $\theta | \vec{x}$.
b) the Bayesian estimator of θ for the squared error loss function.

5.2.3 Let $X_1, \ldots, X_n$ be a sample of iid Geo(θ) random variables with $\Theta = [0, 1]$. If $\theta \sim U[0, 1]$, determine
a) the posterior distribution of $\theta | \vec{x}$.
b) the Bayesian estimator of θ for the squared error loss function.

5.2.4 Let $X_1, \ldots, X_n$ be a sample of iid $\mathrm{Bin}(1, \theta)$ random variables with $\Theta = [0, 1]$. If $\theta \sim \mathrm{Beta}(\alpha, \beta)$, determine
  a) the posterior distribution of $\theta | \vec{x}$.
  b) the Bayesian estimator of θ for the squared error loss function.

5.2.5 Let $X_1, \ldots, X_n$ be a sample of iid $N(\theta, \sigma^2)$ random variables with $\Theta = \mathbb{R}$ and σ^2 known. If $\theta \sim N(\mu, \sigma^2)$, determine
  a) the posterior distribution of $\theta | \vec{x}$.
  b) the Bayesian estimator of θ for the squared error loss function.
  c) the Bayesian estimator of θ for the absolute deviation loss function.

5.2.6 Let $X_1, \ldots, X_{50}$ be a sample of iid observations from a $\mathrm{Bin}(1, \theta)$ random variable and suppose that $\sum_{i=1}^{50} x_i = 12$. Using the posterior distribution in Example 5.14, determine
  a) the value of the Bayesian estimator of θ for the squared error loss function.
  b) the value of the Bayesian estimator of θ for the absolute deviation loss function.

5.2.7 Let $X_1, \ldots, X_{50}$ be a sample of iid observations from a $\mathrm{Geo}(\theta)$ random variable and suppose that $\sum_{i=1}^{50} x_i = 13, \alpha = 4$, and $\beta = 2$. Using the posterior distribution in Example 5.16, determine
  a) the value of the Bayesian estimator of θ for the squared error loss function.
  b) the value of the Bayesian estimator of θ for the absolute deviation loss function.

5.2.8 Let $X_1, \ldots, X_{50}$ be a sample of iid observations from a $\mathrm{Geo}(\theta)$ random variable and suppose that $\sum_{i=1}^{50} x_i = 149, \alpha = 5$, and $\beta = 10$. Using the posterior distribution in Example 5.16, determine
  a) the value of the Bayesian estimator of θ for the squared error loss function.
  b) the value of the Bayesian estimator of θ for the absolute deviation loss function.

5.2.9 Let $X_1, \ldots, X_{30}$ be a sample of iid $\mathrm{Pois}(\theta)$ random variables with $\Theta = \mathbb{R}^+$ and let $\theta \sim \mathrm{Gamma}(\alpha, \beta)$. Determine
  a) the posterior distribution of $\theta | \vec{x}$.
  b) the Bayesian estimator of θ for the squared error loss function.
  c) the Bayesian estimator of θ for the squared error loss function when $\sum_{i=1}^{30} x_i = 173, \alpha = 4$, and $\beta = 2$.

d) the Bayesian estimator of θ for the absolute loss function when $\sum_{i=1}^{30} x_i = 173$, $\alpha = 4$, and $\beta = 2$.

5.2.10 Let $X_1, \ldots, X_{60}$ be a sample of iid random variables with pdf $f(x|\theta) = \theta\, e^{-\theta x}$ for $x \in \mathbb{R}^+$ and $\Theta = \mathbb{R}^+$. If $\theta \sim$ Gamma(α, β), determine
   a) the posterior distribution of $\theta | \vec{x}$.
   b) the Bayesian estimator of θ for the squared error loss function.
   c) the Bayesian estimator of θ for the squared error loss function when $\sum_{i=1}^{60} x_i = 143.1$, $\alpha = 3.5$, and $\beta = 6$.
   d) the Bayesian estimator of θ for the absolute loss function when $\sum_{i=1}^{60} x_i = 143.1$, $\alpha = 3.5$, and $\beta = 6$.

5.2.11 Let $X_1, \ldots, X_n$ be a sample of iid $N(\theta, 1)$ random variables with $\Theta = \mathbb{R}$. If $\theta \sim N(\nu, \tau^2)$, determine
   a) the posterior distribution of $\theta | \vec{x}$.
   b) the Bayesian estimator of θ for the squared error loss function.
   c) the Bayesian estimator of θ for the absolute loss function.
   d) the Bayesian estimator of θ for the squared error loss function when $n = 100$, $\sum_{i=1}^{100} x_i = 876.5$, $\nu = 15$, $\tau^2 = 2$.

5.2.12 Let $X_1, \ldots, X_n$ be a sample of iid Bin$(2, \theta)$ random variables with $\Theta = [0, 1]$. If $\theta \sim U[0, 1]$, determine
   a) the posterior distribution of $\theta | \vec{x}$.
   b) the Bayesian estimator of θ for the squared error loss function.
   c) the Bayesian estimator of θ for the squared error loss function when $n = 10$ and $\sum_{i=1}^{10} x_i = 17$.
   d) the Bayesian estimator of θ for the absolute loss function when $n = 10$ and $\sum_{i=1}^{10} x_i = 17$.

5.2.13 Let $X \sim$ Bin$(\theta, 0.5)$ with $\Theta = \mathbb{W}$ and let $\theta \sim$ Pois(λ). Determine
   a) the posterior distribution of $\theta | \vec{x}$.
   b) the Bayesian estimator of θ for the squared error loss function.
   c) the value of the Bayesian estimator of θ for the squared error loss function when $\lambda = 50$ and $x = 20$.

5.2.14 Let $X \sim$ Bin$(\theta, 0.8)$ with $\Theta = \mathbb{W}$ and let $\theta \sim$ Pois(λ). Determine
   a) the posterior distribution of $\theta | x$.
   b) the Bayesian estimator of θ for the squared error loss function.
   c) the value of the Bayesian estimator of θ for the squared error loss function when $\lambda = 50$ and $x = 20$.

5.2.15 Prove Theorem 5.10 for a continuous prior distribution.

5.3 Interval Estimation

Point estimation produces a single estimate of a parameter θ, determined according to a particular criterion, for which the MSE is used to measure the accuracy of the estimate. An alternative method for estimating θ is *interval estimation*, which produces an interval of plausible estimates of θ.

Definition 5.7 Let $X_1, \ldots, X_n$ be a sample and let $L(\vec{X})$ and $U(\vec{X})$ be statistics with $L(\vec{X}) \leq U(\vec{X})$. An interval $[L(\vec{X}), U(\vec{X})]$ used for estimating θ is called an interval estimator.

Interval estimators are generally based on the sampling distribution of a point estimator. For example, an interval estimator that is often used is $\widehat{\theta} \pm 2\sqrt{\text{MSE}(\widehat{\theta}; \theta)}$, where $\widehat{\theta}$ is a point estimator of θ. Interval estimators considered in this section have random end points $L(\vec{X})$ and $U(\vec{X})$ and use *coverage probability* as a measure of reliability.

Definition 5.8 Let $X_1, \ldots, X_n$ be a sample and let $[L(\vec{X}, U(\vec{X})]$ be an interval estimator of θ. The coverage probability associated with the interval estimator $[L(\vec{X}, U(\vec{X})]$ is $P(\theta \in [L(\vec{X}), U(\vec{X})])$.

It is important to note that the coverage probability refers to the reliability of an interval estimator as a procedure, not the reliability of an interval estimate it produces, that is, before the data are observed, an interval estimator has random end points and a coverage probability; however, once the data have been collected and an interval estimate has been computed, the end points of the interval are numbers, not random variables. Hence, for an observed sample $\vec{x}$, the probability that θ is in $[L(\vec{x}, U(\vec{x})]$ is either 0 or 1.

Example 5.20 Let $X_1, \ldots, X_n$ be a sample of iid random variables and let $\widehat{\theta}$ be an estimator of θ. Then, $\widehat{\theta} \pm 2\sqrt{\text{MSE}(\widehat{\theta}; \theta)}$ is an interval estimator of θ. The coverage probability associated with this interval estimator varies from one probability model to another. At best, Chebyshev's inequality can be used to place a lower-bound on the coverage probability since

$$P\left(|\widehat{\theta} - \theta| \leq 2\text{MSE}(\widehat{\theta}; \theta)\right) \geq 1 - \frac{1}{2^2} = 0.75$$

provided $\text{MSE}(\widehat{\theta}; \theta) < \infty$. Thus, the interval estimator $\widehat{\theta} \pm 2\sqrt{\text{MSE}(\widehat{\theta}; \theta)}$ has a minimum coverage probability of 0.75.

An interval estimator that has a known and prespecified coverage probability is a *confidence interval*.

Definition 5.9 Let $X_1, \ldots, X_n$ be a sample and let $L(\vec{X})$ and $U(\vec{X})$ be statistics with $L(\vec{X}) \le U(\vec{X})$. The interval $[L(\vec{X}), U(\vec{X})]$ is called a $(1 - \alpha) \times 100\%$ confidence interval for the parameter θ when

$$P(L(\vec{X}) \le \theta \le U(\vec{X})) = 1 - \alpha.$$

A $(1 - \alpha) \times 100\%$ lower-bound confidence interval for θ is a confidence interval of the form $[L(\vec{X}), \infty)$, and an $(1 - \alpha) \times 100\%$ upper-bound confidence interval for θ is a confidence interval of the form $(-\infty, U(\vec{X})]$.

The prespecified coverage probability for a confidence interval is called the *confidence level* of the interval. The confidence level refers to the probability that θ lies between the random variables $L(\vec{X})$ and $(\vec{X})$ before the data are observed, and confidence levels of greater than 90% are generally used. Once the data have been observed, $L(\vec{x})$ and $U(\vec{x})$ are numbers, not random variables, and therefore, the probability that $L(\vec{x}) \le \theta \le U(\vec{X})$ is 0 or 1. Again, the confidence level measures the reliability of the confidence interval procedure not a particular interval estimate.

The lower and upper bounds of a $(1 - \alpha) \times 100\%$ confidence interval are by solving $P(L(\vec{X}) < \theta) = \alpha_1$ and $P(U(\vec{X}) > \theta) = \alpha_2$ subject to $\alpha_1 + \alpha_2 = \alpha$. The choice of α_1 and α_2 is not unique, and when $\alpha_1 = \alpha_2 = \frac{\alpha}{2}$ a confidence interval is called an *equal-tailed confidence interval*. In practice, equal-tailed confidence intervals are generally used.

Theorem 5.11 shows that when $\tau(\theta)$ is a strictly monotone function, a confidence for $\tau(\theta)$ can easily be derived from a confidence interval for θ.

Theorem 5.11 *If* $[L(\vec{X}), U(\vec{X})]$ *is a* $(1 - \alpha) \times 100\%$ *confidence interval for* θ *and* $\tau(\theta)$ *is a monotone function, then*

i) $[\tau(L(\vec{X})), \tau(U(\vec{X}))]$ *is a* $(1 - \alpha) \times 100\%$ *confidence interval for* $\tau(\theta)$ *when* τ *is an increasing function.*

ii) $[\tau(U(\vec{X})), \tau(L(\vec{X}))]$ *is a* $(1 - \alpha) \times 100\%$ *confidence interval for* $\tau(\theta)$ *when* τ *is a decreasing function.*

Proof. The proof of Theorem 5.11 is left as an exercise. ∎

Example 5.21 Let $[L(\vec{X}), U(\vec{X})]$ be a $(1 - \alpha) \times 100\%$ confidence interval for σ^2. Then, since $\sqrt{x}$ is an increasing function, $\left[\sqrt{L(\vec{X})}, \sqrt{U(\vec{X})}\right]$ is a $(1 - \alpha) \times 100\%$ confidence interval for σ.

5.3.1 Exact Confidence Intervals

When the sampling distribution of an estimator of θ is known, it is possible to construct a confidence interval having an exact prespecified confidence level. An exact confidence interval can also be constructed when there exists a function of the sample and the parameter that has a known distribution that does not depend upon θ.

First, consider the case where there exists a function of the sample and the unknown parameter, say $g(\vec{X}, \theta)$, that has a known distribution that does not depend upon the value of θ. In this case, the function $g(\vec{X}; \theta)$ is called a *pivotal quantity*.

Definition 5.10 Let $X_1, \ldots, X_n$ be a sample and let $g(\vec{X}; \theta)$ be a function of the sample and the parameter θ. The random variable $g(\vec{X}, \theta)$ is said to be a pivotal quantity when the distribution of $g(\vec{X}, \theta)$ does not depend on θ.

The distributional relationships given in Section 3.3 can be often used to identify a pivotal quantity.

Example 5.22 Let $X_1, \ldots, X_n$ be a sample of iid $N(\theta_1, \theta_2)$ and let $\overline{X} = \frac{1}{n} \sum_{i=1}^{n} X_i$ and $S^2 = \frac{1}{n-1} \sum_{i=1}^{n} (X_i - \overline{X})^2$. Then,

- by Theorem 3.17, $\frac{\sqrt{n}(\overline{X}-\theta_1)}{\sqrt{\theta_2}}$ is a pivotal quantity since $\frac{\sqrt{n}(\overline{X}-\theta_1)}{\sqrt{\theta_2}} \sim N(0, 1)$.
- by Corollary 3.5, $\frac{\sqrt{n}(\overline{X}-\theta_1)}{S}$ is a pivotal quantity since $\frac{\sqrt{n}(\overline{X}-\theta_1)}{S} \sim t_{n-1}$.
- by Theorem 3.18, $\frac{(n-1)S^2}{\sigma^2}$ is a pivotal quantity since $\frac{(n-1)S^2}{\sigma^2} \sim \chi^2_{n-1}$.

Example 5.23 Let $X_1, \ldots, X_n$ be a sample of iid $\text{Exp}(\theta)$. Then, by Theorem 3.16, $\frac{2}{\theta} \sum_{i=1}^{n} X_i$ is a pivotal quantity since $\frac{2}{\theta} \sum_{i=1}^{n} X_i \sim \chi^2_{2n}$.

Theorem 5.12 shows how a pivotal quantity can be used to determine a $(1 - \alpha) \times 100\%$ *confidence region* for a parameter θ; a $(1 - \alpha) \times 100\%$ confidence region is a set B, not necessarily an interval, such that $P(\theta \in B) = 1 - \alpha$.

Theorem 5.12 *Let $X_1, \ldots, X_n$ be a sample, $T(\vec{X}) = T$ a statistic, θ a parameter, $g(T; \theta)$ a pivotal quantity, and a and b be constants. If $P(a \leq g(T; \theta) \leq b) = 1 - \alpha$, then $\{\theta \in \Theta : a \leq g(T; \theta) \leq b\}$ is a $(1 - \alpha) \times 100\%$ confidence region for θ.*

Proof. Let $X_1, \ldots, X_n$ be a sample, $T(\vec{X}) = T$ a statistic, θ a parameter, $g(T; \theta)$ a pivotal quantity, and a and b constants such that $P(a \leq g(T; \theta) \leq b) = 1 - \alpha$. Then, $P(\theta \in \{\theta \in \Theta : a \leq g(T; \theta) \leq b\}) = P(a \leq g(T; \theta) \leq b) = 1 - \alpha$, and therefore, $\{\theta \in \Theta : a \leq g(T; \theta) \leq b\}$ is a $(1 - \alpha) \times 100\%$ confidence region for θ. ∎

Theorem 5.12 provides a confidence region for θ based on a pivotal quantity; however, it may be possible to rewrite the confidence region as a confidence interval by applying probability preserving transformations to the confidence region. In particular, the probability preserving transformations that can be used to transform $\{\theta \in \Theta : a \leq g(T, \theta) \leq b\}$ include adding a real number to each side of the inequality, multiplying each side of the inequality by a positive number, multiplying each side of the inequality by a negative number and reversing inequality signs, or taking the reciprocal of each side of the inequality and reversing inequality signs.

Example 5.24 Let $X_1, \ldots, X_n$ be a sample of iid $N(\theta, \sigma^2)$ random variables with σ^2 known and $\Theta = \mathbb{R}$, and let $\overline{X} = \frac{1}{n} \sum_{i=1}^{n} X_i$. In Example 5.22, $\frac{\sqrt{n}(\overline{X}-\theta)}{\sigma} \sim N(0, 1)$ was shown to be a pivotal quantity. Thus, for $\alpha \in (0, 1)$,

$$P\left(z_{\frac{\alpha}{2}} \leq \frac{\sqrt{n}(\overline{X} - \theta)}{\sigma} \leq z_{1-\frac{\alpha}{2}}\right) = 1 - \alpha,$$

where $z_{\frac{\alpha}{2}}$ and $z_{1-\frac{\alpha}{2}}$ are the $\frac{\alpha}{2}$th and $1 - \frac{\alpha}{2}$th quantiles of an $N(0, 1)$ random variable.

Thus, $\mathcal{R} = \left\{\theta \in \mathbb{R} : z_{\frac{\alpha}{2}} \leq \frac{\sqrt{n}(\overline{X}-\theta)}{\sigma} \leq z_{1-\frac{\alpha}{2}}\right\}$ is a $(1 - \alpha) \times 100\%$ confidence region for θ. Note that $\theta \in \mathcal{R}$ if and only if

$$\theta \in \left\{\theta \in \mathbb{R} : \overline{X} - z_{\frac{\alpha}{2}} \cdot \frac{\sigma}{\sqrt{n}} \leq \theta \leq \overline{X} + z_{1-\frac{\alpha}{2}} \cdot \frac{\sigma}{\sqrt{n}}\right\}.$$

Hence, $\left[\overline{X} - z_{\frac{\alpha}{2}} \cdot \frac{\sigma}{\sqrt{n}}, \overline{X} + z_{1-\frac{\alpha}{2}} \cdot \frac{\sigma}{\sqrt{n}}\right]$ is an exact $(1 - \alpha) \times 100\%$ confidence interval for θ.

Example 5.25 Let $X_1, \ldots, X_n$ be a sample of iid $\text{Exp}(\theta)$ random variables with $\Theta = \mathbb{R}^+$. Then, by Example 5.23, $\frac{2}{\theta} \sum_{i=1}^{n} X_i \sim \chi_{2n}^2$ is a pivotal quantity. Thus, for $\alpha \in (0, 1)$,

$$P\left(\chi_{2n, \frac{\alpha}{2}}^2 \leq \frac{2}{\theta} \sum_{i=1}^{n} X_i \leq \chi_{2n, 1-\frac{\alpha}{2}}^2\right) = 1 - \alpha,$$

and a $(1 - \alpha) \times 100\%$ confidence region for θ is

$$\mathcal{R} = \left\{\theta \in \mathbb{R}^+ : \chi_{2n, \frac{\alpha}{2}}^2 \leq \frac{2}{\theta} \sum_{i=1}^{n} X_i \leq \chi_{2n, 1-\frac{\alpha}{2}}^2\right\}.$$

Since $\theta \in \mathcal{R}$ if and only if $\theta \in \left\{\theta \in \mathbb{R}^+ : \frac{2 \sum_{i=1}^{n} X_i}{\chi_{2n, 1-\frac{\alpha}{2}}^2} \leq \theta \leq \frac{2 \sum_{i=1}^{n} X_i}{\chi_{2n, \frac{\alpha}{2}}^2}\right\}$, $\left[\frac{2 \sum_{i=1}^{n} X_i}{\chi_{2n, 1-\frac{\alpha}{2}}^2}, \frac{2 \sum_{i=1}^{n} X_i}{2n, \chi_{\frac{\alpha}{2}}^2}\right]$ is an exact $(1 - \alpha) \times 100\%$ confidence interval for θ.

Example 5.26 Let $X_1, \ldots, X_n$ be a sample of iid $N(\theta_1, \theta_2)$ with $\Theta = \mathbb{R} \times \mathbb{R}^+$, and let $\overline{X} = \frac{1}{n} \sum_{i=1}^{n} X_i$ and $S^2 = \frac{1}{n-1} \sum_{i=1}^{n} (X_i - \overline{X})^2$. Since $\frac{\sqrt{n}(\overline{X}-\theta_1)}{S} \sim t_{n-1}$

was shown to be a pivotal quantity in Example 5.22, it follows that for $\alpha \in (0, 1)$

$$P\left(\frac{\sqrt{n}(\overline{X} - \theta_1)}{S} \leq t_{1-\alpha,n-1}\right) = 1 - \alpha,$$

where $t_{1-\alpha,n-1}$ is the $1 - \alpha$th quantile of a t_{n-1} random variable.

Thus, a $(1 - \alpha) \times 100\%$ confidence region for θ_1 is

$$\left\{\theta_1 \in \mathbb{R} : \frac{\sqrt{n}(\overline{X} - \theta_1)}{S} \leq t_{1-\alpha,n-1}\right\}.$$

Now,

$$\left\{\theta_1 \in \mathbb{R} : \frac{\sqrt{n}(\overline{X} - \theta_1)}{S} \leq t_{1-\alpha,n-1}\right\}$$

$$= \left\{\theta_1 \in \mathbb{R} : \overline{X} - \theta_1 \leq \frac{S}{\sqrt{n}} t_{1-\alpha,n-1}\right\}$$

$$= \left\{\theta_1 \in \mathbb{R} : -\theta_1 \leq -\overline{X} + \frac{S}{\sqrt{n}} \cdot t_{1-\alpha,n-1}\right\}$$

$$= \left\{\theta_1 \in \mathbb{R} : \theta_1 \geq \overline{X} - \frac{S}{\sqrt{n}} \cdot t_{1-\alpha,n-1}\right\}.$$

Thus, $\left[\overline{X} - \frac{S}{\sqrt{n}} \cdot t_{\alpha,n-1}, \infty\right)$ is a $(1 - \alpha) \times 100\%$ lower-bound confidence interval for θ_1.

When a pivotal quantity is unavailable, it is still possible to determine an exact confidence interval for θ when the sampling distribution of a sufficient statistic for θ or the sampling distribution of an estimator of θ is known. In particular, if $T(\vec{X}) = T$ is a sufficient statistic or an estimator of θ with known pdf, say $f_T(t; \theta)$, and for each $\theta \in \Theta$ there exist $h_1(\theta)$ and $h_2(\theta)$ such that

$$P[h_1(\theta) \leq T \leq h_2(\theta)] = 1 - \alpha,$$

then $\mathcal{R} = \{\theta \in \Theta : h_1(\theta) \leq T \leq h_2(\theta)\}$ is a $(1 - \alpha) \times 100\%$ confidence region for θ. As noted before, it may be possible to transform the confidence region $\mathcal{R}$ into a confidence interval for θ using probability preserving transformations on $\mathcal{R}$.

Example 5.27 Let $X_1, \ldots, X_n$ be a sample of iid $U[0, \theta]$ random variables with $\Theta = \mathbb{R}^+$. Then, $X_{(n)}$ is the MLE of θ, and the pdf of the sampling distribution of $X_{(n)}$ is

$$f_n(x; \theta) = \frac{nx^{n-1}}{\theta^n}, \quad x \in [0, \theta].$$

If $\int_0^a \frac{nx^{n-1}}{\theta^n}\, dx = \frac{\alpha}{2}$ and $\int_0^b \frac{nx^{n-1}}{\theta^n}\, dx = 1 - \frac{\alpha}{2}$, then $a = \theta \sqrt[n]{\frac{\alpha}{2}}$ and $b = \theta \sqrt[n]{1 - \frac{\alpha}{2}}$.
Moreover,

$$1 - \alpha = P\left(\theta \sqrt[n]{\frac{\alpha}{2}} \le X_{(n)} \le \theta \sqrt[n]{1 - \frac{\alpha}{2}} \right)$$

$$= P\left(\sqrt[n]{\frac{\alpha}{2}} \le \frac{X_{(n)}}{\theta} \le \sqrt[n]{1 - \frac{\alpha}{2}} \right) = P\left(\frac{1}{\sqrt[n]{1 - \frac{\alpha}{2}}} \le \frac{\theta}{X_{(n)}} \le \frac{1}{\sqrt[n]{\frac{\alpha}{2}}} \right)$$

$$= P\left(\frac{X_{(n)}}{\sqrt[n]{1 - \frac{\alpha}{2}}} \le \theta \le \frac{X_{(n)}}{\sqrt[n]{\frac{\alpha}{2}}} \right).$$

Hence, $\left[\dfrac{X_{(n)}}{\sqrt[n]{1 - \frac{\alpha}{2}}} \le \theta \le \dfrac{X_{(n)}}{\sqrt[n]{\frac{\alpha}{2}}} \right]$ is an exact $(1 - \alpha) \times 100\%$ confidence interval for θ.

Mathematically, $h_1(\theta)$ and $h_2(\theta)$ are not needed, since for an observed value of $T = t^\star$, solving $\alpha_1 = P(T \le t^\star | \theta)$ and $\alpha_2 = P(T \ge t^\star | \theta)$ for θ is equivalent to solving $P(h_1(\theta) \le T \le h_2(\theta)) = 1 - \alpha$. Also, any choice of α_1 and α_2 are allowed provided $\alpha_i \in [0, 1]$ and $\alpha_1 + \alpha_2 = \alpha$; however, in practice, equal-tailed confidence intervals are generally used.

Example 5.28 In Example 5.27, suppose that $X_{(n)} = t^\star$ is observed, then a $(1 - \alpha) \times 100\%$ confidence interval for θ can be found by solving $\frac{\alpha}{2} = \int_0^{t^\star} \frac{nx^{n-1}}{\theta^n}\, dx$ and $\frac{\alpha}{2} = \int_{t^\star}^{\theta} \frac{nx^{n-1}}{\theta^n}\, dx$ for θ. In particular, the solution to $\frac{\alpha}{2} = \int_0^{t^\star} \frac{nx^{n-1}}{\theta^n}\, dx$ is $\theta = \frac{t^\star}{\sqrt[n]{\frac{\alpha}{2}}}$, and the solution to $\frac{\alpha}{2} = \int_{t^\star}^{\theta} \frac{nx^{n-1}}{\theta^n}\, dx$ is $\theta = \frac{t^\star}{\sqrt[n]{1 - \frac{\alpha}{2}}}$.

Thus, $\left[\dfrac{t^\star}{\sqrt[n]{\frac{\alpha}{2}}}, \dfrac{t^\star}{\sqrt[n]{1 - \frac{\alpha}{2}}} \right]$ is a $(1 - \alpha) \times 100\%$ confidence interval for θ, which is identical to the interval derived in Example 5.27.

This approach can also be used with discrete distributions as shown in Example 5.29; however, in the discrete case, the equations generally must be solved numerically for θ.

Example 5.29 Let $X_1, \ldots, X_n$ be a sample of iid $\text{Pois}(\theta)$ random variables with $\Theta = \mathbb{R}^+$. Then, $S = \sum_{i=1}^n X_i$ is sufficient for θ and $S \sim \text{Pois}(n\theta)$. Suppose that

$S = s$ is observed. Then, a $(1 - \alpha) \times 100\%$ confidence interval for θ can be found by solving the equations

$$\frac{\alpha}{2} = P(S \le s|\theta) = \sum_{i=0}^{s} \frac{e^{-n\theta}(n\theta)^i}{i!}$$

and

$$\frac{\alpha}{2} = P(S \ge s|\theta) = \sum_{i=s}^{\infty} \frac{e^{-n\theta}(n\theta)^i}{i!}.$$

For example, if $\alpha_1 = \alpha_2 = 0.025$, $n = 50$, and $S = 67$, then the equations that must be solved for θ are

$$0.025 = P(S \le 67|\theta) = \sum_{i=0}^{67} \frac{e^{-n\theta}(n\theta)^i}{i!}$$

$$0.025 = P(S \ge 67|\theta) = \sum_{i=67}^{\infty} \frac{e^{-n\theta}(n\theta)^i}{i!}.$$

Figure 5.3 shows a plot of $P(S \le s|\theta)$ and $P(S \ge s|\theta)$ for $\theta \in [0.7, 2.0]$. The values of θ where the functions $P(S \le s|\theta)$ and $P(S \ge s|\theta)$ cross the horizontal line of height 0.025 are the end points of the confidence interval. In particular, the exact 95% confidence for θ when $\alpha_1 = \alpha_2 = 0.025$, $n = 50$, and $S = 67$ is $[1.055, 1.702]$.

5.3.2 Large Sample Confidence Intervals

In some cases, there is no pivotal quantity available nor is there a known sampling distribution that can be used for determining an exact confidence interval

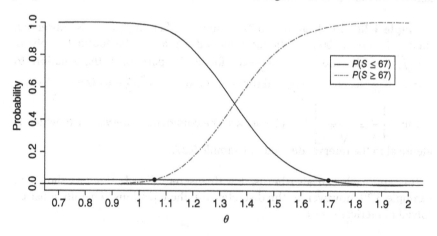

Figure 5.3 A plot of the functions $P(S \le s|\theta)$ and $P(S \ge s|\theta)$ showing the exact confidence interval for a Poisson when $\alpha_1 = \alpha_2 = 0.025$, $n = 50$, and $S = 67$.

for θ. When n is sufficiently large and the underlying pdf of a sample of iid random variables satisfies the regularity conditions given in Definition 4.18, an approximate confidence interval can be computed. In particular, an approximate $(1 - \alpha) \times 100\%$ confidence interval for θ can be based on the MLE of θ.

Recall that when the regularity conditions are satisfied and $\widehat{\theta}$ is the MLE of θ, Theorem 5.6 states that the asymptotic distribution of $\sqrt{n}(\widehat{\theta} - \theta)$ is $N(0, I(\theta)^{-1})$. Thus, for sufficiently large n, $\widehat{\theta}$ is approximately distributed as $N(\theta, \text{AsyVar}(\widehat{\theta}))$, and

$$P\left(\widehat{\theta} - z_{1-\frac{\alpha}{2}} \sqrt{\text{AsyVar}(\widehat{\theta})} \le \theta \le \widehat{\theta} + z_{1-\frac{\alpha}{2}} \sqrt{\text{AsyVar}(\widehat{\theta})} \right) \to 1 - \alpha$$

as $n \to \infty$. Therefore, for sufficiently large samples,

$$\left[\widehat{\theta} - z_{1-\frac{\alpha}{2}} \sqrt{\text{AsyVar}(\widehat{\theta})}, \widehat{\theta} + z_{1-\frac{\alpha}{2}} \sqrt{\text{AsyVar}(\widehat{\theta})} \right]$$

is an approximate $(1 - \alpha) \times 100\%$ confidence interval for θ, where $\text{AsyVar}(\widehat{\theta})$ is the Cramér–Rao lower bound on unbiased estimators of θ evaluated at $\theta = \widehat{\theta}$. Large sample confidence intervals of this type are called *Wald confidence intervals* and due to Wald and Wolfowitz [14].

Example 5.30 Let $X_1, \dots, X_n$ be a sample of iid Pois(θ) random variables with $\Theta = \mathbb{R}^+$. Since the Poisson distribution with $\Theta = \mathbb{R}^+$ is in the regular exponential family of distributions, the regularity conditions are satisfied. The MLE of θ is $\widehat{\theta} = \overline{X}$, and since the Cramér–Rao lower bound for unbiased estimators of θ is CRLB $= \frac{\theta}{n}$, an approximate $(1 - \alpha) \times 100\%$ large sample confidence interval for θ based on the MLE is

$$\left[\widehat{\theta} - z_{1-\alpha} \sqrt{\frac{\widehat{\theta}}{n}}, \widehat{\theta} + z_{1-\alpha} \sqrt{\frac{\widehat{\theta}}{n}} \right] = \left[\overline{X} - z_{1-\alpha} \sqrt{\frac{\overline{X}}{n}}, \overline{X} + z_{1-\alpha} \sqrt{\frac{\overline{X}}{n}} \right].$$

For example, when $n = 50$ and $\overline{x} = 1.34$ as in Example 5.29, an approximate 95% confidence interval for θ is $[1.34 - 1.96 \times \sqrt{0.0268}, 1.34 + 1.96 \times \sqrt{0.268}]$ or $[1.019, 1.661]$. In comparison, the exact 95% confidence interval computed in Example 5.29 was $[1.055, 1.702]$.

Example 5.31 Let $X_1, \dots, X_n$ be a sample of iid Exp(θ) random variables with $\Theta = \mathbb{R}^+$. Since the exponential distribution with $\Theta = \mathbb{R}^+$ is in the regular exponential family of distributions, the regularity conditions are satisfied. The MLE of θ is $\widehat{\theta} = \overline{X}$, the Cramér–Rao lower bound is CRLB $= \frac{\theta^2}{n}$, and the estimated asymptotic variance is $\widehat{\text{CRLB}} = \sqrt{\frac{\widehat{\theta}^2}{n}} = \frac{\overline{x}}{\sqrt{n}}$. Thus, an approximate 96% large sample confidence interval for θ is

$$\left[\overline{X} - 2.05 \frac{\overline{X}}{\sqrt{n}}, \overline{X} + 2.05 \frac{\overline{X}}{\sqrt{n}} \right].$$

For a multiparameter probability model with $\Theta \subset \mathbb{R}^d$ for $d > 1$, large sample approximate confidence intervals for θ_i can still be based on the MLE of θ_i provided the regularity conditions are satisfied. In this case, an approximate $(1 - \alpha) \times 100\%$ confidence interval for θ_i is

$$\left[\widehat{\theta}_i - z_{1-\frac{\alpha}{2}} \sqrt{\text{AsyVar}(\widehat{\theta}_i)}, \quad \widehat{\theta}_i + z_{1-\frac{\alpha}{2}} \sqrt{\text{AsyVar}(\widehat{\theta}_i)} \right],$$

where $\widehat{\theta}_i$ is the MLE of θ_i and $\text{AsyVar}(\widehat{\theta}_i)$ is the Cramér–Rao lower bound for unbiased estimators of θ_i evaluated at $\vec{\theta} = (\widehat{\theta}_1, \dots, \widehat{\theta}_d)$, which is $[I_n(\widehat{\theta})^{-1}]_{ii}$.

Example 5.32 Let $X_1, \dots, X_n$ be a sample of iid Gamma(θ_1, θ_2) random variables with $\Theta = \mathbb{R}^+ \times \mathbb{R}^+$. Since the gamma with $\Theta = \mathbb{R}^+ \times \mathbb{R}^+$ is a regular exponential family distribution, the regularity conditions are satisfied. A large sample confidence interval for θ_i based on the MLE is

$$\left[\widehat{\theta}_i - z_{1-\frac{\alpha}{2}} \times \text{se}(\widehat{\theta}_i), \widehat{\theta}_i + z_{1-\frac{\alpha}{2}} \times \text{se}(\widehat{\theta}_i) \right],$$

where $\text{se}(\widehat{\theta}_i) = \sqrt{\text{AsyVar}(\widehat{\theta}_i)}$ is evaluated at $\vec{\theta} = (\widehat{\theta}_1, \widehat{\theta}_2)$.

Recall that the MLEs of θ_1 and θ_2 must be found numerically. Since the output of the R command provides the asymptotic standard errors of the MLEs of θ_1 and θ_2, for sufficiently large samples, confidence intervals for θ_1 and θ_2 can easily be computed from the R output.

For example, let xg be an object in R containing a sample of $n = 50$ iid observations from a Gamma(θ_1, θ_2) distribution. The following R output can be used to compute large sample confidence intervals for the shape parameter θ_1 and the scale parameter θ_2.

```
> fitdist(xg,"gamma",start=list("scale"=2,"shape"=2))
    Fitting of the distribution 'gamma' by maximum likelihood
    Parameters:
            estimate Std. Error
    scale 1.068205  0.2190830
    shape 3.721112  0.7129044
```

In particular, an approximate 99% confidence interval for θ_1 is

$$[1.068 - 2.576 \times 0.219, 1.068 + 2.576 \times 0.219] = [0.504, 1.633],$$

and an approximate 99% confidence interval for θ_2 is

$$[3.721 - 2.576 \times 0.713, 3.721 + 2.576 \times 0.713] = [1.885, 5.557].$$

Finally, an alternative large sample approach used to determine an approximate confidence interval for θ is a *p% likelihood region*.

Definition 5.11   A $p\%$ likelihood region is the set $\left\{\theta \in \Theta : \frac{L(\theta)}{L(\hat{\theta})} \geq \frac{p}{100}\right\}$, where $\hat{\theta}$ is the MLE of θ.

When $X_1, \ldots, X_n$ is a sample of iid random variables with pdf satisfying the regularity conditions in Definition 4.18 and n is sufficiently large,

$$P\left(\theta \in \Theta : \frac{L(\theta)}{L(\hat{\theta})} \geq \frac{p}{100}\right) = P\left(\theta \in \Theta : \ln\left[\frac{L(\theta)}{L(\hat{\theta})}\right] \geq \ln\left[\frac{p}{100}\right]\right)$$

$$= P\left(\theta \in \Theta : \ell(\theta) - \ell(\hat{\theta}) \geq \ln\left[\frac{p}{100}\right]\right)$$

$$= P\left(\theta \in \Theta : -2[\ell(\theta) - \ell(\hat{\theta})] \leq -2\ln\left[\frac{p}{100}\right]\right).$$

For sufficiently large n, $2[l(\theta) - l(\hat{\theta})] \sim \chi_1^2$, and therefore,

$$P\left(\theta \in \Theta : -2(\ell(\theta) - \ell(\hat{\theta})) \leq -2\ln\left[\frac{p}{100}\right]\right) \approx P\left(\chi_1^2 \leq -2\ln\left[\frac{p}{100}\right]\right).$$

For example, a 14.7% likelihood region produces an approximate 95% confidence interval for θ, provided n is sufficiently large and the regularity conditions are satisfied, since

$$P\left(\theta \in \Theta : \frac{L(\theta)}{L(\hat{\theta})} \geq \frac{14.7}{100}\right) = P(\theta \in \Theta : -2[l(\theta) - l(\hat{\theta})] \leq 3.835)$$

$$\approx \underbrace{P(\chi_1^2 \leq 3.835)}_{\text{for large } n} = 0.9498.$$

Example 5.33   Let $X_1, \ldots, X_{50}$ be a sample of iid $Pois(\theta)$ random variables with $\Theta = \mathbb{R}^+$ and suppose that $\sum_{i=1}^{50} x_i = 67$. Since $Pois(\theta)$ random variable with $\Theta = \mathbb{R}^+$ has regular exponential family pdf, the required regularity conditions are satisfied. The likelihood function is

$$L(\theta) = \frac{e^{-50\theta}\theta^{\sum_{i=1}^{50} x_i}}{\prod_{i=1}^{50} x_i!} \propto e^{-50\theta}\theta^{\sum_{i=1}^{50} x_i},$$

and a plot of the likelihood function is given in Figure 5.4 for $n = 50$ and $\sum_{i=1}^{50} = 67$.

The resulting 14.7% likelihood interval is $[1.045, 1.682]$, differing slightly from the exact 95% confidence interval, $[1.055, 1.702]$, in Example 5.29, and the approximate 95% confidence interval based on the MLE, $[1.019, 1.661]$, in Example 5.30.

5.3.3   Bayesian Credible Intervals

Recall that in Bayesian estimation, θ is a random variable and inferences on θ are based on the posterior distribution, $f(\theta|\vec{X})$. A *Bayesian credible region* is an interval estimator based on the posterior distribution.

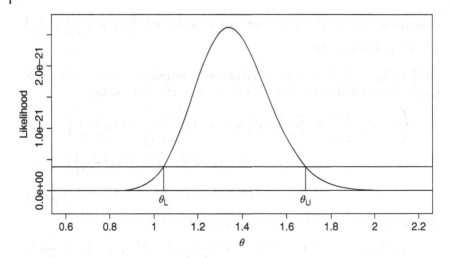

Figure 5.4 A plot of the likelihood function for Example 5.33.

Definition 5.12 Let $X_1, \ldots, X_n$ be a sample and $f(\theta|\vec{X})$ the posterior distribution of θ for a prior distribution $\pi(\theta; \lambda)$. An interval estimator $[L(\vec{X}, \lambda), U(\vec{X}, \lambda)]$ is a $(1 - \alpha) \times 100\%$ Bayesian credible interval for θ when

$$P(L(\vec{X}, \lambda) \leq \theta \leq U(\vec{X} \leq \theta \leq \lambda|\vec{X})) = 1 - \alpha.$$

One-sided credible intervals also exist, and in particular, a $(1 - \alpha) \times 100\%$ lower-bound credible interval for θ is an interval of the form $[L(\vec{X}, \lambda), \infty)$, where $P(\theta \geq L(\vec{X}, \lambda)|\vec{X}) = 1 - \alpha$; a $(1 - \alpha) \times 100\%$ upper-bound credible interval for θ is $[-\infty, U(\vec{X}, \lambda)]$, where $P(\theta \leq U(\vec{X}|\vec{X})) = 1 - \alpha$.

A Bayesian credible interval uses both the prior information about θ and the information in an observed sample in the interval of estimates of θ, while a frequentist's confidence interval is based entirely on the information in the observed sample. Furthermore, because θ is a random variable in the Bayesian approach, the probability that θ is in the resulting $(1 - \alpha) \times 100\%$ credible interval is $1 - \alpha$, even after a sample has been observed. On the other hand, the probability that θ is in a confidence interval is either 0 or 1 after the data have been observed.

Example 5.34 Let $X_1, \ldots, X_n$ be a sample of iid Pois(θ) random variables with $\Theta = \mathbb{R}^+$, and let $\theta \sim \text{Exp}(\lambda)$ be the prior distribution. The posterior distribution was derived in Example 5.15 and is $\theta|\vec{x} \sim \text{Gamma}\left(\sum_{i=1}^{n} x_i + 1, \frac{1}{n+\frac{1}{\lambda}}\right)$. Thus, a $(1 - \alpha) \times 100\%$ credible interval for θ is $[g^\star_{\frac{\alpha}{2}}, g^\star_{1-\frac{\alpha}{2}}]$, where $g^\star_{\frac{\alpha}{2}}$ and

$g^\star_{1-\frac{\alpha}{2}}$ are the $\frac{\alpha}{2}$th and $(1-\frac{\alpha}{2})$th percentiles of a Gamma $\left(\sum_{i=1}^n x_i + 1, \frac{1}{n+\frac{1}{\lambda}}\right)$ distribution.

For example, if $n = 25$, $\sum_{i=1}^{25} x_i = 112$, and $\lambda = 2$, then the posterior is distributed as Gamma$(113, 0.0392)$, and a 95% credible interval for θ is $[3.65, 5.28]$. The R command qgamma was used to determine the 95% credible region for θ.

Example 5.35 Let $X_1, \ldots, X_n$ be a sample of iid $N(\theta_1, \sigma^2)$ random variables with $\Theta = \mathbb{R}$ and σ^2 known. When $\theta \sim N(\mu, \tau^2)$ is the prior distribution of θ, the posterior distribution is $\theta | \vec{x} \sim N\left(\frac{\tau^2}{\frac{\sigma^2}{n}+\tau^2}\bar{x} + \frac{\frac{\sigma^2}{n}}{\frac{\sigma^2}{n}+\tau^2}\mu, \frac{\frac{\sigma^2}{n}\tau^2}{\frac{\sigma^2}{n}+\tau^2}\right)$, and the posterior mean is

$$\frac{\tau^2}{\frac{\sigma^2}{n}+\tau^2}\bar{x} + \frac{\frac{\sigma^2}{n}}{\frac{\sigma^2}{n}+\tau^2}\mu = a\bar{x} + (1-a)\mu,$$

where $a = \frac{\tau^2}{\frac{\sigma^2}{n}+\tau^2} \in (0,1)$. Hence, the posterior mean is a weighted average of the sample mean and the prior mean.

Now, a $(1-\alpha) \times 100\%$ credible interval for θ is

$$\left[\frac{\tau^2}{\frac{\sigma^2}{n}+\tau^2}\bar{x} + \frac{\frac{\sigma^2}{n}}{\frac{\sigma^2}{n}+\tau^2}\mu - z_{1-\frac{\alpha}{2}}\sqrt{V}, \frac{\tau^2}{\frac{\sigma^2}{n}+\tau^2}\bar{x} + \frac{\frac{\sigma^2}{n}}{\frac{\sigma^2}{n}+\tau^2}\mu + z_{1-\frac{\alpha}{2}}\sqrt{V}\right],$$

where $V = \frac{\frac{\sigma^2}{n}\tau^2}{\frac{\sigma^2}{n}+\tau^2}$ is the variance of the posterior distribution.

Problems

5.3.1 Let $X_1, \ldots, X_n$ be a sample of iid $N(\theta, 1)$ random variables with $\Theta = \mathbb{R}$.
a) Show that $T = \bar{X} - \theta$ is a pivotal quantity.
b) Determine an exact $(1-\alpha) \times 100\%$ confidence interval for θ based on T.
c) Determine an exact $(1-\alpha) \times 100\%$ lower-bound confidence interval for θ based on T.

5.3.2 Let $X_1, \ldots, X_n$ be a sample of iid $N(0, \theta)$ random variables with $\Theta = \mathbb{R}$.
a) Show that $T = \frac{1}{\theta}\sum_{i=1}^n X_i^2$ is a pivotal quantity.
b) Determine an exact $(1-\alpha) \times 100\%$ confidence interval for θ based on T.
c) Determine an exact $(1-\alpha) \times 100\%$ upper-bound confidence interval for θ based on T.

d) Determine an exact $(1 - \alpha) \times 100\%$ confidence interval for $SD(X) = \sqrt{\theta}$ based on T.

5.3.3 Let $X_1, \ldots, X_n$ be a sample of iid $U[0, \theta]$ random variables with $\Theta = \mathbb{R}^+$.
   a) Show that $T = \frac{X_{(n)}}{\theta}$ is a pivotal quantity.
   b) Determine an exact $(1 - \alpha) \times 100\%$ confidence interval for θ based on T.
   c) Determine an exact $(1 - \alpha) \times 100\%$ lower-bound confidence interval for θ based on T.

5.3.4 Let $X_1, \ldots, X_n$ be a sample of iid $U[\theta, 1]$ random variables with $\Theta = (-\infty, 1]$.
   a) Show that $T = \frac{1 - X_{(1)}}{1 - \theta}$ is a pivotal quantity.
   b) Determine an exact $(1 - \alpha) \times 100\%$ confidence interval for θ based on T.
   c) Determine an exact $(1 - \alpha) \times 100\%$ upper-bound confidence interval for θ based on T.

5.3.5 Let $X_1, \ldots, X_n$ be a sample of iid $Exp(1, \theta)$ random variables with $\Theta = \mathbb{R}$.
   a) Show that $T = X_{(1)} - \theta$ is a pivotal quantity.
   b) Determine an exact a $(1 - \alpha) \times 100\%$ confidence interval for θ based on T.
   c) Determine an exact $(1 - \alpha) \times 100\%$ upper-bound confidence interval for θ based on T.

5.3.6 Let $X_1, \ldots, X_n$ be a sample of iid $Gamma(4, \theta)$ random variables with $\Theta = \mathbb{R}^+$.
   a) Show that $\frac{2}{\theta} \sum_{i=1}^{n} X_i$ is a pivotal quantity.
   b) Determine an exact $(1 - \alpha) \times 100\%$ confidence interval for θ based on T.
   c) Determine an exact $(1 - \alpha) \times 100\%$ lower-bound confidence interval for θ based on T.
   d) Determine an exact $(1 - \alpha) \times 100\%$ confidence interval for $E(X) = 4\theta$ based on T.

5.3.7 Let $X_1, \ldots, X_n$ be a sample of iid $N(\theta_1, \theta_2)$ random variables.
   a) Show $\frac{(n-1)S^2}{\theta_2}$ is a pivotal quantity.
   b) Determine an exact $(1 - \alpha) \times 100\%$ confidence interval for θ_2.
   c) Determine an exact $(1 - \alpha) \times 100\%$ confidence interval for $SD(X) = \sqrt{\theta_2}$ based on T.

5.3.8 Let $X_1, \ldots, X_n$ be a sample of iid $LN(\theta, 1)$ random variables with $\Theta = \mathbb{R}$.

a) Show that $T = \frac{1}{n} \sum_{i=1}^{n} (\ln(X_i) - \theta)$ is a pivotal quantity.

b) Determine an exact $(1 - \alpha) \times 100\%$ confidence interval for θ.

c) Determine an exact $(1 - \alpha) \times 100\%$ lower-bound confidence interval for θ.

5.3.9 Let $X_1, \ldots, X_n$ be a sample of iid $Bin(1, \theta)$ random variables with $\Theta = [0, 1]$.

a) Determine the asymptotic variance of the MLE $\hat{\theta} = \overline{X}$.

b) For a sufficiently large sample, determine a $(1 - \alpha) \times 100\%$ confidence interval for θ.

c) If $n = 200$ and $\sum_{i=1}^{200} x_i = 86$, determine an approximate 95% confidence interval for θ.

5.3.10 Let $X_1, \ldots, X_n$ be a sample of iid $Geo(\theta)$ random variables with $S = W$ and $\Theta = [0, 1]$.

a) Determine the asymptotic variance of the MLE of θ, $\hat{\theta} = \frac{1}{1+\overline{X}}$.

b) For a sufficiently large sample, determine a $(1 - \alpha) \times 100\%$ confidence interval for θ.

c) If $n = 250$ and $\sum_{i=1}^{250} x_i = 866$, determine an approximate 95% confidence interval for θ.

5.3.11 Let $X_1, \ldots, X_n$ be a sample of iid random variables with common pdf $f(x; \theta) = \theta x^{\theta-1}$ on $(0, 1)$ with $\Theta = \mathbb{R}^+$.

a) Determine the MLE of θ.

b) Determine the asymptotic variance of the MLE $\hat{\theta}$.

c) For a sufficiently large sample, determine a $(1 - \alpha) \times 100\%$ confidence interval for θ.

5.3.12 Suppose that $X_1, \ldots, X_{90}$ is a sample of iid $Gamma(\theta, 5)$ random variables with $\Theta = \mathbb{R}^+$. Use the following R output to answer the questions below.

```
> fitdist(xg,"gamma",fix.arg=list(rate=1/5))
    Fitting of the distribution 'gamma' by maximum likeli-
hood
    Parameters:
        estimate Std. Error
    shape 3.012895  0.1681537
```

a) Determine an approximate 98% confidence interval for θ.

b) Determine an approximate 98% confidence interval for $E(X_i)$.

c) Determine an approximate 98% confidence interval for $Var(X_i)$.

5.3.13 Suppose that $X_1, \ldots, X_{75}$ is a sample of iid Gamma(θ_1, θ_2) random variables with $\Theta = \mathbb{R}^+ \times \mathbb{R}^+$. Use the following R output to answer the questions below.

```
> fitdist(xg,"gamma",start=list(shape=2,scale=1))
  Fitting of the distribution 'gamma' by maximum likelihood
  Parameters:
          estimate Std. Error
  shape 4.945961  0.7807381
  scale 2.344928  0.3895400
```

a) Determine an approximate 98% confidence interval for θ_1.
b) Determine an approximate 98% confidence interval for θ_2.

5.3.14 Suppose that $X_1, \ldots, X_{100}$ is a sample of iid Beta(θ_1, θ_2) random variables with $\Theta = \mathbb{R}^+ \times \mathbb{R}^+$. Use the following R output to answer the questions below.

```
> fitdist(xb,"beta")
  Fitting of the distribution 'beta' by maximum likelihood
  Parameters:
          estimate Std. Error
  shape1 4.531252  0.6769981
  shape2 1.045916  0.1311762
```

a) Determine an approximate 99% confidence interval for θ_1.
b) Determine an approximate 99% confidence interval for θ_2.

5.3.15 Let $X_1, \ldots, X_n$ be a sample of iid random variables with common pdf satisfying the regularity conditions of Definition 4.18. Determine the value of p so that a p%-likelihood region produces an approximate
a) 98% confidence interval for θ.
b) $(1 - \alpha) \times 100$% confidence interval for θ.

5.3.16 Prove Theorem 5.11.

5.3.17 Let $X_1, \ldots, X_{30}$ be a sample of iid Bin$(1, \theta)$ random variables with $\Theta = [0, 1]$ and $\theta \sim U(0, 1)$. Using the posterior distribution derived in Example 5.14, determine a
a) 96% credible interval for θ when $\sum_{i=1}^{30} x_i = 17$.
b) a 96% lower-bound credible interval for θ when $\sum_{i=1}^{30} x_i = 23$.
c) a 96% upper-bound credible interval for θ when $\sum_{i=1}^{30} x_i = 8$.

5.3.18 Let $X_1, \ldots, X_{50}$ be a sample of iid Pois(θ) random variables with $\Theta = \mathbb{R}^+$ and $\theta \sim Exp(\lambda = 5)$. Using the posterior distribution derived in Example 5.15, determine a

a) 98% credible interval for θ when $\sum_{i=1}^{50} x_i = 170$.
b) a 98% lower-bound credible interval for θ when $\sum_{i=1}^{50} x_i = 253$.
c) a 98% upper-bound credible interval for θ when $\sum_{i=1}^{50} x_i = 80$.

5.3.19 Let $X_1, \ldots, X_{60}$ be a sample of iid Geo(θ) random variables with $\mathcal{S} = \mathbb{W}$ and $\Theta = [0, 1]$. If $\theta \sim$ Beta($\alpha = 1.5, \beta = 3.2$) use the posterior distribution derived in Example 5.16 to determine a
a) 95% credible interval for θ when $\sum_{i=1}^{60} x_i = 170$.
b) a 95% lower-bound credible interval for θ when $\sum_{i=1}^{60} x_i = 212$.
c) a 95% upper-bound credible interval for θ when $\sum_{i=1}^{60} x_i = 580$.

5.4 Case Study – Modeling Obsidian Rind Thicknesses

In the paper "Obsidian-Hydration Dating of Fluvially Reworked Sediments in the West Yellowstone Region, Montana" published in the journal *Quaternary Research*, Adams et al. [15] studied obsidian glass, an igneous rock, in an effort to build a model that could be used to date previous glacial events. Obsidian glass is a volcanic rock that can absorb water over time forming a *rind*. The thickness of the rind can be used to date the last fracture of an obsidian grain through a method called *hydration dating*. Prior research has identified three major glacial events in the West Yellowstone region of Montana, and the focus of the authors' research was to date the most recent glacial event in this region.

The authors collected 236 observations on obsidian grains and measured the rind thickness of each grain according to a detailed protocol. The 236 rind thicknesses are given in Table B.1. As a first step in the analysis of the observed rind thicknesses, the 236 rind thicknesses were plotted in a histogram along with a *Gaussian kernel density* (GKD) estimate.

A Gaussian kernel density estimator (GKDE) is a nonparametric density estimator based on a random sample n iid random variables $X_1, \ldots, X_n$ from a common pdf $f(x)$. In particular, a GKDE is a continuous estimator of a pdf that emphasizes the clustering of the observations; a histogram does not produce a continuous pdf estimate, assigns equal weight to each observation within a class interval, and is sensitive to the choice of class intervals.

A GKDE is an estimator of a pdf $f(x)$ of the form

$$\hat{f}(x) = \frac{1}{nh} \sum_{i=1}^{n} \phi\left(\frac{x - X_i}{h}\right),$$

where $\phi(\cdot)$ is the pdf of a $N(0, 1)$ random variable, h is the *smoothing parameter*, and $X_1, \ldots, X_n$ is a sample of iid random variables with pdf $f(x)$. The smoothing parameter h controls the smoothness of a GKDE with larger values of h leading to smoother estimates. The choice of the smoothing parameter is generally based on the observed data, and a commonly used choice of the smoothing parameter is $h = 0.9An^{-0.2}$, where $A = \min\left(\text{standard deviation}, \frac{\text{interquartile range}}{1.34}\right)$. For more information on non-parametric density estimation and GKDEs see *Density Estimation for Statistics and Data Analysis* by Silverman [16].

The histogram and GKDE estimate for the $n = 236$ rind thicknesses are shown in Figure 5.5.

The R commands used to draw Figure 5.5 are

```
> hist(terr,prob=T,xlab="Rind Thickness",ylab="Density",
+ breaks=c(0:20)*1.5,main=" ",axes=F,col="lightgray")
> abline(0,0)
> lines(density(terr,na.rm=T))
> axis(2)
> axis(1,c(0:20)*1.5)
```

Because prior research had identified three major glacial events in the West Yellowstone region of Montana, and the histogram and GKDE suggest a tri-modal distribution, a three-component *finite mixture model* was fit to the data.

5.4.1 Finite Mixture Model

When a histogram or a nonparametric density estimate suggests the underlying distribution of the data is multimodal, a finite mixture model is often used

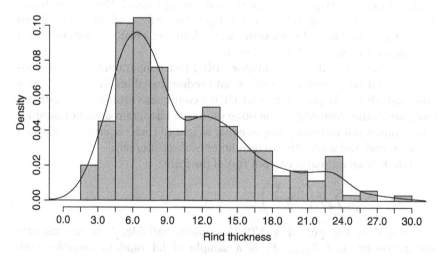

Figure 5.5 A histogram and GKDE for the $n = 236$ rind thickness measurements.

to model the underlying pdf of the data. An M-component continuous finite mixture model is a probability model with pdf of the form

$$f(x; \vec{\theta}) = \sum_{i=1}^{M} p_i \cdot g_i(x; \vec{\theta}_i),$$

where $0 \le p_i \le 1$, $\sum_{i=1}^{M} p_i = 1$, $g_i(x; \vec{\theta}_i)$ is a continuous pdf, and $\vec{\theta}_i$ is the parameter associated with the pdf g_i. Because $p_1, \ldots, p_M$ are nonnegative and sum to 1, and $g_1, g_2, \ldots, g_M$ are continuous pdfs, it follows that $f(x; \vec{\theta})$ is also a continuous pdf. An example of a four-component finite mixture model with normal component pdfs is shown in Figure 5.6. Also, it is important to note that a finite mixture model is not in the exponential family of distributions. For more information on finite mixture models see *Finite Mixture Models* by McLachlan and Peel [17].

In particular, a three-component finite mixture model with normal components has pdf

$$f(x; \vec{p}, \vec{\mu}, \vec{\sigma}) = p_1 \cdot \phi(x; \mu_1, \sigma_1) + p_2 \cdot \phi(x; \mu_2, \sigma_2)$$
$$+ (1 - p_1 - p_2) \cdot \phi(x; \mu_3, \sigma_3),$$

where $\vec{p} = (p_1, p_2)$, $\vec{\mu} = (\mu_1, \mu_2, \mu_3)$, $\vec{\sigma} = (\sigma_1, \sigma_2, \sigma_3)$, and $\phi(\cdot)$ is the pdf of a normal random variable. Note that there are eight parameters to be estimated in a three-component finite mixture model with normal components.

The MLE can be used to find estimates of the eight unknown parameters in the three-component finite mixture model with normal components,

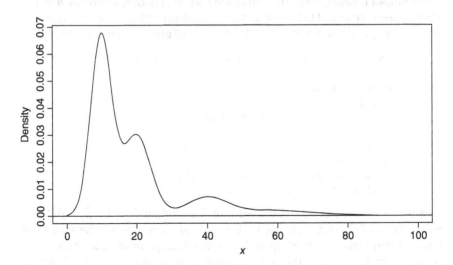

Figure 5.6 A four-component finite mixture model.

and the likelihood function for a sample of n iid observations $x_1, \ldots, x_n$ is

$$L(\vec{p}, \vec{\mu}, \vec{\sigma}) = \prod_{i=1}^{n} f(x_i; \vec{p}, \vec{\mu}, \vec{\sigma})$$

$$= \prod_{i=1}^{n} [p_1 \cdot \phi(x_i; \mu_1, \sigma_1) + p_2 \cdot \phi(x_i; \mu_2, \sigma_2)$$

$$+ (1 - p_1 - p_2) \cdot \phi(x_i; \mu_3, \sigma_3)].$$

The values of $\vec{p}, \vec{\mu}$, and $\vec{\sigma}$ that maximize the likelihood function can also be found by maximizing the log-likelihood function, $\ell(\vec{p}, \vec{\mu}, \vec{\sigma})$, which is differentiable with respect to each of the unknown parameters. The simultaneous solution to the system of eight equations

$$\frac{\partial}{\partial p_i} \ell(\vec{p}, \vec{\mu}, \vec{\sigma}) = 0, \quad i = 1, 2$$

$$\frac{\partial}{\partial \mu_i} \ell(\vec{p}, \vec{\mu}, \vec{\sigma}) = 0, \quad i = 1, 2, 3$$

$$\frac{\partial}{\partial \sigma_i} \ell(\vec{p}, \vec{\mu}, \vec{\sigma}) = 0, \quad i = 1, 2, 3$$

must be found numerically. An R package that can used to fit a finite mixture model is mixtools, and in particular, the mixtools command normalmixEM can be used to find the MLEs of a k-component finite mixture model with normal components.

The output resulting from the command normalmixEM, where the $n = 236$ observations are stored in the object terr, lambda=c(.5,.3,.2) is an initial guess of $\vec{p}$, and mu=c(6,12,20) is an initial guess of $\vec{\mu}$ is shown below.

```
> fit1=normalmixEM(terr,lambda=c(.5,.3,.2),mu=c(6,12,20),
+ k=3)
        number of iterations= 426
> summary.mixEM(fit1)
        summary of normalmixEM object:
                comp 1      comp 2      comp 3
        lambda 0.545       0.305       0.149
        mu     6.208       12.934      20.865
        sigma  1.881       2.406       3.313
        loglik at estimate:  -705
```

In the output, $lambda is the MLE of $\vec{p}$ including $\hat{p}_3 = 1 - \hat{p}_1 - \hat{p}_2$, $mu is the MLE of $\vec{\mu}$, and $sigma is the MLE of $\vec{\sigma}$. A plot of the estimated finite mixture model superimposed over the histogram of rind thicknesses is shown in Figure 5.7. The R commands used to draw Figure 5.7 are as follows:

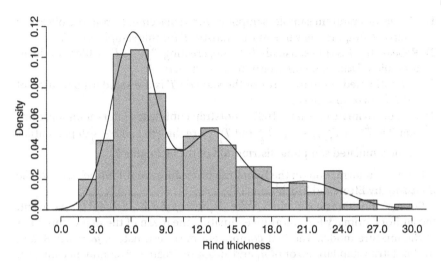

Figure 5.7 A plot of the fitted three-component mixture model superimposed on the histogram of the obsidian grain rind thicknesses.

```
> hist(terr,prob=T,xlab="Rind Thickness",ylab="Density",
+    breaks=c(0:20)*1.5,main=" ",axes=F,col="lightgray",
+    ylim=c(0,.12))
> abline(0,0)
> x=seq(0,30,.01)
> fx=.545*dnorm(x,6.21,1.88)+.305*dnorm(x,12.93,2.41)+
+    +.15*dnorm(x,20.87,3.31)
> lines(x,fx,lty=4)
> axis(2)
> axis(1,c(0:20)*1.5)
```

Because the parameter estimates in a finite mixture model must be found numerically, there are no explicit formulas for the standard errors of the parameter estimates. In this case, the asymptotic standard error of an estimator is often used, but unfortunately, for a finite mixture model, the asymptotic standard errors are based on the information matrix that should only be used when the sample size is extremely large. An alternative method for estimating the standard errors of the parameter estimates is the *bootstrap method*.

The bootstrap method uses the observed sample to simulate the sampling distribution of an estimator and can be used to estimate the standard error of a parameter estimate or to produce an approximate $(1 - \alpha) \times 100\%$ confidence interval for an unknown parameter. In particular, let T be an estimator of θ, and let $x_1, \ldots, x_n$ be an observed sample. The bootstrap procedure for estimating the standard error of a statistic T and generating an approximate $(1 - \alpha) \times 100\%$ confidence interval for the parameter θ is outlined below.

1) Generate a random sample, sampling with replacement, from the observed sample. Compute the value of the statistic T for this sample, say T_1.
2) Repeat step 1 B times, usually $B \geq 500$, creating $T_1, \ldots, T_B$, which produces a simulated sampling distribution of the statistic T.
3) The estimated standard error of the statistic T is the standard deviation of the B bootstrap values $T_1, \ldots, T_B$.
4) An approximate $(1 - \alpha) \times 100\%$ bootstrap confidence interval for a parameter θ is $\widehat{T}^\star_{\frac{\alpha}{2}}$ to $\widehat{T}^\star_{1-\frac{\alpha}{2}}$, where $\widehat{T}^\star_{\frac{\alpha}{2}}$ and $\widehat{T}^\star_{1-\frac{\alpha}{2}}$ are the $\frac{\alpha}{2}$th and $1 - \frac{\alpha}{2}$th percentiles of the simulated sampling distribution of the estimator T.

For more information on the Bootstrap procedure see *An Introduction to the Bootstrap* by Efron and Tibshirani [18].

Because the thinnest rind thicknesses represent the most recent glacial event, the authors' primary focus was on estimating the mean of the first component of the mixture model. The first component mean estimate is $\widehat{\mu}_1 = 6.208$, and the bootstrap standard error of $\widehat{\mu}_1$ and an approximate 95% confidence interval for μ_1 are $\text{se}(\widehat{\mu}_1) = 0.296$ and an approximate 95% confidence interval for μ_1 of 6.02 to 6.80. The `mixtools` command `boot.se` was used to compute the bootstrap standard error. The R commands used to compute the standard error and confidence interval are

```
> b1=boot.se(fit1,B=500)
> b1$mu.se[1]
> quantile(b1$mu[1,],c(0.25,.975))
```

Finally, a hydration dating formula for dating glacial events based on a mean rind thickness is $473 \times (\widehat{\mu}_1)^2$. Thus, the most recent glacial event near West Yellowstone, Montana, is estimated to have occurred $473 \times (6.21)^2 = 18\,240.8$ years ago.

Problems

5.4.1 Determine the number of parameters that need to be estimated in
   a) A five-component finite mixture model with normal components.
   b) A four-component finite mixture model with gamma components.

5.4.2 Plot the following finite mixture models.
   a) A two-component finite mixture model with normal components with $\vec{p} = (0.5, 0.5)$, $\vec{\mu} = (150, 200)$, and $\vec{\sigma} = (20, 25)$.
   b) A three-component finite mixture model with normal components with $\vec{p} = (0.6, 0.3, 0.1)$, $\vec{\mu} = (10, 15, 25)$, and $\vec{\sigma} = (2, 1.5, 5)$.
   c) A two-component finite mixture model with gamma components with $\vec{p} = (0.4, 0.6)$, $\vec{\alpha} = (4, 12)$, and $\vec{\beta} = (1.5, 1.2)$.
   d) A three-component finite mixture model with gamma components with $\vec{p} = (0.25, 0.20, 0.55)$, $\vec{\alpha} = (2.5, 12, 15)$, and $\vec{\beta} = (1, 0.8, 1.6)$.

5.4.3 Using the R package `mixtools` and the command `normalmixEM`
  a) fit a two-component finite mixture model with normal components to the data in Table B.1.
  b) fit a four-component finite mixture model with normal components to the data in Table B.1.

5.4.4 Use the R package `mixtools` with the command `rnormmix` to generate a random sample of $n = 300$ observations from the finite mixture model

$$f(x; \vec{p}, \vec{\mu}, \vec{\sigma}) = 0.5 \cdot \phi(x, 25, 3) + 0.4 \cdot \phi(x, 35, 4) + 0.1 \cdot \phi(x, 50, 5).$$

  a) Fit a three-component finite mixture model with normal components to the sample data generated using the `rnormmix` command.
  b) Plot the fitted finite mixture model and the underlying finite mixture model and compare.
  c) Using the command `boot.se` determine the bootstrap standard errors of the eight parameter estimates using $B = 500$.

5.4.5 Using the formula $\widehat{age} = 473 \times (\hat{\mu}_1)^2$ to estimate the age of the most recent glacial event,
  a) Use the Delta method to determine the approximate standard error formula for the estimated age of the most recent glacial.
  b) Determine the approximate standard error of the age estimate when $\hat{\mu}_1 = 6.21$ and $se(\hat{\mu}_1) = 0.296$.

5.4.6 Using the data given in Table B.1, plot a GKDE estimate for the bandwidths $h = 0.1, 0.5, 1.0.1.5, 2, 2.5$.

5.4.7 Verify a GKDE is a continuous pdf.

5.4.8 Verify an M-component continuous finite mixture model is a continuous pdf.

6

Hypothesis Testing

Another method of likelihood-based inference is *hypothesis testing*. In most statistical analyses, there is a research claim that is under investigation, and a hypothesis test is often used to determine whether or not the information in the observed sample strongly supports this research hypothesis. In a hypothesis test, the research claim must be stated in the form of a *statistical hypothesis*.

Definition 6.1 A statistical hypothesis is any statement about the distribution of one or more random variables. A statistical hypothesis is said to be a simple hypothesis when the hypothesis completely specifies the distribution of the random variables, and a statistical hypothesis is said to be a composite hypothesis when the hypothesis does not completely specify the distribution of the random variables.

Example 6.1 $H: X \sim \text{Pois}(\lambda = 4)$ is a simple hypothesis because it completely specifies the distribution of X.

Example 6.2 $H: \mu = 0$ is a composite hypothesis since there are many distributions having $\mu = 0$, and thus, H does not completely specify the distribution of X.

Example 6.3 Let $X_1, \ldots, X_n$ be a sample of random variables. $H: X_i \sim N(\mu_i, \sigma^2)$ is a composite hypothesis since μ_i and σ^2 are not specified.

In *parametric hypothesis testing*, a statistical hypothesis is a statement about the unknown parameters of a given probability distribution. For example, the statistical hypotheses in Examples 6.1 and 6.3 are parametric statistical hypotheses.

Mathematical Statistics: An Introduction to Likelihood Based Inference, First Edition. Richard J. Rossi.
© 2018 John Wiley & Sons, Inc. Published 2018 by John Wiley & Sons, Inc.

6.1 Components of a Hypothesis Test

In a hypothesis test, there are two hypotheses that are being tested against each other, namely the *null hypothesis* and the *alternative hypothesis*. The null hypothesis is denoted by H_0, the alternative hypothesis is denoted by H_A, H_0, and H_A are mutually exclusive statements, and a hypothesis test is a test of H_0 versus H_A.

In parametric hypothesis testing, H_0 and H_A can be written in terms of the parameter space. In particular, the null and alternative hypotheses are written as $H_0 : \theta \in \Theta_0$ and $H_A : \theta \in \Theta_A$, where Θ_0 is called the *null parameter space* and Θ_A is called the *alternative parameter space*.

A parametric hypothesis test is a test of $H_0 : \theta \in \Theta_0$ versus $H_A : \theta \in \Theta_A$, where H_0 and H_A are both simple hypotheses, H_0 is a simple hypothesis and H_A is a composite hypothesis, or both H_0 and H_A are composite hypotheses. The four most common forms of a parametric hypothesis test about a single parameter θ are

$$H_0 : \theta = \theta_0 \text{ versus } H_A : \theta = \theta_1$$
$$H_0 : \theta \leq \theta_0 \text{ versus } H_A : \theta > \theta_0$$
$$H_0 : \theta \geq \theta_0 \text{ versus } H_A : \theta < \theta_0$$
$$H_0 : \theta = \theta_0 \text{ versus } H_A : \theta \neq \theta_0.$$

A *hypothesis test* of H_0 versus H_A is a decision rule for rejecting a null hypothesis H_0. In particular, a hypothesis test results in one of the two decisions, namely, *reject* H_0 or *fail to reject* H_0. Specifically, a hypothesis test is a function of a sample $X_1, X_2, \ldots, X_n$ of the form

$$\varphi(\vec{X}) = \begin{cases} \text{Reject } H_0 & \text{when } \vec{X} \in \mathcal{C} \\ \text{Fail to reject } H_0 & \text{otherwise} \end{cases},$$

where $\mathcal{C}$ is called the *critical region*.

Definition 6.2 The critical region associated with a hypothesis test φ is the subset of the sample space of $\vec{X} = (X_1, X_2, \ldots, X_n)$ for which H_0 is rejected.

Example 6.4 Let $X_1, \ldots, X_n$ be iid Bin$(25, \theta)$ random variables with $\Theta = [0, 1]$. A critical region that could be used for testing $H_0 : \theta = 0.5$ versus $H_A : \theta = 0.75$ is $\mathcal{C} = \left\{ \vec{x} : \sum_{i=1}^{25} x_i > 17 \right\}$.

Given a critical region $\mathcal{C}$, the two possible errors that could occur in the hypothesis test are (i) rejecting a true H_0 and (ii) failing to reject a false H_0, and these errors are called *Type I* and *Type II* errors, respectively.

Definition 6.3 A Type I error is made by rejecting H_0 when H_0 is true, and a Type II error is made by failing to reject H_0 when H_0 is false.

Table 6.1 Possible errors in a hypothesis test.

	Truth	
Decision	H_0 is true	H_A is true
Reject H_0	Type I error	No error
Fail to reject H_0	No error	Type II error

The possible errors in a hypothesis test are summarized in Table 6.1. The measures of reliability of a hypothesis test are the probabilities of making a Type I error or a Type II error, and in a well-designed hypothesis test, the probability of making either type of testing error is small. Specifically, the measures of reliability associated with a hypothesis test are called the *size* of the test and the *power* of the test.

Definition 6.4 For testing $H_0 : \theta \in \Theta_0$ versus $H_A : \theta \in \Theta_A$ with critical region $\mathcal{C}$, the size of the hypothesis test is $\alpha = \sup_{\theta \in \Theta_0} P(\vec{X} \in \mathcal{C}|\theta)$.

Definition 6.5 For testing $H_0 : \theta \in \Theta_0$ versus $H_A : \theta \in \Theta_A$ with critical region $\mathcal{C}$, the power function for the hypothesis test is defined to be $\pi(\theta) = P(\vec{X} \in \mathcal{C}|\theta)$, for $\theta \in \Theta$. The power of the test at $\theta_A \in \Theta_A$ is $\pi(\theta_A)$.

The size of a test is also referred to as the *significance level* of the test. It is important to note that the size of the test is the largest probability of making a Type I error, and the power of the test is the probability of rejecting H_0 for a given value of $\theta \in \Theta$.

The power function gives the probability of correctly rejecting H_0 when $\theta \in \Theta_A$, and thus, the power function measures the ability of the test to distinguish values in Θ_A from values in Θ_0. A well-designed hypothesis test will have large power for values of $\theta \in \Theta_A$ that are different in a practical sense from those in Θ_0. For $\theta \in \Theta_A$, the probability of making a Type II error is a function of power and is $\beta(\theta) = P(\vec{X} \notin \mathcal{C}|\theta) = 1 - \pi(\theta)$. Thus, it is unlikely that a Type II error will be made when the power at θ is large. The optimal choice of the critical region maximizing the power depends on the underlying probability model and the structure of the hypotheses being tested and is discussed in Sections 6.2–6.4.

Example 6.5 Let $X_1, \dots, X_{10}$ be a sample of iid Poisson random variables. Consider the critical region $\mathcal{C} = \left\{ \vec{x} : \sum_{i=1}^{10} x_i \geq 51 \right\}$ for testing $H_0 : \lambda = 4$ versus $H_A : \lambda = 5$. Then, the decision function for this test is

$$\varphi(\vec{x}) = \begin{cases} \text{Reject } H_0 & \text{when } \sum_{i=1}^{10} x_i \geq 51 \\ \text{Fail to reject } H_0 & \text{otherwise} \end{cases}.$$

Since $\sum_{i=1}^{10} X_i \sim \text{Pois}(10\lambda)$ and H_0 is a simple hypothesis, it follows that the size of the critical region $\mathcal{C}$ is

$$\alpha = P\left(\sum_{i=1}^{10} X_i \geq 51 \mid \lambda = 4\right) = 0.0526.$$

Since H_A is also a simple hypothesis, the power of the test is

$$\pi(5) = P\left(\sum_{i=1}^{10} X_i \geq 51 \mid \lambda = 5\right) = 0.462,$$

and thus, the probability of making a Type II error is $\beta(5) = 1 - 0.462 = 0.538$.

Example 6.6 Let $X_1, \ldots, X_{25}$ be a sample of iid $N(\theta, 25)$ random variables. For testing $H_0 : \theta \leq 0$ versus $H_A : \theta > 0$, let the critical region be $\mathcal{C} = \{\vec{x} : \bar{x} > 1.645\}$. Now, since $\bar{X} \sim N(\theta, 1)$, it follows that

$$\alpha = \sup_{\theta \leq 0} P(\mathcal{C} \mid \theta) = \sup_{\theta \leq 0} P(\bar{X} > 1.645 \mid \theta).$$

Consider, $P(\bar{X} > 1.645 \mid \theta)$.

$$P(\bar{X} > 1.645 \mid \theta) = P(\bar{X} - \theta > 1.645 - \theta \mid \theta) = \underbrace{P(Z > 1.645 - \theta \mid \theta)}_{Z \sim N(0,1)}.$$

Now, for $\theta \leq 0$, $P(Z > 1.645 - \theta)$ is increasing in θ, and therefore,

$$\alpha = \sup_{\theta \leq 0} P(\mathcal{C} \mid \theta) = P(\bar{X} > 1.645 \mid \theta = 0) = 0.05.$$

The power function for $\mathcal{C}$ is shown in Figure 6.1. Note that the power function is an increasing function, and for $\theta \geq 2.5$, $\pi(\theta) \geq 0.80$.

Common sense dictates that it should be more likely to reject H_0 for $\theta \in \Theta_A$ than it is to reject H_0 for any value of $\theta \in \Theta_0$, and a critical region having this property is called an *unbiased critical region*.

Definition 6.6 A critical region $\mathcal{C}$ of size α with power function $\pi(\theta)$ for testing $H_0 : \theta \in \Theta_0$ versus $H_A : \theta \in \Theta_A$ is said to be an unbiased critical when $\pi(\theta_1) \geq \pi(\theta_0)$ for every $\theta_1 \in \Theta_A$ and $\theta_0 \in \Theta_0$.

Example 6.7 The critical region mentioned in Example 6.6 is an unbiased critical region since the power function is an increasing function on Θ.

The final component of a hypothesis test is a measure of the strength of the information in an observed sample against the null hypothesis. The most commonly used measure of the strength of the sample evidence against the null hypothesis is the *p-value*.

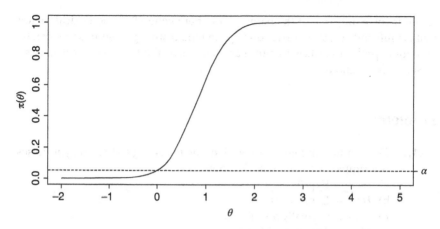

Figure 6.1 A plot of the power function for $\mathcal{C} = \{\vec{x} : \bar{x} > 1.645\}$ in Example 6.6.

Definition 6.7 The p-value associated with a sample and a critical region $\mathcal{C}$ is the smallest value of α for which H_0 is rejected.

The smaller the p-value, the stronger the evidence in the observed sample is against H_0. When an observed sample $\vec{x}$ falls in the critical region $\mathcal{C}$, the p-value is less than α, and conversely, when $\vec{x}$ is not in $\mathcal{C}$, the p-value is larger than α. Thus, a hypothesis test based on a critical region $\mathcal{C}$ amounts to rejecting H_0 when the p-value is less than α and failing to reject H_0 when p-value is greater than α.

For a critical region of the form $\{\vec{x} : T(\vec{x}) \geq k\}$ and an observed sample $\vec{x}$, the p-value associated with $T_{\text{obs}} = T(\vec{x})$ is $p = \sup_{\theta \in \Theta_0} P(T(\vec{X}) \geq T_{\text{obs}} | \theta)$. Similarly, for a critical region of the form $\{\vec{x} : T(\vec{x}) \leq k\}$ and an observed sample $\vec{x}$, the p-value is $p = \sup_{\theta \in \Theta_0} P(T(\vec{X} \leq T_{\text{obs}} | \theta)$.

Example 6.8 Let $X_1, \ldots, X_{25}$ be a sample of iid Pois(θ) random variables, and let $\mathcal{C} = \{\vec{x} : \sum_{i=1}^{25} x_i \geq 171\}$ be a size $\alpha = 0.049$ critical region for testing $H_0 : \theta \leq 6$ versus $H_A : \theta > 6$. If the observed sample results in $\sum_{i=1}^{25} x_i = 188$, then the p-value is $P\left(\sum_{i=1}^{25} X_i \geq 188 | \theta = 6\right) = 0.0015$. Since the p-value is less than $\alpha = 0.049$, there is sufficient evidence in the observed sample for rejecting H_0.

The goal of parametric hypothesis testing is to determine the critical region having the largest power at each value of $\theta \in \Theta_A$. Methods of determining the best critical region are discussed in Sections 6.2 and 6.3; however, it is not always possible to create a critical region having the largest power on Θ_A. In particular, when θ is a vector of unknown parameters, and the hypotheses being tested involve only one of the parameters, there generally is no critical region that is uniformly most powerful on Θ_A. General testing procedures that can be used for hypotheses tests with multiparameter probability models are discussed in Sections 6.4 and 6.5.

Problems

6.1.1 Determine whether or not each of the following statistical hypotheses is a simple or a composite hypothesis.
a) $H: X \sim \text{Exp}(\theta)$.
b) $H: X \sim \text{Exp}(\theta = 4)$.
c) $H: X$ is normally distributed.
d) $H: X \sim U(0, \theta)$ and $\theta < 3$.

6.1.2 Let $X_1, \ldots, X_{10}$ be a sample of iid $\text{Bin}(1, \theta)$ random variables, and let $\mathcal{C} = \{\vec{x}: \sum_{i=1}^{10} x_i \geq 9\}$ be a critical region for testing $H_0: \theta = 0.6$ versus $H_A: \theta = 0.8$. Determine
a) the size of this critical region.
b) the power of this critical region for $\theta = 0.8$.

6.1.3 Let $X_1, \ldots, X_{16}$ be a sample of iid $\text{Pois}(\theta)$ random variables, and let $\mathcal{C} = \{\vec{x}: \sum_{i=1}^{16} x_i \leq 20\}$ be a critical region for testing $H_0: \theta = 2$ versus $H_A: \theta = 1$. Determine
a) the size of this critical region.
b) the power of this critical region for $\theta = 1$.

6.1.4 Let $X_1, \ldots, X_{10}$ be a sample of iid $\text{Exp}(\theta)$ random variables, and let $\mathcal{C} = \{\vec{x}: \sum_{i=1}^{10} x_i \leq 19\}$ be a critical region for testing $H_0: \theta \geq 3.5$ versus $H_A: \theta < 3.5$. Using the fact that $\sum_{i=1}^{10} X_i \sim \text{Gamma}(10, \theta)$,
a) determine the size of this critical region.
b) determine the power of this critical region for $\theta = 2$.
c) plot the power function for $\theta \in (0, 3.5)$.

6.1.5 Let $X_1, \ldots, X_{10}$ be a sample of iid $U(0, \theta)$ random variables, and let $\mathcal{C} = \{\vec{x}: x_{(10)} > 0.995\}$ be a critical region for testing $H_0: \theta \leq 1$ versus $H_A: \theta > 1$. Using the fact that the pdf of $X_{(10)}$ is $f_n(x; \theta) = \frac{10x^9}{\theta^{10}}$ for $x \in (0, \theta)$,
a) determine the size of this critical region.
b) determine the power of this critical region for $\theta = 1.25$.
c) plot the power function for $\theta \in (1, 5)$.
d) determine the value of θ such that $\pi(\theta) = 0.95$.

6.1.6 Let $X_1, \ldots, X_{16}$ be a sample of iid $N(\theta, 16)$ random variables and let $\mathcal{C}_1 = \{\vec{x} : \bar{x} < -1.96 \text{ or } \bar{x} > 1.96\}$ and $\mathcal{C}_2 = \{\vec{x} : \bar{x} < -2.33 \text{ or } \bar{X} > 1.75\}$ be two critical regions for testing $H_0 : \theta = 0$ versus $\theta \neq 0$. Using the fact that $\bar{X} \sim N(\theta, 1)$,
   a) determine the size of the critical region $\mathcal{C}_1$.
   b) determine the size of the critical region $\mathcal{C}_2$.
   c) determine which critical region has higher power for $\theta = 1$.

6.1.7 Let $X_1, \ldots, X_{20}$ be a sample of iid $\text{Bin}(1, \theta)$ random variables with $\Theta = \{0.4, 0.6\}$. For testing $H_0 : \theta = 0.4$ versus $H_A : \theta = 0.6$, the critical region $\mathcal{C} = \{\vec{x} : \sum_{i=1}^{20} x_i \geq k\}$ will be used. Determine
   a) the value of k so that $\alpha \approx 0.05$
   b) $\pi(0.6)$ for $\mathcal{C}$ using the value of k found in Problem 6.1.7(a).

6.1.8 Let $X_1, \ldots, X_{20}$ be a sample of iid $\text{Pois}(\theta)$ random variables with $\Theta = \{2, 4\}$. For testing $H_0 : \theta = 4$ versus $H_A : \theta = 5$, the critical region $\mathcal{C} = \{\vec{x} : \sum_{i=1}^{20} x_i \geq k\}$ will be used. Determine
   a) the value of k so that $\alpha \approx 0.02$
   b) $\pi(5)$ for $\mathcal{C}$ using the value of k found in Problem 6.1.8(a).

6.1.9 Let $X_1, \ldots, X_{36}$ be a sample of iid $N(\theta, 108)$ random variables with $\Theta = \mathbb{R}$. For testing $H_0 : \theta \geq 0$ versus $H_A : \theta < 0$, the critical region $\mathcal{C} = \{\vec{X} : \bar{X} < k\}$ will be used.
   a) Determine the value of k so that $\alpha = 0.01$.
   b) Determine $\pi(-5)$ for $\mathcal{C}$ using the value of k found in Problem 6.1.9(a).
   c) Create a plot of the power function, $\pi(\theta)$, using the value of k found in Problem 6.1.9(a) for $\theta \in (-10, 0)$.

6.1.10 Let $X_1, \ldots, X_{25}$ be a sample of iid $\text{Bin}(1, \theta)$ random variables, and let $\mathcal{C} = \{\vec{x} : \sum_{i=1}^{25} x_i \geq 20\}$ be a size α critical region for testing $H_0 : \theta \leq 0.6$ versus $H_A : \theta > 0.6$. Determine the value of
   a) α.
   b) the p-value when $\sum_{i=1}^{10} x_i = 21$.

6.1.11 Let $X_1, \ldots, X_{20}$ be a sample of iid $\text{Exp}(\theta)$ random variables, and let $\mathcal{C} = \{\vec{x} : \sum_{i=1}^{20} x_i \leq 12.9\}$ be a size α critical region for testing $H_0 : \theta \geq 1$ versus $H_A : \theta < 1$. Determine the
   a) sampling distribution of $\sum_{i=1}^{20} X_i$ when $\theta = 1$.
   b) value of α.
   c) $\pi(0.5)$.
   d) value of the p-value when $\sum_{i=1}^{20} x_i = 12.1$.

6.1.12 Let $X_1, \ldots, X_{25}$ be a sample of iid $N(\theta, 125)$ random variables, and let $\mathcal{C} = \{\vec{x} : \bar{x} \geq 3.8\}$ be a size α critical region for testing $H_0 : \theta \leq 0$ versus $H_A : \theta > 0$. Determine the

a) sampling distribution of $\overline{X}$ when $\theta = 0$.
b) value of α.
c) $\pi(3.5)$.
d) value of the p-value when $\bar{x} = 2.9$.

6.2 Most Powerful Tests

The goal of a hypothesis test is to find, when it exists, the critical region with the largest power for each value of the parameter in Θ_A. Moreover, when testing a simple null hypothesis versus a simple alternative hypothesis, a critical region that maximizes the power of the test always exists. A critical region of size α having the largest power is called a *most powerful critical region*.

Definition 6.8 A critical region $\mathcal{C}^\star$ for testing a simple null hypothesis $H_0 : \theta = \theta_0$ against a simple alternative hypothesis $H_A : \theta = \theta_A$ is said to be a most powerful (MP) critical region of size α when

i) $\pi_{\mathcal{C}^*}(\theta_0) = \alpha$.
ii) $\pi_{\mathcal{C}^*}(\theta_A) \geq \pi_{\mathcal{C}}(\theta_A)$ for any other critical region $\mathcal{C}$ of size α.

Note that in a test of a simple null hypothesis $H_0 : \theta = \theta_0$ versus a simple alternative hypothesis $H_A : \theta = \theta_A$, the size of a critical region $\mathcal{C}$ is $\alpha = P(\mathcal{C}|\theta_0)$ and power of the critical region is $\pi_{\mathcal{C}}(\theta_A) = P(\vec{X} \in \mathcal{C}|\theta_A)$.

The form of the most powerful critical region for testing a simple null hypothesis against a simple alternative hypothesis was given by Neyman and Pearson [19] in the *Neyman–Pearson Lemma*, Theorem 6.1. Specifically, the Neyman–Pearson Lemma shows that the most powerful critical region for testing $H_0 : \theta = \theta_0$ versus $H_A : \theta = \theta_A$ is based on the ratio of the likelihood functions evaluated at θ_0 and θ_A.

Theorem 6.1 (Neyman–Pearson Lemma) *Let $X_1, \ldots, X_n$ be a sample of random variables with joint pdf $f(x_1, \ldots, x_n; \theta)$ on $\Theta = \{\theta_0, \theta_A\}$, let $L(\theta)$ be the likelihood function, and let $\Lambda(\theta_0, \theta_A) = \frac{L(\theta_0)}{L(\theta_A)}$. Then, $\mathcal{C}^\star = \{\vec{x} : \Lambda(\theta_0, \theta_A) \leq k_\alpha\}$, where k_α is a constant chosen so that $\pi_{\mathcal{C}^*}(\theta_0) = \alpha$, is an MP critical region of size α for testing $H_0 : \theta = \theta_0$ versus $H_A : \theta = \theta_A$.*

Proof. The proof will be given for continuous random variables $X_1, \ldots, X_n$ since the proof is similar for the discrete case with integration replaced by summation.

Let $X_1, \ldots, X_n$ be a sample of continuous random variables with joint pdf $f(\vec{x}; \theta)$ on $\Theta = \{\theta_0, \theta_A\}$, let $\Lambda(\theta_0, \theta_A) = \frac{L(\theta_0)}{L(\theta_A)}$, and let $\mathcal{C}^\star = \{\vec{x} : \Lambda(\theta_0, \theta_A) \leq k_\alpha\}$, where k_α is a positive constant chosen so that $\pi_{\mathcal{C}^\star}(\theta_0) = \alpha$.

First, on $\mathcal{C}^\star$, $\Lambda(\theta_0, \theta_A) = \frac{L(\theta_0)}{L(\theta_A)} \leq k_\alpha$ which means $f(\vec{x}; \theta_0) \leq k_\alpha f(\vec{x}; \theta_A)$ on $\mathcal{C}^\star$. Also, on $\mathcal{C}^{\star c}$, $f(\vec{x}; \theta_0) \geq k_\alpha f(\vec{x}; \theta_A)$. Now, let $\mathcal{C}$ be any other critical region of size α for testing $H_0 : \theta = \theta_0$ versus $H_A : \theta = \theta_A$.

Then, since both $\mathcal{C}^\star$ and $\mathcal{C}$ are size α critical regions, it follows that

$$
\begin{aligned}
0 = \pi_{\mathcal{C}^\star}(\theta_0) - \pi_{\mathcal{C}}(\theta_0) &= \int_{\mathcal{C}^\star} f(\vec{x}; \theta_0) \, d\vec{x} - \int_{\mathcal{C}} f(\vec{x}; \theta_0) \, d\vec{x} \\
&= \int_{\mathcal{C}^\star \cap \mathcal{C}} f(\vec{x}; \theta_0) \, d\vec{x} + \int_{\mathcal{C}^\star \cap \mathcal{C}^c} f(\vec{x}; \theta_0) \, d\vec{x} - \int_{\mathcal{C} \cap \mathcal{C}^\star} f(\vec{x}; \theta_0) \, d\vec{x} \\
&\quad - \int_{\mathcal{C} \cap \mathcal{C}^{\star c}} f(\vec{x}; \theta_0) \, d\vec{x} \\
&= \int_{\mathcal{C}^\star \cap \mathcal{C}^c} f(\vec{x}; \theta_0) \, d\vec{x} - \int_{\mathcal{C} \cap \mathcal{C}^{\star c}} f(\vec{x}; \theta_0) \, d\vec{x}.
\end{aligned}
$$

Now, $k_\alpha f(\vec{x}; \theta_A) \geq f(\vec{x}; \theta_0)$ on $\mathcal{C}^\star$ and any subset of $\mathcal{C}^\star$, and thus,

$$
\int_{\mathcal{C}^\star \cap \mathcal{C}^c} f(\vec{x}; \theta_0) \, d\vec{x} \leq k_\alpha \int_{\mathcal{C}^\star \cap \mathcal{C}^c} f(\vec{x}; \theta_A) \, d\vec{x}.
$$

Also, $k_\alpha f(\vec{x}; \theta_A) \leq f(\vec{x}; \theta_0)$ on $\mathcal{C}^{\star c}$ and its subsets, and therefore,

$$
\int_{\mathcal{C} \cap \mathcal{C}^{\star c}} f(\vec{x}; \theta_0) \, d\vec{x} \geq k_\alpha \int_{\mathcal{C} \cap \mathcal{C}^{\star c}} f(\vec{x}; \theta_A) \, d\vec{x}.
$$

Now,

$$
\begin{aligned}
0 &= \int_{\mathcal{C}^\star \cap \mathcal{C}^c} f(\vec{x}; \theta_0) \, d\vec{x} - \int_{\mathcal{C} \cap \mathcal{C}^{\star c}} f(\vec{x}; \theta_0) \, d\vec{x} \\
&\leq k_\alpha \int_{\mathcal{C}^\star \cap \mathcal{C}^c} f(\vec{x}; \theta_A) \, d\vec{x} - k_\alpha \int_{\mathcal{C} \cap \mathcal{C}^{\star c}} f(\vec{x}; \theta_A) \, d\vec{x} \\
&= k_\alpha \int_{\mathcal{C}^\star \cap \mathcal{C}^c} f(\vec{x}; \theta_A) \, d\vec{x} + \underbrace{k_\alpha \int_{\mathcal{C}^\star \cap} f(\vec{x}; \theta_A) \, d\vec{x} - k_\alpha \int_{\mathcal{C}^\star \cap} f(\vec{x}; \theta_A) \, d\vec{x}}_{0} \\
&\quad - k_\alpha \int_{\mathcal{C} \cap \mathcal{C}^{\star c}} f(\vec{x}; \theta_A) \, d\vec{x} \\
&= k_\alpha \left[\int_{\mathcal{C}^\star} f(\vec{x}; \theta_A) \, d\vec{x} - \int_{CR} f(\vec{x}; \theta_A) \, d\vec{x} \right] \\
&= k_\alpha [\pi_{CRS}(\theta_A) - \pi_{\mathcal{C}}(\theta_A)].
\end{aligned}
$$

Thus, $k_\alpha [\pi_{CRS}(\theta_A) - \pi_{\mathcal{C}}(\theta_A)] \geq 0$, and since $k_\alpha > 0$, $\pi_{CRS}(\theta_A) \geq \pi_{\mathcal{C}}(\theta_A)$.

Hence, since $\mathcal{C}$ was an arbitrary size α critical region for testing $H_0 : \theta = \theta_0$ versus $H_A : \theta = \theta_A$, $\mathcal{C}^\star$ is the most powerful critical region of size α. ∎

Moreover, not only is the critical region given in the Neyman–Pearson Lemma an MP critical region, Corollaries 6.1 and 6.2 show that it is also an unbiased critical region based entirely on a sufficient statistic for θ.

Corollary 6.1 *If $X_1, \ldots, X_n$ is a sample of iid random variables and $\mathcal{C}^\star$ is the Neyman–Pearson MP critical region for testing $H_0 : \theta = \theta_0$ versus $H_A : \theta = \theta_A$, then $\mathcal{C}^\star$ is an unbiased critical region.*

Proof. The proof is given in *Testing Statistical Hypotheses* by Lehmann [20]. ∎

Corollary 6.2 *If $X_1, \ldots, X_n$ is a sample of iid random variables with common pdf $f(x; \theta)$ for $\Theta = \{\theta_0, \theta_A\}$, $\mathcal{C}^\star$ is the Neyman–Pearson MP critical region of size α for testing $H_0 : \theta = \theta_0$ versus $H_A : \theta_A$, and S is a sufficient statistic for θ, then $\mathcal{C}^\star$ depends on $X_1, \ldots, X_n$ only through S.*

Proof. Let $X_1, \ldots, X_n$ be a sample of iid random variables with common pdf $f(x; \theta)$ for $\Theta = \{\theta_0, \theta_A\}$, $\mathcal{C}^\star$ be an MP critical region of size α for testing $H_0 : \theta = \theta_0$ versus $H_A : \theta = \theta_A$, and S be a sufficient statistic for θ. Then,

$$\Lambda(\theta_0, \theta_A) = \frac{\prod_{i=1}^n f(x_i; \theta_0)}{\prod_{j=1}^n f(x_j; \theta_A)} = \frac{g(S; \theta_0)h(\vec{x})}{g(S; \theta_A)h(\vec{x})} = \frac{g(S; \theta_0)}{g(S; \theta_A)}.$$

Thus, the MP critical region depends on $\vec{X}$ only through the sufficient statistic S. ∎

The Neyman–Pearson Lemma is used in Example 6.9 to find an MP critical region for testing a simple null versus a simple alternative.

Example 6.9 Let $X_1, \ldots, X_n$ be a sample of iid Pois(θ) random variables with $\Theta = \{4, 5\}$. Then, for testing $H_0 : \theta = 4$ versus $H_A : \theta = 5$, the MP critical region of size α is $\mathcal{C}^\star = \{\vec{x} : \Lambda(4, 5) \le k_\alpha\}$. Now,

$$\Lambda(4, 5) = \frac{e^{-4n} 4^{\sum_{i=1}^n x_i} / \prod_{i=1}^n x_i}{e^{-5n} 5^{\sum_{i=1}^n x_i} / \prod_{i=1}^n x_i} = \frac{e^{-4n} 4^{\sum_{i=1}^n x_i}}{e^{-5n} 5^{\sum_{i=1}^n x_i}} = e^n \left[\frac{4}{5}\right]^{\sum_{i=1}^n x_i},$$

and therefore, the MP critical region for testing $H_0 : \theta = 4$ versus $H_A : \theta = 5$ is

$$\mathcal{C}^\star = \left\{ \vec{x} : e^n \left[\frac{4}{5}\right]^{\sum_{i=1}^n x_i} \le k_\alpha \right\}.$$

In order to find the value of k_α, it is important to find an equivalent critical region based on a sufficient statistic with known sampling distribution. Now, using a sequence of set-preserving transformations, an equivalent form of $\mathcal{C}^\star$

is $\mathcal{C}^\star = \{\vec{x}: \sum_{i=1}^{n} x_i \geq k\}$ since

$$\Lambda(4,5) \leq k \quad \text{if and only if} \quad \underbrace{\left[\frac{4}{5}\right]^{\sum x_i} \leq k'}_{\text{since } e^{-n} > 0}$$

$$\text{if and only if} \quad \underbrace{\ln\left[\frac{4}{5}\right] \sum x_i \leq k''}_{\text{taking ln both sides}}$$

$$\text{if and only if} \quad \underbrace{\sum x_i \geq k'''}_{\text{since } \ln\left[\frac{4}{5}\right] < 0}.$$

Since the sampling distribution of $\sum_{i=1}^{n} X_i$ is Pois($n\theta$), the values of k_α and $\pi(5)$ can be found using the Poisson distribution. For example, if $n = 25$ and $\alpha \approx 0.05$, the value of k_α is found using $\sum_{i=1}^{25} X_i \sim \text{Pois}(100)$. In particular, since $0.05 = P(\vec{x}: \sum_{i=1}^{25} x_i \geq k_\alpha | \theta = 4)$ the value of $k_{0.05}$ can be found using the command qpois(0.95,100), which yields $k_{0.052} = 117$; because the distribution of $\sum_{i=1}^{25} X_i$ is discrete, an exact size of $\alpha = 0.05$ is not possible in this case.

Finally, the power associated with $\mathcal{C}^\star$ when $n = 25$ is

$$\pi(5) = P\left(\sum_{i=1}^{25} X_i \geq 117 \middle| \lambda = 5\right) = 1 - P\left(\sum_{i=1}^{25} X_i \leq 116 \middle| \lambda = 5\right),$$

which can be found using the R command 1-ppois(116,125). Hence, for $n = 25$ and $\alpha = 0.052$, $\pi(5) = 0.775$.

Theorem 6.2 provides convenient forms for the Neyman–Pearson MP critical regions when $f(x; \theta)$ is an exponential family of distribution.

Theorem 6.2 *If $X_1, \ldots, X_n$ is a sample of iid random variables with pdf $f(x; \theta) = c(\theta)h(x)e^{q(\theta)t(x)}$ in the exponential family of distributions, then the MP critical region for testing $H_0: \theta = \theta_0$ versus $H_A: \theta = \theta_A$ is*

i) $\mathcal{C}^\star = \{\vec{x}: \sum_{i=1}^{n} t(x_i) \geq k_\alpha\}$ *when $q(\theta)$ is increasing and $\theta_0 < \theta_A$.*
ii) $\mathcal{C}^\star = \{\vec{x}: \sum_{i=1}^{n} t(x_i) \leq k_\alpha\}$ *when $q(\theta)$ is decreasing and $\theta_0 < \theta_A$.*
iii) $\mathcal{C}^\star = \{\vec{x}: \sum_{i=1}^{n} t(x_i) \leq k_\alpha\}$ *when $q(\theta)$ is increasing and $\theta_0 > \theta_A$.*
iv) $\mathcal{C}^\star = \{\vec{x}: \sum_{i=1}^{n} t(x_i) \geq k_\alpha\}$ *when $q(\theta)$ is decreasing and $\theta_0 > \theta_A$.*

Proof. The proof of Theorem 6.2 is left as an exercise. ∎

Note that when $X_1, \ldots, X_n$ are iid one-parameter exponential family random variables, the Neyman–Pearson critical regions are based entirely on the sufficient statistic $\sum_{i=1}^{n} t(X_i)$.

Example 6.10 Let $X_1, \ldots, X_n$ be a sample of iid Bin$(1, \theta)$ random variables. Since the pdf of Bin$(1, \theta)$ is an exponential family distribution, for testing $H_0 : \theta = 0.5$ versus $H_A : \theta = 0.8$, Theorem 6.2 can be used to find the form of the MP critical region.

First, $f(x; \theta) = \theta^x (1 - \theta)^{1-x} = (1 - \theta) e^{x \ln \left[\frac{\theta}{1-\theta} \right]}$, and thus, $f(x; \theta)$ is in the exponential family of distributions with $q(\theta) = \frac{\theta}{1-\theta}$ and $t(x) = x$. Thus, $q(\theta)$ is increasing and $\theta_0 < \theta_A$, and by Theorem 6.2, the MP critical region for testing $H_0 : \theta = 0.5$ versus $H_A : \theta = 0.8$ is $\mathcal{C}^\star = \{\vec{x} : \sum_{i=1}^{n} x_i \geq k_\alpha\}$.

For example, if $n = 30$ and $\alpha = 0.02$, $\sum_{i=1}^{30} X_i \sim$ Bin$(30, 0.5)$ under H_0, and the MP critical region for testing $H_0 : \theta = 0.5$ versus $H_A : \theta = 0.8$ is $\mathcal{C}^\star = \{\vec{X} : \sum_{i=1}^{n} X_i \geq 21\}$. The value of $k_{0.02} = 21$ was found using the R command qbinom(0.98,30,0.5).

Also, for $n = 30$ and $\theta = 0.8$, $\sum_{i=1}^{30} X_i \sim$ Bin$(30, 0.8)$. Thus, for $n = 30$ and $\alpha = 0.02$, the power of $\mathcal{C}^\star$ for $\theta = 0.8$ is

$$\pi_{\mathcal{C}^\star}(0.8) = P(\mathcal{C}^\star) = P\left(\sum_{i=1}^{30} X_i \geq 21 \,\middle|\, \theta = 0.8\right)$$

$$= 1 - P\left(\sum_{i=1}^{30} X_i \leq 20\} \,\middle|\, \theta = 0.8\right)$$

$$= 1\text{-pbinom}(20,30,0.8) = 0.939.$$

Example 6.11 Let $X_1, \ldots, X_n$ be a sample of iid $N(0, \theta)$ random variables with $\Theta = \{5, 10\}$. Then, for testing $H_0 : \theta = 10$ versus $H_A : \theta = 5$, Theorem 6.2 can again be used to find the form of the MP critical region since $f(x; \theta)$ is in the exponential family of distributions with $q(\theta) = -\frac{1}{2\theta}$ and $t(x) = x^2$. Now, since $q(\theta)$ is increasing and since $\theta_0 = 10 > \theta_A = 5$, by Theorem 6.2, the MP critical region for testing $H_0 : \theta = 10$ versus $H_A : \theta = 5$ is $\mathcal{C}^\star = \{\vec{x} : \sum_{i=1}^{n} x_i^2 \leq k_\alpha\}$.

Now, $\frac{1}{\theta} \sum_{i=1}^{n} X_i^2 \sim \chi_n^2$ when $X_1, X_2, \ldots, X_n$ are iid $N(0, \theta)$ random variables. Thus,

$$P(\mathcal{C}^\star | \theta) = P\left(\sum_{i=1}^{n} X_i^2 \leq k_\alpha \,\middle|\, \theta\right) = P\left(\frac{1}{\theta} \sum_{i=1}^{n} X_i^2 \leq \frac{k_\alpha}{\theta} \,\middle|\, \theta\right)$$

$$= P\left(\chi_n^2 \leq \frac{k_\alpha}{\theta}\right).$$

Hence, for $\theta = 10$,

$$P(\mathcal{C}^\star | \theta = 10) = P\left(\sum_{i=1}^{n} X_i^2 \le k_\alpha \middle| \theta = 10 \right) = P\left(\frac{1}{10} \sum_{i=1}^{n} X_i^2 \le \frac{k_\alpha}{10} \middle| \theta = 10 \right)$$

$$= P\left(\chi_n^2 \le \frac{k_\alpha}{10} \right).$$

In particular, for $n = 50$ and $\alpha = 0.01$, $\frac{1}{10} \sum_{i=1}^{50} X_i^2 \sim \chi_{50}^2$, and therefore, the value of $k_{0.01} = 10 \cdot \text{qchisq}(0.01, 50) = 297.07$. Thus, the most powerful critical region of size $\alpha = 0.01$ for testing $H_0 : \theta = 10$ versus $H_A : \theta = 5$ is

$$\mathcal{C}^\star = \left\{ \vec{X} : \sum_{i=1}^{50} X_i^2 \le 297.07 \right\}.$$

The power of $\mathcal{C}^\star$ when $n = 50$ and $\alpha = 0.01$ is

$$P(\mathcal{C}^\star | \theta = 5) = P\left(\sum_{i=1}^{n} X_i^2 \le 2978.07 \middle| \theta = 5 \right)$$

$$= P\left(\frac{1}{5} \sum_{i=1}^{n} X_i^2 \le \frac{297.07}{5} \middle| \theta = 5 \right)$$

$$= P(\chi_n^2 \le 59.414) = 0.830.$$

Problems

6.2.1 For the Neyman–Pearson MP critical region for testing $H_0 : \theta = 4$ versus $H_A : \theta = 5$ in Example 6.9, determine
a) the value of k_α for $\alpha \approx 0.02$ when $n = 10$.
b) $\pi(5)$ for $\alpha = 0.02$ when $n = 10$.

6.2.2 For the Neyman–Pearson MP critical region for testing $H_0 : \theta = 0.5$ versus $H_A : \theta = 0.8$ in Example 6.10, determine
a) the value of k_α for $\alpha \approx 0.025$ when $n = 15$.
b) $\pi(0.8)$ for $\alpha \approx 0.025$ when $n = 15$.

6.2.3 For the Neyman–Pearson MP critical region for testing $H_0 : \theta = 10$ versus $H_A : \theta = 5$ in Example 6.11, determine
a) the value of k_α for $\alpha = 0.05$ when $n = 10$.
b) $\pi(5)$ for $\alpha \approx 0.05$ when $n = 10$.

6.2.4 Let $X_1, \ldots, X_n$ be a sample of iid $\text{Geo}(\theta)$ random variables with $\Theta = \{0.2, 0.5\}$. For testing $H_0 : \theta = 0.2$ versus $H_A : \theta = 0.5$, determine
   a) the form of the Neyman–Pearson MP critical region for a size α test.
   b) the sampling distribution of $\sum_{i=1}^{n} X_i$.
   c) the value of k_α for $\alpha \approx 0.05$ when $n = 10$.
   d) $\pi(0.5)$ for $\alpha \approx 0.05$ when $n = 10$.

6.2.5 Let $X_1, \ldots, X_n$ be a sample of iid $\text{Bin}(2, \theta)$ random variables with $\Theta = \{0.6, 0.8\}$. For testing $H_0 : \theta = 0.6$ versus $H_A : \theta = 0.8$, determine
   a) the form of the Neyman–Pearson MP critical region for a size α test.
   b) the sampling distribution of $\sum_{i=1}^{n} X_i$.
   c) the value of k_α for $\alpha \approx 0.05$ when $n = 20$.
   d) $\pi(0.8)$ for $\alpha \approx 0.05$ when $n = 20$.

6.2.6 Let $X_1, \ldots, X_n$ be a sample of iid $\text{Exp}(\theta)$ random variables with $\Theta = \{5, 8\}$. For testing $H_0 : \theta = 5$ versus $H_A : \theta = 8$, determine
   a) the form of the Neyman–Pearson MP critical region for a size α test.
   b) the sampling distribution of $\sum_{i=1}^{n} X_i$.
   c) the value of k_α for $\alpha = 0.05$ when $n = 25$.
   d) $\pi(8)$ for $\alpha = 0.05$ when $n = 25$.

6.2.7 Let $X_1, \ldots, X_n$ be a sample of iid $\text{Exp}(\theta)$ random variables with $\Theta = \{4, 7\}$. For testing $H_0 : \theta = 7$ versus $H_A : \theta = 4$, determine
   a) the form of the Neyman–Pearson MP critical region for a size α test.
   b) the sampling distribution of $\sum_{i=1}^{n} X_i$.
   c) the value of k_α for $\alpha = 0.01$ when $n = 20$.
   d) $\pi(4)$ for $\alpha = 0.01$ when $n = 20$.

6.2.8 Let $X_1, \ldots, X_n$ be a sample of iid $N(\theta, 25)$ random variables with $\Theta = \{10, 15\}$. For testing $H_0 : \theta = 15$ versus $H_A : \theta = 10$, determine
   a) the form of the Neyman–Pearson MP critical region for a size α test.
   b) the sampling distribution of $\sum_{i=1}^{n} X_i$.
   c) the value of k_α for $\alpha = 0.02$ when $n = 30$.
   d) $\pi(10)$ for $\alpha = 0.02$ when $n = 30$.

6.2.9 Let $X_1, \ldots, X_n$ be a sample of iid $\text{Gamma}(2, \theta)$ random variables with $\Theta = \{5, 10\}$. For testing $H_0 : \theta = 5$ versus $H_A : \theta = 10$, determine
   a) the form of the Neyman–Pearson MP critical region for a size α test.

  b) the sampling distribution of $\sum_{i=1}^{n} X_i$.

  c) the value of k_α for $\alpha = 0.02$ when $n = 10$.

  d) $\pi(10)$ for $\alpha = 0.02$ when $n = 10$.

6.2.10 Let $X_1, \ldots, X_n$ be a sample of iid $N(10, \theta)$ random variables with $\Theta = \{1, 2\}$. For testing $H_0 : \theta = 2$ versus $H_A : \theta = 1$, determine

  a) the form of the Neyman–Pearson MP critical region for a size α test.

  b) the sampling distribution of $\sum_{i=1}^{n} \frac{(X_i - 10)^2}{\theta}$.

  c) the value of k_α for $\alpha = 0.04$ when $n = 12$.

  d) $\pi(1)$ for $\alpha = 0.04$ when $n = 12$.

6.2.11 Let $X_1, \ldots, X_n$ be a sample of iid random variables with common pdf $f(x; \theta) = \theta x^{\theta-1}$ for $x \in (0, 1)$ and $\Theta = \{1, 3\}$. For testing $H_0 : \theta = 1$ versus $H_A : \theta = 3$, determine

  a) the form of the Neyman–Pearson MP critical region for a size α test.

  b) the value of k_α when $n = 10$ and $\alpha = 0.04$ using the sampling distribution $-\sum_{i=1}^{n} \ln(X_i) \sim \text{Gamma}\left(n, \frac{1}{\theta}\right)$.

6.2.12 Let $X_1, \ldots, X_n$ be a sample of iid $N(0, \theta^2)$ random variables with $\Theta = \{\theta_0, \theta_A\}$. For testing $H_0 : \theta = \theta_0$ versus $H_A : \theta = \theta_A$, determine

  a) the form of the Neyman–Pearson MP critical region for a size α test when $\theta_0 < \theta_A$.

  b) the form of the Neyman–Pearson MP critical region for a size α test when $\theta_0 > \theta_A$.

6.2.13 Let $X_1, \ldots, X_n$ be a sample of iid Gamma$(\theta, 1)$ random variables with $\Theta = \{2, 4\}$. For testing $H_0 : \theta = 2$ versus $H_A : \theta = 4$, determine the form of the Neyman–Pearson MP critical region for a size α test.

6.2.14 Let $X_1, \ldots, X_n$ be a sample of iid random variables with common pdf $f(x; \theta) = \frac{1}{\theta} e^{-\frac{1}{\theta}(x-5)}$ on $\Theta = \{1, 4\}$. For testing $H_0 : \theta = 1$ versus $H_A : \theta = 4$, determine the form of the Neyman–Pearson MP critical region for a size α test.

6.2.15 Let $X_1, \ldots, X_n$ be a sample of iid $N(\theta, \theta)$ random variables with $\Theta = \{\theta_0, \theta_A\}$. For testing $H_0 : \theta = \theta_0$ versus $H_A : \theta = \theta_A$, determine the form of the Neyman–Pearson MP critical region for a size α test when $0 < \theta_0 < \theta_A$.

6.2.16 Prove Theorem 6.2.

6.3 Uniformly Most Powerful Tests

In most hypothesis testing scenarios, a composite null hypothesis is being tested against a composite alternative hypothesis. The most commonly tested hypotheses are $H_0 : \theta \leq \theta_0$ versus $H_A : \theta > \theta_0$, $H_0 : \theta \geq \theta_0$ versus $H_A : \theta < \theta_0$, and $H_0 : \theta = \theta_0$ versus $H_A : \theta \neq \theta_0$. The tests $H_0 : \theta \leq \theta_0$ versus $H_A : \theta > \theta_0$ and $H_0 : \theta = \theta_0$ versus $H_A : \theta < \theta_0$ are called *one-sided tests*, and $H_0 : \theta = \theta_0$ versus $H_A : \theta \neq \theta_0$ is called a *two-sided test*.

For a one-parameter probability model, testing a composite null versus a composite alternative, say $H_0 : \theta \in \Theta_0$ versus $H_A : \theta \in \Theta_A$, the size and power of a critical region $\mathcal{C}$ are $\alpha = \sup_{\theta \in \Theta_0} P(\mathcal{C}|\theta)$ and $\pi(\theta) = P(\mathcal{C}|\theta)$. Furthermore, when the alternative hypothesis is a composite hypothesis, the best critical region is the one having the largest power at each value of $\theta \in \Theta_A$. A critical region of size α having the largest power uniformly over Θ_A is called a *uniformly most powerful* critical region of size α.

Definition 6.9 A critical region $\mathcal{C}^\star$ for testing $H_0 : \theta \in \Theta_0$ versus $H_A : \theta \in \Theta_A$ is said to be a uniformly most powerful (UMP) critical region of size α when

i) $\sup_{\theta \in \Theta_0} \pi_{\mathcal{C}^\star}(\theta) = \alpha$.

ii) $\pi_{\mathcal{C}^\star}(\theta) \geq \pi_{\mathcal{C}}(\theta)$ for all $\theta \in \Theta_A$ for any other critical region $\mathcal{C}$ of size α.

A UMP critical region may or may not exist for one-parameter probability model; however, under certain conditions, a UMP critical region for a one-sided test does exist. The Karlin–Rubin theorem, Theorem 6.4, shows that a UMP critical region exists for a one-sided test when the likelihood function has *monotone likelihood ratio* in θ.

Definition 6.10 For a random sample $X_1, \ldots, X_n$, a likelihood function $L(\theta)$ is said to have monotone likelihood ratio (MLR) in a statistic T when for every $\theta_1 < \theta_2$, the likelihood ratio $\Lambda(\theta_1, \theta_2) = \frac{L(\theta_1)}{L(\theta_2)}$ is a nondecreasing or nonincreasing function of T.

Example 6.12 Let $X_1, \ldots, X_n$ be a sample of iid $U(0, \theta)$ random variables with $\Theta = \mathbb{R}^+$. Then,

$$L(\theta) = \prod_{i=1}^{n} \frac{1}{\theta} I_{(0,\theta)}(x_i) = \begin{cases} \theta^{-n} & \text{when } 0 < x_i < \theta, \quad \forall \ i \\ 0 & \text{otherwise} \end{cases}$$
$$= \theta^{-n} I_{(0,\theta)}(x_{(n)}).$$

Thus, when $\theta_2 > \theta_1 > 0$,

$$\Lambda(\theta_1, \theta_2) = \frac{L(\theta_1)}{L(\theta_2)} = \frac{\theta_1^{-n} I_{(0,\theta_1)}(x_{(n)})}{\theta_2^{-n} I_{(0,\theta_2)}(x_{(n)})}$$

$$= \begin{cases} \left[\frac{\theta_2}{\theta_1}\right]^n & \text{when } 0 < x_{(n)} < \theta_1 \\ 0 & \text{when } \theta_1 \leq x_{(n)} < \theta_2 \end{cases}.$$

Hence, $\Lambda(\theta_1, \theta_2)$ is a nonincreasing function of $x_{(n)}$, and therefore, $L(\theta)$ has MLR in $X_{(n)}$.

Theorem 6.3 *If $X_1, \ldots, X_n$ is a sample of iid random variable with pdf $f(x; \theta) = c(\theta)h(x)e^{q(\theta)t(x)}$ in the one-parameter exponential family, then $L(\theta)$ has monotone likelihood ratio in the sufficient statistic $\sum_{i=1}^{n} t(X_i)$.*

Proof. Let $X_1, \ldots, X_n$ be a sample of iid random variable with pdf $f(x; \theta) = c(\theta)h(x)e^{q(\theta)t(x)}$ in the one-parameter exponential family. Then, for $\theta \in \Theta$,

$$L(\theta) = \prod_{i=1}^{n} c(\theta)h(x_i)e^{q(\theta)t(x_i)} = c(\theta)^n e^{q(\theta)\sum_{i=1}^{n} t(x_i)} \prod_{i=1}^{n} h(x_i).$$

Hence, for $\theta_1, \theta_2 \in \Theta$ with $\theta_1 < \theta_2$,

$$\Lambda(\theta_1, \theta_2) = \frac{c(\theta_1)^n e^{q(\theta_1)\sum_{i=1}^{n} t(x_i)} \prod_{i=1}^{n} h(x_i)}{c(\theta_2)^n e^{q(\theta_2)\sum_{i=1}^{n} t(x_i)} \prod_{i=1}^{n} h(x_i)}$$

$$= \left[\frac{c(\theta_1)}{c(\theta_2)}\right]^n e^{[q(\theta_1)-q(\theta_2)]\sum_{i=1}^{n} t(x_i)}.$$

Let $t = \sum_{i=1}^{n} t(x_i)$. Then,

$$\frac{\partial}{\partial t}\Lambda(\theta_1, \theta_2) = [q(\theta_1) - q(\theta_2)]e^{[q(\theta_1)-q(\theta_2)]t}$$

$$\begin{cases} > 0 & \text{when } q(\theta) \text{ is decreasing} \\ < 0 & \text{when } q(\theta) \text{ is increasing} \end{cases}.$$

Hence, $\Lambda(\theta_1, \theta_2)$ is decreasing in $\sum_{i=1}^{n} t(X_i)$ when $q(\theta)$ is increasing, and $\Lambda(\theta_1, \theta_2)$ is a increasing function of $\sum_{i=1}^{n} t(X_i)$ when $q(\theta)$ is decreasing. Therefore, $L(\theta)$ has monotone likelihood ratio in $\sum_{i=1}^{n} t(X_i)$ when $f(x; \theta)$ is in the one-parameter exponential family. ∎

Example 6.13 Let $X_1, \ldots, X_n$ be a sample of iid $\text{Exp}(\theta)$ random variables with $\Theta = \mathbb{R}^+$. Then, $f(x; \theta)$ is a one-parameter exponential family pdf with sufficient statistic $T = \sum_{i=1}^{n} X_i$, and therefore, $L(\theta)$ has MLR in $\sum_{i=1}^{n} X_i$. Moreover, since $q(\theta) = -\frac{1}{\theta}$ is an increasing function, it follows that $L(\theta)$ has MLR decreasing in $\sum_{i=1}^{n} X_i$.

Theorem 6.4, the Karlin–Rubin theorem, shows that UMP critical regions exist for the one-sided tests when the likelihood function of a one-parameter probability model having monotone likelihood ratio in a statistic T.

Theorem 6.4 (Karlin–Rubin) *Let* $X_1, \ldots, X_n$ *be iid random variables with one-parameter pdf* $f(x; \theta)$ *and parameter space* Θ, *and let* $L(\theta)$ *have MLR in a statistic* $T(\vec{X})$.

i) If $L(\theta)$ *has MLR nondecreasing in* T, *then a UMP critical region of size* α *for testing* $H_0 : \theta \leq \theta_0$ *versus* $H_A : \theta > \theta_0$ *is* $\mathcal{C}^* = \{\vec{x} : T(\vec{x}) \leq k_\alpha\}$.

ii) If $L(\theta)$ *has MLR nonincreasing in* T, *then a UMP critical region of size* α *for testing* $H_0 : \theta \leq \theta_0$ *versus* $H_A : \theta > \theta_0$ *is* $\mathcal{C}^* = \{\vec{x} : T(\vec{x}) \geq k_\alpha\}$.

Proof. The proof of the Karlin–Rubin theorem is given in *Statistical Inference* by Casella and Berger [13]. ∎

The Karlin–Rubin theorem can also be used for the lower-tail tests, $H_0 : \theta \geq \theta_0$ versus $H_A : \theta < \theta_0$, with the inequalities reversed in the UMP critical regions. For example, for testing the lower-tail test $H_0 : \theta \geq \theta_0$ versus $H_A : \theta < \theta_0$ when the likelihood function has MLR nonincreasing in a statistic T, the UMP critical region is $\mathcal{C}\{\vec{x} : T(\vec{x}) \leq k_\alpha\}$.

Example 6.14 Let $X_1, \ldots, X_n$ be a sample of iid $N(\theta, 1)$ random variables. Since $f(x; \theta)$ is an exponential family pdf and $\sum_{i=1}^{n} X_i$ is a sufficient statistic, it follows that $L(\theta)$ has MLR in $\sum_{i=1}^{n} X_i$. Furthermore, since $q(\theta) = \theta$ is an increasing function, $L(\theta)$ has nonincreasing MLR in $\sum_{i=1}^{n} X_i$.

Hence, by the Karlin–Rubin theorem the UMP critical region of size α for testing $H_0 : \theta \leq 0$ versus $H_A : \theta > 0$ is $\mathcal{C}^* = \{\vec{x} : \sum x_i \geq k_\alpha\}$.

Example 6.15 Let $X_1, \ldots, X_n$ be a sample of iid $\mathrm{NegBin}(4, \theta)$ random variables with $\Theta = [0, 1]$. Then,

$$f(x; \theta) = \binom{3 + x}{x} \theta^4 (1 - \theta)^x = \theta^4 \binom{3 + x}{x} e^{x \ln(1-\theta)}.$$

Thus, $f(x; \theta)$ is in the exponential family with $q(\theta) = \ln(1 - \theta)$ and sufficient statistic $\sum_{i=1}^{n} X_i$. Since $q(\theta)$ is a decreasing function, it follows that $L(\theta)$ has nondecreasing MLR in $\sum_{i=1}^{n} X_i$. Thus, by the Karlin–Rubin theorem, the UMP critical region for testing $H_0 : \theta \leq 0.3$ versus $H_A : \theta > 0.3$ is $\mathcal{C}^* = \{\vec{x} : \sum x_i \leq k_\alpha\}$.

Example 6.16 Let $X_1, \ldots, X_n$ be a sample of iid $U(0, \theta)$ random variables with $\Theta = \mathbb{R}^+$. In Example 6.12, $L(\theta)$ was shown to have nonincreasing MLR in $X_{(n)}$, and therefore, by the Karlin–Rubin theorem, the UMP critical region for testing $H_0: \theta \geq 1$ versus $H_A: \theta < 1$ is $\mathcal{C}^\star = \{\vec{x} : x_{(n)} \leq k_\alpha\}$.

The pdf of the sampling distribution of $X_{(n)}$ is $f_n(x; \theta) = \frac{nx^{n-1}}{\theta^n}$ for $x \in (0, \theta)$. Hence, the value of k_α is found by solving $\alpha = \int_0^{k_\alpha} \frac{nx^{n-1}}{\theta^n} \, dx$ for k_α. In particular, $k_\alpha = \sqrt[n]{\alpha}$.

For example, if $\alpha = 0.05$ and $n = 20$, then $k_{0.05} = 0.8609$, and the power function associated with $\mathcal{C}^\star = \{\vec{x} : x_{(n)} \leq 0.869\}$ is

$$\pi(\theta) = \int_0^{\min(\theta, 0.8609)} \frac{20x^{19}}{\theta^{20}} \, dx = \left(\frac{\min(\theta, 0.8609)}{\theta} \right)^{20}$$

The power function for $\mathcal{C}^\star$ when $\alpha = 0.05$ and $n = 20$ is shown in Figure 6.2. Note that $\pi(\theta) = 1$ for $\theta < 0.8609$, and the power increases rapidly as θ moves away from $\theta_0 = 1$.

6.3.1 Uniformly Most Powerful Unbiased Tests

In general, there is no UMP critical region for the two-sided test $H_0: \theta = \theta_0$ versus $H_A: \theta \neq \theta_0$ because the most powerful critical region for testing $H_0: \theta = \theta_0$ versus $H_A: \theta = \theta_A$ depends on whether $\theta_A < \theta_0$ or $\theta_A > \theta_0$. For example, when $X_1, \ldots, X_n$ is a sample of iid $N(\theta, 1)$ random variables, the most powerful critical region for testing $H_0: \theta = 0$ versus $H_A: \theta = \theta_A$ is

$$\mathcal{C} = \begin{cases} \mathcal{C}^- = \{\vec{x} : \sum_{i=1}^n x_i \leq k_\alpha\} & \text{when } \theta_A < 0 \\ \mathcal{C}^+ = \{\vec{x} : \sum_{i=1}^n x_i \geq k_\alpha\} & \text{when } \theta_A > 0 \end{cases}.$$

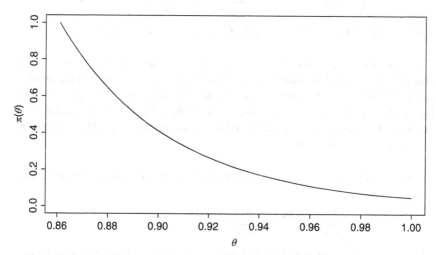

Figure 6.2 A plot of the power function for $\mathcal{C}^\star = \{\vec{x} : x_{(n)} \leq 0.8609\}$ for $n = 20$.

While no UMP critical region exists for the two-sided test, when only unbiased critical regions are considered, there may be a size α critical region that is uniformly most powerful within the class of unbiased critical regions, called a *uniformly most powerful unbiased* (UMPU) critical region.

Definition 6.11 A critical region $\mathcal{C}^{\star}$ is a UMPU critical region of size α for testing $H_0 : \theta \in \Theta_0$ versus $H_A : \theta \in \Theta_A$ when

i) $\sup_{\theta \in \Theta_0} \pi_{\mathcal{C}^{\star}}(\theta) = \alpha$.
ii) $\mathcal{C}^{\star}$ is an unbiased critical region.
iii) $\pi_{\mathcal{C}^{\star}}(\theta) \geq \pi_{\mathcal{C}}(\theta)$ for every $\theta \in \Theta_A$ and for any other unbiased critical region $\mathcal{C}$ of size α.

Because a UMP critical region is unbiased, it follows that every UMP critical region is a UMPU critical region. On the other hand, there may exist a critical region more powerful than a UMPU critical when critical regions outside of the class of unbiased critical regions are considered. Theorem 6.5 shows that a UMPU critical region does exist for a two-sided test when sampling from a one-parameter canonical form exponential family distribution.

Theorem 6.5 *If $X_1, \ldots, X_n$ is a sample of iid random variables with $f(x; \theta) = c(\theta)h(x)e^{\theta t(x)}$ for $\theta \in \Theta$, then for testing $H_0 : \theta = \theta_0$ versus $H_A : \theta \neq \theta_0$ there exists a size α UMPU critical region of the form*

$$\mathcal{C}^{\star} = \left\{ \vec{x} : \sum_{i=1}^{n} t(x_i) \leq k_{\alpha_1} \text{ or } \sum_{i=1}^{n} t(x_i) \geq k_{\alpha_2} \right\},$$

where $P\left(\sum_{i=1}^{n} t(x_i) \leq k_{\alpha_1} | \theta_0\right) = \alpha_1$, $P\left(\sum_{i=1}^{n} t(x_i) \geq k_{\alpha_2} | \theta_0\right) = \alpha_2$, and $\alpha_1 + \alpha_2 = \alpha$.

It is important to note that Theorem 6.5 only applies to canonical form exponential family distributions, and Theorem 6.5 only provides the existence of the UMPU critical region. That is, Theorem 6.5 does not explicitly state how α should be allocated to α_1 and α_2 to produce the UMPU critical region. Theorem 6.5 can also be generalized for tests of the form $H_0 : \theta_1 \leq \theta \leq \theta_2$ versus $H_A : \theta < \theta_1$ or $\theta > \theta_2$.

Example 6.17 Let $X_1, \ldots, X_n$ be a sample of iid $N(\theta, 1)$ random variables with $\Theta = \mathbb{R}$. Then,

$$f(x; \theta) = \frac{1}{\sqrt{2\pi}} e^{-\frac{1}{2}(x-\theta)^2} = \frac{1}{\sqrt{2\pi}} e^{-\frac{\theta^2}{2} - \frac{x}{2} + \theta x},$$

which is an exponential family distribution in canonical form with sufficient statistic $\sum_{i=1}^{n} X_i$.

Hence, there exists a size α UMPU critical region of the form

$$\mathcal{C}^\star = \left\{ \vec{x} : \sum_{i=1}^n x_i \le k_{\alpha_1} \text{ or } \sum_{i=1}^n x_i \ge k_{\alpha_2} \right\}$$

for testing $H_0 : \theta = \theta_0$ versus $H_A : \theta \ne \theta_0$.

Example 6.18 Let $X_1, \ldots, X_n$ be a sample of iid $\mathrm{Bin}(1, \theta)$ random variables with $\Theta = [0, 1]$. Then, with the parameterization $\eta = \ln\left[\frac{\theta}{1-\theta}\right]$, the binomial pdf becomes

$$f(x; \eta) = \binom{n}{x} \left[\frac{e^\eta}{1 + e^\eta}\right]^n e^{\eta x},$$

which is an exponential family distribution in canonical form with sufficient statistic $\sum_{i=1}^n X_i$.

Hence, there exists a size α UMPU critical region of the form

$$\mathcal{C}^\star = \left\{ \vec{x} : \sum_{i=1}^n x_i \le k_{\alpha_1} \text{ or } \sum_{i=1}^n x_i \ge k_{\alpha_2} \right\}$$

for testing $H_0 : \eta = \eta_0$ versus $H_A : \eta \ne \eta_0$.

Note that testing $H_0 : \theta = 0.5$ versus $H_A : \theta \ne 0.5$ is equivalent to testing $H_0 : \eta = 0$ versus $H_A : \eta \ne 0$ since $\eta = \ln\left[\frac{\theta}{1-\theta}\right]$. Thus, the UMPU test of $H_0 : \eta = 0$ versus $H_A : \eta \ne 0$ is also a UMPU test of $H_0 : \theta = 0.5$ versus $H_A : \theta \ne 0.5$.

Problems

6.3.1 Let $X_1, \ldots, X_n$ be a sample of iid $\mathrm{Geo}(\theta)$ random variables with $\Theta = [0, 1]$. Show that $L(\theta)$ has MLR in $\sum_{i=1}^n X_i$.

6.3.2 Let $X_1, \ldots, X_n$ be a sample of iid $\mathrm{Pois}(\theta)$ random variables with $\Theta = \mathbb{R}^+$. Show that $L(\theta)$ has MLR in $\sum_{i=1}^n X_i$.

6.3.3 Let $X_1, \ldots, X_n$ be a sample of iid $N(0, \theta)$ random variables with $\Theta = \mathbb{R}^+$. Show that $L(\theta)$ has MLR in $\sum_{i=1}^n X_i^2$.

6.3.4 Let $X_1, \ldots, X_n$ be a sample of iid random variables with common pdf $f(x; \theta) = \theta x^{\theta-1}$ for $x \in (0, 1)$ and $\Theta = \mathbb{R}^+$. Show that $L(\theta)$ has MLR in $\prod_{i=1}^n X_i$.

6.3.5 Let $X_1, \ldots, X_n$ be a sample of iid $U(1, \theta)$ random variables with $\Theta = (1, \infty)$. Show that $L(\theta)$ has MLR in $X_{(n)}$.

6.3.6 Let $X_1, \ldots, X_n$ be a sample of iid random variables with common pdf $f(x; \theta) = e^{-(x-\theta)}$ for $x > \theta$ and $\Theta = \mathbb{R}$. Show that $L(\theta)$ has MLR in $X_{(1)}$.

6.3.7 Let $X_1, \ldots, X_n$ be a sample of iid $U(\theta, 1)$ random variables with $\Theta = (-\infty, 1)$. Show that $L(\theta)$ has MLR in $X_{(1)}$.

6.3.8 Let $X_1, \ldots, X_n$ be a sample of iid random variables with pdf $f(x; \theta) = \theta x^{\theta-1}$ for $x \in (0, 1)$ and $\Theta = \mathbb{R}^+$.
   a) Determine the form of the UMP critical region for testing $H_0 : \theta \leq 2$ versus $H_A : \theta > 2$.
   b) Given that $-\sum_{i=1}^{n} \ln(X_i) \sim \text{Gamma}(n, \frac{1}{\theta})$, determine the value of k_α in the UMP critical region for testing $H_0 : \theta \leq 2$ versus $H_A : \theta > 2$ for $n = 20$ and $\alpha = 0.04$.

6.3.9 Let $X_1, \ldots, X_n$ be a sample of iid random variables with common pdf $f(x; \theta) = \theta^2 x e^{-\theta x}$ for $x \in \mathbb{R}^+$ and $\Theta = \mathbb{R}^+$.
   a) Show that $L(\theta)$ has MLR in $\sum_{i=1}^{n} X_i$.
   b) Determine the form of the UMP critical region for testing $H_0 : \theta \geq 3$ versus $H_A : \theta < 3$.
   c) Determine the distribution of $\sum_{i=1}^{n} X_i$.
   d) For $n = 10$ and $\alpha = 0.01$, determine the value of k_α in the UMP critical region for testing $H_0 : \theta \geq 3$ versus $H_A : \theta < 3$.
   e) Determine the power of the UMP critical region for $\theta = 1.5$ when $n = 10$ and $\alpha = 0.01$.

6.3.10 Let $X_1, \ldots, X_n$ be a sample of iid $\text{Bin}(1, \theta)$ random variables with $\Theta = [0, 1]$. Determine
   a) the form of the UMP critical region for testing $H_0 : \theta \leq \theta_0$ vs $H_A : \theta > \theta_0$.
   b) the form of the UMP critical region for testing $H_0 : \theta \geq \theta_0$ vs $H_A : \theta < \theta_0$.
   c) the sampling distribution of the statistic defining the UMP critical region.

6.3.11 Let $X_1, \ldots, X_n$ be a sample of iid $\text{Pois}(\theta)$ random variables with $\Theta = \mathbb{R}^+$. Determine
   a) the form of the UMP critical region for testing $H_0 : \theta \leq \theta_0$ vs $H_A : \theta > \theta_0$.
   b) the form of the UMP critical region for testing $H_0 : \theta \geq \theta_0$ vs $H_A : \theta < \theta_0$.
   c) the sampling distribution of the statistic defining the UMP critical region.

6.3.12 Let $X_1, \ldots, X_n$ be a sample of iid $\text{Exp}(\theta)$ random variables with $\Theta = \mathbb{R}^+$. Determine
  a) the form of the UMP critical region for testing $H_0 : \theta \leq \theta_0$ vs $H_A : \theta > \theta_0$.
  b) the form of the UMP critical region for testing $H_0 : \theta \geq \theta_0$ vs $H_A : \theta < \theta_0$.
  c) the sampling distribution of the statistic defining the UMP critical region.

6.3.13 Let $X_1, \ldots, X_n$ be a sample of iid $\text{Geo}(\theta)$ random variables with $\Theta = [0, 1]$. Determine
  a) the form of the UMP critical region for testing $H_0 : \theta \leq \theta_0$ vs $H_A : \theta > \theta_0$.
  b) the form of the UMP critical region for testing $H_0 : \theta \geq \theta_0$ vs $H_A : \theta < \theta_0$.
  c) the sampling distribution of the statistic defining the UMP critical region.

6.3.14 Let $X_1, \ldots, X_n$ be a sample of iid $N(0, \theta)$ random variables with $\Theta = \mathbb{R}^+$. Determine
  a) the form of the UMP critical region for testing $H_0 : \theta \leq \theta_0$ vs $H_A : \theta > \theta_0$.
  b) the form of the UMP critical region for testing $H_0 : \theta \geq \theta_0$ vs $H_A : \theta < \theta_0$.
  c) the sampling distribution of the statistic defining the UMP critical region.

6.3.15 Let $X_1, \ldots, X_n$ be a sample of iid $N(\mu, \theta)$ random variables with $\Theta = \mathbb{R}^+$ and μ known. Determine
  a) the form of the UMP critical region for testing $H_0 : \theta \leq \theta_0$ vs $H_A : \theta > \theta_0$.
  b) the form of the UMP critical region for testing $H_0 : \theta \geq \theta_0$ vs $H_A : \theta < \theta_0$.
  c) the sampling distribution of the statistic defining the UMP critical region.

6.3.16 Let $X_1, \ldots, X_n$ be a sample of iid $N(\theta, \sigma^2)$ random variables with $\Theta = \mathbb{R}$ and σ known. Determine
  a) the form of the UMP critical region for testing $H_0 : \theta \leq \theta_0$ vs $H_A : \theta > \theta_0$.
  b) the form of the UMP critical region for testing $H_0 : \theta \geq \theta_0$ vs $H_A : \theta < \theta_0$.
  c) the sampling distribution of the statistic defining the UMP critical region.

6.3.17 Let $X_1, \ldots, X_n$ be a sample of iid Gamma(α, θ) random variables with α known and $\Theta = \mathbb{R}^+$. Determine
a) the form of the UMP critical region for testing $H_0 : \theta \leq \theta_0$ vs $H_A : \theta > \theta_0$.
b) the form of the UMP critical region for testing $H_0 : \theta \geq \theta_0$ vs $H_A : \theta < \theta_0$.
c) the sampling distribution of the statistic defining the UMP critical region.

6.3.18 $X_1, \ldots, X_n$ be a sample of iid random variables with $f(x; \theta) = \frac{\theta}{x^{\theta+1}}$ for $x \in (1, \infty)$ and $\Theta = \mathbb{R}^+$. Determine the form of the UMP critical region for testing $H_0 : \theta \leq \theta_0$ vs $H_A : \theta > \theta_0$.

6.3.19 Let $X_1, \ldots, X_n$ be a sample of iid Gamma(θ, β) random variables with β known and $\Theta = \mathbb{R}^+$. Determine the form of the UMP critical region for testing
a) $H_0 : \theta \leq \theta_0$ vs $H_A : \theta > \theta_0$.
b) $H_0 : \theta \geq \theta_0$ vs $H_A : \theta < \theta_0$.

6.3.20 Let $X_1, \ldots, X_n$ be a sample of iid Beta(θ, β) random variables with β known and $\Theta = \mathbb{R}^+$. Determine the form of the UMP critical region for testing
a) $H_0 : \theta \leq \theta_0$ vs $H_A : \theta > \theta_0$.
b) $H_0 : \theta \geq \theta_0$ vs $H_A : \theta < \theta_0$.

6.3.21 Let $X_1, \ldots, X_n$ be a sample of iid Beta(α, θ) random variables with α known and $\Theta = \mathbb{R}^+$. Determine the form of the UMP critical region for testing
a) $H_0 : \theta \leq \theta_0$ vs $H_A : \theta > \theta_0$.
b) $H_0 : \theta \geq \theta_0$ vs $H_A : \theta < \theta_0$.

6.3.22 Let $X_1, \ldots, X_n$ be a sample of iid Exp(θ, η) random variables with η known and $\Theta = \mathbb{R}^+$. Determine the form of the UMP critical region for testing
a) $H_0 : \theta \leq \theta_0$ vs $H_A : \theta > \theta_0$.
b) $H_0 : \theta \geq \theta_0$ vs $H_A : \theta < \theta_0$.

6.3.23 Let $X_1, \ldots, X_n$ be a sample of iid Exp(β, θ) random variables with β known and $\Theta = \mathbb{R}$. Determine the form of the UMP critical region for testing
a) $H_0 : \theta \leq \theta_0$ vs $H_A : \theta > \theta_0$.
b) $H_0 : \theta \geq \theta_0$ vs $H_A : \theta < \theta_0$.

6.3.24 Let $X_1, \ldots, X_n$ be a sample of iid $U(\theta, 1)$ random variables with $\Theta = (-\infty, 1)$. Determine the form of the UMP critical region for testing
a) $H_0 : \theta \leq \theta_0$ vs $H_A : \theta > \theta_0$.
b) $H_0 : \theta \geq \theta_0$ vs $H_A : \theta < \theta_0$.

6.3.25 Let $X_1, \ldots, X_n$ be a sample of iid $\text{Pois}(\theta)$ random variables with $\Theta = \mathbb{R}^+$. Determine the form of the UMPU critical region for testing $H_0 : \theta = 0.5$ vs $H_A : \theta \neq 0.5$.

6.3.26 Let $X_1, \ldots, X_n$ be a sample of iid exponential random variables with pdf $f(x; \theta) = \theta e^{-\theta x}$ and $\Theta = \mathbb{R}^+$. Determine the form of the UMPU critical region for testing $H_0 : \theta = 5$ vs $H_A : \theta \neq 5$.

6.3.27 Let $X_1, \ldots, X_n$ be a sample of iid random variables with pdf $f(x; \theta) = \theta x^{\theta-1}$ for $x \in (0, 1)$ and $\Theta = \mathbb{R}^+$. Determine the form of the UMPU critical region for testing $H_0 : \theta = 10$ vs $H_A : \theta \neq 10$.

6.3.28 Let $X_1, \ldots, X_n$ be a sample of iid $\text{Pois}(\theta)$ random variables with $\Theta = \mathbb{R}^+$.
a) Show that with the parameterization $\eta = \ln(\theta)$, the Poisson distribution is in canonical exponential family form.
b) Determine the form of the UMPU critical region for testing $H_0 : \eta = \eta_0$ vs $H_A : \eta \neq \eta_0$.

6.3.29 Let $X_1, \ldots, X_n$ be a sample of iid Weibull random variables with pdf given by $f(x; \theta) = \frac{3x^2}{\theta^3} e^{-\left(\frac{x}{\theta}\right)^3}$ for $x \in \mathbb{R}^+$ and $\Theta = \mathbb{R}^+$. Determine the form of the UMP critical region for testing
a) $H_0 : \theta \leq \theta_0$ vs $H_A : \theta > \theta_0$.
b) $H_0 : \theta \geq \theta_0$ vs $H_A : \theta < \theta_0$.

6.3.30 If $L(\theta)$ has MLR increasing in T, prove that $L(\theta)$ will have MLR in any increasing function of T.

6.3.31 If $\mathcal{C}^\star$ is a most powerful critical region of size α for testing $H_0 : \theta = \theta_0$ versus $H_A : \theta = \theta_A$ for every $\theta_A \in \Theta_A$, prove that $\mathcal{C}^\star$ is a UMP critical region of size α for testing $H_0 : \theta = \theta_0$ versus $H_A : \theta \in \Theta_A$.

6.4 Generalized Likelihood Ratio Tests

While it is often possible to find a best critical region for testing hypotheses about the unknown parameter in a one-parameter probability model, there is generally no uniformly best critical region for tests concerning a single parameter in a multiparameter probability model. For example, when $X_1, \ldots, X_n$ is a sample of iid $N(\mu, \sigma^2)$ random variables with both μ and σ unknown, there is no UMP critical region for testing $H_0 : \mu \leq 0$ versus $H_A : \mu > 0$ because σ is unknown. In this example, σ is called a *nuisance parameter*.

Definition 6.12   Let $f(x; \theta)$ be the pdf of a multiparameter probability model. Any parameter that is not of immediate interest and must be accounted for in the analysis of those parameters that are of interest is called a nuisance parameter.

Example 6.19   The multiple regression model is $Y = \beta_0 + \beta_1 x_1 + \cdots + \beta_p x_p + \epsilon$, where $x_1, \ldots, x_p$ are known fixed (i.e. not random) variables, Y is a random variable with variance σ^2, $\beta_0, \ldots, \beta_p$ are unknown parameters, and $\epsilon \sim N(0, \sigma^2)$. In the multiple regression model, the mean of the random variable Y is a linear function of the fixed variables $x_1, \ldots, x_p$.

A common test on a multiple regression model is a test of $H_0 : \beta_i = 0$ versus $H_A : \beta_i \neq 0$, and in this test, the unknown variance σ^2 is a nuisance parameter.

Nuisance parameters often are present when testing parametric hypotheses about the parameters in a multiparameter statistical model and require alternative approaches to determine good critical regions. A general testing approach that can be used for two-sided tests and tests involving a nuisance parameter is *generalized likelihood ratio testing* (GLRT).

A generalized likelihood ratio test of $H_0 : \theta \in \Theta_0$ versus $H_A : \theta \in \Theta_A$ is based on the *generalized likelihood ratio* statistic, which can be used for testing any type of null hypothesis against any alternative hypothesis.

Definition 6.13   Let $X_1, \ldots, X_n$ be a sample of random variables with likelihood function $L(\theta)$. For testing $H_0 : \theta \in \Theta_0$ versus $H_A : \theta \in \Theta_A$, the generalized likelihood ratio is defined to be $\lambda(\vec{x}) = \dfrac{\sup_{\theta \in \Theta_0} L(\theta)}{\sup_{\theta \in \Theta} L(\theta)}$.

The generalized likelihood ratio can also be written as $\lambda(\vec{x}) = \frac{L(\widehat{\theta}_0)}{L(\widehat{\theta})}$, where $\widehat{\theta}_0$ is the maximum likelihood estimation (MLE) of θ restricted to Θ_0 and $\widehat{\theta}$ is the MLE on Θ. Also, since the unrestricted MLE, $\widehat{\theta}$, maximizes the likelihood function, it follows that $L(\widehat{\theta}_0) \leq L(\widehat{\theta})$, and therefore, $0 \leq \lambda(\vec{x}) \leq 1$. In a generalized likelihood ratio test, H_0 is rejected when the observed data agrees with $\widehat{\theta}$ much better than it does with $\widehat{\theta}_0$.

When $X_1, \ldots, X_n$ is a sample of iid random variables with likelihood function $L(\theta)$, the generalized likelihood ratio critical region of size α for testing $H_0 : \theta \in \Theta_0$ versus $H_A : \theta \in \Theta_A$ be $\mathcal{C}_{GLR} = \{\vec{x} : \lambda(\vec{x}) \leq k_\alpha\}$. In practice, to determine the value of k_α, it will be necessary to find a critical region equivalent to $\mathcal{C}_{\text{GLR}}$ based on a statistic $T(\vec{X})$ with known sampling distribution.

Theorem 6.6 shows that for testing a simple null hypothesis versus a simple alternative hypothesis, the generalized likelihood ratio critical region is equivalent to the Neyman–Pearson MP critical region. Moreover, a generalized likelihood ratio test also often produces a UMP critical region when one exists.

Theorem 6.6 *If* $X_1, \dots, X_n$ *is a sample of iid random variables with one-parameter pdf, then the generalized likelihood ratio critical region for testing* $H_0 : \theta = \theta_0$ *versus* $H_A : \theta = \theta_1$ *is equivalent to the Neyman–Pearson MP critical region.*

Proof. The proof of Theorem 6.6 is left as an exercise. ∎

Example 6.20 Let $X_1, \dots, X_n$ be a sample of iid $\text{Exp}(\theta)$ random variables with $\Theta = \mathbb{R}^+$. For testing $H_0 : \theta = 1$ versus $H_A : \theta \neq 1$, the generalized likelihood ratio critical region is $\mathcal{C}_{\text{GLR}} = \{\vec{x} : \lambda(\vec{x}) \leq k_\alpha\}$.

Now, the unrestricted MLE of θ is $\bar{x}$ and the MLE of θ on $\Theta_0 = \{1\}$ is $\widehat{\theta}_0 = 1$. Thus,

$$\lambda(\vec{x}) = \frac{L(\widehat{\theta}_0)}{L(\bar{x})} = \frac{\prod_{i=1}^{n} e^{-x_i}}{\prod_{j=1}^{n} \frac{1}{\bar{x}} e^{-\frac{x_j}{\bar{x}}}}$$

$$= \bar{x}^n e^{-n\bar{x}+n}.$$

Thus, the generalized likelihood ratio critical region of size α for testing $H_0 : \theta = 1$ versus $H_A : \theta \neq 1$ is $\mathcal{C}_{\text{GLR}} = \{\vec{x} : \bar{x}^n e^{-n\bar{x}+n} \leq k_\alpha\}$. Let $t = \bar{x}$ so that $\lambda(t) = g(t) = t^n e^{-nt+n}$. Then,

$$g'(t) = nt^{n-1} e^{-nt+n} + t^n(-n)e^{-nt+n} = nt^{n-1} e^{-nt+n}(1 - t)$$

Hence, $g'(t) \geq 0$ for $t \in (0, 1)$, $g'(t) = 0$ for $t = 1$, and $g'(t) \leq 0$ for $t > 1$, and hence, $\lambda(\vec{x})$ is increasing on $(0, 1)$ and decreasing on $(1, \infty)$ with a maximum at $t = 1$. Hence, $\lambda(\vec{x}) \leq k_\alpha$ if and only if $\bar{x} \leq k_{\alpha_1}$ or $\bar{x} \geq k_{\alpha_2}$.

An equivalent form of the size α GLRT critical region for testing $H_0 : \theta = 1$ versus $H_A : \theta \neq 1$ is

$$\mathcal{C}_{\text{GLR}} = \{\vec{x} : \bar{x} \leq k_{\alpha_1} \text{ or } \bar{x} \geq k_{\alpha_2}\},$$

where $P(\bar{x} \leq k_{\alpha_1} | \theta = 1) = \alpha_1$, $P(\bar{x} \geq k_{\alpha_2} | \theta = 1) = \alpha_2$, and $\alpha_1 + \alpha_2 = \alpha$.

Example 6.21 Let $X_1, \dots, X_n$ be a sample of iid $\text{Pois}(\theta)$ random variables with $\Theta = \mathbb{R}^+$. For testing $H_0 : \theta \leq 1$ versus $H_A : \theta > 1$, the unrestricted MLE of θ is $\bar{x}$, and the MLE of θ on $\Theta_0 = (0, 1]$ is

$$\widehat{\theta}_0 = \begin{cases} \bar{x} & \text{when } \bar{x} \leq 1 \\ 1 & \text{when } \bar{x} > 1 \end{cases}.$$

Thus,

$$\lambda(\vec{x}) = \begin{cases} 1 & \text{when } \bar{x} \leq 1 \\ e^{-n+n\bar{x}-n\bar{x}\ln(\bar{x})} & \text{when } \bar{x} > 1 \end{cases}.$$

Now, for $\bar{x} > 1$, $\lambda(\vec{x})$ is decreasing since $\frac{d}{dx}\ln[\lambda(\bar{x})] = -n\ln(\bar{x}) < 0$. Hence, $\lambda(\vec{x}) \leq k$ if and only if $\bar{x} \geq k'$, and therefore, the GLRT critical region of size α for testing $H_0 : \theta \leq \theta_0$ versus $H_A : \theta > \theta_0$ is

$$\mathcal{C}_{\text{GLR}} = \{\vec{x} : \lambda(\vec{x}) \leq k_\alpha\} = \{\vec{x} : \bar{x} \geq k_\alpha\}.$$

The UMP critical region for testing $H_0 : \theta \leq 1$ versus $H_A : \theta > 1$ is $\mathcal{C}^\star = \{\vec{x} : \sum_{i=1}^n x_i \geq k_\alpha\}$, which is equivalent to the GLRT critical region.

Example 6.22 Let $X_1, \ldots, X_n$ be iid $N(\theta_1, \theta_2)$ random variables with both θ_1 and θ_2 unknown and $\Theta = \mathbb{R} \times \mathbb{R}^+$. A generalized likelihood ratio test can be used to test $H_0 : \theta_1 = 0$ versus $H_A : \theta_1 \neq 0$ when θ_2 is a nuisance parameter.

First, the unrestricted MLEs of θ_1 and θ_2 are $\hat{\theta}_1 = \bar{x}$ and $\hat{\theta}_2 = s_n^2$, and on $\Theta_0 = \{0\}$, the MLE of θ_1 is $\hat{\theta}_{1_0} = 0$, and the MLE of θ_2 is the value of θ_2 maximizing $L(0, \theta_2)$, which is $\hat{\theta}_{2_0} = s_0^2 = \frac{1}{n}\sum_{i=1}^n x_i^2$.

Since $L(\theta) = (2\pi\theta_2)^{-\frac{n}{2}} e^{-\frac{1}{2\theta_2}\sum_{i=1}^n (x_i - \theta_1)^2}$, it follows that

$$\lambda(\vec{x}) = \frac{(2\pi s_0^2)^{-\frac{n}{2}} e^{-\frac{1}{2s_0^2}\sum x_i^2}}{(2\pi s_n^2)^{-\frac{n}{2}} e^{-\frac{1}{2s_n^2}\sum (x_i - \bar{x})^2}}$$

$$= \left(\frac{s_0^2}{s_n^2}\right)^{-\frac{n}{2}}$$

since $\frac{1}{2s_0^2}\sum x_i^2 = \frac{\sum x_i^2}{2\frac{1}{n}\sum x_i^2} = \frac{n}{2}$ and $\frac{1}{2s_n^2}\sum (x_i - \bar{x})^2 = \frac{ns_n^2}{2s_n^2} = \frac{n}{2}$.

Thus, the generalized likelihood ratio critical region of size α for testing $H_0 : \theta_1 = 0$ versus $H_A : \theta_1 \neq 0$ is

$$\mathcal{C}_{\text{GLR}} = \left\{\vec{x} : \left(\frac{s_0^2}{s_n^2}\right)^{-\frac{n}{2}} \leq k_\alpha\right\}.$$

Furthermore, $\vec{x} \in \mathcal{C}_{\text{GLR}}$ if and only if $\left(\frac{s_0^2}{s_n^2}\right)^{-\frac{n}{2}} \leq k$, or equivalently $\frac{s_0^2}{s_n^2} \geq k$, and

$$\frac{s_0^2}{s_n^2} = \frac{\frac{1}{n}\sum_{i=1}^n x_i^2}{\frac{1}{n}\sum_{j=1}^n (x_j - \bar{x})^2} = \frac{\sum_{i=1}^n x_i^2}{\sum_{j=1}^n (x_j - \bar{x})^2} = \frac{\sum_{i=1}^n (x_i - \bar{x} + \bar{x})^2}{\sum_{j=1}^n (x_j - \bar{x})^2}$$

$$= \frac{\sum_{i|1}^{n}(x_i - \bar{x})^2 + 2\bar{x}\overbrace{\sum(x_i - \bar{x})}^{0} + n\bar{x}^2}{\sum_{j=1}^{n}(x_i - \bar{x})^2} = \frac{(n-1)s^2 + n\bar{x}^2}{(n-1)s^2}$$

$$= 1 + \frac{1}{n-1}\frac{n\bar{x}^2}{s^2}.$$

Hence, the generalized likelihood ratio critical region for testing $H_0 : \theta_1 = 0$ versus $H_A : \theta_1 \neq 0$ can be expressed as $\mathcal{C}_{\text{GLR}} = \left\{1 + \frac{1}{n-1}\frac{n\bar{x}^2}{s^2} \geq k\right\}$, which is equivalent to

$$\mathcal{C}_{\text{GLR}} = \left\{\frac{\sqrt{n}\bar{x}}{s} \geq k_{\alpha_1} \text{ or } \frac{\sqrt{n}\bar{x}}{s} \leq k_{\alpha_2}\right\},$$

where $P\left(\frac{\sqrt{n}\bar{x}}{s} \leq k_{\alpha_1}|\theta = 1\right) = \alpha_1, P\left(\frac{\sqrt{n}\bar{x}}{s} \geq k_{\alpha_2}|\theta = 1\right) = \alpha_2$, and $\alpha_1 + \alpha_2 = \alpha$.

Since the sampling distribution of $\frac{\sqrt{n}\bar{x}}{s} \sim t_{n-1}$ for $X_i \sim N(0, \theta_2)$ (i.e. under H_0), the values of k_{α_1} and k_{α_2} are found by solving $P(t_{n-1} \leq k_{\alpha_1}) = \alpha_2$ and $P(t_{n-1} \geq k_{\alpha_2}) = \alpha_2$ subject to $\alpha_1 + \alpha_2 = \alpha$. For example, when $\alpha = 0.05$ and $n = 25$, then a size $\alpha = 0.05$ generalized likelihood ratio critical region for testing $H_0 : \theta_1 = 0$ versus $H_A : \theta_1 \neq 0$ with $\alpha_1 = \alpha_2 = 0.025$ is $\mathcal{C}_{\text{GLR}} = \left\{\frac{\sqrt{n}\bar{x}}{s} \leq -2.064 \text{ or } \frac{\sqrt{n}\bar{x}}{s} \geq 2.064\right\}$.

A generalized likelihood ratio test can also be used in multisample problems. For example, when $X_1, \ldots, X_n$ is a random sample from $f_x(x; \theta_x)$, $Y_1, \ldots, Y_m$ is a random sample from $f_y(y; \theta_y)$, a generalized likelihood ratio test can be used to test claims about θ_x and θ_y. The most common two-sample tests are $H_0 : \theta_x = \theta_y$ versus $H_A : \theta_x \neq \theta_y$, $H_0 : \theta_x \leq \theta_y$ versus $H_A : \theta_x > \theta_y$, and $H_0 : \theta_x \geq \theta_y$ versus $H_A : \theta_x < \theta_y$. A generalized likelihood ratio test for testing the parameters of two Poisson distributions is given in Example 6.23.

Example 6.23 Let $X_1, \ldots, X_n$ be a sample of iid Pois(θ_x) random variables with $\Theta_x = \mathbb{R}^+$, let $Y_1, \ldots, Y_m$ be a sample of iid Pois(θ_y) random variables with $\Theta_y = \mathbb{R}^+$, and suppose that $X_1, \ldots, X_n, Y_1, \ldots, Y_m$ is a collection of independent random variables.

Note that for testing $H_0 : \theta_x = \theta_y$ versus $H_A : \theta_x \neq \theta_y$, there is a hidden nuisance parameter, namely under $H_0 : \theta_x = \theta_y$, the common value of θ_x and θ_y. The unrestricted MLEs of θ_x and θ_y are $\bar{x}$ and $\bar{y}$, respectively. Under H_0, the random variables $X_1, \ldots, X_n, Y_1, \ldots, Y_m$ are a sample of iid Pois(θ_0) random variables, and therefore, the MLE of θ_0 is $\hat{\theta}_0 = \frac{\sum_{i=1}^{n}x_i + \sum_{j=1}^{m}y_j}{n+m} = \frac{n\bar{x} + m\bar{y}}{n+m}$.

The generalized likelihood ratio is

$$\lambda(\vec{x}, \vec{y}) = \frac{L(\widehat{\theta}_0)}{L(\overline{x}, \overline{y})} = \frac{\prod_{i=1}^{n} e^{-\widehat{\theta}_0} \widehat{\theta}_0^{x_i} \prod_{j=1}^{m} e^{-\widehat{\theta}_0} \widehat{\theta}_0^{y_j}}{\prod_{i=1}^{n} e^{-\overline{x}} \overline{x}^{x_i} \prod_{j=1}^{m} e^{-\overline{y}} \overline{y}^{y_j}}$$

$$= \frac{e^{-(n+m)\widehat{\theta}_0} \times \widehat{\theta}_0^{\sum_{i=1}^{n} x_i + \sum_{j=1}^{m} y_j}}{e^{-n\overline{x} - m\overline{y}} \times \overline{x}^{\sum_{i=1}^{n} x_i} \overline{y}^{\sum_{j=1}^{m} y_j}} = \frac{\left(\frac{n\overline{x} + m\overline{y}}{n+m} \right)^{\sum_{i=1}^{n} x_i + \sum_{j=1}^{m} y_j}}{\overline{x}^{\sum_{i=1}^{n} x_i} \overline{y}^{\sum_{j=1}^{m} y_j}}.$$

Thus, the generalized likelihood ratio critical region of size α for testing $H_0 : \theta_x = \theta_y$ versus $H_A : \theta_x \neq \theta_y$ is

$$\mathcal{C}_{\mathrm{GLR}} = \left\{ (\vec{x}, \vec{y}) : \frac{\left(\frac{n\overline{x} + m\overline{y}}{n+m} \right)^{\sum_{i=1}^{n} x_i + \sum_{j=1}^{m} y_j}}{\overline{x}^{\sum_{i=1}^{n} x_i} \overline{y}^{\sum_{j=1}^{m} y_j}} \leq k_\alpha \right\};$$

however, determining the exact sampling distribution of $\lambda(\vec{x}, \vec{y})$ or an equivalent statistic is difficult in this case.

In many problems, the sampling distribution of $\lambda(\vec{x})$ or an equivalent statistic is difficult to determine; however, for large sample sizes, Theorem 6.6 provides a solution to this problem.

Theorem 6.7 *Let $X_1, \ldots, X_n$ be a sample of iid random variables with pdf satisfying the regularity conditions in Definition 4.18. For testing $H_0 : \theta \in \Theta_0$ versus $H_A : \theta \in \Theta_A$, if $\theta \in \Theta_0$, then $-2\ln[\lambda_n(\vec{x})] \xrightarrow{d} \chi_r^2$, where $r = \dim(\Theta) - \dim(\Theta_0)$, and $\lambda_n(\vec{x})$ is the generalized likelihood ratio statistic for a sample of size n.*

When the regularity conditions are met and n is sufficiently large, the large sample generalized likelihood ratio critical region for testing $H_0 : \theta \in \Theta_0$ versus $H_A : \theta \in \Theta_A$ is $\mathcal{C}_{\mathrm{GLRA}} = \{\vec{x} : -2\ln[\lambda(\vec{x})] \geq k\}$. The degrees of freedom of the large sample χ^2 test are based on the difference in the number of free parameters in Θ and Θ_0.

Since the regularity conditions are satisfied for regular exponential family distributions, the large sample critical region is often used for exponential family distributions where the MLEs must be found numerically.

Example 6.24 Let $X_1, \ldots, X_n$ be a sample of iid Gamma(θ_1, θ_2) random variables with θ_1 and θ_2 unknown and $\Theta = \mathbb{R}^+ \times \mathbb{R}^+$. Then, for testing $H_0 : \theta_1 = 2$ versus $H_0 : \theta_1 \neq 2$, the unrestricted MLEs of θ_1 and θ_2 must found numerically. Since the gamma distribution is in the regular exponential family, with a sufficiently large sample, the generalized likelihood ratio test can be based on the critical region $\mathcal{C}_{\mathrm{GLRA}} = \{\vec{x} : -2\ln[\lambda(\vec{x})] \geq k_\alpha\}$, where $P(\chi_1^2 \geq k_\alpha) = \alpha$. Under H_0, the asymptotic distribution of $-2\ln[\lambda(\vec{x})] \sim \chi_1^2$ since $\dim(\Theta) = 2$ and

$\dim(\Theta_0) = 1$; there are two free parameters in Θ but only one free parameter in H_0.

Example 6.25 Let $X_1, \dots, X_{200}$ be a sample of iid $\text{Bin}(1, \theta)$ random variables with $\Theta = [0, 1]$ and suppose that $\sum_{i=1}^{200} x_i = 116$. For testing $H_0 : \theta = 0.5$ versus $H_A : \theta \neq 0.5$ at the $\alpha = 0.02$ level, the large sample generalized likelihood ratio critical region can be used since the binomial distribution is in the regular exponential family and the sample size is large.

Under H_0, the asymptotic distribution of $-2 \ln[\lambda(\vec{x})] \sim \chi_1^2$ and the large sample critical region is $\mathcal{C}_{\text{GLRA}} = \{\vec{x} : -2 \ln[\lambda(\vec{x})] \geq 5.41\}$, where $P(\chi_1^2 \geq 5.41) = 0.02$.

For $\sum_{i=1}^{200} x_i = 116$, the likelihood function is

$$\lambda(0.5, \overline{x}) = \frac{0.5^{200}}{0.58^{116} \times 0.42^{84}} = 0.0765.$$

Thus, $-2 \ln[\lambda(\vec{x})] = -2 \ln[0.0765] = 5.142$, which is not in $\mathcal{C}_{\text{GLRA}}$, and therefore, there is not sufficient evidence in the observed sample for rejecting $H_0 : \theta = 0.5$.

Finally, Theorem 6.6 can be generalized to handle samples consisting of a collection of independent variables rather than iid random variables. Moreover, the generalized version of Theorem 6.6 is often used in testing hypotheses about the parameters in *generalized linear models*, which are discussed in Chapter 7.

Problems

6.4.1 Let $X_1, \dots, X_n$ be a sample of iid $\text{Bin}(1, \theta)$ random variables with $\Theta = [0, 1]$. For the generalized likelihood ratio test of $H_0 : \theta = 0.25$ versus $H_A : \theta \neq 0.25$, determine
   a) the unrestricted MLE of θ.
   b) the MLE of θ on Θ_0.
   c) $\lambda(\vec{x})$.

6.4.2 Let $X_1, \dots, X_n$ be a sample of iid $\text{Exp}(\theta)$ random variables with $\Theta = \mathbb{R}^+$. For the generalized likelihood ratio test of $H_0 : \theta = 1$ versus $H_A : \theta \neq 1$, determine
   a) the unrestricted MLE of θ.
   b) the MLE of θ on Θ_0.
   c) $\lambda(\vec{x})$.

6.4.3 Let $X_1, \dots, X_n$ be a sample of iid $N(0, \theta)$ random variable with $\Theta = \mathbb{R}^+$. For the generalized likelihood ratio test of $H_0 : \theta \geq 1$ versus $H_A : \theta < 1$,
   a) determine the unrestricted MLE of θ.
   b) determine the MLE of θ on Θ_0.

c) determine $\lambda(\vec{x})$.

d) show that the generalized likelihood ratio critical region of size α is equivalent to $\{\vec{x}: \sum_{i=1}^{n} x_i^2 \leq k_\alpha\}$.

6.4.4 Let $X_1, \ldots, X_n$ be a sample of iid $U(0, \theta)$ random variables with $\Theta = \mathbb{R}^+$. For the generalized likelihood ratio test of $H_0: \theta \leq 1$ versus $H_A: \theta > 1$,

a) determine the unrestricted MLE of θ.

b) determine the MLE of θ on Θ_0.

c) determine $\lambda(\vec{x})$.

d) show that the generalized likelihood ratio critical region of size α is equivalent to $\{\vec{x}: x_{(n)} \geq k_\alpha\}$.

6.4.5 Let $X_1, \ldots, X_n$ be a sample of iid $N(\theta_1, \theta_2)$ random variables with $\Theta = \mathbb{R} \times \mathbb{R}^+$. For the generalized likelihood ratio test of $H_0: \theta_2 = 1$ versus $H_A: \theta_2 \neq 1$,

a) determine the unrestricted MLEs of θ_1 and θ_2.

b) determine the MLEs of θ_1 and θ_2 on Θ_0.

c) determine $\lambda(\vec{x})$.

d) show that the generalized likelihood ratio critical region of size α is equivalent to $\{\vec{x}: \sum_{i=1}^{n} (x_i - \overline{x})^2 \leq k_{\alpha_1} \text{ or } \sum_{i=1}^{n} (x_i - \overline{x})^2 \geq k_{\alpha_2}\}$.

6.4.6 Let $X_1, \ldots, X_n$ be a sample of iid Gamma(α, θ) random variables with $\Theta = \mathbb{R}^+$ and α known. For the generalized likelihood ratio test of $H_0: \theta = 4$ versus $H_A: \theta \neq 4$,

a) determine the unrestricted MLE of θ.

b) determine $\lambda(\vec{x})$.

c) show that the generalized likelihood ratio critical region of size α is equivalent to $\{\vec{x}: \overline{x} \leq k_{\alpha_1} \text{ or } \overline{x} \geq k_{\alpha_2}\}$.

6.4.7 Let $X_1, \ldots, X_n$ be a sample of iid random variables with pdf $f(x: \theta_1, \theta_2) = \frac{1}{\theta_1} e^{-\frac{(x-\theta_2)}{\theta_1}}$ for $x > \theta_2$ and $\Theta = \mathbb{R}^+ \times \mathbb{R}$. For the generalized likelihood ratio test of $H_0: \theta_1 \geq 1$ versus $H_A: \theta_1 < 1$, the MLE of the nuisance parameter θ_2 is $X_{(1)}$ on both Θ and Θ_0.

a) Determine the unrestricted MLE of θ_1.

b) Determine the MLE of θ_1 on Θ_0.

c) Determine $\lambda(\vec{x})$.

d) Show that the generalized likelihood ratio critical region of size α is equivalent to $\{\vec{x}: \sum_{i=1}^{n} x_i - n x_{(1)} \leq k_\alpha\}$.

6.4.8 Let $X_1, \ldots, X_{300}$ be a sample of iid Pois(θ) random variables with $\Theta = \mathbb{R}^+$ and $\sum_{i=1}^{300} x_i = 659$. For testing $H_0: \theta = 2$ versus $H_A: \theta \neq 2$,

a) determine the unrestricted value of the MLE of θ.

b) determine the value of $\lambda(\vec{x})$.

c) determine the large sample critical region for the likelihood ratio test for $\alpha = 0.04$.

d) perform a $\alpha = 0.04$ level test of $H_0 : \theta = 2$ versus $H_A : \theta \neq 2$ using the large sample generalized likelihood ratio critical region.

e) compute the p-value associated with the large sample likelihood ratio test.

6.4.9 Let $X_1, \ldots, X_{150}$ be a sample of iid $Exp(\theta)$ random variables with $\Theta = \mathbb{R}^+$ and $\sum_{i=1}^{150} x_i = 869.6$. For testing $H_0 : \theta = 5$ versus $H_A : \theta \neq 5$,

a) determine the unrestricted value of the MLE of θ.

b) determine the value of $\lambda(\vec{x})$.

c) determine the large sample critical region for the likelihood ratio test for $\alpha = 0.05$.

d) perform a $\alpha = 0.05$ level test of $H_0 : \theta = 5$ versus $H_A : \theta \neq 5$ using the large sample generalized likelihood ratio critical region.

e) compute the p-value associated with the large sample likelihood ratio test.

6.4.10 Let $X_1, \ldots, X_{250}$ be a sample of iid $Beta(4, \theta)$ random variables with $\Theta = \mathbb{R}^+$.

a) Determine the form of the log-likelihood function $\ell(\theta)$.

b) If $\sum_{i=1}^{250} \ln(x_i) = -159.58$ and $\sum_{i=1}^{250} \ln(1 - x_i) = -229.25$, determine the value of $l(3)$.

c) For testing $H_0 : \theta = 3$ versus $H_A : \theta \neq 3$ at the $\alpha = 0.05$ level, determine the large sample critical region for the likelihood ratio test.

d) For testing $H_0 : \theta = 3$ versus $H_A : \theta \neq 3$ at the $\alpha = 0.05$ level, compute $-2\ln[\lambda(\vec{x})]$ when $l(\hat{\theta}) = 86.92$.

e) Test $H_0 : \theta = 3$ versus $H_A : \theta \neq 3$ at the $\alpha = 0.05$ level.

6.4.11 Let $X_1, \ldots, X_{250}$ be a sample of iid $Gamma(\theta, 10)$ random variables with $\Theta = \mathbb{R}^+$.

a) Determine the form of the log-likelihood function $\ell(\theta)$.

b) If $\sum_{i=1}^{250} \ln(x_i) = 992.18$ and $\sum_{i=1}^{250} x_i = 14473.7$, determine the value of $l(5)$.

c) For testing $H_0 : \theta = 5$ versus $H_A : \theta \neq 5$ at the $\alpha = 0.01$ level, determine the large sample critical region for the likelihood ratio test.

d) For testing $H_0 : \theta = 5$ versus $H_A : \theta \neq 5$ at the $\alpha = 0.01$ level, compute $-2\ln[\lambda(\vec{x})]$ when $l(\hat{\theta}) = -1136.12$.

e) Test $H_0 : \theta = 5$ versus $H_A : \theta \neq 5$ at the $\alpha = 0.01$ level.

6.4.12 Prove Theorem 6.6.

6.5 Large Sample Tests

Two other commonly used large sample testing approaches are also based on the likelihood function. In particular, large sample tests based on the asymptotic distribution of the MLE or the Score function can be used for testing $H_0 : \theta \in \Theta_0$ versus $H_A : \theta \in \Theta_A$. As is the case with a large sample likelihood ratio test, hypothesis tests based on an MLE or the Score function are only valid when the regularity conditions given in Definition 4.18 are satisfied and n is sufficiently large.

6.5.1 Large Sample Tests Based on the MLE

Recall that when $X_1, \ldots, X_n$ is a sample of iid random variables from a one-parameter pdf $f(x; \theta)$ satisfying the regularity conditions given in Definition 4.18, Theorem 5.6 applies, and it follows that $\dfrac{\widehat{\theta} - \theta}{\sqrt{V}} \xrightarrow{d} N(0, 1)$. The asymptotic variance of the MLE of θ is $V = I_n^{-1}(\theta)$. Moreover, it is also true that $\dfrac{\widehat{\theta} - \theta}{\sqrt{\widehat{V}}} \xrightarrow{d} N(0, 1)$ where $\widehat{V} = I_n^{-1}(\widehat{\theta})$.

Similarly, for a multiparameter distribution with $\theta = (\theta_1, \ldots, \theta_k)$ satisfying the regularity conditions in Definition 4.18, $\dfrac{\widehat{\theta}_i - \theta_i}{\sqrt{\widehat{V}_i}} \xrightarrow{d} N(0, 1)$, where the asymptotic variance of the MLE of θ_i is $\widehat{V}_i = I_n^{-1}(\widehat{\theta})_{ii}$ and $I_n(\theta)$ is the Fisher Information matrix.

A test based on the asymptotic distribution of the MLE was first proposed by Wald [21] and, hence, is referred to as a *Wald test*. The Wald test can be used with a sufficiently large sample and a probability model satisfying the necessary regularity conditions, and in this case, $\dfrac{\sqrt{n}(\widehat{\theta} - \theta_0)}{\sqrt{\widehat{V}}}$ is approximately $N(0, 1)$.

In particular, for testing $H_0 : \theta \leq \theta_0$ versus $H_A : \theta > \theta_0$, the critical region for the Wald test is

$$\mathcal{C}_W = \left\{ \vec{x} : \frac{\widehat{\theta} - \theta_0}{\sqrt{\widehat{V}}} > z_{1-\alpha} \right\},$$

for testing $H_0 : \theta \geq \theta_0$ versus $H_A : \theta < \theta_0$, the Wald test critical region is

$$\mathcal{C}_W = \left\{ \vec{x} : \frac{\widehat{\theta} - \theta_0}{\sqrt{\widehat{V}}} < z_{\alpha} \right\},$$

and for testing for $H_0 : \theta = \theta_0$ versus $H_A : \theta \neq \theta_0$, the Wald test critical region is

$$\mathcal{C}_W = \left\{ \vec{x} : \left| \frac{\widehat{\theta} - \theta_0}{\sqrt{\widehat{V}}} \right| > z_{1-\frac{\alpha}{2}} \right\},$$

where $\hat{V} = I_n^{-1}(\hat{\theta})$ for a one-parameter pdf and z_α is the $100 \cdot \alpha$th percentile of a $N(0,1)$ random variable. When $\theta = (\theta_1, \dots, \theta_k)$, the critical regions for a Wald test on the parameter θ_i are the similar using $\hat{V}_i = I_n^{-1}(\hat{\theta})_{ii}$.

Example 6.26 Let $X_1, \dots, X_n$ be iid $N(\theta_1, \theta_2)$ random variables with θ_1 and θ_2 unknown. Recall that the MLEs of θ_1 and θ_2 are $\bar{x}$ and s_n^2, respectively, and the Fisher's Information matrix is $I_n(\theta_1, \theta_2) = \begin{bmatrix} \frac{n}{\theta_2} & 0 \\ 0 & \frac{n}{2\theta_2^2} \end{bmatrix}$. Thus, the asymptotic variance of $\hat{\theta}_1$ is $\hat{V} = I_n^{-1}(\hat{\theta}_1, \hat{\theta}_2)_{11} = \frac{s_n^2}{n}$.

Hence, for testing $H_0 : \theta_1 = \theta_{10}$ versus $H_A : \theta_1 \neq \theta_{10}$, the critical region for the Wald test is

$$\mathcal{C}_W = \left\{ \left| \frac{\bar{x} - \theta_{10}}{\frac{s_n}{\sqrt{n}}} \right| > z_{1-\frac{\alpha}{2}} \right\}.$$

Note that the critical region in the Wald test is based on $\frac{\bar{x}-\theta_{10}}{\frac{s_n}{\sqrt{n}}}$, which is slightly different from the t-test developed in Example 6.22 based on $\frac{\bar{x}-\theta_{10}}{\frac{s}{\sqrt{n}}}$. However, the Wald test and the generalized likelihood ratio test are asymptotically equivalent and should produce nearly the same results for large samples.

For example, if $n = 120$, $\bar{x} = 12.3$, and $s_n^2 = 25.2$, then for testing $H_0 : \theta_1 = 10$ versus $\theta_1 \neq 10$, the value of the Wald test statistic is

$$\frac{12.3 - 10}{\sqrt{\frac{25.2}{120}}} = 5.019$$

with p-value $p = 0.0000$. The value of the generalized likelihood ratio t statistic for this sample is $\frac{12.3-10}{\sqrt{\frac{24.99}{120}}} = 5.04$ with p-value $p = 0.000$.

Wald tests are easily performed using a maximum likelihood fitting procedure that provides the MLE and its asymptotic standard error such as R's `fitdist`. Example 6.27 illustrates using the output from R's `fitdist` command to perform a Wald test when sampling from a beta distribution.

Example 6.27 Let $X_1, \dots, X_{100}$ be iid Beta(θ_1, θ_2) random variables. Suppose that the R output in Table 6.2 results from fitting the MLEs to the observed sample. In the `fitdist` output for fitting a beta distribution with MLE, `shape1` corresponds to θ_1 and `shape2` corresponds to θ_2.

The information in Table 6.2, from the R command `fitdist(x,dbeta)`, can easily be used to test $H_0 : \theta_1 \leq 1$ versus $H_A : \theta_1 > 1$ with a Wald test.

Table 6.2 The R output for Example 6.27.

```
Fitting of the distribution "beta" by maximum likelihood
Parameters:
        estimate Std. Error
shape1 1.154192  0.1467954
shape2 2.828578  0.4055633
```

In particular, based on the information in Table 6.2, the value of the Wald test statistic is

$$\frac{\hat{\theta}_1 - 1}{\sqrt{V}} = \frac{1.154 - 1}{0.1468} = 1.04.$$

The size $\alpha = 0.05$ critical region for the Wald test of $H_0 : \theta_1 \leq 1$ versus $H_A : \theta_1 > 1$ is $\mathcal{C}_W = \left\{ \frac{\hat{\theta}_1 - 1}{\sqrt{V}} > 1.645 \right\}$. Therefore, there is insufficient evidence in the observed sample to support rejecting H_0. The p-value associated for testing $H_0 : \theta_1 \leq 1$ versus $H_A : \theta_1 > 1$ with a Wald test is $p = 1\text{-pnorm}(1.049, 0, 1) = 0.1471$.

6.5.2 Score Tests

A large sample alternative to the Wald test for testing $H_0 : \theta = \theta_0$ versus $H_A : \theta \neq \theta_0$ is *Rao's Score test* due to Rao [22]. Rao's Score test is based on the information about θ contained in the likelihood function through the Score function.

Definition 6.14   Let $X_1, \ldots, X_n$ be a sample with log-likelihood function $\ell(\theta)$ for $\theta \in \Theta$. The Score function $\dot{\ell}(\theta)$ is defined to be $\dot{\ell}(\theta) = \frac{\partial}{\partial \theta} \ln[L(\theta)]$ when $\dim(\Theta) = 1$, and for $\dim(\Theta) = k > 1$, the Score function is the vector

$$\vec{\ell} = \left(\frac{\partial}{\partial \theta_1} \ln[L(\theta)], \ldots, \frac{\partial}{\partial \theta_k} \ln[L(\theta)] \right).$$

Recall that for a sample of iid random variables with a one-parameter pdf satisfying the regularity conditions, $E[\dot{\ell}(\theta)] = 0$ and $\text{Var}[\dot{\ell}(\theta)] = I_n(\theta)$. Also, when $X_1, \ldots, X_n$ are iid random variables with pdf $f(x; \theta)$, the score function is $\dot{\ell}(\theta) = \sum_{i=1}^{n} \frac{\partial \ln[f(x_i; \theta)]}{\partial \theta}$, and by Theorem 5.3, the Score function is asymptotically normal. Theorem 6.8 shows that the asymptotic distribution of the square of the Score function divided by its variance is asymptotically distributed as a χ^2 random variable.

Theorem 6.8 *If $X_1, \ldots, X_n$ is a sample of iid random variables with pdf $f(x; \theta)$ satisfying the regularity conditions in Definition 4.18, then $\frac{\dot{\ell}(\theta)^2}{I_n(\theta)} \xrightarrow{d} \chi_1^2$.*

When $\hat{\theta}$ is the MLE of θ, $\dot{\ell}(\hat{\theta}) = 0$, and therefore, in a test of $H_0 : \theta = \theta_0$ versus $H_A : \theta \neq \theta_0$, H_0 should be rejected when $\dot{\ell}(\theta_0)$ is far from zero. When $X_1, \ldots, X_n$ are iid random variables with pdf satisfying the regularity conditions and n is sufficiently large, Theorem 6.8 applies and under $H_0 : \theta = \theta_0$, $\frac{\dot{\ell}(\theta_0)^2}{I_n(\theta_0)} \approx \chi_1^2$. Thus, the large sample critical region of size α for the Score test of $H_0 : \theta = \theta_0$ versus $H_A : \theta \neq \theta_0$ is

$$
\mathcal{C}_S = \left\{ \vec{x} : \frac{\dot{\ell}(\theta_0)^2}{I_n(\theta_0)} \geq k_\alpha \right\},
$$

where $P(\chi_1^2 \geq k_\alpha) = \alpha$.

Example 6.28 Let $X_1, \ldots, X_n$ be a sample of iid random variables with pdf $f(x; \theta) = \theta e^{-\theta x}$ for $x \in \mathbb{R}^+$ and $\Theta = \mathbb{R}^+$. Then, $f(x; \theta)$ is in the exponential family of distributions and and the regularity conditions of the Score test are satisfied. The log-likelihood function is

$$
\ell(\theta) = n \ln(\theta) - \theta \sum_{i=1}^{n} x_i,
$$

and therefore, the Score function is

$$
\dot{\ell}(\theta) = \frac{n}{\theta} - \sum_{i=1}^{n} x_i
$$

and

$$
I_n(\theta) = -E \left[\frac{\partial^2}{\partial \theta^2} \dot{\ell}(\theta) \right] = -E \left[\frac{-n}{\theta^2} \right] = \frac{n}{\theta^2}.
$$

Hence, for a sufficiently large sample, the size α critical region for the Score test of $H_0 : \theta = 2$ versus $H_A : \theta \neq 2$ is $\mathcal{C}_S \left\{ \frac{\dot{\ell}(2)^2}{I_n(2)} \geq k_\alpha \right\}$, which is

$$
\mathcal{C}_S = \left\{ \vec{x} : \frac{\left(\frac{n}{2} - \sum_{i=1}^{n} x_i \right)^2}{\frac{n}{4}} \geq k_\alpha \right\},
$$

where $P(\chi_1^2 \geq k_\alpha) = \alpha$.

For example, if $n = 100$ and $\sum_{i=1}^{100} x_i = 61.2$, then

$$
\frac{\dot{\ell}(2)^2}{I_n(2)} = \frac{(50 - 61.2)^2}{25} = 5.0176
$$

and the size $\alpha = 0.02$ critical region for the Score test is $\left\{ \frac{\dot{\ell}(2)^2}{I_n(2)} \geq 5.412 \right\}$. Thus, when $\sum_{i=1}^{100} x_i = 61.2$, there is not sufficient evidence to reject $H_0 : \theta = 2$ in a Score test.

When $X_1, \ldots, X_n$ is a sample of iid random variables from a k-parameter pdf $f(x; \theta)$ satisfying the regularity conditions on Definition 4.18, the Score function is a vector of functions, Fisher's Information is a matrix, and Score test of $H_0: \theta = \theta_0$ versus $H_A: \theta \neq \theta_0$ is based on the test statistic $\dot{\ell}^{\mathrm{T}}(\hat{\theta}_0) I_n^{-1}(\hat{\theta}_0) \dot{\ell}(\hat{\theta}_0)$, where $\hat{\theta}_0$ is the MLE of θ under the null hypothesis. Under the standard regularity conditions, $\dot{\ell}^{\mathrm{T}}(\hat{\theta}_0) I_n^{-1}(\hat{\theta}_0) \dot{\ell}(\hat{\theta}_0) \xrightarrow{d} \chi_p^2$, where p is the number of constraints on θ in H_0. In this case, for large samples the size α critical region for the Score test is $\mathcal{C}_S = \{\vec{x}: \dot{\ell}^{\mathrm{T}}(\hat{\theta}_0) I_n^{-1}(\hat{\theta}_0) \dot{\ell}(\hat{\theta}_0) \geq k_\alpha\}$, where $P(\chi_p^2 \geq k_\alpha) = \alpha$.

Example 6.29   Let $X_1, \ldots, X_n$ be a sample of iid Gamma(θ_1, θ_2) random variables with $\Theta = \mathbb{R}^+ \times \mathbb{R}^+$. For testing $H_0: \theta_1 = 4$ versus $H_A: \theta_1 \neq 4$, θ_2 is a nuisance parameter, and in this case, Rao's Score test can be used to deal with the nuisance parameter θ_2 with a sufficiently large sample.

The likelihood function is

$$L(\theta_1, \theta_2) = \prod_{i=1}^{n} \frac{1}{\Gamma(\theta_1)\theta_2^{\theta_1}} x_i^{\theta_1 - 1} e^{-\frac{x_i}{\theta_2}} = \Gamma(\theta_1)^{-n} \theta_2^{-n\theta_1} \prod_{i=1}^{n} x_i^{\theta_1 - 1} e^{-\frac{\sum_{i=1}^{n} x_i}{\theta_2}},$$

and therefore, the log-likelihood function is

$$\ell(\theta_1, \theta_2) = -n \ln[\Gamma(\theta_1)] - n\theta_1 \ln(\theta_2) + (\theta_1 - 1) \sum_{i=1}^{n} \ln(x_i) - \frac{\sum_{i=1}^{n} x_i}{\theta_2}.$$

For the gamma distribution, the Score function is the two-component vector of partial derivatives of $\ell(\theta_1, \theta_2)$ taken with respect to θ_1 and θ_2. Hence,

$$\dot{\ell}(\theta_1, \theta_2) = \begin{pmatrix} \dot{\ell}_1(\theta_1, \theta_2) \\ \dot{\ell}_2(\theta_1, \theta_2) \end{pmatrix} = \begin{pmatrix} \frac{\partial}{\partial \theta_1} \ell(\theta_1, \theta_2) \\ \frac{\partial}{\partial \theta_2} \ell(\theta_1, \theta_2) \end{pmatrix}$$

$$= \begin{pmatrix} -n \frac{\Gamma'(\theta_1)}{\Gamma(\theta_1)} - n \ln(\theta_2) + \sum_{i=1}^{n} \ln(x_i) \\ -\frac{n\theta_1}{\theta_2} + \frac{\sum_{i=1}^{n} x_i}{\theta_2^2} \end{pmatrix}.$$

Fisher's Information matrix for the gamma distribution is

$$I_n(\theta_1, \theta_2) = \begin{bmatrix} n\psi_1(\theta_1) & \frac{n}{\theta_2} \\ \frac{n}{\theta_2} & \frac{n\theta_1}{\theta_2^2} \end{bmatrix},$$

where $\psi_1(\theta_1) = \frac{\partial^2}{\partial \theta_1^2} \ln[\Gamma(\theta_1)]$, which is called the *trigamma function*.

Now, the MLE of $\theta = (\theta_1, \theta_2)$ under H_0 is $\widehat{\theta}_0 = (\widehat{\theta}_{10}, \widehat{\theta}_{20}) = \left(4, \frac{\bar{x}}{4}\right)$, and the Score function evaluated at $\widehat{\theta}_0$ is

$$
\ell\left(4, \frac{\bar{x}}{4}\right) = \begin{pmatrix} -n\frac{\Gamma'(4)}{\Gamma(4)} - n\ln\left[\frac{\bar{x}}{4}\right] + \sum_{i=1}^{n}\ln(x_i) \\ 0 \end{pmatrix}
$$

$$
= \begin{pmatrix} -n\psi(4) - n\ln\left[\frac{\bar{x}}{4}\right] + \sum_{i=1}^{n}\ln(x_i) \\ 0 \end{pmatrix},
$$

where $\psi(\theta) = \frac{\partial}{\partial\theta}\ln[\Gamma(\theta)]$, which is called the *digamma function*.

The test statistic for the Score test is $\ell^{\mathrm{T}}(\widehat{\theta}_0)I_n^{-1}(\widehat{\theta}_0)\ell(\widehat{\theta}_0)$, and the size $\alpha = 0.01$ critical region for the Score test is

$$
\mathcal{C}_S = \left\{ \bar{x} : \ell^{\mathrm{T}}(\widehat{\theta}_0)I_n^{-1}(\widehat{\theta}_0)\ell(\widehat{\theta}_0) \geq 6.635 \right\}.
$$

Consider testing $H_0 : \theta_1 = 4$ versus $H_A : \theta_1 \neq 4$, with a random sample of size $n = 100$, $\sum_{i=1}^{100} x_i = 1377.493$, and $\sum_{i=1}^{100}\ln(x_i) = 254.425$. Then, the MLE under H_0 is $\widehat{\theta}_0 = (4, 3.444)$, the value of the Score function evaluated at $\widehat{\theta}_0$ is $\ell(4, 3.322) = \begin{pmatrix} 5.158 \\ 0 \end{pmatrix}$, and the inverse of Fisher's Information matrix evaluated at $\widehat{\theta}_0$ is

$$
I_n(\widehat{\theta}_0)^{-1} = \begin{bmatrix} 28.38230 & 29.03826 \\ 29.03826 & 33.72883 \end{bmatrix}^{-1} = \begin{bmatrix} 0.2956572 & -0.254541 \\ -0.2545410 & 0.248791 \end{bmatrix}.
$$

Now, the observed value of the Score test statistic is

$$
\begin{pmatrix} 5.158 \\ 0 \end{pmatrix}^{\mathrm{T}} \begin{bmatrix} 0.2956572 & -0.254541 \\ -0.2545410 & 0.248791 \end{bmatrix} \begin{pmatrix} 5.158 \\ 0 \end{pmatrix} = 7.865.
$$

Thus, there is sufficient evidence in the sample to reject $H_0 : \theta_1 = 4$, and the approximate p-value for the Score test is $p = P(\chi_1^2 \geq 7.865) = 0.0050$.

Finally, Rao's Score test, Wald's test, and the large sample likelihood ratio test are asymptotically equivalent. In practice, when there is strong evidence against the null hypothesis, all three tests will reject H_0; however, when the evidence against the null hypothesis is weak, these three tests may result in conflicting decisions.

Problems

6.5.1 Let $X_1, \ldots, X_{100}$ be a sample of iid Pois(θ) random variables with $\Theta = \mathbb{R}^+$.
   a) If $\sum_{i=1}^{100} x_i = 158$, determine the MLE of θ.
   b) Determine $I(\theta)$.
   c) Determine the large sample MLE critical region of size $\alpha = 0.05$ for testing $H_0: \theta \geq 2$ versus $H_0: \theta < 2$.
   d) Test $H_0: \theta \geq 2$ versus $H_0: \theta < 2$ using the large sample MLE test with $\alpha = 0.05$.

6.5.2 Let $X_1, \ldots, X_{100}$ be a sample of iid Exp(θ) random variables with $\Theta = \mathbb{R}^+$.
   a) If $\sum_{i=1}^{100} x_i = 536$, determine the MLE of θ.
   b) Determine $I(\theta)$.
   c) Determine the large sample MLE critical region of size $\alpha = 0.02$ for testing $H_0: \theta \leq 5$ versus $H_0: \theta > 5$.
   d) Test $H_0: \theta \leq 5$ versus $H_0: \theta > 5$ using the large sample MLE test with $\alpha = 0.02$.

6.5.3 Let $X_1, \ldots, X_{100}$ be a sample of iid random variables with $f(x; \theta) = \theta x^{\theta-1}$ for $x \in (0, 1)$ and $\Theta = \mathbb{R}^+$.
   a) If $\sum_{i=1}^{100} \ln(x_i) = -66.51$, determine the MLE of θ.
   b) Determine $I(\theta)$.
   c) Determine the large sample MLE critical region of size $\alpha = 0.01$ for testing $H_0: \theta \leq 1$ versus $H_0: \theta > 1$.
   d) Test $H_0: \theta \leq 1$ versus $H_0: \theta > 1$ using the large sample MLE test with $\alpha = 0.01$.

6.5.4 Let $X_1, \ldots, X_{100}$ be a sample of iid $N(0, \theta)$ random variables with $\Theta = \mathbb{R}^+$.
   a) If $\sum_{i=1}^{100} x_i^2 = 148.2$, determine the MLE of θ.
   b) Determine $I(\theta)$.
   c) Determine the large sample MLE critical region of size $\alpha = 0.01$ for testing $H_0: \theta = 3$ versus $H_0: \theta \neq 3$.
   d) Test $H_0: \theta = 3$ versus $H_0: \theta \neq 3$ using the large sample MLE test with $\alpha = 0.01$.
   e) Compute the p-value associated with the large sample MLE test of $H_0: \theta = 3$ versus $H_0: \theta \neq 3$.

6.5.5 Let $X_1, \ldots, X_{250}$ be a sample of iid Gamma($\theta, 3$) random variables with shape parameter θ and $\Theta = \mathbb{R}^+$. Use the R output give in Table 6.3 to answer the following questions.

a) Determine the large sample MLE critical region of size $\alpha = 0.04$ for testing $H_0: \theta \leq 1$ versus $H_0: \theta > 1$.
b) Test $H_0: \theta \leq 1$ versus $H_0: \theta > 1$ using the large sample MLE test with $\alpha = 0.04$.
c) Compute the p-value associated with the large sample MLE test of $H_0: \theta \leq 1$ versus $H_0: \theta > 1$.

6.5.6 Let $X_1, \ldots, X_{250}$ be a sample of iid Gamma(θ_1, θ_2) random variables with shape parameter θ_1, scale parameter θ_2, and $\Theta = \mathbb{R}^+ \times \mathbb{R}^+$. Use the R output give in Table 6.4 to answer the following questions.

a) Determine the large sample MLE critical region of size $\alpha = 0.04$ for testing $H_0: \theta_1 \leq 3.5$ versus $H_0: \theta_1 > 3.5$.
b) Test $H_0: \theta_1 \leq 3.5$ versus $H_0: \theta_1 > 3.5$ using the large sample MLE test with $\alpha = 0.04$.
c) Compute the p-value associated with the large sample MLE test of $H_0: \theta_1 \leq 3.5$ versus $H_0: \theta_1 > 3.5$.
d) Determine the large sample MLE critical region of size $\alpha = 0.05$ for testing $H_0: \theta_2 = 1$ versus $H_0: \theta_2 \neq 1$.
e) Test $H_0: \theta_2 = 1$ versus $H_0: \theta_2 \neq 1$ using the large sample MLE test with $\alpha = 0.05$.
f) Compute the p-value associated with the large sample MLE test of $H_0: \theta_2 = 2$ versus $H_0: \theta_2 \neq 2$.

Table 6.3 R output for Problem 6.5.5.

```
Fitting of the distribution 'gamma' by maximum likelihood
Parameters:
      estimate Std. Error
shape 1.309736 0.05968956
Fixed parameters:
         value
scale 3
```

Table 6.4 R output for Problem 6.5.6.

```
Fitting of the distribution 'gamma' by maximum likelihood
Parameters:
      estimate Std. Error
shape 4.016050  0.4459281
scale 2.019409  0.2388624
```

Table 6.5 R output for Problem 6.5.7.

```
Fitting of the distribution 'beta' by maximum likelihood
Parameters:
       estimate Std. Error
shape1 5.120072  0.5878650
shape2 2.579221  0.2822914
```

6.5.7 Let $X_1, \ldots, X_{150}$ be a sample of iid Beta(θ_1, θ_2) random variables where θ_1 is the first scale parameter, θ_2 is the second scale parameter, and $\Theta = \mathbb{R}^+ \times \mathbb{R}^+$. Use the R output give in Table 6.5 to answer the following questions.

a) Determine the large sample MLE critical region of size $\alpha = 0.03$ for testing $H_0 : \theta_1 \geq 6$ versus $H_0 : \theta_1 < 6$.

b) Test $H_0 : \theta_1 \geq 6$ versus $H_0 : \theta_1 < 6$ using the large sample test MLE test with $\alpha = 0.03$.

c) Compute the p-value associated with the large sample MLE test of $H_0 : \theta_1 \geq 6$ versus $H_0 : \theta_1 < 6$.

d) Determine the large sample MLE critical region of size $\alpha = 0.05$ for testing $H_0 : \theta_2 = 2$ versus $H_0 : \theta_2 \neq 2$.

e) Test $H_0 : \theta_2 = 2$ versus $H_0 : \theta_2 \neq 2$ using the large sample MLE test with $\alpha = 0.05$

f) Compute the p-value associated with the large sample MLE test of $H_0 : \theta_2 = 2$ versus $H_0 : \theta_2 \neq 2$.

6.5.8 Let $X_1, \ldots, X_{60}$ be a sample of iid Pois(θ) random variables with $\Theta = \mathbb{R}^+$.

a) Determine the form of the Score function.

b) Evaluate the Score function at $\theta = 5$ when $\sum_{i=1}^{60} x_i = 361$.

c) Using $I(\theta) = \frac{1}{\theta}$, test $H_0 : \theta = 5$ versus $H_0 : \theta \neq 5$ using the large sample score test with $\alpha = 0.02$.

d) Compute the p-value associated with the large sample score test of $H_0 : \theta = 5$ versus $H_0 : \theta \neq 5$.

6.5.9 Let $X_1, \ldots, X_{50}$ be a sample of iid Exp(θ) random variables with $\Theta = \mathbb{R}^+$.

a) Determine the form of the Score function.

b) Evaluate the Score function at $\theta = 8$ when $\sum_{i=1}^{50} x_i = 470.8$.

c) Using $I(\theta) = \frac{1}{\theta^2}$, test $H_0 : \theta = 8$ versus $H_0 : \theta \neq 8$ using the large sample score test with $\alpha = 0.01$.

d) Compute the p-value associated with the large sample score test of $H_0 : \theta = 8$ versus $H_0 : \theta \neq 8$.

6.5.10 Let $X_1, \ldots, X_{100}$ be a sample of iid random variables with common pdf
$f(x; \theta) = \theta x^{\theta-1}$ for $x \in (0, 1)$ and $\Theta = \mathbb{R}^+$.
  a) Determine the form of the Score function.
  b) Evaluate the Score function at $\theta = 1$ when $\sum_{i=1}^{100} \ln(x_i) = -66.51$.
  c) Using $I(\theta) = \frac{1}{\theta^2}$, test $H_0 : \theta = 1$ versus $H_0 : \theta \neq 1$ using the large sample score test with $\alpha = 0.05$.
  d) Compute the p-value associated with the large sample score test of $H_0 : \theta = 1$ versus $H_0 : \theta \neq 1$.

6.5.11 Let $X_1, \ldots, X_{100}$ be a sample of iid $N(0, \theta)$ random variables with $\Theta = \mathbb{R}^+$.
  a) Determine the form of the Score function.
  b) Evaluate the Score function at $\theta = 5$ when $\sum_{i=1}^{100} x_i^2 = 419.84$.
  c) Using $I(\theta) = \frac{1}{2\theta^2}$, test $H_0 : \theta = 5$ versus $H_0 : \theta \neq 5$ using the large sample score test with $\alpha = 0.02$.
  d) Compute the p-value associated with the large sample score test of $H_0 : \theta = 5$ versus $H_0 : \theta \neq 5$.

6.5.12 Let $X_1, \ldots, X_{100}$ be a sample of iid Gamma$(5, \theta)$ random variables with $\Theta = \mathbb{R}^+$.
  a) Determine the form of the Score function.
  b) Evaluate the Score function at $\theta = 1$ when $\sum_{i=1}^{100} x_i = 620.2$.
  c) Using $I(\theta) = \frac{5}{\theta^2}$, test $H_0 : \theta_1 = 1$ versus $H_0 : \theta \neq 1$ using the large sample score test with $\alpha = 0.01$.
  d) Compute the p-value associated with the large sample score test of $H_0 : \theta = 1$ versus $H_0 : \theta \neq 1$.

6.5.13 Let $X_1, \ldots, X_{100}$ be a sample of iid $N(\theta, 25)$ random variables with $\Theta = \mathbb{R}^+$.
  a) Determine the form of the Score function.
  b) Evaluate the Score function at $\theta = 10$ when $\sum_{i=1}^{100} x_i = 1128.6$.
  c) Using $I(\theta) = \frac{1}{\sigma^2}$, test $H_0 : \theta_1 = 10$ versus $H_0 : \theta \neq 10$ using the large sample score test with $\alpha = 0.04$.
  d) Compute the p-value associated with the large sample score test of $H_0 : \theta = 10$ versus $H_0 : \theta \neq 10$.

6.6 Case Study – Modeling Survival of the Titanic Passengers

The RMS Titanic was a British passenger liner that was deemed unsinkable; however, the Titanic sank on its maiden voyage after colliding with an iceberg in the North Atlantic Ocean on April 15, 1912. There were 2224 passengers and crew aboard the Titanic of which over 1500 died. Records on 1309 passengers on the Titanic will be used to explore the chance of survival for a Titanic passenger. In particular, the probability of surviving the Titanic's sinking will be modeled as a function of a passenger's age and gender.

The data used to model the probability of survival consists of records of 1046 Titanic passengers having complete records on the variables survival, age, and gender. Furthermore, the model will be based on a random sample of $n = 600$ records from the 1046; the remaining records can be used as a model validation dataset.

The variables of interest in this data set are

$$Y = \text{Survival} = \begin{cases} 0 & \text{Died} \\ 1 & \text{Survived} \end{cases}$$

$$A = \text{Age of a passenger}$$

$$G = \text{Gender} = \begin{cases} 0 & \text{Female} \\ 1 & \text{Male} \end{cases},$$

where Y is a random variable and A and G are fixed variables; Y is the *response variable*, and A and G are *explanatory variables* in the model, and the data for the ith passenger is a vector of the form (Y_i, A_i, G_i).

Definition 6.15 In a statistical model, the response variable is the random variable of interest and the explanatory variables are the variables believed to cause changes in the response variable.

The response variable is sometimes called the dependent variable because it is believed to depend on the explanatory variables.

6.6.1 Exploring the Data

R was used to generate a random sample of $n = 600$ observations from the 1046 complete passenger records. Table 6.6 summarizes the data according

Table 6.6 A table showing for survival of $n = 600$ Titanic passengers broken down by gender.

Gender	Died	Survived
Female	55	167
Male	299	79
Total	354	246

Table 6.7 A table showing for survival of $n = 600$ Titanic passengers broken down by gender and age.

Age	Females			Males		
	Died	Survived	Percentage survived (%)	Died	Survived	Percentage survived (%)
0–10	9	17	65.4	8	19	70.4
10–20	8	29	78.4	42	6	12.5
20–30	23	53	69.7	111	25	18.4
30–40	11	33	75	66	15	18.5
40–50	2	17	89.5	42	9	17.6
50–60	1	12	92.3	16	3	15.8
60–70	1	5	83.3	12	1	7.7
70–80	0	1	100	2	1	33.3

to survival and gender. Note that the percentage of sampled females surviving the Titanic sinking was 75%, whereas the percentage of males surviving was only 21%.

Table 6.7 summarizes the data according to survival by age and gender. Note that in every age class except for the youngest passengers, females have a much higher survival rate than males; survival rates in the age class of 0–10 are roughly the same for each gender. Clearly, the maritime code of conduct to save women and children first seemed to apply in the case of the Titanic sinking.

6.6.2 Modeling the Probability of Survival

Using the $n = 600$ randomly selected observations to build a model for the probability of survival, it is assumed that the response variable $Y_i \sim \text{Bin}(1, \theta_i)$, where θ_i is the probability of survival for $i = 1, \dots, 600$. Furthermore, θ_i is a function of a passenger's age and gender according to

$$\theta_i = \frac{e^{\beta_0 + \beta_1 A_i + \beta_2 G_i + \beta_3 A_i \cdot G_i}}{1 + e^{\beta_0 + \beta_1 A_i + \beta_2 G_i + \beta_3 A_i \cdot G_i}},$$

which is a *generalized linear model* in the unknown parameters $\beta_0, \beta_1, \beta_2$, and β_3.

A generalized linear model is a model where the response variable Y is assumed to follow an exponential family distribution with mean $\mu(\vec{X})$ for a set of explanatory variables $\vec{X}$, and there exists a linear predictor $\eta = \vec{X}^T \vec{\beta}$ with a one-to-one link function $g(\mu)$ such that $E(Y) = \mu = g^{-1}(\eta)$. Generalized linear models are discussed in detail in Chapter 7.

The model being fit to the response variable Y in this case is called a *logistic regression model*; a logistic regression model is a generalized linear model where the response variable follows a binomial distribution and the *logit* link function is used. For the Titanic data, a logistic regression model with

$$\text{logit}(\theta_i) = \ln\left[\frac{\theta_i}{1 - \theta_i}\right] = \beta_0 + \beta_1 A_i + \beta_2 G_i + \beta_3 A_i \cdot G_i$$

will be fit.

For convenience, let

$$\vec{\beta} = \begin{pmatrix} \beta_0 \\ \beta_1 \\ \beta_2 \\ \beta_3 \end{pmatrix}, X_i = \begin{pmatrix} 1 \\ A_i \\ G_i \\ A_i \cdot G_i \end{pmatrix}$$

for $i = 1, \ldots, 600$. Then,

$$\text{logit}(\theta_i) = X_i^T \vec{\beta} = \beta_0 + \beta_1 A_i + \beta_2 G_i + \beta_3 A_i \cdot G_i.$$

The parameters $\beta_0, \beta_1, \beta_2,$ and β_3 are estimated using MLE, and the likelihood function for the observed values $y_1, y_2, \ldots, y_{600}$ is

$$L(\vec{\beta}) = \prod_{i=1}^{n} \theta_i^{y_i}(1 - \theta_i)^{1-y_i} = \prod_{i=1}^{n} \left(\frac{e^{X_i^T \vec{\beta}}}{1 + e^{X_i^T \vec{\beta}}}\right)^{y_i} \left(\frac{1}{1 + e^{X_i^T \vec{\beta}}}\right)^{1-y_i}$$

$$= \prod_{i=1}^{n} \frac{e^{y_i X_i^T \vec{\beta}}}{1 + e^{X_i^T \vec{\beta}}}.$$

Thus, the log-likelihood function is

$$\mathit{l}(\vec{\beta}) = \sum_{i=1}^{n} y_i X_i^T \vec{\beta} - \sum_{i=1}^{n} \ln[1 + e^{X_i^T \vec{\beta}}].$$

The MLEs of $\beta_0, \beta_1, \beta_2,$ and β_3 must be found numerically and can be found using the R command glm command; the command glm is used to fit a generalized linear model and produces the MLEs of the unknown parameters along with their asymptotic standard errors.

Now, provided the sample size is sufficiently large, a Wald test can be used to test hypotheses about the unknown parameters of the model (i.e. $\vec{\beta}$). In particular, a test of $H_0 : \beta_i = 0$ versus $H_A : \beta_i \neq 0$ for $i = 1, 2,$ or 3 can be performed to investigate the importance of the ith explanatory variable since the variable X_i drops out of the model when $\beta_i = 0$.

The Wald test for testing $H_0 : \beta_i = 0$ versus $H_A : \beta_i \neq 0$ is based on the test statistic $Z = \dfrac{\widehat{\beta}_i}{\sqrt{\text{asyvar}(\widehat{\beta}_i)}}$, which is approximately normally distributed for large samples.

6.6.3 Analysis of the Fitted Survival Model

The statistical software R was used to fit the logistic regression model to the 600 sampled observations using the commands

```
> out_titanic=glm(survived age+sex+age*sex,
+ family="binomial",data=titanic_s)
> summary(out_titanic).
```

The resulting R output includes summary information on the fitted model including the maximum likelihood estimate of $\vec{\beta}$, the asymptotic standard errors of the individual MLEs, the value of the Wald test statistic, and the p-value associated with the Wald test of $H_0 : \beta_i = 0$ versus $H_A : \beta_i \neq 0$. The R output for the Wald tests for each parameter in the model is shown in Table 6.8.

In the model $\text{logit}(\theta_i) = X_i^T \vec{\beta} = \beta_0 + \beta_1 A_i + \beta_2 G_i + \beta_3 A_i \cdot G_i$, β_3 is first tested to determine whether or not there is an interaction effect between the explanatory variables age and gender. The p-value for testing $H_0 : \beta_3 = 0$ versus $H_A : \beta_3 \neq 0$ is $p = 0.0002$, and therefore, there is strong evidence that β_3 is significantly different from 0. Hence, the survival probability depends on both age and gender through an interaction effect. A large sample 95% confidence interval for β_3 based on the MLE is $\hat{\beta}_3 \pm 1.96\text{se}(\hat{\beta}_3)$, which results in the interval -0.0842 to -0.0260.

The effects of the explanatory variables age and gender are not tested because the interaction term is significant meaning both age and gender are important in the model. Thus, the fitted model for the probability of surviving the Titanic sinking is

$$\widehat{P}(Y = 1|A, G) = \frac{e^{0.543+0.021A-0.893G-0.055AG}}{1 + e^{0.543+0.021A-0.893G-0.055AG}}.$$

Specifically, for males ($G = 1$), the estimated probability of survival for a given age is

$$\widehat{P}(Y = 1|A, G = 1) = \frac{e^{-0.350-0.034A}}{1 + e^{-0.350-0.034A}},$$

Table 6.8 Partial R output for the logistic regression model fit to the Titanic data.

```
Coefficients:
              Estimate Std. Error z value Pr(>|z|)
(Intercept)   0.54290   0.32703    1.660  0.096899
age           0.02097   0.01106    1.896  0.057999
sexmale      -0.89324   0.44122   -2.024  0.042920
age:sexmale  -0.05513   0.01485   -3.712  0.000206
```

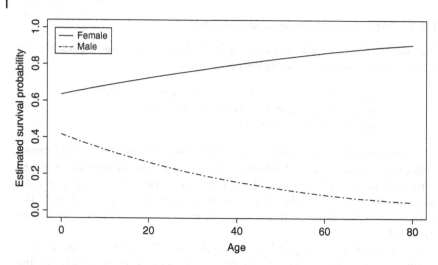

Figure 6.3 Estimated survival probabilities for the Titanic passengers.

and for females ($G = 0$), the estimated probability of survival for a given age is

$$\widehat{P}(Y = 1|A, G = 0) = \frac{e^{0.543+0.021A}}{1 + e^{0.543+0.021A}}.$$

For example, a 30-year-old male passenger has an estimated survival probability of

$$\widehat{P}(Y = 1|A = 30, G = 1) = \frac{e^{-0.350-0.034\cdot30}}{1 + e^{-0.350-0.034\cdot30}} = 0.2026.$$

On the other hand, a 30-year-old female passenger has an estimated survival probability of

$$\widehat{P}(Y = 1|A = 30, G) = \frac{e^{0.543+0.021\cdot30}}{1 + e^{0.543+0.021\cdot30}} = 0.7637.$$

The estimated survival probabilities for males and females are shown in Figure 6.3. Note that the estimated survival probabilities for females are larger than the estimated survival probabilities for males over the entire range of ages. Also, the survival probabilities slightly increase over age for females, whereas the survival probabilities decrease over age for males.

Problems

6.6.1 Using the information in Table 6.8,
a) test $H_0 : \beta_1 = 0$ versus $H_A : \beta_1 = 0$ using $\alpha = 0.05$.
b) test $H_0 : \beta_2 = 0$ versus $H_A : \beta_2 = 0$ using $\alpha = 0.05$.
c) estimate the probability of survival for a 20-year-old male.

   d) estimate the probability of survival for a 20-year-old female.
   e) determine the age at which the estimated survival probability for a
      male is 0.10.

6.6.2 Using the information in Table 6.8,
   a) compute a large sample 95% confidence interval for β_1.
   b) compute a large sample 95% confidence interval for β_4.

6.6.3 For $Y_1, \ldots, Y_n \overset{\text{ind}}{\sim} \text{Bin}(1, \theta_i)$ and the model $\text{logit}(\theta_i) = \beta x_i$, determine
   a) likelihood function $L(\beta)$.
   b) the log-likelihood function $l(\beta)$.
   c) the Score function $\dot{\ell}(\beta)$.

6.6.4 Let $Y_1, \ldots, Y_n \overset{\text{ind}}{\sim} \text{Bin}(1, \theta_i)$ be a random sample, and suppose that
   $\text{logit}(\theta_i) = \beta_0 + \beta_1 x_i$ for known constants $x_1, \ldots, x_n$. Determine
   a) the likelihood function $L(\beta_0, \beta_1)$.
   b) the log-likelihood function $l(\beta_0, \beta_1)$.
   c) the Score function $\dot{\ell}(\beta_0, \beta_1)$.

7

Generalized Linear Models

This final chapter deals with likelihood-based inferences for a wide class of commonly used statistical models known as *generalized linear models*. Generalized linear models were introduced by Nelder and Wedderburn [23] and are parametric models based on an underlying exponential family distribution. Furthermore, generalized linear models cover a wide range of commonly used modeling approaches including classical linear models, logistic regression, probit regression, and Poisson regression.

Generalized linear models require a prespecified exponential family distribution; a random component Y called the response variable; a systematic component $\vec{X}$, which is a vector of explanatory variables; and a model for some function of the parameters of the distribution of Y. In particular, in a generalized linear model, the mean of the response variable Y is modeled as a function of $\vec{X}$ and an unknown parameter $\vec{\beta}$ by $E(Y|\vec{X}) = g(\mu(\vec{X}))$, where $g(\cdot)$ is a one-to-one real-valued function.

The earliest statistical models fit to data were the classical linear models. A classical linear model is a model for a continuous variable Y, where $Y \sim N(X^T\vec{\beta}, \sigma^2)$ for a set of explanatory variables $\vec{X} = (1, X_1, \ldots, X_p)^T$. In a linear model of this nature, the mean of the response variable is modeled as a linear function of the parameters $\beta_0, \beta_1, \ldots, \beta_p$, namely, $\mu = \vec{X}^T\vec{\beta}$, or equivalently, $\mu = \beta_0 + \beta_1 X_1 + \cdots + \beta_p X_p$.

Nelder and Wedderburn generalized the setting for the classical linear model to response variables that are not normally distributed but are in the exponential family of distributions. Furthermore, in this generalized setting, the mean is not necessarily modeled as a linear function of $\vec{\beta}$, but rather, a specific function of the mean is modeled as a linear function of $\vec{\beta}$. Because the classical linear model falls under Nelder and Wedderburn's generalized setting, generalized linear models are discussed first in this chapter, and the classical linear model is covered as a special type of generalized linear model in a Section 7.4.

Mathematical Statistics: An Introduction to Likelihood Based Inference, First Edition. Richard J. Rossi.
© 2018 John Wiley & Sons, Inc. Published 2018 by John Wiley & Sons, Inc.

7.1 Generalized Linear Models

The form of the exponential family distributions that Nelder and Wedderburn based their generalized linear model on is

$$f(y; \theta, \phi) = e^{\frac{y\theta - b(\theta)}{a(\phi)} + c(y, \phi)}, \tag{7.1}$$

where θ and ϕ are scalar parameters and $a(\cdot), b(\cdot)$, and $c(\cdot)$ are known functions. A random variable with an exponential family pdf of the form in (7.1) is denoted by $Y \sim EF(b(\theta), \phi)$. Note that pdfs of the form (7.1) are exponential family distributions when ϕ is known, but in general, when ϕ is unknown, a pdf of the form (7.1) may not be an exponential family distribution.

Example 7.1   Let $Y \sim N(\mu, \sigma^2)$. Then, $f(y; \mu, \sigma^2) = \frac{1}{\sqrt{2\pi\sigma^2}} e^{-\frac{(y-\mu)^2}{2\sigma^2}}$, which can be written as

$$f(y; \mu, \sigma^2) = e^{-\frac{y^2}{2\sigma^2} + \frac{y\mu}{\sigma^2} - \frac{\mu^2}{2\sigma^2} - \ln[\sqrt{2\pi\sigma^2}]} = e^{\frac{y\mu - \frac{\mu^2}{2}}{\sigma^2} + \left(-\frac{y^2}{2\sigma^2} - \ln[\sqrt{2\pi\sigma^2}]\right)}.$$

Letting $\theta = \mu$ and $\phi = \sigma^2$, $Y \sim EF(\frac{\mu^2}{2}, \sigma^2)$ with $b(\mu) = \frac{\mu^2}{2}, a(\sigma^2) = \sigma^2$, and $c(y, \sigma^2) = \left(-\frac{y^2}{2\sigma^2} - \ln\left[\sqrt{2\pi\sigma^2}\right]\right)$, it follows that $Y \sim EF(\frac{\mu^2}{2}, \sigma^2)$.

Example 7.2   Let $Y \sim \text{Bin}(n, p)$ with n known. Then, $f(y; p) = \binom{n}{y} p^y (1 - p)^{n-y}$, which can be written as

$$f(y; p) = e^{\ln\binom{n}{y} + y\ln(p) + (n-y)\ln(1-p)} = e^{y\ln\left[\frac{p}{1-p}\right] + n\ln(1-p) + \ln\binom{n}{y}}.$$

Letting $\theta = \ln\left[\frac{p}{1-p}\right], \phi = 1, b(\theta) = n\ln(1 + e^\theta), a(\phi) = 1$, and $c(y, \phi) = \ln\binom{n}{y}$, it follows that $Y \sim EF(n\ln(1 + e^\theta), 1)$.

Theorem 7.1 shows that when $Y \sim EF(b(\theta), \phi)$, the mean and variance are completely determined by the functions $b(\theta)$ and $a(\phi)$.

Theorem 7.1   *If* $Y \sim EF(b(\theta), \phi)$, *then*

i) $E(Y) = b'(\theta)$.
ii) $\text{Var}(Y) = a(\phi)b''(\theta)$.

Proof. Let $Y \sim EF(b(\theta), \phi)$.

i) By Lemma 4.1, $E\left[\frac{\partial}{\partial\theta}\ln(f(y; \theta, \phi))\right] = 0$. Since $Y \sim EF(b(\theta), \phi)$,

$$\frac{\partial}{\partial\theta}\ln[f(y; \theta, \phi)] = \frac{y - b'(\theta)}{a(\phi)}.$$

Hence, $E\left[\frac{Y - b'(\theta)}{a(\phi)}\right] = 0$, and therefore, $E(Y) = b'(\theta)$.

ii) The proof of part (ii) is left as an exercise. ∎

When $Y \sim EF(b(\theta), \phi)$, $V(\mu) = b''(\theta)$ is called the variance function, ϕ is called the *dispersion parameter*, and $\text{Var}(Y) = b''(\theta)a(\phi)$. For generalized linear models based on the binomial and the Poisson distributions, the dispersion parameter ϕ and $a(\phi)$ are constant; however, in the classical linear model based on the normal distribution, the dispersion parameter $\phi = a(\phi) = \sigma^2$ is generally unknown.

Example 7.3 Let $Y \sim B(n, p)$. By Example 7.2, $Y \sim EF(n \ln(1 + e^{\theta}), 1)$. Hence, by Theorem 7.1,

$$E(Y) = b'(\theta) = \frac{ne^{\theta}}{1 + e^{\theta}} = np$$

and

$$\text{Var}(Y) = a(\phi)b''(\theta) = 1 \cdot \left[\frac{n\, e^{\theta}}{e^{\theta} + 1} - \frac{n\, e^{2\theta}}{(e^{\theta} + 1)^2} \right] = np - np^2 = np(1 - p).$$

Since the dispersion parameter is $\phi = 1$, the variance function is

$$V(\mu) = \text{Var}(Y) = np(1 - p) = \mu \left(1 - \frac{\mu}{n} \right).$$

Example 7.4 Let $Y \sim \text{Gamma}(\alpha, \beta)$. Then, letting $\theta = -\frac{1}{\alpha\beta} = -\frac{1}{\mu}$ and $\phi = \frac{1}{\alpha}$, it follows that $Y \sim EF\left(\ln\left[-\frac{1}{\theta} \right], \frac{1}{\alpha} \right)$ (see Problem 7.1.10). Thus, by Theorem 7.1,

$$E(Y) = b'(\theta) = \frac{\frac{1}{\theta^2}}{-\frac{1}{\theta}} = -\frac{1}{\theta} = -\frac{1}{-\frac{1}{\alpha\beta}} = \alpha\beta$$

and

$$\text{Var}(Y) = a(\phi)b''(\theta) = \frac{1}{\alpha} \cdot \frac{1}{\theta^2} = \frac{1}{\alpha} \cdot \frac{1}{\frac{1}{\theta^2}} = \frac{1}{\alpha}\alpha^2\beta^2 = \alpha\beta^2.$$

Thus, for $Y \sim \text{Gamma}(\alpha, \beta)$, the dispersion parameter is $\frac{1}{\alpha}$, and the variance function is $V(\mu) = \alpha^2\beta^2 = \mu^2$.

When $Y_1, \ldots, Y_n$ are independent random variables, $\mathbf{X}_i$ is a $(p + 1)$-dimensional vector of explanatory variables associated with Y_i, and $Y_i \sim EF(b(\theta_i), \phi)$, a generalized linear model, consists of four components, namely, a random component, which is the response variable; a systematic component, which is the vector of explanatory variables; a one-to-one parametric function $g(\mu)$, which links the mean to the explanatory variables; and a linear function $\mathbf{X}^T \vec{\beta}$, where $g(\mu_i) = \mathbf{X}_i^T \vec{\beta}$ and $\vec{\beta} = (\beta_0, \beta_1, \ldots, \beta_p)^T$ is a vector of unknown parameters. The function $g(\cdot)$ is called the *link function*, and $\mathbf{X}_i^T \vec{\beta}$ is called the *linear predictor*.

Also, the values of ϕ and the functions $b(\cdot)$ and $a(\phi)$ must be the same for each random variable Y_i in a generalized linear model.

One of the most important aspects of a generalized linear model is the link function, and since the link function is a one-to-one function, a model for the mean of the response variable is $\mu_i = g^{-1}(\mathbf{X}_i^T \vec{\beta})$. The choice of the link function is purely a modeling choice, and in particular, significant lack of fit within a model suggests that a different link function should be used. Choosing a link simply amounts to choosing a transformation of the mean, and thus, two different link functions correspond to different models for the mean. In most cases, it makes sense to match the domain of the link function to the range of the means. Finally, when $g(\mu_i) = \theta_i = \mathbf{X}_i^T \vec{\beta}$, the link function is called the *canonical link function*. A canonical link function always exists when $Y_i \sim EF(b(\theta_i), \phi)$.

Example 7.5 Let $Y_1, \ldots, Y_n$ be a sample of independent $N(\mathbf{X}_i^T \vec{\beta}, \sigma^2)$ random variables with vector of explanatory variables $\mathbf{X}_i = (1, X_{1i}, \ldots, X_{pi})$. In Example 7.1, it was shown that $Y_i \sim EF(\frac{\mu_i^2}{2}, \sigma^2)$ with $\theta_i = \mu_i$. Therefore, the canonical link function is the identity link function $\theta_i = \mu_i$, and the linear predictor is $\mu_i = \mathbf{X}_i^T \vec{\beta}$.

The generalized linear model for $Y_i \sim N(\mu_i, \sigma^2)$ is

$$\mu_i = \beta_0 + \beta_1 X_{1i} + \cdots + \beta_p X_{pi},$$

which can also be written as a model for Y_i. In particular, the model can be stated as $Y_i | \mathbf{X}_i \sim N(\mathbf{X}_i^T \vec{\beta}, \sigma^2)$, or equivalently, $Y = \beta_0 + \beta_1 X_1 + \cdots + \beta_p X_p + \epsilon$, where $\epsilon \sim N(0, \sigma^2)$.

Example 7.6 Let $Y_1, \ldots, Y_n$ be a sample of independent $B(n, p_i)$ random variables. Example 7.2 shows that $Y_i \sim EF(n \ln[1 + e^{\theta_i}], 1)$ with $\theta_i = \ln\left[\frac{p}{1-p}\right]$. Thus, the canonical link function is $g(p_i) = \ln\left[\frac{p_i}{1-p_i}\right]$, which is called the *logit function*. A generalized linear model for $Y_i \sim \text{Bin}(n, p_i)$ with the logit link function is $\text{logit}(p_i) = \mathbf{X}_i^T \vec{\beta}$ and is called a *logistic regression model*. Note that the logistic regression model for $\text{logit}(p_i)$ is equivalent to a modeling p_i as $p_i = \frac{e^{\mathbf{X}_i^T \vec{\beta}}}{1 + e^{\mathbf{X}_i^T \vec{\beta}}}$.

Problems

7.1.1 Let $Y \sim \text{Pois}(\lambda)$.
   a) Show that the pdf of Y can be written in the form of (7.1).
   b) Determine $E(Y)$ using Theorem 7.1.
   c) Determine $\text{Var}(Y)$ using Theorem 7.1.
   d) Identify the canonical link function.

7.1.2 Let $Y \sim \text{Geo}(p)$.
   a) Show that the pdf of Y can be written in the form of (7.1).
   b) Determine $E(Y)$ using Theorem 7.1.
   c) Determine $\text{Var}(Y)$ using Theorem 7.1.
   d) Identify the canonical link function.

7.1.3 Let Y be a random variable with pdf $f(y; \eta) = \eta e^{-\eta y}$ for $y > 0$ and $\Theta = \mathbb{R}^+$.
   a) Show that the pdf of Y can be written in the form of (7.1).
   b) Determine $E(Y)$ using Theorem 7.1.
   c) Determine $\text{Var}(Y)$ using Theorem 7.1.
   d) Identify the canonical link function.

7.1.4 Let $Y \sim \text{NegBin}(r, p)$.
   a) Show that the pdf of Y can be written in the form of (7.1).
   b) Determine $E(Y)$ using Theorem 7.1.
   c) Determine $\text{Var}(Y)$ using Theorem 7.1.
   d) Identify the canonical link function.

7.1.5 Let $Y \sim \text{Exp}(\beta)$.
   a) Show that the pdf of Y can be written in the form of (7.1).
   b) Determine $E(Y)$ using Theorem 7.1.
   c) Determine $\text{Var}(Y)$ using Theorem 7.1.
   d) Identify the canonical link function.

7.1.6 Let $Y \sim \text{Gamma}(4, \beta)$.
   a) Show that the pdf of Y can be written in the form of (7.1).
   b) Determine $E(Y)$ using Theorem 7.1.
   c) Determine $\text{Var}(Y)$ using Theorem 7.1.
   d) Identify the canonical link function.

7.1.7 Let $Y_1, \ldots, Y_n$ be independent random variables with $Y_i \sim \text{Bin}(1, p_i)$.
   a) Show that $a(\phi) = 1$.
   b) Determine the variance function.

7.1.8 Let $Y_1, \ldots, Y_n$ be independent random variables with $Y_i \sim \text{Bin}(n_i, p_i)$. Show that $a(\phi) = 1$.

7.1.9 Let $Y_1, \ldots, Y_n$ be independent random variables with $Y_i \sim N(\mu_i, \sigma^2)$. Show that $a(\phi) = \sigma^2$.

7.1.10 Let $Y \sim \text{Gamma}(\alpha, \beta)$. Show that the pdf of Y can be written in the form of (7.1) with $\theta = -\frac{1}{\alpha\beta}$ and $\phi = \frac{1}{\alpha}$.

7.2 Fitting a Generalized Linear Model

Because the underlying probability model in a generalized linear model is an exponential family member, maximum likelihood estimation is generally used for estimating the parameters of the model. Moreover, when the pdf is in the exponential family of distributions, the regularity conditions required for using Fisher's Information, the asymptotic standard errors, and large sample hypothesis tests are satisfied.

7.2.1 Estimating $\vec{\beta}$

When $Y_1, \ldots, Y_n$ are independent random variables with $Y_i \sim EF(b(\theta_i), \phi)$, the likelihood function is

$$L(\theta_1, \ldots, \theta_n, \phi) = \prod_{i=1}^{n} e^{\frac{y_i\theta_i - b(\theta_i)}{a(\phi)} + c(y_i, \phi)} = e^{\sum_{i=1}^{n} \frac{y_i\theta_i - b(\theta_i)}{a(\phi)}} e^{\sum_{i=1}^{n} c(y_i, \phi)}.$$

When the dispersion parameter ϕ is known, and the canonical link $\theta_i = g(\mu_i)$ and canonical parameter $\eta_i = \theta_i$ are used, the likelihood function is

$$L(\vec{\beta}, \phi) = e^{\sum_{i=1}^{n} \frac{y_i\theta_i - b(\theta_i)}{a(\phi)} + c(y_i, \phi)} = e^{\sum_{i=1}^{n} \frac{y_i \mathbf{X}_i^{\mathrm{T}} \vec{\beta} - b(\theta_i)}{a(\phi)} + c(y_i, \phi)}$$

$$= e^{\frac{\sum_{i=1}^{n} y_i \mathbf{X}_i^{\mathrm{T}} \vec{\beta}}{a(\phi)} - \frac{b(\mathbf{X}_i^{\mathrm{T}} \vec{\beta})}{a(\phi)} + c(y_i, \phi)} = g\left(\sum_{i=1}^{n} y_i \mathbf{X}_i^{\mathrm{T}}, \vec{\beta}\right) h(\vec{y}),$$

where $\mathbf{X}_i = (1, x_{i1}, \ldots, x_{ip})^{\mathrm{T}}$. Thus, $\sum_{i=1}^{n} y_i \mathbf{X}_i^{\mathrm{T}}$ is a sufficient statistic for $\vec{\beta}$ in a generalized linear model with canonical link function when ϕ is known.

Example 7.7 Let $Y_1, \ldots, Y_n$ be independent random variables with $Y_i \sim \mathrm{Pois}(\lambda_i)$. Then, $Y_i \sim EF(e^{\lambda}, 1)$ and the canonical link is $g(\lambda_i) = \ln(\lambda_i)$ (see Exercise 7.2.1). Since $\phi = 1$, the sufficient statistic for $\vec{\beta}$ is $\sum_{i=1}^{n} y_i \mathbf{X}_i^{\mathrm{T}}$.

The log-likelihood function for a generalized linear model with $Y_i \sim EF(b(\theta_i), \phi)$ is

$$\ell(\theta_1, \ldots, \theta_n, \phi) = \sum_{i=1}^{n} \frac{y_i\theta_i - b(\theta_i)}{a(\phi)} + \sum_{i=1}^{n} c(y_i, \phi).$$

Moreover, in a generalized linear model with link function $g(\cdot)$, $\eta_i = g(\mu_i) = \mathbf{X}_i^{\mathrm{T}} \vec{\beta}$, and $\theta_i = h(\mu_i)$, the log-likelihood function in terms of $\vec{\beta}$ is

$$\ell(\vec{\beta}, \phi) = \sum_{i=1}^{n} \frac{y_i h(\mu_i) - b(\mu_i)}{a(\phi)} + c(y_i, \phi) = \sum_{i=1}^{n} l_i(\vec{\beta}, \phi),$$

where $l_i(\vec{\beta}, \phi) = \frac{y_i h(\mu_i) - b(\mu_i)}{a(\phi)} + c(y_i, \phi)$ for $1 = 1, \ldots, n$.

For a generalized linear model with the canonical link function $g(\mu_i) = \theta(\mu_i)$, or equivalently $\theta_i = \mathbf{X}_i^{\mathrm{T}} \vec{\beta}$, the log-likelihood function is

$$\ell(\vec{\beta}, \phi) = \sum_{i=1}^{n} \frac{y_i \mathbf{X}_i^{\mathrm{T}} \vec{\beta} - b(\mathbf{X}_i^{\mathrm{T}} \vec{\beta})}{a(\phi)} + \sum_{i=1}^{n} c(y_i, \phi).$$

The maximum likelihood estimate of $\vec{\beta}$ is found by differentiating $\ell(\vec{\beta}, \phi)$ with respect to β_j for $j = 0, \ldots, p$ and then setting each partial derivative equal to zero and simultaneously solving this system of equations. Equivalently, the MLE of $\vec{\beta}$ is found by setting the Score vector function equal to the zero vector and solving for $\vec{\beta}$.

The Score function is $l(\vec{\beta}) = \left(\frac{\partial \ell(\vec{\beta}, \phi)}{\partial \beta_1}, \ldots, \frac{\partial \ell(\vec{\beta}, \phi)}{\partial \beta_p} \right)^{\mathrm{T}}$, and since $\ell(\vec{\beta}, \phi) = \sum_{i=1}^{n} l_i(\vec{\beta}, \phi)$, it follows that $l(\vec{\beta}, \phi) = \sum_{i=1}^{n} l_i(\vec{\beta}, \phi)$. Moreover, since $\frac{\partial l_i}{\partial \beta_j} = \frac{\partial l_i}{\partial \theta_i} \frac{\partial \theta_i}{\partial \mu_i} \frac{\partial \mu_i}{\partial \eta_i} \frac{\partial \eta_i}{\partial \beta_j}$, where

$$\frac{\partial l_i}{\partial \theta_i} = \frac{y_i - b'(\theta_i)}{a(\phi)} = \frac{y_i - \mu_i}{a(\phi)}$$

$$\frac{\partial \theta_i}{\partial \mu_i} = \underbrace{b''(\theta_i)}_{\mu_i = b'(\theta_i)} = \underbrace{\frac{\mathrm{Var}(Y_i)}{a(\phi)}}_{\mathrm{Var}(Y_i) = a(\phi)b''(\theta_i)}$$

$$\frac{\partial \mu_i}{\partial \theta_i} = \frac{1}{\frac{\partial \theta_i}{\partial \mu_i}} = \frac{a(\phi)}{\mathrm{Var}(Y_i)}$$

$$\frac{\partial \eta_i}{\partial \beta_j} = x_{ij},$$

it follows that $\frac{\partial l_i}{\partial \beta_j} = \frac{y_i - \mu_i}{a(\phi)} \frac{a(\phi)}{\mathrm{Var}(Y_i)} \frac{\partial \mu_i}{\partial \eta_i} x_{ij}$ and $\frac{\partial}{\partial \beta_j} \ell(\vec{\beta}, \phi) = \sum_{i=1}^{n} \frac{(y_i - \mu_i)x_{ij}}{\mathrm{Var}(Y_i)} \frac{\partial \mu_i}{\partial \eta_i}$. Note that $\frac{\partial \mu_i}{\partial \eta_i}$ depends on the link function and, therefore, is model specific.

Example 7.8 Let $Y_1, \ldots, Y_n$ be independent random variables with $Y_i \sim \mathrm{Pois}(\lambda_i)$, and $\theta_i = \ln(\lambda_i)$. Then, $Y_i \sim EF(\ln(\lambda_i), 1)$, and when the canonical link $g(\mu_i) = \ln(\mu_i)$ is used, $\mu_i = \mathrm{Var}(Y_i) = e^{\mathbf{X}_i \vec{\beta}}$. Thus,

$$\frac{\partial}{\partial \beta_j} \ell(\vec{\beta}, \phi) = \sum_{i=1}^{n} \frac{(y_i - \mu_i)x_{ij}}{\mu_i} \frac{\partial \mu_i}{\partial \eta_i} = \sum_{i=1}^{n} \frac{(y_i - e^{\mathbf{X}_i^{\mathrm{T}} \vec{\beta}})x_{ij}}{e^{\mathbf{X}_i^{\mathrm{T}} \vec{\beta}}} e^{\mathbf{X}_i^{\mathrm{T}} \vec{\beta}}$$

$$= \sum_{i=1}^{n} (y_i - e^{\mathbf{X}_i^{\mathrm{T}} \vec{\beta}}) x_{ij}.$$

In general, the MLE of $\vec{\beta}$ in a generalized linear model must be found numerically, and for sufficiently large samples, the asymptotic standard error of the

MLE of β_i can be used in confidence intervals and hypothesis tests based on the MLE. Software packages such as R have commands for estimating the parameters in a generalized linear model for a specific underlying distribution, link function, and linear predictor. For example, in R, a generalized linear model can be fit using the command glm(formula, family, data), where formula is the linear predictor, family is the probability model with link function, and data is the data set being fit. The R command glm produces the MLE of $\vec{\beta}$ along with the asymptotic standard error of $\widehat{\beta}_i$ and a multitude of other useful information on the model's fit. Examples where R is used to fit a generalized linear model are given in Sections 7.4–7.6.

Charnes et al. [24] showed that the numerical solution to simultaneous system of equations $\dot{l}(\vec{\beta}, \phi) = \vec{0}$ can be solved using the method of *iteratively weighted least squares*. For more information on the method of iteratively weighted least squares, see *Generalized Linear Models* by Nelder and McCullagh [25].

7.2.2 Model Deviance

Once a generalized linear model has been fit, and before it is used for making inferences, the fit of the model to the observed data must be assessed. The first step in assessing the fit of a generalized linear model is to compare the fitted model with the *saturated model*.

Definition 7.1 A model is saturated when it contains as many unknown parameters as there are data points.

Under the saturated model, the fitted values are equal to the observed data values. In particular, for a generalized linear model, the fitted values under the saturated model are $\widehat{\mu}_i = y_i$. The *deviance*, sometimes referred to as the *scaled deviance*, associated with a model is the logarithm of the likelihood ratio evaluated at the MLE $\widehat{\beta}$ and the estimate of $\vec{\beta}$ under the saturated model. That is, when $\widehat{\beta}$ is the MLE of $\vec{\beta}$, and $\widetilde{\beta}$ is the estimate of $\vec{\beta}$ under the saturated model, the deviance, denoted by $\mathcal{D}$, is

$$\mathcal{D} = -2[\ell(\widehat{\beta}, \phi) - \ell(\widetilde{\beta}, \phi)] = 2[\ell(\widetilde{\beta}, \phi) - \ell(\widehat{\beta}, \phi)].$$

Also, since each log-likelihood is sum over the observed values, the deviance can also be written as $\mathcal{D} = \sum_{i=1}^{n} \mathcal{D}_i$, where $\mathcal{D}_i = 2[\ell_i(\widetilde{\beta}, \phi) - \ell_i(\widehat{\beta}, \phi)]$.

Example 7.9 Let $Y_1, \ldots, Y_n$ be independent Poisson random variables with $Y_i \sim \text{Pois}(\lambda_i)$. Then, $Y_i \sim EF(\ln(\mu_i), 1)$. The likelihood function is $L(\vec{\lambda}) = \prod_{i=1}^{n} \frac{e^{-\lambda_i} \lambda_i^{y_i}}{y_i!}$, which as a function of $\vec{\mu} = (\mu_1, \ldots, \mu_n)$ is $L(\vec{\mu}) = \prod_{i=1}^{n} \frac{e^{-\mu_i} \mu_i^{y_i}}{y_i!}$.

Thus, the log-likelihood function is $\ell(\vec{\mu}) = \sum_{i=1}^{n} y_i \ln(\mu_i) - \sum_{i=1}^{n} \mu_i$.

Hence, the deviance is

$$\mathcal{D} = -2 \sum_{i=1}^{n} [(y_i \ln(\widehat{\mu}_i) - \widehat{\mu}_i) - (y_i \ln(y_i) - y_i)]$$

$$= 2 \sum_{i=1}^{n} \left[y_i \ln \left(\frac{y_i}{\widehat{\mu}_i} \right) + \widehat{\mu}_i - y_i \right].$$

The deviance is used for assessing a models fit, comparing nested models, and estimating an unknown dispersion parameter ϕ. A model M_2 is nested in a larger model M_1 when all of the explanatory variables in model M_2 are in model M_1. That is, when the vector of explanatory variables $\mathbf{X}$ is partitioned into mutually disjoint subvectors, say $\mathbf{X}_1$ and $\mathbf{X}_2$, and the *full model* is denoted by $g_F(\mu) = \beta_0 + \mathbf{X}_1^{\mathrm{T}} \vec{\beta}_1 + \mathbf{X}_2^{\mathrm{T}} \vec{\beta}_2$, the model $g_R(\mu_i) = \beta_0 + \mathbf{X}_1^{\mathrm{T}} \vec{\beta}_1$ is nested in the full model and is called a *reduced model*.

Example 7.10 For the four-variable explanatory vector $\mathbf{X} = (1, X_1, X_2, X_3, X_4)^{\mathrm{T}}$, the full model is $g(\mu) = \mathbf{X}^{\mathrm{T}} \vec{\beta} = \beta_0 + \beta_1 X_1 + \beta_2 X_2 + \beta_3 X_3 + \beta_4 X_4$. Examples of reduced models nested in the full model include

$$g(\mu) = \beta_0 + \beta_1 X_1 + \beta_2 X_2 + \beta_3 X_3$$
$$g(\mu) = \beta_0 + \beta_2 X_2 + \beta_3 X_3$$
$$g(\mu) = \beta_0 + \beta_1 X_1 + \beta_3 X_3$$
$$g(\mu) = \beta_0 + \beta_1 X_1.$$

A generalized linear model containing only the intercept β_0 is called the *null model* and corresponds to $Y_1, \ldots, Y_n$ being iid random variables. Thus, in the null model, $Y_1, \ldots, Y_n$ all have the same mean.

Let the deviance for the full model be $\mathcal{D}_F$, and let the deviance of a reduced model be $\mathcal{D}_R$. Then, $\mathcal{D}_F = 2[\ell_S - \ell(\widehat{\beta}_1, \widehat{\beta}_2, \phi)]$ and $\mathcal{D}_R = 2[\ell_S - \ell(\widehat{\beta}_1', \phi)]$, where ℓ_S is the log-likelihood for the saturated model, which is constant with respect to $\widehat{\beta}$. The difference of the deviances for the full and reduced model is

$$\mathcal{D}_F - \mathcal{D}_R = 2[\ell_S - \ell(\widehat{\beta}_1, \widehat{\beta}_2, \phi)] - 2[\ell_S - \ell(\widehat{\beta}_1', \phi)]$$
$$= 2[\ell(\widehat{\beta}_1', \phi) - \ell(\widehat{\beta}_1, \widehat{\beta}_2, \phi)],$$

where $\widehat{\beta}_1'$ is the MLE of β_1 under the reduced model and $\widehat{\beta}_1$ is the MLE of β_1 under the full model. A hypothesis testing approach for comparing the fit of a reduced model with the full model based on the difference of the deviances is discussed in Section 7.3.

The deviance is also used to measure how well the model fits a particular observation through the use of the deviance residuals. The ith *deviance residual* is defined to be

$$e_{\mathcal{D}_i} = \mathrm{sign}(y_i - \widehat{\mu}_i) \sqrt{\mathcal{D}_i},$$

where $\text{sign}(y_i - \widehat{\mu}_i) = \begin{cases} 1 & \text{when } y_i > \widehat{\mu}_i \\ -1 & \text{when } y_i < \widehat{\mu}_i \end{cases}$ and $\mathcal{D}_i$ is the contribution of the ith observation to the deviance. An observation with a large deviance residual suggests that the model does not fit this observed value very well, and a model with several large deviance residuals would suggest that the model does not explain the response variable very well.

Problems

7.2.1 Let $Y_1, \ldots, Y_n$ be independent random variables with $f(y_i; \eta_i) = \eta_i e^{-\eta_i y_i}$ for $y_i \in \mathbb{R}^+$.
  a) Determine the likelihood function as a function of $\mu_i = \frac{1}{\eta_i}$.
  b) Determine the log-likelihood function as a function of μ_i.
  c) Determine the Score function as a function of μ_i.
  d) Determine the deviance as a function of μ_i.

7.2.2 Let $Y_1, \ldots, Y_n$ be independent random variables with $Y_i \sim \text{Exp}(\beta_i)$.
  a) Determine the likelihood function as a function of $\mu_i = \beta_i$.
  b) Determine the log-likelihood function as a function of μ_i.
  c) Determine the Score function as a function of μ_i.
  d) Determine the deviance as a function of μ_i.

7.2.3 Let $Y_1, \ldots, Y_n$ be independent random variables with $Y_i \sim \text{Gamma}(4, \beta_i)$.
  a) Determine the likelihood function as a function of $\mu_i = 4\beta_i$.
  b) Determine the log-likelihood function as a function of μ_i.
  c) Determine the Score function as a function of μ_i.
  d) Determine the deviance as a function of μ_i.

7.2.4 Let $Y_1, \ldots, Y_n$ be independent random variables with $Y_i \sim N(\mu_i, 1)$.
  a) Determine the likelihood function as a function of μ_i.
  b) Determine the log-likelihood function as a function of μ_i.
  c) Determine the Score function as a function of μ_i.
  d) Determine the deviance as a function of μ_i.

7.2.5 Let $Y_1, \ldots, Y_n$ be independent random variables with $Y_i \sim N(\mu_i, 1)$ with $\mu_i = \beta x_i$ for known constants $x_1, \ldots, x_n$.
  a) Determine the likelihood function as a function of β.
  b) Determine the log-likelihood function as a function of β.
  c) Determine the MLE of β.
  d) Determine the deviance for the generalized linear model with $\theta_i = \mu_i$ and link function $g(\mu_i) = \mu_i = \beta x_i$.

7.2.6 Let $Y_1, \ldots, Y_n$ be independent random variables with $Y_i \sim N(\mu_i, 1)$ with $\mu_i = \beta_0 + \beta_1 x_i$ for known constants $x_1, \ldots, x_n$.
a) Determine the likelihood function as a function of β_0 and β_1.
b) Determine the log-likelihood function as a function of β_0 and β_1.
c) Determine the MLEs of β_0 and β_1.
d) Determine the deviance for the generalized linear model with $\theta_i = \mu_i$ and link function $g(\mu_i) = \mu_i = \beta_0 + \beta_1 x_i$.

7.2.7 Let $Y_1, \ldots, Y_n$ be a sample of independent random variables with $Y_i \sim \text{Pois}(\lambda_i)$. Using the deviance function in Example 7.9, determine the deviance residual for $y_i = 12$ and $\widehat{\mu}_i = 7.8$.

7.2.8 Let $Y_1, \ldots, Y_n$ be a sample of independent random variables with $Y_i \sim \text{Exp}(\beta_i)$. Using the deviance function in Problem 7.2.2, determine the deviance residual for $y_i = 8.8$ and $\widehat{\mu}_i = 11.2$.

7.3 Hypothesis Testing in a Generalized Linear Model

In general, the exact distributions of the MLE of $\vec{\beta}$ and the likelihood ratio statistic are unknown in a generalized linear model; however, because the underlying distributions are exponential family distributions, the regularity conditions for using large sample testing procedures are satisfied. Thus, for sufficiently large samples, tests and confidence intervals concerning the parameters of a generalized linear model and tests comparing different models can be based on large sample procedures. In particular, in the generalized linear models setting, tests and confidence intervals for β_i are based on the Wald test, and tests comparing different models are based on the large sample likelihood ratio test.

7.3.1 Asymptotic Properties

If $\widehat{\vec{\beta}}$ is the MLE of $\vec{\beta}$ and $\mathcal{D}$ is the deviance in a generalized linear model where ϕ is known, then from the asymptotic properties of the MLE and the likelihood ratio statistic, it follows that

- $\widehat{\vec{\beta}}$ is a consistent estimator of $\vec{\beta}$;
- $\widehat{\vec{\beta}} - \vec{\beta} \xrightarrow{d} N_p(0, I_n^{-1}(\vec{\beta}))$, where $I_n(\vec{\beta})$ is Fisher's Information matrix;
- the asymptotic standard error of $\widehat{\beta}_i$ for $i = 0, 1, \ldots, p$ is $I_n^{-1}(\vec{\beta})_{ii}$, where $I_n(\vec{\beta})$ is the $(p+1) \times (p+1)$ Fisher's Information matrix with $\vec{\beta} = (\beta_0, \ldots, \beta_p)$;
- $\mathcal{D} \xrightarrow{d} \chi^2_{n-p}$ when the dispersion parameter is known.

In most cases, asymptotic properties are used for making inferences about the parameters in a generalized linear model because the exact distributions of the

MLEs and the likelihood ratio statistic are generally unknown; an exception is the classical linear model where the exact distributions are known.

7.3.2 Wald Tests and Confidence Intervals

Provided the sample is sufficiently large, a hypothesis test for β_i, with null value β_{i0}, is based on the test statistic

$$Z = \frac{\widehat{\beta}_i - \beta_{i0}}{\text{ase}(\widehat{\beta}_i)},$$

which is approximately distributed as a $N(0, 1)$ random variable. For example, an α level large sample test of $H_0 : \beta_i = \beta_{i0}$ versus $H_A : \beta_i \neq \beta_{i0}$ has critical region

$$\mathcal{C} = \left\{ \vec{y} : \left| \frac{\widehat{\beta}_i - \beta_{i0}}{\text{ase}(\widehat{\beta}_i)} \right| > z_{1-\frac{\alpha}{2}} \right\},$$

where $z_{1-\frac{\alpha}{2}}$ is the $1 - \frac{\alpha}{2}$th quantile of a standard normal random variable. Note that testing $\beta_i = 0$ versus $\beta_i \neq 0$ is an important test because when $\beta_i = 0$, then the explanatory variable X_i drops out of the model. One-sided tests for β_i can also be performed using the Wald test procedure using the critical regions given in Chapter 6.

A $(1 - \alpha) \times 100\%$ large sample Wald confidence interval for β_i is

$$\widehat{\beta}_i \pm z_{1-\frac{\alpha}{2}} \cdot \text{ase}(\widehat{\beta}_i).$$

Example 7.11   The output listed in Table 7.1 resulted from a fitting a generalized linear model to the historical data on the Donner Party. The response variable is Survival and the explanatory variables are Age and Gender. It is assumed that the random variable Survival follows a Bernoulli distribution with $Y_i \sim \text{Bin}(1, p_i)$, where p_i is the probability of surviving. The generalized linear model fit used the canonical link $g(p_i) = \ln\left[\frac{p_i}{1-p_i}\right]$ and linear predictor $\ln\left[\frac{p_i}{1-p_i}\right] = \beta_0 + \beta_1$ Age $+ \beta_2$ Gender $+ \beta_3$ Age*Gender; this model is a *logistic regression model*, which is discussed in detail in Section 7.5.

Based on the information in Table 7.1, the fitted model is

$$\ln\left[\frac{\widehat{p}}{1-p}\right] = 7.246 - 0.194 \text{ Age} - 6.928 \text{ Gender} + 0.162 \text{ Age*Gender}.$$

Testing whether or not the explanatory variables Age and Gender interact involves testing $H_0 : \beta_3 = 0$ versus $H_A : \beta_3 \neq 0$. The observed value of the Wald test statistic for testing $\beta_3 = 0$ is $\frac{0.1616}{0.9426} = 1.714$ with p-value $p = 0.0865$. Thus, it appears that there is insufficient evidence to conclude that $\beta_3 \neq 0$, and therefore, the variables Age and Gender do not appear to interact. Note that

Table 7.1 Summary output for the coefficients in the logistic regression model on the survival of Donner Party members.

```
Coefficients:
                  Estimate Std. Error z value Pr(>|z|)
(Intercept)        7.24638    3.20517   2.261   0.0238
Age               -0.19407    0.08742  -2.220   0.0264
Gender MALE       -6.92805    3.39887  -2.038   0.0415
Age:Gender MALE    0.16160    0.09426   1.714   0.0865
```

the information for the Wald tests of the coefficients in the model is given in Table 7.1 under the columns `z value` and `Pr(>|z|)`.

7.3.3 Likelihood Ratio Tests

When ϕ is known, large sample likelihood ratio tests can be used to test the goodness of fit associated with a fitted model, the utility of the model, and to compare nested models. Furthermore, because the deviance is directly related to the logarithm of the likelihood ratio for comparing the saturated model with the fitted model, model tests can be based on the deviance when the sample size is sufficiently large.

The first test that is generally performed in the analysis of a fitted generalized linear model is the *goodness-of-fit test*. When the model fits the data reasonably well, the log-likelihoods of the saturated and fitted model should be roughly the same, and therefore, the deviance is used to test the goodness of fit of the model.

Since $D \xrightarrow{d} \chi^2_{K-p}$, where K is the degrees of freedom of the saturated model and p is the degrees of freedom of the fitted model, when ϕ is known and n is sufficiently large, the critical region for a the goodness-of-fit test is

$$\mathcal{C}_{\text{GF}} = \{\vec{y} : D > \chi^2_{K-p,1-\alpha}\},$$

where $\chi^2_{K-p,1-\alpha}$ is the $(1 - \alpha)$th percentile of a chi-square random variable with $K - p$ degrees of freedom.

The fitted model is only rejected when there is sufficiently strong evidence that it does not fit the data adequately. Thus, a significant goodness-of-fit test suggests that the model does not fit the data very well and an alternative model needs to be fit. Two reasons for a significant lack of fit are (i) an incorrect link function has been used and (ii) the underlying distribution was misspecified.

The second test that is generally performed in the analysis of a generalized linear model is the model utility test, which compares the fitted model with the null model. The model utility test should only be performed when the goodness-of-fit test indicates that there is no evidence of significant lack of fit. In the model utility test, the null hypothesis is $H_0 : g(\mu_i) = \beta_0$ and the alternative hypothesis is $H_A : g(\mu_i) = \mathbf{X}_i^T \vec{\beta}$. Equivalently, the model utility test is a test of the null hypothesis $H_0 : (\beta_1, \ldots, \beta_p) = \vec{0}$ versus $H_A : (\beta_1, \ldots, \beta_p) \neq \vec{0}$.

The test statistic for the model utility test is the difference of the deviance for the null model and the deviance for the full model. Specifically, if $\mathcal{D}_F$ is the deviance for the full model and $\mathcal{D}_0$ is the deviance for the null model, then the model utility test statistic is

$$G = \mathcal{D}_0 - \mathcal{D}_F = 2[l_{Sat} - l_{Null}] - 2[l_{Sat} - l_{Full}] = 2[l_{Full} - l_{Null}].$$

For a sufficiently large sample size, G is approximately distributed as a χ_p^2 random variable. Thus, the large sample critical region for testing $H_0 : (\beta_1, \ldots, \beta_p) = \vec{0}$ versus $H_A : (\beta_1, \ldots, \beta_p) \neq \vec{0}$ is $\mathcal{C}_{Util} = \{\vec{y} : G > \chi_{p,1-\alpha}^2\}$.

The deviance for the full model is referred to as the *residual deviance*, and the deviance for the null model is called the *null deviance*. The null and residual deviances are both reported as part of the glm command in R; however, R does not report the p-values for the goodness of fit nor the model utility test in its output.

Example 7.12 Using the data and generalized linear model for the Donner Party data introduced in Example 7.11, the full output for the model

$$\ln\left[\frac{p_i}{1 - p_i}\right] = \beta_0 + \beta_1 \text{ Age} + \beta_2, \text{Gender} + \beta_3 \text{ Age*Gender}$$

is given in Table 7.2.

From the output in Table 7.2, the residual deviance is 47.346 with 41 degrees of freedom. The p-value for the goodness-of-fit test is

$$p = P(\chi_{41}^2 > 47.346) = 0.2295.$$

Hence, there is insufficient evidence to conclude that the model has significant lack of fit.

The value of the test statistic for the model utility test is the difference of the null and residual deviances, which is $G = 61.827 - 47.346 = 14.481$ with 3 degrees of freedom and p-value $p = 0.0023$. Thus, there is sufficient evidence to reject $H_0 : (\beta_1, \beta_2, \beta_3) = \vec{0}$, and therefore, the model with the explanatory variables Age, Gender, and Age*Gender fits the data significantly better than does the model with no explanatory variables at all (i.e. the null model).

Finally, it is possible to compare the fit of the full model and any submodel nested in the full model. Let the full model be $g(\mu) = \beta_0 + \mathbf{X}_1^T\beta_1 + \mathbf{X}_2^T\beta_2$ and the reduced model be $g(\mu) = \beta_0 + \mathbf{X}_1^T\beta_1$, where, without loss of generality, $\mathbf{X}_1 = (X_1, \ldots, X_k)^T$, $\beta_1 = (\beta_1, \ldots, \beta_k)^T$, $\mathbf{X}_2 = (X_{k+1}, \ldots, X_p)^T$, and $\beta_2 = (\beta_{k+1}, \ldots, \beta_p)^T$.

The large sample test of $H_0 : g(\mu) = \beta_0 + \mathbf{X}_1^T\beta_1$ versus $H_A : g(\mu) = \beta_0 + \mathbf{X}_1^T\beta_1 + \mathbf{X}_2^T\beta_2$ is based on the difference of the residual deviance for the reduced

Table 7.2 R output for the model ln $\left[\frac{p_i}{1-p_i}\right] = \beta_0 + \beta_1$ Age $+ \beta_2$, Gender $+ \beta_3$ Age*Gender fit to the Donner Party data in Example 7.11.

```
glm(formula = Survival ~ AGE + Gender + AGE * Gender,
    family = binomial, data = donner)

Deviance Residuals:
   Min        1Q     Median        3Q        Max
-2.2279   -0.9388   -0.5550    0.7794     1.6998

Coefficients:
                Estimate  Std. Error  z value  Pr(>|z|)
(Intercept)      7.24638     3.20517    2.261    0.0238 *
AGE             -0.19407     0.08742   -2.220    0.0264 *
GenderMALE      -6.92805     3.39887   -2.038    0.0415 *
AGE:GenderMALE   0.16160     0.09426    1.714    0.0865 .

Signif. codes:  0 '***' 0.001 '**' 0.01 '*' 0.05 '.' 0.1 ' ' 1

(Dispersion parameter for binomial family taken to be 1)

    Null deviance: 61.827  on 44  degrees of freedom
Residual deviance: 47.346  on 41  degrees of freedom
```

model and the residual deviance for the full model and is referred to as a *extra deviance test*. In particular, if $\mathcal{D}_F$ is the residual deviance for the full model, $\mathcal{D}_R$ is the residual deviance for the reduced model, $\widetilde{\beta}_0$ and $\widetilde{\beta}_1$ are the MLEs of β_0 and β_1 under the reduced model, and $\widehat{\beta}_0, \widehat{\beta}_1,$ and $\widehat{\beta}_2$ are the MLEs of $\beta_0, \beta_1,$ and β_2 under the full model, then the test statistic for the extra deviance test is

$$\Delta\mathcal{D} = \mathcal{D}_F - \mathcal{D}_R = 2[\ell_{\text{Sat}} - \ell(\widetilde{\beta}_0, \widetilde{\beta}_1)] - 2[\ell_{\text{Sat}} - \ell(\widehat{\beta}_0, \widehat{\beta}_1, \widehat{\beta}_2)]$$
$$= 2[\ell(\widehat{\beta}_0, \widehat{\beta}_1, \widehat{\beta}_2) - \ell(\widetilde{\beta}_0, \widetilde{\beta}_1)].$$

Furthermore, under $H_0, \Delta\mathcal{D} \xrightarrow{d} \chi^2_{p-k}.$

Thus, the large sample critical region for testing $H_0 : g(\mu) = \beta_0 + \mathbf{X}_1^T\beta_1$ versus $H_A : g(\mu) = \beta_0 + \mathbf{X}_1^T\beta_1 + \mathbf{X}_2^T\beta_2$ is $\mathcal{C}_{\Delta\mathcal{D}} = \{\vec{y} : \Delta\mathcal{D} > \chi^2_{p-k,1-\alpha}\}.$ Note that in order to carry out the test comparing the full and reduced models, both models must be fit separately in order to obtain their respective residual deviances.

Example 7.13 In order to test the importance of the explanatory variable Gender in the generalized linear model fit to the Donner Party data in Example 7.11, the reduced model without the explanatory variables Gender and Age*Gender will be compared with the full model. The R output for the reduced model is given in Table 7.3.

Table 7.3 R output for the model ln $\left[\frac{p_i}{1-p_i}\right] = \beta_0 + \beta_1$ Age fit to the Donner Party data.

```
glm(formula = Survival ~ AGE,family=binomial,data=donner)

Deviance Residuals:
   Min       1Q    Median        3Q       Max
-1.5401  -1.1594   -0.4651    1.0842    1.7283

Coefficients:
             Estimate  Std. Error  z value  Pr(>|z|)
(Intercept)   1.81852     0.99937    1.820    0.0688.
AGE          -0.06647     0.03222   -2.063    0.0391 *

Signif. codes:  0 '***' 0.001 '**' 0.01 '*' 0.05 '.' 0.1 ' ' 1

(Dispersion parameter for binomial family taken to be 1)

    Null deviance: 61.827  on 44  degrees of freedom
Residual deviance: 56.291  on 43  degrees of freedom
```

Now, using the information on the full model given in Table 7.2 and the information on the reduced model given in Table 7.3, the extra deviance test statistic is

$$\Delta\mathcal{D} = \mathcal{D}_F - \mathcal{D}_R = 56.291 - 47.346 = 8.945$$

with 2 degrees of freedom and p-value $p = 0.0114$. Thus, there is sufficient evidence that $(\beta_2, \beta_3) \neq \vec{0}$, and the full model is significantly better than the reduced model. Therefore, the explanatory variable Gender is an important explanatory variable in the model for the survival of the members of the Donner Party.

Problems

7.3.1 Using the information in Table 7.2,
  a) test the importance of explanatory variable Age by testing $H_0 : \beta_1 = 0$ versus $H_a : \beta_1 \neq 0$ at the $\alpha = 0.05$ level;
  b) compute a 98% confidence interval for β_1.

7.3.2 Use the information in Table 7.4 for the generalized linear model with $Y_i \sim \text{Bin}(1, p_i)$, link $g(p_i) = \ln\left[\frac{p_i}{1-p_i}\right]$, and linear predictor $\beta_0 + \beta_1 X_1 + \beta_2 X_2 + \beta_3 X_3$ to
  a) test $H_0 : \beta_1 = 0$ versus $H_A : \beta_1 \neq 0$ at the $\alpha = 0.05$ level.

b) test $H_0 : \beta_2 = 0$ versus $H_A : \beta_2 \neq 0$ at the $\alpha = 0.05$ level.
c) test $H_0 : \beta_3 = 0$ versus $H_A : \beta_3 \neq 0$ at the $\alpha = 0.05$ level.
d) compute a 95% confidence interval for β_2.

7.3.3 Use the information in Table 7.4 for the generalized linear model with $Y_i \sim \text{Bin}(1, p_i)$, link $g(p_i) = \ln\left[\frac{p_i}{1-p_i}\right]$, and $\ln\left[\frac{p_i}{1-p_i}\right] = \beta_0 + \beta_1 X_1 + \beta_2 X_2 + \beta_3 X_3$ to
a) test the goodness of fit of the model at the $\alpha = 0.05$ level.
b) test $H_0 : \beta_1 = \beta_2 = \beta_3 = 0$ at the $\alpha = 0.05$ level.

7.3.4 Use the information in Table 7.5 and the information in Table 7.4 to
a) test $H_0 : \beta_1 = \beta_3 = 0$ at the $\alpha = 0.05$ level.
b) compute a 95% confidence interval for β_2 for the better fitting of the two models.

Table 7.4 R output for the model $\ln\left[\frac{p_i}{1-p_i}\right] = \beta_0 + \beta_1 X_1 + \beta_2 X_2 + \beta_3 X_3$ in Problem 7.3.2.

```
Coefficients:
              Estimate   Std. Error
(Intercept)   -12.3445     2.3412
X1             -1.2317     0.9529
X2              3.4367     1.3319
X3              0.8901     0.7438

(Dispersion parameter for binomial family taken to be 1)

    Null deviance: 67.8903 on 74   degrees of freedom
Residual deviance: 54.2223 on 71   degrees of freedom
```

Table 7.5 R output for the model $\ln\left[\frac{p_i}{1-p_i}\right] = \beta_0 + \beta_2 X_2$ in Problem 7.3.2.

```
Coefficients:
              Estimate   Std. Error
(Intercept)   -16.6881     3.4545
X2              5.1181     2.0098

(Dispersion parameter for binomial family taken to be 1)

    Null deviance: 67.8903 on 74   degrees of freedom
Residual deviance: 56.5211 on 73   degrees of freedom
```

7.4 Generalized Linear Models for a Normal Response Variable

Statistical models for a normal underlying probability model were commonly used before the development of generalized linear models. A *classical linear model* is a generalized linear model where

- the underlying distribution of the response variable is normally distributed;
- the mean is modeled by a linear predictor in the nonrandom explanatory variables $X_1, \ldots, X_p$;
- the variance of the response variable is constant over all experimental conditions.

Thus, the basis for a classical linear model is a collection of independent random variables $Y_1, \ldots, Y_n$ with $Y_i \sim N(\mathbf{X}_i^T \vec{\beta}, \sigma^2)$, where $\mathbf{X}_i$ is a vector of explanatory variables associated with Y_i.

Classical linear models are used in a wide range of statistical procedures including simple linear regression, multiple regression, and models for designed experiments with a quantitative response variable. Prior to the development of generalized linear models, in modeling scenarios where the underlying distribution was not normal, the response variable was generally transformed to approximate normality and the model fit to the transformed data.

Let $Y_1, \ldots, Y_n$ be independent random variables with $Y_i \sim N(\mu_i, \sigma^2)$ so that the variance is the same for each Y_i and $Y_i \sim \text{EF}(\frac{\mu_i^2}{2}, \sigma^2)$. The components of the classical linear model are

- a set of nonrandom explanatory variables $X_1, \ldots, X_p$ called *covariates*;
- the identity link function $g(\mu_i) = \mu_i$;
- a linear predictor $\mu_i = \mathbf{X}_i \vec{\beta}$;
- an unknown dispersion parameter σ^2.

In the standard notation used in a classical linear model, the vector of response random variables is denoted by $\mathbf{Y} = (Y_1, \ldots, Y_n)^T$, the vector of observed values of the response variables denoted by $\mathbf{y} = (y_1, \ldots, y_p)^T$, the vector of explanatory variables associated with Y_i denoted by $\mathbf{X}_i = (1, x_{1i}, \ldots, x_{pi})^T$, and the vector of unknown parameters in the linear predictor denoted by $\vec{\beta} = (\beta_0, \beta_1, \ldots, \beta_p)^T$. The matrix with the ith row $\mathbf{X}_i$ is called the *design matrix*, which is denoted by $\mathbf{X}$. In particular, the design matrix is

$$\mathbf{X} = \begin{pmatrix} 1 & X_{11} & X_{21} & \cdots & X_{p1} \\ 1 & X_{12} & X_{22} & \cdots & X_{p2} \\ 1 & \vdots & \vdots & \cdots & \vdots \\ 1 & X_{1n} & X_{2n} & \cdots & X_{pn} \end{pmatrix},$$

where the ith row of the design matrix is X_i^T. The design matrix plays an important role in many aspects fitting a classical linear model. For example, the mean of the response vector Y is $E(Y) = X\vec{\beta}$.

7.4.1 Estimation

In a classical linear model, the likelihood function in terms of the unknown parameters $\vec{\beta}$ and σ^2 is

$$L(\vec{\beta}, \sigma^2) = (2\pi\sigma^2)^{-\frac{n}{2}} e^{-\frac{1}{2\sigma^2} \sum_{i=1}^{n} (y_i - X_i^T \vec{\beta})^2}$$

$$= \sigma^{-n} e^{-\frac{1}{2\sigma^2} \sum_{i=1}^{n} [y_i^2 - 2y_i X_i^T \vec{\beta} + (X_i^T \vec{\beta})^2]} (2\pi)^{-\frac{n}{2}}$$

$$= \underbrace{\sigma^{-n} e^{-\frac{1}{2\sigma^2} [y^T y - 2y^T X \vec{\beta} + \sum_{i=1}^{n} (X_i^T \vec{\beta})^2]}}_{g(S;\vec{\beta},\sigma^2)} \underbrace{(2\pi)^{-\frac{n}{2}}}_{h(y)},$$

and therefore, $\vec{S} = (y^T y, y^T X)$ is a jointly sufficient statistic for $(\vec{\beta}, \sigma^2)$. Furthermore, when the parameter space is $\mathbb{R}^{p+1} \times \mathbb{R}^+$, $\vec{S}$ is a complete sufficient statistic for $(\vec{\beta}, \sigma^2)$.

The log-likelihood function for the classical linear model is

$$\ell(\vec{\beta}, \sigma^2) = -\frac{n}{2} \ln(2\pi\sigma^2) - \frac{1}{2\sigma^2} \sum_{i=1}^{n} (y_i - X_i^T \vec{\beta})^2,$$

and for $j = 0, 1, \ldots, p$, the jth component of the Score function corresponding to $\frac{\partial \ell(\vec{\beta},\sigma^2)}{\partial \beta_j}$ is

$$\dot{\ell}_j = \frac{\partial \ell(\vec{\beta}, \sigma^2)}{\partial \beta_j} = \frac{1}{\sigma^2} \sum_{i=1}^{n} (y_i - X_i \vec{\beta}) x_{ij} = \frac{1}{\sigma^2} \sum_{i=1}^{n} (y_i x_{ij} - x_{ij} X_i \vec{\beta})$$

$$= \frac{1}{\sigma^2} \left[X_i^T y - X_i^T X_i \vec{\beta} \right].$$

Since σ^2 is not important in solving for the MLE of $\vec{\beta}$, only the partial derivatives $\dot{\ell}_0, \ldots, \dot{\ell}$ are used to determine $\hat{\vec{\beta}}$; however, the partial derivative of $\ell(\vec{\beta}, \sigma^2)$ with respect to σ^2 is needed for determining the MLE of σ^2.

The system of simultaneous equations $\begin{pmatrix} \dot{\ell}_0(\vec{\beta}, \sigma^2) \\ \vdots \\ \dot{\ell}_p(\vec{\beta}, \sigma^2) \end{pmatrix} = \vec{0}$ are called the *normal equations*. In terms of the design matrix, the normal equations are $X^T y = X^T X \vec{\beta}$. Note that $X^T X$ is a symmetric $(p+1) \times (p+1)$ matrix.

When $X^T X$ is nonsingular (i.e. invertible), the solution to the normal equations determines the MLE of $\vec{\beta}$, which is $\hat{\vec{\beta}} = (X^T X)^{-1} X^T y$. On the other hand, when $X^T X$ is singular, the model must be reparameterized by constraining the parameters in the linear predictor so that $\mathbf{X}'^T\mathbf{X}'$ is nonsingular, where $\mathbf{X}'$ is the constrained design matrix. A nonsingular matrix $\mathbf{X}^T\mathbf{X}$ occurs when $\mathbf{X}$ does not have full rank and often occurs in designed experiments where the covariates are qualitative variables (see Example 7.15).

When $\mathbf{X}^T\mathbf{X}$ is nonsingular, the MLE of $\vec{\beta}$ is equal to the *least squares estimator* of $\vec{\beta}$, which minimizes $SS(\vec{\beta}) = (\mathbf{y} - \mathbf{X}^T\vec{\beta})^T(\mathbf{y} - \mathbf{X}^T\vec{\beta})$.

The MLE of σ^2 is found by solving $\frac{\partial\ell(\widehat{\beta},\sigma^2)}{\partial\sigma^2} = 0$, where

$$\frac{\partial\ell(\widehat{\beta},\sigma^2)}{\partial\sigma^2} = -\frac{n}{2\sigma^2} + \frac{1}{2\sigma^4}\sum_{i=1}^{n}(y_i - \mathbf{X}_i^T\widehat{\beta})^2.$$

Hence, the MLE of σ^2 is $\widehat{\sigma}^2 = \frac{1}{n}\sum_{i=1}^{n}(y_i - \mathbf{X}_i^T\widehat{\beta})^2 = \frac{1}{n}(\mathbf{y} - \mathbf{X}^T\widehat{\beta})^T(\mathbf{y} - \mathbf{X}^T\widehat{\beta})$.

Example 7.14 Let $Y_1, \ldots, Y_n$ be independent random variables with $Y_i \sim N(\mu_i, \sigma^2)$, and let $\mu_i = \beta_0 + \beta_1 x_i$ for $i = 1, \ldots, n$ and know constants $x_1, \ldots, x_n$. This is a generalized linear model with $Y_i \sim EF(\frac{\mu_i}{2}, \sigma^2)$ with the identity link, and linear predictor $\mu_i = \beta_0 + \beta_1 x_i$. This classical linear model is referred to as a *simple linear regression model*.

The log-likelihood function for the simple linear regression model is

$$\ell(\beta_0, \beta_1, \sigma^2) = -\frac{n}{2}\ln[2\pi\sigma^2] - \frac{1}{2\sigma^2}\sum_{i=1}^{n}(y_i - \beta_0 - \beta_1 x_i)^2,$$

and the partial derivatives $\dot{\ell}_0 = \frac{\partial\ell(\beta_0,\beta_1,\sigma^2)}{\partial\beta_0}$ and $\dot{\ell}_1 = \frac{\partial\ell(\beta_0,\beta_1,\sigma^2)}{\partial\beta_1}$ are

$$\dot{\ell}_0 = \frac{1}{\sigma^2}\sum_{i=1}^{n}(y_i - \beta_0 - \beta_1 x_i)$$

$$\dot{\ell}_1 = \frac{1}{\sigma^2}\sum_{i=1}^{n}(y_i - \beta_0 - \beta_1 x_i)x_i.$$

Thus, the normal equations, ignoring σ^2, that must be solved to find the MLEs of β_0 and β_1 are

$$\sum_{i=1}^{n}(y_i - \beta_0 - \beta_1 x_i) = 0$$

$$\sum_{i=1}^{n}(y_i - \beta_0 - \beta_1 x_i)x_i = 0,$$

which is a linear system of two equations in two unknowns.

Solving the normal equations, the MLEs of β_0 and β_1 are

$$\widehat{\beta}_1 = \frac{\sum_{i=1}^{n}x_i y_i - n\bar{x}\bar{y}}{\sum_{i=1}^{n}x_i^2 - n\bar{x}^2}$$

$$\widehat{\beta}_0 = \bar{y} - \widehat{\beta}_1\bar{x}.$$

Furthermore, the MLEs of β_0 and β_1 for the simple linear regression model are unbiased estimators since

$$E(\widehat{\beta_1}) = E\left(\frac{\sum_{i=1}^{n} x_i Y_i - n\bar{x}\bar{Y}}{\sum_{i=1}^{n} x_i^2 - n\bar{x}^2}\right) = \frac{n\sum_{i=1}^{n} x_i E(Y_i) - n\bar{x}E(\bar{Y})}{\sum_{i=1}^{n} x_i^2 - n\bar{x}^2}$$

$$= \frac{n\sum_{i=1}^{n} x_i(\beta_0 + \beta_1 x_i) - n\bar{x}\sum_{j=1}^{n}(\beta_0 + \beta_1 x_i)}{\sum_{i=1}^{n} x_i^2 - n\bar{x}^2}$$

$$= \frac{n\bar{x}\beta_0 + \beta_1 n\sum_{i=1}^{n} x_i^2 - (n\bar{x}\beta_0 + n\beta_1\bar{x}^2)}{\sum_{i=1}^{n} x_i^2 - n\bar{x}^2}$$

$$= \frac{\beta_1(\sum_{i=1}^{n} x_i^2 - n\bar{x}^2)}{\sum_{i=1}^{n} x_i^2 - n\bar{x}^2} = \beta_1$$

and

$$E(\widehat{\beta_0}) = E(\bar{Y} - \widehat{\beta_1}\bar{x}) = E(\bar{Y}) - \bar{x}E(\widehat{\beta_1})$$
$$= \beta_0 + \beta_1\bar{x} - \beta_1\bar{x} = \beta_0.$$

Finally, the MLE of σ^2 in the simple linear regression model is

$$\widehat{\sigma}^2 = \frac{1}{n}\sum_{i=1}^{n}(y_i - \widehat{\beta_0} - \widehat{\beta_1}x_i)^2.$$

Example 7.15 Let $Y_1, \ldots, Y_n$ be independent random variables with $Y_i \sim N(\mu_i, \sigma^2)$. Furthermore, suppose that the random variables $Y_1, \ldots, Y_n$ are associated with three treatments. Let $\mu_i = \mu + \tau_1 T_{1i} + \tau_2 T_{2i} + \tau_3 T_{3i}$, where T_{1i}, T_{2i}, and T_{3i} are indicator functions identifying which treatment Y_i is associated with. That is, when Y_i is associated with treatment j, $T_{ji} = 1$ and the indicator functions for the other two treatments are 0.

The design matrix for this model is an $n \times 4$ matrix with columns representing the parameters μ, τ_1, τ_2, and τ_3. Assuming that $Y_1, \ldots, Y_{n_1}$ are associated with treatment 1, $Y_{n_1+1}, \ldots, Y_{n_1+n_2}$ are associated with treatment 2, and $Y_{n_1+n_2+1}, \ldots, Y_n$ are associated with treatment 3, the design matrix is

$$\mathbf{X} = \begin{pmatrix} 1 & 1 & 0 & 0 \\ 1 & 1 & 0 & 0 \\ \vdots & \vdots & \vdots & \vdots \\ 1 & 1 & 0 & 0 \\ 1 & 0 & 1 & 0 \\ \vdots & \vdots & \vdots & \vdots \\ 1 & 0 & 1 & 0 \\ 1 & 0 & 0 & 1 \\ \vdots & \vdots & \vdots & \vdots \\ 1 & 0 & 0 & 1 \end{pmatrix}.$$

Note that the sum of columns 2, 3, and 4 is equal to the column of 1s, and therefore, $\mathbf{X}$ is not a full-rank matrix and $\mathbf{X}^T\mathbf{X}$ is singular. Because $\mathbf{X}$ is not full rank, the model must be reparameterized so that the constrained design matrix, say $\mathbf{X}'$, is full rank.

One of many possible reparameterizations of the model is to set $\tau_3 = 0$. No information will be lost with this reparameterization since Y_i is associated with treatment 3 if and only if $T_{1i} = T_{2i} = 0$. Thus, the reparameterized model is $\mu_i = \mu' + \tau_1' T_{1i} + \tau_2' T_{2i}$, and the new design matrix is

$$\mathbf{X}' = \begin{pmatrix} 1 & 1 & 0 \\ 1 & 1 & 0 \\ \vdots & \vdots & \vdots \\ 1 & 1 & 0 \\ 1 & 0 & 1 \\ \vdots & \vdots & \vdots \\ 1 & 0 & 1 \\ 1 & 0 & 0 \\ \vdots & \vdots & \vdots \\ 1 & 0 & 0 \end{pmatrix}.$$

Since $\mathbf{X}'$ is full rank, the MLE of $\vec{\beta} = (\mu', \tau_1', \tau_2')^T$ is $\hat{\beta} = (\mathbf{X}'^T\mathbf{X}')^{-1}\mathbf{X}'\mathbf{Y}$; however, the parameters μ', τ_1', and τ_2' are not *estimable*.

Definition 7.2 A linear function of $\vec{\beta}$, say $C\beta$, is estimable if and only if there exists a linear function of $\mathbf{Y}$, say $A\mathbf{Y}$, that is an unbiased estimator of $C\vec{\beta}$. Otherwise, $C\vec{\beta}$ is said to be nonestimable.

Estimable functions are unique regardless of the reparameterization used. In this example, while there is no linear combination of $\mathbf{Y}$ that has expectation τ_i, the difference between treatment effects, $\tau_1 - \tau_2$, is estimable since

$$E(\overline{Y}_1 - \overline{Y}_2) = E(\overline{Y}_1) - E(\overline{Y}_2) = (\mu + \tau_1) - (\mu + \tau_2) = \tau_1 - \tau_2,$$

where $\overline{Y}_j$ is the mean of the Y values associated with the jth treatment.

Finally, there are other reparameterizations of the model that also will produce alternative full-rank design matrices. More importantly, any valid reparameterization of the model will produce the same estimates of all of the estimable quantities. For example, a second reparameterization of the model is to constrain the τ's so that $\sum_{i=1}^{3} \tau_i = 0$. In this parameterization, $\tau_3 = -\tau_1 - \tau_2$, and the model is $\mu_i = \mu'' + \tau_1'' T_{1i} + \tau_2'' T_{2i}$ with design matrix

$$\mathbf{X}'' = \begin{pmatrix} 1 & 1 & 0 \\ 1 & 1 & 0 \\ \vdots & \vdots & \vdots \\ 1 & 1 & 0 \\ 1 & 0 & 1 \\ \vdots & \vdots & \vdots \\ 1 & 0 & 1 \\ 1 & -1 & -1 \\ \vdots & \vdots & \vdots \\ 1 & -1 & -1 \end{pmatrix}.$$

In Example 7.15, the parameters of the model were not estimable, and in general, this is a problem for experimental design models when the explanatory variables are qualitative variables. On the other hand, when the explanatory variables are all quantitative variables and the design matrix is full rank, then for every vector $\mathbf{a} = (a_0, \ldots, a_p)^\mathrm{T}$, $\mathbf{a}^\mathrm{T} \vec{\beta}$ is estimable, since $\mathbf{a}^\mathrm{T} \hat{\beta} = \mathbf{a}^\mathrm{T} (\mathbf{X}^\mathrm{T} \mathbf{X})^{-1} \mathbf{X}^\mathrm{T} \mathbf{Y}$ is a linear combination of $\mathbf{Y}$ with $E(\mathbf{a}^\mathrm{T} \hat{\beta}) = \mathbf{a}^\mathrm{T} \vec{\beta}$.

7.4.2 Properties of the MLEs

Because the MLEs in a classical linear model can be determined analytically, many of the important properties of the MLEs can also be determined. For example, the mean and variance of the sampling distribution of $\hat{\beta}$ are given in Theorem 7.2, and the sampling distribution of $\hat{\beta}$ is given in Theorem 7.3.

Theorem 7.2 *If* $Y_1, \ldots, Y_n$ *are independent* $N(\mu_i, \sigma^2)$ *random variables with* $\mu_i = \mathbf{X}_i \vec{\beta}$, $\mathbf{X}^\mathrm{T} \mathbf{X}$ *is nonsingular, and* $(\hat{\beta}, \hat{\sigma}^2)$ *is the MLE of* $(\vec{\beta}, \sigma^2)$, *then*

i) $E(\hat{\beta}) = \beta$.
ii) $\mathrm{Var}(\hat{\beta}) = \sigma^2 (\mathbf{X}^\mathrm{T} \mathbf{X})^{-1}$.
iii) $E(\mathbf{x}^\mathrm{T} \hat{\beta}) = \mathbf{x}^\mathrm{T} \vec{\beta}$ *for* $\mathbf{x} = (1, x_1, \ldots, x_p)^\mathrm{T}$.
iv) $\mathrm{Var}(\mathbf{x}^\mathrm{T} \hat{\beta}) = \sigma^2 \mathbf{x}^\mathrm{T} (\mathbf{X}^\mathrm{T} \mathbf{X})^{-1} \mathbf{x}$ *for* $\mathbf{x} = (1, x_1, \ldots, x_p)^\mathrm{T}$.

Proof. Let $Y_1, \ldots, Y_n$ be independent $N(\mu_i, \sigma^2)$ random variables with $\mu_i = \mathbf{X}_i \vec{\beta}$ where $\mathbf{X}^\mathrm{T} \mathbf{X}$ is nonsingular, and let $(\hat{\beta}, \hat{\sigma}^2)^\mathrm{T}$ be the MLE of $(\vec{\beta}, \sigma^2)^\mathrm{T}$.

i) Since $\widehat{\beta} = (\mathbf{X}^T\mathbf{X})^{-1}\mathbf{X}^T\mathbf{Y}$, it follows that

$$E(\widehat{\beta}) = E[(\mathbf{X}^T\mathbf{X})^{-1}\mathbf{X}^T\mathbf{Y}] = (\mathbf{X}^T\mathbf{X})^{-1}\mathbf{X}^T \underbrace{E(\mathbf{Y})}_{\mathbf{X}\vec{\beta}}$$

$$= (\mathbf{X}^T\mathbf{X})^{-1}\mathbf{X}^T\mathbf{X}\vec{\beta} = \vec{\beta}.$$

ii) Since $\widehat{\beta} = (\mathbf{X}^T\mathbf{X})^{-1}\mathbf{X}^T\mathbf{Y}$, it follows that

$$V(\widehat{\beta}) = V[(\mathbf{X}^T\mathbf{X})^{-1}\mathbf{X}^T\mathbf{Y}]$$

$$= (\mathbf{X}^T\mathbf{X})^{-1}\mathbf{X}^T \underbrace{V(\mathbf{Y})}_{\sigma^2 \mathbf{I}_{n \times n}} ((\mathbf{X}^T\mathbf{X})^{-1}\mathbf{X}^T)^T$$

$$= (\mathbf{X}^T\mathbf{X})^{-1}\mathbf{X}^T\sigma^2\mathbf{I}_{n \times n}\mathbf{X}(\mathbf{X}^T\mathbf{X})^{-1}$$

$$= \sigma^2 \underbrace{(\mathbf{X}^T\mathbf{X})^{-1}\mathbf{X}^T\mathbf{X}(\mathbf{X}^T\mathbf{X})^{-1}}_{\mathbf{I}_{p+1 \times p+1}}$$

$$= \sigma^2(\mathbf{X}^T\mathbf{X})^{-1}.$$

iii) The proof of part (iii) is left as an exercise.
iv) The proof of part (iv) is left as an exercise. ∎

Corollary 7.1 shows that the MLE $\widehat{\beta}$ is a uniformly minimum variance unbiased estimator (UMVUE) of $\vec{\beta}$ when $\mathbf{X}$ is full rank and $\Theta = \mathbb{R}^{p+1} \times \mathbb{R}^+$.

Corollary 7.1 *Let* $Y_1, \dots, Y_n$ *be independent* $N(\mu_i, \sigma^2)$ *random variables with* $\mu_i = \mathbf{X}_i\vec{\beta}$ *and* $\Theta = \mathbb{R}^{p+1} \times \mathbb{R}^+$. *If* $\mathbf{X}^T\mathbf{X}$ *is nonsingular and* $(\widehat{\beta}, \widehat{\sigma}^2)^T$ *is the MLE of* $(\vec{\beta}, \sigma^2)$, *then*

i) $\widehat{\beta}_j$ *is a UMVUE of* β_j, $j = 0, \dots, p$.
ii) $S^2 = \dfrac{n}{n-p-1}\widehat{\sigma}^2$ *is a UMVUE of* σ^2.

Proof. Let $Y_1, \dots, Y_n$ be independent $N(\mu_i, \sigma^2)$ random variables with $\mu_i = \mathbf{X}_i\vec{\beta}$ and $\Theta = \mathbb{R}^{p+1} \times \mathbb{R}^+$, and suppose that $\mathbf{X}^T\mathbf{X}$ is nonsingular and $(\widehat{\beta}, \widehat{\sigma}^2)$ the MLE of $(\vec{\beta}, \sigma^2)^T$.

i) $\widehat{\beta}_j$ is an unbiased estimator of β_j for $j = 0, \dots, p$ based completely on a complete sufficient statistic, and therefore, by the Lehmann–Scheffé theorem, it follows that $\widehat{\beta}_j$ is a UMVUE of β_j.
ii) S^2 is an unbiased estimator of σ^2 based completely on a complete sufficient statistic, and therefore, by the Lehmann–Scheffé theorem, it follows that S^2 is a UMVUE of σ^2. ∎

Example 7.16 In the simple linear regression problem, Example 7.14,

$$(\mathbf{X}^T\mathbf{X})^{-1} = \frac{1}{n\sum_{i=1}^{n} x_i^2 - \left(\sum_{i=1}^{n} x_i\right)^2} \begin{pmatrix} \sum_{i=1}^{n} x_i^2 & -\sum_{i=1}^{n} x_i \\ -\sum_{i=1}^{n} x_i & n \end{pmatrix}.$$

Since $n\sum_{i=1}^{n} x_i^2 - \left(\sum_{i=1}^{n} x_i\right)^2 = n\sum_{i=1}^{n} (x_i - \overline{x})^2$, the exact standard errors of the MLEs are

$$\text{se}(\widehat{\beta}_0) = \sqrt{\frac{\sigma^2 \sum_{i=1}^{n} x_i^2}{n\sum_{i=1}^{n} (x_i - \overline{x})^2}}$$

and

$$\text{se}(\widehat{\beta}_1) = \sqrt{\frac{n\sigma^2}{n\sum_{i=1}^{n} (x_i - \overline{x})^2}} = \sqrt{\frac{\sigma^2}{\sum_{i=1}^{n} (x_i - \overline{x})^2}}.$$

By the invariance property of MLEs, the MLE of the mean of the response variable when $\mathbf{X}^\star = (1, x^\star)^T$ is $\widehat{\mu}(\mathbf{X}^\star) = \widehat{\beta}_0 + \widehat{\beta}_1 x^\star$. Since $\widehat{\beta}_0$ and $\widehat{\beta}_1$ are unbiased estimators, $\widehat{\mu}(\mathbf{X}^\star) = \widehat{\beta}_0 + \widehat{\beta}_1 x^\star$ is an unbiased estimator of $\mu(x^\star) = \beta_0 + \beta_1 x^\star$. The standard error of the $\widehat{\mu}(x^\star)$ is

$$\text{se}[\widehat{\mu}(\mathbf{X}^\star)] = \sigma\sqrt{\frac{1}{n} + \frac{(x^\star - \overline{x})^2}{\sum_{i=1}^{n} (x_i - \overline{x})^2}}.$$

The sampling distributions of the maximum likelihood estimators are given Theorem 7.3, and the distribution of S^2 is given in Corollary 7.2.

Theorem 7.3 *If $Y_1, \ldots, Y_n$ are independent $N(\mathbf{X}_i^T\vec{\beta}, \sigma^2)$ random variables, $\mathbf{X}^T\mathbf{X}$ is nonsingular, and $(\widehat{\beta}, \widehat{\sigma}^2)^T$ the MLE of $(\vec{\beta}, \sigma^2)$, then*

i) $\widehat{\beta} \sim N_p(\vec{\beta}, \sigma^2(\mathbf{X}^T\mathbf{X})^{-1})$.

ii) $\widehat{\beta}_j \sim N(\beta_j, \sigma^2(\mathbf{X}^T\mathbf{X})_{jj}^{-1})$, *for $j = 0, \ldots, p$.*

iii) $\frac{(\mathbf{Y} - \mathbf{X}\widehat{\beta})^T(\mathbf{Y} - \mathbf{X}\widehat{\beta})}{\sigma^2} \sim \chi_{n-p-1}^2$.

iv) $\widehat{\sigma}^2 \sim n\sigma^2\chi_{n-p-1}^2$.

Proof. Let $Y_1, \ldots, Y_n$ be independent random variables with $Y_i \sim N(\mathbf{X}_i^T\vec{\beta}, \sigma^2)$, $\mathbf{X}^T\mathbf{X}$ is nonsingular, and let $(\widehat{\beta}, \widehat{\sigma}^2)^T$ be the MLE of $(\vec{\beta}, \sigma^2)$. First, note that $\mathbf{Y}$ is distributed as a p-dimensional multivariate normal with $\mathbf{Y} \sim N_p(\mathbf{X}\vec{\beta}, \sigma^2\mathbf{I}_{n\times n})$.

i) Since $\widehat{\beta} = (\mathbf{X}^T\mathbf{X})^{-1}\mathbf{X}^T\mathbf{Y}$, it follows from properties of the multivariate normal that $\widehat{\beta}$ follows a p-dimensional multivariate normal with mean and variance given in Theorem 7.2. Therefore, $\widehat{\beta} \sim N_p(\beta, \sigma^2(\mathbf{X}^T\mathbf{X})^{-1})$.

ii) Part (ii) follows directly from part (i) since the univariate marginals of a multivariate normal are normally distributed.

iii) Part (iii) follows from properties of the multivariate normal and quadratic forms. See Graybill [26] for a detailed proof of part (iii).

iv) Since $\hat{\sigma}^2 = \frac{1}{n}(\mathbf{Y} - \mathbf{X}\hat{\beta})^{\mathrm{T}}(\mathbf{Y} - \mathbf{X}\hat{\beta})$, it follows from part (iii) that $\hat{\sigma}^2 \sim n\sigma^2\chi^2_{n-p-1}$.

∎

Corollary 7.2 *If $Y_1, \ldots, Y_n$ are independent $N(\mathbf{X}_i^{\mathrm{T}}\vec{\beta}, \sigma^2)$ random variables with, $\mathbf{X}_i = (1, X_{1i}, \ldots, X_{pi})^{\mathrm{T}}$, $\mathbf{X}^{\mathrm{T}}\mathbf{X}$ is nonsingular, and $S^2 = \frac{(\mathbf{y} - \mathbf{X}\hat{\beta})^{\mathrm{T}}(\mathbf{y} - \mathbf{X}\hat{\beta})}{n-p-1}$, then*

i) $S^2 \sim \frac{\sigma^2}{n-p-1}\chi^2_{n-p-1}$.

ii) S^2 *is an unbiased estimator of σ^2.*

iii) $\dfrac{\hat{\beta}_j - \beta_j}{\sqrt{S^2(\mathbf{X}^{\mathrm{T}}\mathbf{X})_{jj}^{-1}}} \sim t_{n-p-1}$.

Proof.

i) The proof of part (i) is left as an exercise.

ii) The proof of part (ii) is left as an exercise.

iii) By Theorem 7.3, $\hat{\beta}_j \sim N(\beta_j, \sigma^2(\mathbf{X}^{\mathrm{T}}\mathbf{X})_{jj}^{-1})$ and $S^2 \sim \frac{\sigma^2}{n-p-1}\chi^2_{n-p-1}$, and therefore,

$\dfrac{\hat{\beta}_j - \beta_j}{\sqrt{\sigma^2(\mathbf{X}^{\mathrm{T}}\mathbf{X})_{jj}^{-1}}} \sim N(0, 1)$. Now,

$$\frac{\hat{\beta}_j - \beta_j}{\sqrt{S^2(\mathbf{X}^{\mathrm{T}}\mathbf{X})^{-1}}} = \frac{\frac{1}{\sqrt{\sigma^2(\mathbf{X}^{\mathrm{T}}\mathbf{X})^{-1}}}}{\frac{1}{\sqrt{\sigma^2(\mathbf{X}^{\mathrm{T}}\mathbf{X})^{-1}}}} \frac{\hat{\beta}_j - \beta_j}{\sqrt{S^2(\mathbf{X}^{\mathrm{T}}\mathbf{X})^{-1}}}$$

$$= \frac{\frac{\hat{\beta}_j - \beta_j}{\sqrt{\sigma^2(\mathbf{X}^{\mathrm{T}}\mathbf{X})^{-1}}}}{\sqrt{\frac{S^2}{\sigma^2}}},$$

which is the ratio of $\dfrac{\hat{\beta}_j - \beta_j}{\sqrt{\sigma^2(\mathbf{X}^{\mathrm{T}}\mathbf{X})^{-1}}} \sim N(0, 1)$ and $\sqrt{\dfrac{S^2}{\sigma^2}} \sim \sqrt{\dfrac{\chi^2_{n-p-1}}{n-p-1}}$. Hence,

$\dfrac{\hat{\beta}_j - \beta_j}{\sqrt{S^2(\mathbf{X}^{\mathrm{T}}\mathbf{X})^{-1}}} \sim t_{n-p-1}$.

∎

Theorem 7.2 shows that the variance–covariance matrix for $\hat{\beta}$ (i.e. $\mathrm{Var}(\hat{\beta})$) can be computed exactly, and therefore, the asymptotic standard errors need not be used in a classical linear model. In practice, σ^2 is unknown and $S^2(\mathbf{X}^{\mathrm{T}}\mathbf{X})^{-1}$ is an unbiased estimator that is used for estimating $\mathrm{Var}(\hat{\beta})$.

Example 7.17 In the simple linear regression setting, the exact standard errors of $\hat{\beta}_0$ and $\hat{\beta}_1$ were given in Example 7.16. In particular,

$$\mathrm{se}(\hat{\beta}_0) = \sqrt{\frac{\sigma^2 \sum_{i=1}^{n} x_i^2}{n \sum_{i=1}^{n} (x_i - \bar{x})^2}}$$

and

$$\text{se}(\widehat{\beta}_1) = \sqrt{\frac{\sigma^2}{\sum_{i=1}^{n}(x_i - \bar{x})^2}}.$$

Now, in practice, σ^2 is unknown, and in this case, S^2 is used in place of σ^2 in the standard errors. Therefore, in practice, the standard errors of $\widehat{\beta}_0$ and $\widehat{\beta}_1$ are

$$\text{se}(\widehat{\beta}_0) = \sqrt{\frac{S^2 \sum_{i=1}^{n} x_i^2}{n \sum_{i=1}^{n}(x_i - \bar{x})^2}} = S\sqrt{\frac{\sum_{i=1}^{n} x_i^2}{n \sum_{i=1}^{n}(x_i - \bar{x})^2}}$$

and

$$\text{se}(\widehat{\beta}_1) = \sqrt{\frac{S^2}{\sum_{i=1}^{n}(x_i - \bar{x})^2}} = S\sqrt{\frac{1}{\sum_{i=1}^{n}(x_i - \bar{x})^2}}.$$

7.4.3 Deviance

The MLE of μ_i under the saturated model is $\widetilde{\mu}_i = y_i$, and hence,

$$\ell_{\text{Sat}} = -\frac{n}{2}\ln(2\pi\sigma^2) - \frac{1}{2\sigma^2}\sum_{i=1}^{n}(y_i - y_i)^2 = -\frac{n}{2}\ln(2\pi\sigma^2).$$

Thus, the residual deviance in a classical linear model is

$$\begin{aligned}
\mathcal{D}_R &= 2[\ell_{\text{sat}} - \ell(\widehat{\beta}, \sigma^2)] \\
&= 2\left[-\frac{n}{2}\ln(2\pi\sigma^2) - \left[-\frac{n}{2}\ln(2\pi\sigma^2) - \frac{1}{2\sigma^2}\sum_{i=1}^{n}(y_i - \mathbf{X}_i^{\mathsf{T}}\widehat{\beta})^2\right]\right] \\
&= \frac{1}{\sigma^2}\sum_{i=1}^{n}(y_i - \mathbf{X}_i^{\mathsf{T}}\widehat{\beta})^2 \\
&= \frac{1}{\sigma^2}(\mathbf{y} - \mathbf{X}^{\mathsf{T}}\widehat{\beta})^{\mathsf{T}}(\mathbf{y} - \mathbf{X}^{\mathsf{T}}\widehat{\beta}).
\end{aligned}$$

The residual deviance in a classical linear model is a sum of squares involving the nuisance parameter σ^2, and therefore, the unscaled deviance $(\mathbf{y} - \mathbf{X}^{\mathsf{T}}\widehat{\beta})^{\mathsf{T}}(\mathbf{y} - \mathbf{X}^{\mathsf{T}}\widehat{\beta})$ is used in a classical linear model. The unscaled deviance can also be computed using $\mathbf{y}^{\mathsf{T}}\mathbf{y} - \widehat{\beta}^{\mathsf{T}}\mathbf{X}^{\mathsf{T}}\mathbf{y}$ since

$$\begin{aligned}
(\mathbf{y} - \mathbf{X}^{\mathsf{T}}\widehat{\beta})^{\mathsf{T}}(\mathbf{y} - \mathbf{X}^{\mathsf{T}}\widehat{\beta}) &= \mathbf{y}^{\mathsf{T}}\mathbf{y} - 2\widehat{\beta}\mathbf{X}^{\mathsf{T}}\mathbf{y} + \widehat{\beta}\mathbf{X}^{\mathsf{T}}\mathbf{X}\widehat{\beta} \\
&= \mathbf{y}^{\mathsf{T}}\mathbf{y} - 2\widehat{\beta}\mathbf{X}^{\mathsf{T}}\mathbf{y} + \widehat{\beta}\underbrace{\mathbf{X}^{\mathsf{T}}\mathbf{X}(\mathbf{X}^{\mathsf{T}}\mathbf{X})^{-1}}_{\mathbf{I}_p}\mathbf{X}^{\mathsf{T}}\mathbf{y} \\
&= \mathbf{y}^{\mathsf{T}}\mathbf{y} - \widehat{\beta}^{\mathsf{T}}\mathbf{X}^{\mathsf{T}}\mathbf{y}.
\end{aligned}$$

The unscaled residual deviance in a classical linear model is called the *sum of squares due to error* and is denoted by *SSE*.

The null model is $\mu_i = \beta_0$, and under the null model, the MLE of β_0 is $\bar{y}$ and the null deviance is

$$\mathcal{D}_N = 2[\ell_{sat} - \ell(\bar{y}, \sigma^2)]$$

$$= 2\left[-\frac{n}{2}\ln(2\pi\sigma^2) - \left[-\frac{n}{2}\ln(2\pi\sigma^2) - \frac{1}{2\sigma^2}\sum_{i=1}^{n}(y_i - \bar{y})^2 \right] \right]$$

$$= \frac{1}{\sigma^2}\sum_{i=1}^{n}(y_i - \bar{y})^2 = \frac{1}{\sigma^2}(\mathbf{y} - \bar{\mathbf{y}})^{\mathsf{T}}(\mathbf{y} - \bar{\mathbf{y}}),$$

where $\bar{\mathbf{y}}$ is a n-vector with each component equal to $\bar{y}$. The unscaled null deviance is $\mathbf{y}^{\mathsf{T}}\mathbf{y} - n\bar{y}^2$ and is called the *total sum of squares* and is denoted by *SSTot*.

The *sum of squares due to the model* is the difference between the unscaled null deviance and the unscaled residual deviance, which is

$$\mathcal{D}_M = \mathbf{y}^{\mathsf{T}}\mathbf{y} - n\bar{y}^2 - (\mathbf{y}^{\mathsf{T}}\mathbf{y} - \hat{\beta}^{\mathsf{T}}\mathbf{X}^{\mathsf{T}}\mathbf{y}) = \hat{\beta}^{\mathsf{T}}\mathbf{X}^{\mathsf{T}}\mathbf{y} - n\bar{y}^2.$$

The sum of squares due to the model is denoted by *SSM*.

The exact sampling distributions of *SSM* and *SSE* in a classical linear model are given in Theorem 7.4.

Theorem 7.4 *If* $Y_1, \dots, Y_n$ *are independent* $N(\mathbf{X}_i^{\mathsf{T}}\vec{\beta}, \sigma^2)$ *random variables,* $\mathbf{X}^{\mathsf{T}}\mathbf{X}$ *is nonsingular, and* $(\hat{\beta}, \hat{\sigma}^2)^{\mathsf{T}}$ *is the MLE of* $(\vec{\beta}, \sigma^2)$, *then*

i) $SSM \sim \chi_p^2$.

ii) $SSM \sim \chi_{n-p-1}^2$.

iii) $\dfrac{\frac{SSM}{p}}{\frac{SSE}{n-p-1}} \sim F_{p, n-p-1}$.

Proof. The proof of Theorem 7.4 can be found in Graybill [26]. ∎

A breakdown of the sum of squares for a classical linear model is summarized in Table 7.6, which is called an *Analysis of Variance table* or ANOVA table. The F ratio given in Table 7.6 is used for testing the utility of the model and is discussed in Section 7.4.4. Also, $MSE = S^2$ is an unbiased estimator of σ^2.

Finally, the ith deviance residual, which is called an *ordinary residual* in a classical linear model, is $e_i = y_i - \mathbf{X}_i^{\mathsf{T}}\hat{\beta}$. The vector of ordinary residuals is $\mathbf{e} = \mathbf{y} - \mathbf{X}\hat{\beta}$, and the ordinary residuals are used in estimating σ^2, assessing the assumptions, and assessing the fit of a classical linear model. A detailed discussion of the model diagnostics can be found in Chatterjee and Hadi's [27].

Table 7.6 ANOVA table for a classical linear model.

Source	df	SS	MS	F
Model	p	$\text{SSM} = \hat{\beta}^T X^T y - n\bar{y}^2$	$\text{MSM} = \frac{\text{SSM}}{p}$	$F = \frac{\text{MSM}}{\text{MSE}}$
Error	$n - p - 1$	$\text{SSE} = y^T y - \hat{\beta}^T X^T y$	$\text{MSE} = \frac{\text{SSE}}{n-p-1}$	
Total	$n - 1$	$\text{SSTot} = y^T y - n\bar{y}^2$		

7.4.4 Hypothesis Testing

Because the exact sampling distributions of the MLEs and the sums of squares are known in a classical linear model, exact testing procedures can be used for testing the model utility and claims about the unknown parameters of the model.

Recall that for the model $\mu_i = X_i^T \vec{\beta}$, the model utility test is a test of $H_0 : (\beta_1, \dots, \beta_p) = \vec{0}$ versus $H_A : (\beta_1, \dots, \beta_p) \neq \vec{0}$. The test statistic for the model utility test in a classical linear model is $F_{\text{util}} = \frac{\text{MSM}}{\text{MSE}}$, which is referred to as an F *ratio*. The F ratio follows an F distribution with p and $n - p - 1$ degrees of freedom, and therefore, a size α critical region for testing $H_0 : (\beta_1, \dots, \beta_p) = \vec{0}$ versus $H_A : (\beta_1, \dots, \beta_p) \neq \vec{0}$ is

$$\mathcal{C} = \left\{ \vec{y} : \frac{\text{MSM}}{\text{MSE}} > F_{p,n-p-1,1-\alpha} \right\},$$

where $F_{p,n-p-1,1-\alpha}$ is the $(1 - \alpha)$th quantile of an F distribution with p and $n - p - 1$ degrees of freedom.

A significant model utility test suggests that the fitted model fits the observed data better than the null model does; however, it does not suggest that the fitted model is a good fitting model. On the other hand, a nonsignificant model utility test suggests that the fitted model fits the observed data no better than the null model, and hence, the full model is a poor model for explaining the response variable.

Hypothesis tests and confidence intervals for β_j are based on $\frac{\hat{\beta}_j - \beta_{j0}}{\text{se}(\hat{\beta}_j)}$, which follows a t distribution with $n - p$ degrees of freedom. In particular, the size α region for testing $H_0 : \beta_j \leq \beta_{j0}$ versus $H_A : \beta_j > \beta_{j0}$ is

$$\mathcal{C} = \left\{ \vec{y} : \frac{\hat{\beta}_j - \beta_{j0}}{\text{se}(\hat{\beta}_j)} > t_{n-p-1,1-\alpha} \right\},$$

the size α region for testing $H_0 : \beta_j \geq \beta_{j0}$ versus $H_A : \beta_j < \beta_{j0}$ is

$$\mathcal{C} = \left\{ \vec{y} : \frac{\hat{\beta}_j - \beta_{j0}}{\text{se}(\hat{\beta}_j)} < -t_{n-p-1,1-\alpha} \right\},$$

and the size α region for testing $H_0 : \beta_j = \beta_{j0}$ versus $H_A : \beta_j \neq \beta_{j0}$ is

$$
\mathcal{C} = \left\{ \vec{y} : \left| \frac{\widehat{\beta}_j - \beta_{j0}}{\mathrm{se}(\widehat{\beta}_j)} \right| > t_{n-p-1,1-\frac{\alpha}{2}} \right\}.
$$

In practice, the most commonly used test about β_j is $H_0 : \beta_j = 0$ versus $H_A : \beta_j \neq 0$, since a nonsignificant test suggests that the explanatory variable X_j is not important in the model. A $(1 - \alpha) \times 100\%$ confidence interval for β_j is $\widehat{\beta}_j \pm t_{n-p-1,1-\frac{\alpha}{2}} \times \mathrm{se}(\widehat{\beta}_j)$.

For comparing nested models, an *extra sum of squares F test* is used. Let the full model be $\mu_i = \beta_0 + X_1\vec{\beta}_1 + X_2\vec{\beta}_2$ with $p+1$ parameters and model sum of squares SSM_F, and let the reduced model be $\mu_i = \beta_0 + X_1\vec{\beta}_1$ with $k+1$ parameters and model sum of squares SSM_R. The extra sum of squares test of $H_0 : \vec{\beta}_2 = \vec{0}$ versus $H_A : \vec{\beta}_2 \neq \vec{0}$ is based on the F-ratio

$$
F = \frac{\frac{\mathrm{SSM}_F - \mathrm{SSM}_R}{p-k}}{\mathrm{MSE}_F},
$$

which follows an F distribution with $p - k$ and $n - p - 1$ degrees of freedom. Thus, a size α critical region for testing $H_0 : \vec{\beta}_2 = \vec{0}$ versus $H_A : \vec{\beta}_2 \neq \vec{0}$ is

$$
\mathcal{C} = \left\{ \vec{y} : \frac{\frac{\mathrm{SSM}_F - \mathrm{SSM}_R}{p-k}}{\mathrm{MSE}_F} > F_{p-k,n-p-1,1-\alpha} \right\}.
$$

A significant extra sum of squares F-ratio suggests that the full model fits the observed data better than does the reduced model, and a nonsignificant extra sum of squares F-ratio suggests that the reduced model fits the observed data at least as well as the full model.

Example 7.18 Using the body-fat data set given in *Applied Biostatistics for the Health Sciences* by Rossi [28], a model for the mean of an adult male's percentage body fat (PCTBF) based on the explanatory variables age, height, neck circumference, chest circumference, hip circumference, knee circumference, ankle circumference, biceps circumference, forearm circumference, and wrist circumference has been fit.

The primary objectives for this model are to (i) test the model utility, (ii) determine the importance of the explanatory variable age in the model, and (iii) compare the full model with a reduced model without the explanatory variables biceps, forearm, and wrist (i.e. the arm variables).

The full model is

$$
\mu_i = \beta_0 + \beta_1 \text{ Age} + \beta_2 \text{ Height} + \beta_3 \text{ Neck} + \beta_4 \text{ Chest} + \beta_5 \text{ Hip}
$$
$$
+ \beta_6 \text{ Knee} + \beta_7 \text{ Ankle} + \beta_8 \text{ Biceps} + \beta_9 \text{ Forearm} + \beta_{10} \text{ Wrist},
$$

Table 7.7 The R output for the full model in Example 7.18.

```
Coefficients:
               Estimate Std. Error  t value  Pr(>|t|)
(Intercept)   -28.12700    8.58611   -3.276  0.001208
Age             0.18740    0.03312    5.658  4.34e-08
Height         -0.21772    0.10384   -2.097  0.037068
Neck           -0.25808    0.27179   -0.950  0.343284
Chest           0.50391    0.08814    5.717  3.19e-08
Hip             0.43908    0.11491    3.821  0.000169
Knee            0.09717    0.27653    0.351  0.725604
Ankle          -0.03628    0.26960   -0.135  0.893076
Biceps          0.12271    0.20017    0.613  0.540456
Forearm         0.33710    0.24666    1.367  0.173011
Wrist          -2.59257    0.64984   -3.990  8.79e-05

Analysis of Variance Table

              Df  Sum Sq   Mean Sq   F value    Pr(>F)
Model         10  10705.3  1070.53    37.53    < 2.2e-16
Residuals    241   6873.7    28.5
Total        251  17579
```

and the reduced model is

$$\mu_i = \beta_0 + \beta_1 \text{ Age} + \beta_2 \text{ Height} + \beta_3 \text{ Neck} + \beta_4 \text{ Chest} + \beta_5 \text{ Hip}$$
$$+ \beta_6 \text{ Knee} + \beta_7 \text{ Ankle.}$$

The R output for the full model is given in Table 7.7, and the R output for the reduced model is given in Table 7.8. Based on the ANOVA output in Table 7.7, the F-ratio for the model utility test is $F = 37.53$ with p-value $p = 0.0000$, and therefore, the full model is significantly better at explaining the response variable PCTBF than is the null model.

Testing the importance of the explanatory variable Age in the model is carried out by testing $H_0 : \beta_1 = 0$ versus $H_A : \beta_1 \neq 0$. Based on the coefficient output in Table 7.7, there is sufficient evidence to conclude that Age is an important explanatory variable in the model for PCTBF since β_1 is significantly different from zero ($t = 5.658, p = 0.0000$).

Finally, an extra sum of squares F test can be used to test the importance of the arm variables in the model. Using the information in the ANOVA tables in Tables 7.7 and 7.8, the extra sum of squares F-ratio for testing $H_0 : \beta_8 = \beta_9 = \beta_{10} = 0$ is

$$F = \frac{\frac{\text{SSM}_F - \text{SSM}_R}{10 - 7}}{\text{MSE}_F} = \frac{\frac{10705.3 - 10229}{3}}{28.5} = 5.571$$

Table 7.8 The R output for the full model in Example 7.18.

```
Coefficients:
             Estimate Std. Error t value Pr(>|t|)
(Intercept) -34.06832    8.31575  -4.097 5.70e-05
Age           0.13458    0.03095   4.348 2.02e-05
Height       -0.30351    0.10430  -2.910 0.003949
Neck         -0.54255    0.24534  -2.211 0.027937
Chest         0.53336    0.08819   6.048 5.47e-09
Hip           0.41999    0.11560   3.633 0.000341
Knee          0.02593    0.27875   0.093 0.925956
Ankle        -0.33127    0.26525  -1.249 0.212906

Analysis of Variance Table

              Df   Sum Sq   Mean Sq   F value    Pr(>F)
Model          7    10229   1461.29     48.51  < 2.2e-16
Residuals    244     7350      30.1
```

with p-value $p = P(F_{3241} > 5.571) = 0.0010$. Thus, since the extra sum of squares F test of $H_0 : \beta_8 = \beta_9 = \beta_{10} = 0$ is significant, there is sufficient evidence that at least one of β_8, β_9, or β_{10} is not zero, and therefore, at least one of these variables is important in the model.

Problems

7.4.1 Let $Y_1, \ldots, Y_n$ be independent $N(\mu_i, \sigma^2)$ random variables with $\mu_i = \beta_0 + \beta_1 x_i$ for known constants $x_1, \ldots, x_n$.
a) Show that $\hat{\mu}(x^\star) = \hat{\beta}_0 + \hat{\beta}_1 x^\star$ can be written as a linear combination of $Y_1, \ldots, Y_n$ (i.e. $\hat{\beta}_0 + \hat{\beta}_1 x^\star = \sum_{i=1}^{b} a_i Y_i$). Hint: Substitute $\bar{y} - \hat{\beta}_1 \bar{x}$ for $\hat{\beta}_0$.
b) Show that $\text{Var}(\hat{\beta}_0 + \hat{\beta}_1 x^\star) = \sigma^2 \left(\frac{1}{n} + \frac{(x^\star - \bar{x})^2}{\sum_{i=1}^{n} (x_i - \bar{x})^2} \right)$.

7.4.2 Let $Y_1, \ldots, Y_n$ be independent $N(\mu_i, \sigma^2)$ random variables with $\mu_i = \beta x_i$ for known constants $x_1, \ldots, x_n$. Determine
a) the MLE of β.
b) the expected value of the MLE of β.
c) the variance of the MLE of β.
d) the variance of $\hat{\mu}(x^\star) = \hat{\beta} x^\star$.

7.4.3 Let $Y_1, \ldots, Y_n$ be independent $N(\mu_i, \sigma^2)$ random variables with $\mu_i = \beta_0 + \beta_1 x_{1i} + \beta_2 x_{2i}$.
a) Determine the normal equations for this model.
b) Show that $\widehat{\beta}_0 = \bar{y} - \widehat{\beta}_1 \bar{x}_1 - \widehat{\beta}_2 \bar{x}_2$.

7.4.4 Let $Y_1, \ldots, Y_n$ be independent $N(\mu_i, \sigma^2)$ random variables with $\mu_i = \beta_1 x_{1i} + \beta_2 x_{2i}$.
a) Determine the normal equations for this no-intercept model.
b) Determine the MLE of $\vec{\beta} = (\beta_1, \beta_2)^T$ when $\sum_{i=1}^n y_i x_{1i} = 500$, $\sum_{i=1}^n x_{1i}^2 = 500$, $\sum_{i=1}^n x_{1i} x_{2i} = 300$, $\sum_{i=1}^n y_i x_{2i} = 200$, and $\sum_{i=1}^n x_{2i}^2 = 100$.

7.4.5 Show that $(\mathbf{y} - \mathbf{X}\widehat{\beta})^T (\mathbf{y} - \mathbf{X}\widehat{\beta}) = \mathbf{y}^T \mathbf{y} - \widehat{\beta}^T \mathbf{X}^T \mathbf{y}$.

7.4.6 Let $Y_1, \ldots, Y_6$ be independent random variables with $Y_i \sim N(\mu_i, \sigma^2)$ and $\mu_i = \beta_0 + \beta_1 T_{1i} + \beta_2 T_{2i} + \beta_3 T_{3i}$, where T_{ji} is the indicator function associated with the jth treatment. That is,

$$T_{ji} = \begin{cases} 1 & \text{when } Y_i \text{ is associated with treatment } j \\ 0 & \text{otherwise} \end{cases}.$$

Suppose that Y_1 and Y_2 are associated with treatment 1, Y_3 and Y_4 are associated with treatment 2, and Y_5 and Y_6 are associated with treatment 3.
Let the design matrix for the reparameterization with $\beta_3 = 0$ be $\mathbf{X}'$, the design matrix for the reparameterization with $\sum_{i=1}^3 \beta_i = 0$ be $\mathbf{X}''$, and let $\mathbf{y}$ be the observed values of $Y_1, \ldots, Y_6$.

$$\mathbf{X}' = \begin{pmatrix} 1 & 1 & 0 \\ 1 & 1 & 0 \\ 1 & 0 & 1 \\ 1 & 0 & 1 \\ 1 & 0 & 0 \\ 1 & 0 & 0 \end{pmatrix} \qquad \mathbf{X}'' = \begin{pmatrix} 1 & 1 & 0 \\ 1 & 1 & 0 \\ 1 & 0 & 1 \\ 1 & 0 & 1 \\ 1 & -1 & -1 \\ 1 & -1 & -1 \end{pmatrix} \qquad \mathbf{y} = \begin{pmatrix} 10 \\ 12 \\ 16 \\ 19 \\ 11 \\ 9 \end{pmatrix}$$

a) Determine the MLEs of μ, τ_1, τ_2 under the model with $\beta_3 = 0$.
b) Determine the MLEs of μ, τ_1, τ_2 under the model with $\sum_{i=1}^3 \beta_i = 0$.
c) Determine the MLEs of $\mu_1 = \mu + \beta_1$ and $\mu_2 = \mu + \beta_2$ under the model with $\beta_3 = 0$.
d) Determine the MLE of $\mu_1 = \mu + \beta_1$ and $\mu_2 = \mu + \beta_2$ under the model with $\sum_{i=1}^3 \beta_i = 0$.

e) Determine the MLE of μ_3 using $\hat{\mu}_3 = \hat{\mu}$ under the model with $\beta_3 = 0$.

f) Determine the MLEs of μ_3 using $\hat{\mu}_3 = \hat{\mu} - \hat{\beta}_1 - \hat{\beta}_2$ under the model with $\sum_{i=1}^{3} \beta_i = 0$.

7.4.7 Prove

a) part(i) of Corollary 7.2.

b) part(ii) of Corollary 7.2.

7.4.8 Let $Y_1, Y_2, \ldots, Y_{252}$ be independent $N(\mu_i, \sigma^2)$ random variables with $\mu_i = \beta_0 + \beta_1 x_i$ for known constants $x_1, \ldots, x_n$, and suppose that

$$\mathbf{X}^{\mathrm{T}}\mathbf{X} = \begin{pmatrix} 252 & 4826 \\ 4826 & 110000.7 \end{pmatrix} \quad \mathbf{X}^{\mathrm{T}}\mathbf{y} = \begin{pmatrix} 45088.95 \\ 901295.6 \end{pmatrix}$$

$$\mathbf{y}^{\mathrm{T}}\mathbf{y} = 8284308 \qquad \bar{y} = 178.9244.$$

a) Determine the MLE of $\vec{\beta} = (\beta_0, \beta_1)^{\mathrm{T}}$.

b) Determine the ANOVA table associated with this model.

c) Test the utility of the model.

d) Determine the standard error of $\hat{\beta}_1$.

e) Test $H_0 : \beta_1 = 0$ versus $H_A : \beta_1 \neq 0$ at the $\alpha = 0.05$ level.

7.4.9 Let $Y_1, Y_2, \ldots, Y_{252}$ be independent $N(\mu_i, \sigma^2)$ random variables with $\mu_i = \beta_0 + \beta_1 X_{1i} + \beta_2 X_{2i} + \beta_3 X_{3i}$ for known constants $x_1, \ldots, x_n$, and suppose that

$$\mathbf{X}^{\mathrm{T}}\mathbf{X} = \begin{pmatrix} 252.0 & 11311.0 & 17677.5 & 25176 \\ 11311.0 & 547555.0 & 791464.5 & 1128882 \\ 17677.5 & 791464.5 & 1243423.1 & 1767188.7 \\ 25176.0 & 1128882.2 & 1767188.7 & 2528085 \end{pmatrix}$$

$$\mathbf{X}^{\mathrm{T}}\mathbf{y} = \begin{pmatrix} 266.00 \\ 11922.88 \\ 18661.62 \\ 26554.27 \end{pmatrix} \qquad \mathbf{y}^{\mathrm{T}}\mathbf{y} = 280.8784 \qquad \bar{y} = 1.05557.$$

a) Determine the MLE of $\vec{\beta} = (\beta_0, \beta_1, \beta_2, \beta_3)^{\mathrm{T}}$.

b) Determine the ANOVA table associated with this model.

c) Test the utility of the model.

d) Determine the standard error of $\hat{\beta}_2$.

e) Test $H_0 : \beta_2 = 0$ versus $H_A : \beta_2 \neq 0$ at the $\alpha = 0.05$ level.

7.4.10 Show that SSE $= \mathbf{e}^{\mathrm{T}}\mathbf{e}$.

7.4.11 Let the full model be $\mu = \beta_0 + X_1\vec{\beta_1} + X_2\vec{\beta_2}$ and the reduced model $\mu = \beta_0 + X_1\vec{\beta_1}$. Show that the extra sums of squares F ratio for testing $H_0 : \vec{\beta_2} = \vec{0}$ can be written as $F = \frac{\frac{SSE_R - SSE_F}{p-k}}{MSE_F}$.

7.4.12 Let $Y_1, \ldots, Y_n$ be independent $N(\mu_i, \sigma^2)$ random variables with $\mu_i = \beta_0 + \beta_1 x_i$ for known constants $x_1, \ldots, x_n$. Show that $E(\overline{Y}) = \beta_0 + \beta_1 \overline{x}$.

7.4.13 Show that $\sum_{i=1}^{n} y_i(x_i - \overline{x}) = \sum_{i=1}^{n} x_i y_i - n\overline{x}\,\overline{y}$ for real numbers $y_1, \ldots, y_n$ and $x_1, \ldots, x_n$.

7.4.14 Let $Y_1, \ldots, Y_n$ be independent $N(\beta_0 + \beta_1 x_i, \sigma^2)$ random variables for constants $x_1, \ldots, x_n$. Show that $\widehat{\beta_1} \sim N\left(\beta_1, \frac{\sigma^2}{\sum_{j=1}^{n} x_j^2 - n\overline{x}^2}\right)$. Hint: Use Problem 7.4.13.

7.4.15 Let $Y_1, \ldots, Y_n$ be independent $N(\mu_i, \sigma^2)$ random variables with $\mu_i = \beta_0 + \beta_1 x_i$ for known constants $x_1, \ldots, x_n$. Show that the solutions to the normal equations in Example 7.14 are $\widehat{\beta_0} = \overline{y} - \widehat{\beta_1}\overline{x}$ and $\widehat{\beta_1} = \frac{\sum_{i=1}^{n} x_i y_i - n\overline{x}\overline{Y}}{\sum_{i=1}^{n} x_i^2 - n\overline{x}^2}$.

7.4.16 Let $Y_1, \ldots, Y_n$ be independent $N(\mu_i, \sigma^2)$ random variables with $\mu_i = X_i^T \vec{\beta}$, $X^T X$ nonsingular, and $(\hat{\beta}, \hat{\sigma}^2)^T$ the MLE of $(\vec{\beta}, \sigma^2)^T$.
a) Prove part (iii) of Theorem 7.2.
b) Prove part (iv) of Theorem 7.2.

7.5 Generalized Linear Models for a Binomial Response Variable

When the response variable is a binary variable, the underlying distribution of the response variable is often assumed to follow a binomial distribution. Commonly used generalized linear models for a binomial response variable include the logistic and probit regression models.

Let $Y_1, \ldots, Y_n$ be independent random variables with $Y_i \sim \text{Bin}(n_i, p_i)$, so that $Y_i \sim \text{EF}(n_i \ln(1 + e^{p_i}), 1)$. Note that the dispersion parameter for the binomial distribution is $\phi = 1$, and therefore, the only parameters in a generalized linear model for a binomial are $\beta_0, \beta_1, \ldots, \beta_p$.

The notation used in a generalized linear model for a binomial response variable is the same as that for the classical linear model or, for that matter, every generalized linear model. That is, the response vector is $Y = (Y_1, \ldots, Y_n)^T$, the observed response vector is $y = (y_1, \ldots, y_p)^T$, the vector of explanatory variables associated with Y_i is $X_i = (1, x_{1i}, \ldots, x_{pi})^T$, and the vector of unknown parameters in the linear predictor is $\vec{\beta} = (\beta_0, \beta_1, \ldots, \beta_p)^T$. The $n \times (p+1)$ design matrix is

$$\mathbf{X} = \begin{pmatrix} 1 & X_{11} & X_{21} & \cdots & X_{p1} \\ 1 & X_{12} & X_{22} & \cdots & X_{p2} \\ 1 & \vdots & \vdots & \cdots & \vdots \\ 1 & X_{1n} & X_{2n} & \cdots & X_{pn} \end{pmatrix},$$

where the ith row of the design matrix is $\mathbf{X}_i$.

The design matrix $\mathbf{X}$ is linked to $p_i = P(Y_i = 1)$ by the link function. In particular, with a link function $g(p)$, the model for the mean of the response variable is $g(p_i) = \mathbf{X}_i \vec{\beta}$. Commonly used links for a modeling binomial response variable include the *logit link*, the *probit link*, and the *complementary log–log link*.

The logit link function is $g_L(p_i) = \ln\left[\frac{p_i}{1-p_i}\right]$, and a generalized linear model for binomial response with logit link function is called a *logistic regression model*. In a logistic regression model the linear predictor is

$$\text{logit}(p_i) = \ln\left[\frac{p_i}{1-p_i}\right] = \mathbf{X}_i^T \vec{\beta},$$

and the model for p_i is $p_i = \frac{e^{\mathbf{X}_i^T \vec{\beta}}}{1+e^{\mathbf{X}_i^T \vec{\beta}}}$.

The probit link function is $g_P(p_i) = \Phi^{-1}(p_i)$, where $\Phi(\cdot)$ is the cumulative distribution function (CDF) of an $N(0,1)$ random variable, and a generalized linear model with the probit link function is called a *probit regression model*. In a probit regression model, the model for p_i is $p_i = \Phi(\mathbf{X}_i^T \vec{\beta})$.

The complementary log–log link function is $g_C(p_i) = \ln[-\ln(1 - p_i)]$, and when the complementary log–log link function is used, the model for p_i is

$$p_i = 1 - e^{-e^{\mathbf{X}_i^T \vec{\beta}}}.$$

The logit, probit, and complementary log–log link functions are plotted in Figure 7.1. Note that the differences between these link functions increase as p moves away from 0.5, and for values of p near 0 or 1, the complementary log–log function approaches $\pm\infty$ more slowly than do the logit or probit functions.

The choice of link function is generally not important unless p is near 0 or 1, and therefore, only the logit link and logistic regression are discussed in the following sections.

7.5.1 Estimation

Let $Y_1, \ldots, Y_n$ be independent $\text{Bin}(n_i, p_i)$ random variables, and let the link function be the logit link. Then, the generalized linear model is $\ln\left[\frac{p_i}{1-p_i}\right] = \mathbf{X}_i^T \vec{\beta}$.

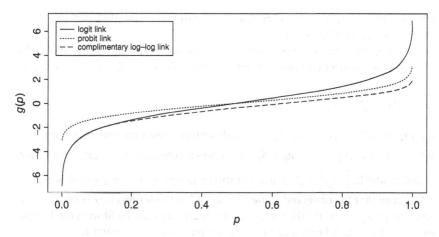

Figure 7.1 A plot of the logit, probit, and complementary log–log link functions.

Maximum likelihood estimation is used to estimate $\vec{\beta}$ since the MLE has important asymptotic properties that can be used in making inferences about $\vec{\beta}$ when the sample size is large.

Since $Y_i \sim \text{Bin}(n_i, p_i)$ and the canonical link is used in logistic regression, it follows that $\mathbf{y}^{\mathsf{T}}\mathbf{X}$ is a sufficient statistic for $\vec{\beta}$, and MLEs are based entirely on the information in $\mathbf{y}^{\mathsf{T}}\mathbf{X}$. The pdf of Y_i is

$$f(y_i; p_i) = \binom{n_i}{y_i} p_i^{y_i}(1 - p_i)^{n_i - y_i} = e^{y_i \ln\left[\frac{p_i}{1 - p_i}\right] + n_i \ln(1 - p_i) + \ln\binom{n_i}{y_i}}$$

$$= e^{y_i \mathbf{X}_i^{\mathsf{T}} \vec{\beta} - n_i \ln(1 + e^{\mathbf{X}_i^{\mathsf{T}} \vec{\beta}}) + \ln\binom{n_i}{y_i}},$$

and thus, the log-likelihood function in terms of $\vec{\beta}$ is

$$\ell(\vec{\beta}) = \sum_{i=1}^{n} \left[y_i \mathbf{X}_i^{\mathsf{T}} \vec{\beta} - n_i \ln(1 + e^{\mathbf{X}_i^{\mathsf{T}} \vec{\beta}}) + \ln\binom{n_i}{y_i} \right].$$

The MLE is found by setting the Score function equal to $\vec{0}$ and solving for $\vec{\beta}$. The jth component of the Score function is

$$\frac{\partial \ell(\vec{\beta})}{\partial \beta_j} = \sum_{i=1}^{n} \left[y_i x_{ji} - n_i x_{ji} \frac{e^{\mathbf{X}_i^{\mathsf{T}} \vec{\beta}}}{1 + e^{\mathbf{X}_i^{\mathsf{T}} \vec{\beta}}} \right].$$

Since there is no analytical solution to this system of equations, the MLE of $\vec{\beta}$ must be found numerically. The R command used to fit a model with the logit link is `glm(model,family=binomial,data)`; a model with the probit link can be fit with the command `glm(model, family = binomial(probit), data)`, and a model with the complementary log–log link can be fit with the command $\texttt{glm(model, family = binomial(cloglog), data)}$.

Applying the invariance property of the maximum likelihood estimate, the MLE of p at a particular vector of explanatory values, say $\widetilde{\mathbf{X}} = (\tilde{x}_1, \ldots, \tilde{x}_p)^{\mathrm{T}}$, is

$$\hat{p}_{\widetilde{X}} = \frac{e^{\widetilde{X}^{\mathrm{T}}\hat{\beta}}}{1 + e^{\widetilde{X}^{\mathrm{T}}\hat{\beta}}}.$$

Example 7.19 Let $Y_1, \ldots, Y_{100}$ be independent random variables with $Y_i \sim \text{Bin}(1, p_i)$ with $\ln\left[\frac{p_i}{1-p_i}\right] = \beta_0 + \beta_1 x_i$ for known constants $x_1, \ldots, x_n$. The generalized model $\ln\left[\frac{p_i}{1-p_i}\right] = \beta_0 + \beta_1 x_i$ is called a *simple logistic regression model*.

Suppose that the observed values are stored in the R objects $\mathtt{y} = (y_1, \ldots, y_n)^{\mathrm{T}}$ and $\mathtt{x} = (1, x_1, \ldots, x_n)^{\mathrm{T}}$. Then, this logistic regression can be fit with the R command $\texttt{glm(y}\sim\texttt{x,binomial())}$. The output of this command is

```
Call:  glm(formula = y~ x, family = binomial())

Coefficients:
(Intercept)              x
   -5.3095         0.1109.
```

Note that $\texttt{glm(y}\sim\texttt{x,binomial())}$ only returns the coefficients (i.e. the MLEs) in its output; however, much more information about the model can be extracted from the command $\texttt{glm(y}\sim\texttt{x,binomial())}$ (see Example 7.20).

Using the MLEs of β_0 and β_1, the estimated probability that $Y = 1$ for a given value of x is $\hat{p}_x = \frac{e^{-5.3095+0.1109\,x}}{1+e^{-5.3095+0.1109\,x}}$. In particular, the estimated probability that $Y = 1$ when $x = 50$ is $\hat{p}_{50} = \frac{e^{-5.3095+0.1109\cdot50}}{1+e^{-5.3095+0.1109\cdot50}} = 0.5589$.

7.5.2 Properties of the MLEs

The exact sampling distribution of $\hat{\beta}$ is hard to determine in the logistic regression setting, and therefore, the large sample properties of MLE are used for confidence intervals and tests about $\vec{\beta}$. The large sample properties of the MLEs include

- $\hat{\beta}$ is asymptotically unbiased for estimating $\vec{\beta}$;
- $\hat{\beta}$ is a consistent estimator of $\vec{\beta}$;
- $\hat{\beta} - \vec{\beta} \xrightarrow{d} N_p(\vec{0}, I_n(\vec{\beta})^{-1})$, where $I_n(\vec{\beta})$ is Fisher's Information matrix;
- $\hat{\beta}_j - \beta_j \xrightarrow{d} N(0, I_n(\vec{\beta})_{jj}^{-1})$, where $I_n(\vec{\beta})$ is Fisher's Information matrix;
- and the asymptotic standard error of $\hat{\beta}_j$ is $I_n(\vec{\beta})_{jj}^{-1}$.

In practice, the asymptotic standard error of $\hat{\beta}_j$ is $\text{ase}(\hat{\beta}_j) = \sqrt{I_n(\hat{\beta})_{jj}^{-1}}$, where $I_n(\hat{\beta})$ is Fisher's Information matrix evaluated at the MLE $\hat{\beta}$. For large samples, an approximate $(1 - \alpha) \times 100\%$ confidence interval for β_j is

$$\hat{\beta}_j \pm z_{1-\frac{\alpha}{2}} \times \text{ase}(\hat{\beta}_j),$$

where $z_{1-\frac{\alpha}{2}}$ is the $(1 - \frac{\alpha}{2})$th quantile of a $N(0, 1)$ random variable.

The asymptotic standard error of $\hat{p}_{\widetilde{X}}$ can be computed using the delta method. In particular, for $\hat{p}_{\widetilde{X}} = \frac{e^{\widetilde{X}^T\hat{\beta}}}{1+e^{\widetilde{X}^T\hat{\beta}}}$, the approximate variance of $\hat{p}_{\widetilde{X}}$ is

$$\hat{V}\text{ar}(\hat{p}_{\widetilde{X}}) \approx \hat{V}\text{ar}(\widetilde{X}^T\hat{\beta})\left[\frac{e^{\widetilde{X}^T\hat{\beta}}}{(1 + e^{\widetilde{X}^T\hat{\beta}})^2}\right]^2 = \hat{V}\text{ar}(\widetilde{X}^T\hat{\beta})[\hat{p}_{\widetilde{X}}(1 - \hat{p}_{\widetilde{X}})]^2,$$

and hence, the approximate asymptotic standard error is

$$\text{ase}(\hat{p}_{\widetilde{X}}) = \sqrt{\text{Var}(\widetilde{X}^T\hat{\beta})\left[\frac{e^{\widetilde{X}^T\hat{\beta}}}{(1 + e^{\widetilde{X}^T\hat{\beta}})^2}\right]^2} = \sqrt{\widetilde{X}^T I_n(\hat{\beta})^{-1}\widetilde{X}\left[\frac{e^{\widetilde{X}^T\hat{\beta}}}{(1 + e^{\widetilde{X}^T\hat{\beta}})^2}\right]^2}.$$

A $(1 - \alpha) \times 100\%$ large sample confidence interval for p at $\widetilde{X}$ can be computed in one of two ways. First, a confidence interval for $\text{logit}(p) = \widetilde{X}^T\vec{\beta}$ can be computed using

$$\widetilde{X}^T\hat{\beta} \pm z_{1-\frac{\alpha}{2}}\sqrt{\widetilde{X}^T I_n(\hat{\beta})^{-1}\widetilde{X}},$$

where $\hat{V}\text{ar}(\widetilde{X}^T\hat{\beta}) = \widetilde{X}^T I_n(\hat{\beta})^{-1}\widetilde{X}$. Then, since $\frac{e^*}{1+e^*}$ is a one-to-one function, and hence, a probability preserving transformation, it follows that a $(1 - \alpha) \times 100\%$ large sample confidence interval for p at $\widetilde{X}$ is $\frac{e^L}{1+e^L}$ to $\frac{e^U}{1+e^U}$, where L to U is the confidence interval for $p_{\widetilde{X}}$.

A second approach is to directly estimate $p_{\widetilde{X}}$ and $\text{ase}(\hat{p}_{\widetilde{X}})$. In this case, a $(1 - \alpha) \times 100\%$ large sample confidence interval for $p_{\widetilde{X}}$ is

$$\hat{p}_{\widetilde{X}} \pm z_{1-\frac{\alpha}{2}}\sqrt{\widetilde{X}^T I_n(\hat{\beta})^{-1}\widetilde{X}\,[\hat{p}_{\widetilde{X}}(1 - \hat{p}_{\widetilde{X}})]^2}.$$

Both approaches are asymptotically equivalent and generally produce similar confidence intervals for $p_{\widetilde{X}}$.

Example 7.20   For the simple linear logistic regression model mentioned in Example 7.19, the asymptotic standard errors of the MLEs can be extracted from the output of the object out_x=glm(y~ x,binomial()), which contains all of the relevant information on the fitted model. The information on $\hat{\beta}$ can extracted from out_x with the command summary(out_x) and is given in Table 7.9.

A 95% large sample confidence interval for β_1 is $0.11092 \pm 1.96 \cdot 0.02406$ or 0.062 to 0.156.

The MLE of p when $x = 50$ is

$$\hat{p}_{50} = \frac{e^{-5.30945+0.11092\cdot50}}{1 + e^{-5.30945+0.11092\cdot50}} = 0.5589.$$

Table 7.9 Summary information on the coefficients in the model fit in Example 7.19.

```
Coefficients:
            Estimate Std. Error z value Pr(>|z|)
(Intercept) -5.30945    1.13365  -4.683 2.82e-06
       x     0.11092    0.02406   4.610 4.02e-06
```

The variance of $\widehat{\beta}$ can be extracted from the R output using the vcov command. In particular, since the output is stored in the object out_x, the R command vcov(out_x) produces the variance of $\widehat{\beta}$ (i.e. $I_n(\widehat{\beta})^{-1}$), which is shown as follows:

```
            (Intercept)           x
(Intercept)  1.28517059 -0.0266769747
       x    -0.02667697  0.0005788748.
```

The approximate standard error of $\widehat{p}_{50}$ is

$$\sqrt{(1\ \ 50)\begin{pmatrix} 1.2852 & -0.02667 \\ -0.02667 & 0.00058 \end{pmatrix}\begin{pmatrix} 1 \\ 50 \end{pmatrix}}\ [0.5589(1-0.5589)]^2 = 0.064.$$

Hence, a 95% large sample confidence interval for $\widehat{p}_{50}$ is $0.5589 \pm 1.96 \cdot 0.064$ or 0.433 to 0.684.

Alternatively, basing the confidence interval for $\widehat{p}_{50}$ on a confidence interval for $\ln\left[\frac{p_i}{1-p_i}\right] = \beta_0 + \beta_1(50)$, the confidence interval for $\widehat{p}_{50}$ is 0.434 to 0.676.

7.5.3 Deviance

Recall that the residual deviance is twice the difference between the log-likelihood of the saturated and fitted models. The log-likelihood for the saturated model is based on $\widehat{\mu}_i = y_i$, and hence,

$$\ell_{\text{Sat}} = \sum_{i=1}^{n}\left[y_i\ln(y_i) - (n_i - y_i)\ln(1-y_i) + \ln\binom{n_i}{y_i}\right].$$

Thus, the residual deviance is

$$\mathcal{D} = 2[\ell_{\text{Sat}} - \ell(\widehat{\beta})] = 2\sum_{i+1}^{n}\left[y_i\ln\left(\frac{y_i}{\widehat{\mu}_i}\right) + (n_i - y_i)\ln\left(\frac{n_i - y_i}{n_i - \widehat{\mu}_i}\right)\right],$$

where $\widehat{\mu}_i = n_i\frac{e^{X_i^T\beta}}{1+e^{X_i^T\beta}}$. Note that when $Y_1, \ldots, Y_n$ are independent Bin$(1, p_i)$ random variables with $\ln\left[\frac{p_i}{1-p_i}\right] = X_i^T\vec{\beta}$, the residual deviance is

$$\mathcal{D} = 2\sum_{i+1}^{n}\left[y_i\ln\left(\frac{y_i}{\widehat{p}_i}\right) + (1-y_i)\ln\left(\frac{1-y_i}{1-\widehat{p}_i}\right)\right],$$

where $\widehat{p}_i = \frac{e^{x_i^T \beta}}{1+e^{x_i^T \beta}}$. The residual deviance is used for testing the goodness of fit of the model and comparing nested models, and these tests are discussed in Section 7.5.4.

The ith deviance residual is

$$d_i = \text{sign}(y_i - \widehat{\mu}_i)\sqrt{2\left[y_i \ln\left(\frac{y_i}{\widehat{\mu}_i}\right) + (n_i - y_i)\ln\left(\frac{n_i - y_i}{n_i - \widehat{\mu}_i}\right)\right]},$$

and again, a large deviance residual indicates that the model does not fit this particular observation very well. Determining whether or not a deviance residual is large is generally based on the *standardized deviance residual*. The standardized deviance residual is the ratio of the deviance residual to its standard error. In particular,

$$d_{si} = \frac{d_i}{\sqrt{1 - H_i}},$$

where H_i is the ith diagonal element of the *hat matrix*

$$H = V^{\frac{1}{2}}X(X^TVX)^{-1}X^TV^{\frac{1}{2}}$$

and where V is a diagonal matrix with $V_{ii} = n_i\widehat{p}_i(1 - \widehat{p}_i)$. A standardized deviance residual is generally considered large when $|d_{si}| > 3$.

Example 7.21 For the simple linear logistic model mentioned in Example 7.19, the 97th observation is $y = 0$ with predicted value $\widehat{p} = 0.8567$ and $H_{97} = 0.0296$. Thus, the deviance residual for the 97th observation is

$$d_{97} = \text{sign}(0 - 0.8567)\sqrt{2\left[0 \cdot \ln\left(\frac{0}{0.8567}\right) + (1 - 0)\ln\left(\frac{1 - 0}{1 - 0.8567}\right)\right]}$$
$$= -1.971,$$

and the standardized deviance residual for the 97th observation is

$$d_{s,97} = \frac{d_{97}}{\sqrt{1 - H_{97}}} = \frac{-1.971}{\sqrt{1 - 0.0296}} = -2.001.$$

Hence, the deviance residual for the 97th observation is not an unusually large residual.

7.5.4 Hypothesis Testing

Because the exact sampling distributions of the MLE and the deviance are generally unknown in logistic regression, large sample tests are used for testing the goodness of fit of a model, the model utility, the parameters in the model, and for comparing nested models.

The goodness-of-fit test is based on the residual deviance, and for a sufficiently large sample size, the size α critical region is

$$\mathcal{C}_{GF} = \{\vec{y} : \mathcal{D} > \chi^2_{K-p,1-\alpha}\},$$

where K is the degrees of freedom of the saturated model and p is the degrees of freedom of the model fit; $K = n$ when there are no identical observations.

A significant goodness-of-fit test indicates that there is a significant lack of fit with the model, and therefore, the model is not fitting the observed data very well. On the other hand, a nonsignificant goodness-of-fit test suggests that there is insufficient evidence to conclude that there is a lack of fit between the model and the observed data.

The model utility test of $H_0 : (\beta_1, \ldots, \beta_p) = \vec{0}$ versus $H_A : (\beta_1, \ldots, \beta_p) \neq \vec{0}$ is based on the difference between the null and residual deviance. The test statistic for the model utility test is

$$\mathcal{D}_{Null} - \mathcal{D}_{Res} = 2[\ell_{Sat} - \ell(\tilde{\beta}_0)] - 2[l_{Sat} - \ell(\hat{\beta})] = 2[\ell(\hat{\beta}) - \ell(\tilde{\beta}_0)].$$

The large sample critical region for the model utility test is

$$\mathcal{C}_{MU} = \{\vec{y} : \mathcal{D} > \chi^2_{p,1-\alpha}\}.$$

A significant model utility test indicates that the model fit is significantly better at explaining the response variable than a model containing no explanatory variables, and a nonsignificant model utility test suggests that the model fit is no better than the model without the explanatory variables for explaining the response variable.

Tests of the parameters in the model are based on the large sample distribution of the MLEs of the parameters. In particular, Wald tests are used for testing claims about β_j and are based on the test statistic $\frac{\hat{\beta}_j - \beta_j}{ase(\hat{\beta}_j)}$, which for a sufficiently large sample size is approximately distributed as a $N(0, 1)$ random variable. Thus, the large sample critical region for testing $H_0 : \beta_j \leq \beta_{j0}$ versus $H_A : \beta_j > \beta_{j0}$ is

$$\mathcal{C} = \left\{\vec{y} : \frac{\hat{\beta}_j - \beta_{j0}}{ase(\hat{\beta}_j)} > z_{1-\alpha}\right\},$$

the size α region for testing $H_0 : \beta_j \geq \beta_{j0}$ versus $H_A : \beta_j < \beta_{j0}$ is

$$\mathcal{C} = \left\{\vec{y} : \frac{\hat{\beta}_j - \beta_{j0}}{ase(\hat{\beta}_j)} < -z_\alpha\right\},$$

and the size α region for testing $H_0 : \beta_j = \beta_{j0}$ versus $H_A : \beta_j \neq \beta_{j0}$ is

$$\mathcal{C} = \left\{\vec{y} : \left|\frac{\hat{\beta}_j - \beta_{j0}}{ase(\hat{\beta}_j)}\right| > z_{1-\frac{\alpha}{2}}\right\}.$$

Again, in practice, the most common test about β_j is $H_0 : \beta_j = 0$ versus $H_A : \beta_j \neq 0$ since a nonsignificant test suggests that the explanatory variable X_j is not important in the model.

For simultaneously testing the importance of two or more variables in the full model $\text{logit}(p) = \beta_0 + \beta_1 X_1 + \cdots + \beta_p X_p$, an *extra deviance test* can be used. The extra deviance test is based on the difference between the residual deviance for full model and the residual deviance for the model without the variables being investigated.

Let the full model be $\ln\left[\frac{p_i}{1-p_i}\right] = \beta_0 + X_1\vec{\beta}_1 + X_2\vec{\beta}_2$ and the reduced model be $\ln\left[\frac{p_i}{1-p_i}\right] = \beta_0 + X_1\vec{\beta}_1$, and let $\mathcal{D}_F$ and $\mathcal{D}_R$ be the residual deviances for the full and reduced models. The extra deviance test statistic for testing $H_0 : \vec{\beta}_2 = \vec{0}$ versus $H_A : \vec{\beta}_2 \neq \vec{0}$ is $\Delta\mathcal{D} = \mathcal{D}_R - \mathcal{D}_F$, and the large sample critical region is $\mathcal{C} = \{\vec{y} : \Delta\mathcal{D} > \chi^2_{k,1-\alpha}\}$, where $\vec{\beta}_2$ has k parameters.

A significant extra deviance test suggests that there is sufficient evidence supporting the use of the full model, and a nonsignificant extra deviance test suggests that the reduced model is at least as good as the full model for explaining the response variable.

Example 7.22 Suppose that the residual deviance for the full model

$$\text{logit}(p) = \beta_0 + \beta_1 X_1 + \cdots + \beta_9 X_9$$

is 152.245 with 140 degrees of freedom, and the residual deviance for the reduced model,

$$\text{logit}(p) = \beta_0 + \beta_1 X_1 + \cdots + \beta_5 X_5,$$

is 158.695 with 144 degrees of freedom.

First, there is no evidence of a lack of fit of the model since the residual deviance of the full model is 152.245 with 140 degrees of freedom, and the p-value for the goodness-of-fit test is $p = 0.2263$.

Now, for testing $H_0 : (\beta_6, \beta_7, \beta_8, \beta_9) = \vec{0}$, the extra deviance test statistic is $\mathcal{D}_R - \mathcal{D}_F = 158.695 - 152.245 = 6.450$ with 4 degrees freedom and p-value $p = 0.1680$. Thus, based on the extra deviance test, the data supports the use of the reduced model over the full model, and hence, the variables $X_6, X_7, X_8,$ and X_9 do not appear to be important for explaining the response variable.

Problems

7.5.1 Let $Y_1, \ldots, Y_n$ be independent $\text{Bin}(1, p_i)$ random variables with $\ln\left[\frac{p_i}{1-p_i}\right] = \beta x_i$ for known constants $x_1, \ldots, x_n$.
  a) Determine the log-likelihood function as a function of β.
  b) Determine the Score function as a function of β.

c) Determine Fisher's Information as a function of β.

d) Determine the asymptotic standard error of $\hat{\beta}$.

7.5.2 Using the information in Example 7.20,

a) estimate the probability that $Y = 1$ when $x = 30$.

b) compute an approximate 95% confidence interval for the probability $Y = 1$ when $x = 30$ based on the confidence interval for logit(p).

c) compute an approximate 95% confidence interval for the probability $Y = 1$ when $x = 30$ based on $\hat{p}$ and ase($\hat{p}$).

7.5.3 Use the output given in Table 7.10 for the logistic regression model $\ln\left[\frac{p_i}{1-p_i}\right] = \beta_0 + \beta_1 X_1 + \beta_2 X_2 + \beta_3 X_3 + \beta_4 X_4$ to

a) test the goodness of fit of the model.

b) test the utility of the model.

c) test $H_0 : \beta_1 = 0$ versus $H_A : \beta_1 \neq 0$.

d) estimate the probability that $Y = 1$ when $X_1 = 35, X_2 = 100, X_3 = 95, X_4 = 30$.

7.5.4 Use the output given in Table 7.11 for the logistic regression model $\ln\left[\frac{p_i}{1-p_i}\right] = \beta_0 + \beta_2 X_2 + \beta_3 X_3$, a subset of the model fit in Problem 7.5.3, to

a) test $H_0 : \beta_1 = \beta_4 = 0$.

b) estimate the probability that $Y = 1$ when $X_2 = 100, X_3 = 100$ for the model with the explanatory variables $X2$ and $X3$.

7.5.5 Let $Y_1, \ldots, Y_n$ be independent Bin($1, p_i$) random variables. For a generalized linear model with the complementary log–log link function and linear predictor $\ln[-\ln(1 - p)] = \beta_0 + \beta_1 x$,

a) express p_i as a function of $\vec{\beta}$.

b) estimate p when $x = 10$ if $\hat{\beta} = (12.11, -1.29)^{\mathrm{T}}$.

Table 7.10 Summary information for Problem 7.5.3.

```
Coefficients:
              Estimate   Std. Error  z value   Pr(>|z|)
(Intercept)  -73.45040   10.80990    -6.795    1.09e-11
      X1       0.10958    0.18098     0.605     0.5449
      X2       0.46189    0.09212     5.014     5.34e-07
      X3       0.17792    0.08016     2.220     0.0264
      X4       0.17067    0.13907     1.227     0.2197

    Null deviance: 349.33  on 251  degrees of freedom
Residual deviance: 116.66  on 247  degrees of freedom
```

Table 7.11 Summary information for Problem 7.5.4.

```
Coefficients:
            Estimate Std. Error z value Pr(>|z|)
(Intercept) -72.37606   10.51586  -6.883 5.88e-12
    X2        0.50803    0.08732   5.818 5.96e-09
    X3        0.21760    0.07737   2.812 0.00492
    Null deviance: 349.33  on 251  degrees of freedom
Residual deviance: 119.14  on 249  degrees of freedom
```

7.5.6 Let $Y_1, \dots, Y_n$ be independent $\text{Bin}(1, p_i)$ random variables. For a generalized linear model with the logit link, show that for

a) $y = 0$, the deviance residual is $-\sqrt{2 \ln \left[\frac{1}{1-\hat{p}} \right]}$.

b) $y = 1$, the deviance residual is $\sqrt{2 \ln \left[\frac{1}{\hat{p}} \right]}$.

7.5.7 Let $Y_1, \dots, Y_n$ be independent $\text{Bin}(1, p_i)$ random variables. For a generalized linear model with the logit link and $y_1 = 0$ and $\hat{p}_1 = 0.74$,
a) compute the value of the deviance residual.
b) compute the value of the standardized residual if $H_1 = 0.232$.

7.5.8 Suppose that the residual deviance for the full logistic regression model with explanatory variables $X_1, \dots, X_7$ is 183.335 with 204 degrees of freedom, and the residual deviance for the logistic regression model with explanatory variables X_1, X_2, X_3, and X_4 is 197.867 with 207 degrees of freedom.
a) Test the goodness of fit of the full model at the $\alpha = 0.05$ level.
b) Test $H_0 : \beta_5 = \beta_6 = \beta_7$ at the $\alpha = 0.01$ level.

7.5.9 If H is the hat matrix, show that $H_i = n_i \hat{p}_i (1 - \hat{p}_i) \mathbf{X}_i^T (\mathbf{X}^T \mathbf{V} \mathbf{X})^{-1} \mathbf{X}_i$.

7.6 Case Study – IDNAP Experiment with Poisson Count Data

In the article "Iron-Doped Apatite Nanoparticle Influence on Viral Infection Examined in Prokaryotic and Eukaryotic Systems" by Andriolo et al. [29], the authors studied the potential of using iron-doped apatite nanoparticles (IDANP) for treating antibiotic-resistant bacteria. Two experiments were run on IDANP-exposed Staphylococcus aureus bacterium. An exploratory experiment with nine treatments was first run, followed up by a second experiment investigating the four best treatments from the first experiment.

Table 7.12 Treatments used in the Staphylococcus aureus phage nanoparticle experiment.

Treatments		
1: Cells only (control)	4: 25C Cit 30%Fe	7: 65C Cit 30%Fe
2: 25C Cit 0%Fe	5: 25C Cit 40%Fe	8: 45C no Cit 30%Fe
3: 25C Cit 20%Fe	6: 45C Cit 30%Fe	9: 45C 2XCit 30%Fe

The nine treatments listed in Table 7.12 were first studied to determine the influence of IDANP-exposed cells on viral infection in a completely randomized design experiment with each treatment replicated three times. The response variable was the death zone counts, which are typically referred to as plaques in prokaryotic host/virus plating experiments. The four treatments that end up having the significantly largest number of plaques are to be compared in the second experiment; however, only the analysis of the first experiment is discussed in detail here since a similar analysis was used in the second experiment.

7.6.1 The Model

The response variable in each experiment is assumed to follow a Poisson distribution with $Y_{ij} \sim \text{Poisson}(\lambda_{ij})$, where Y_{ij} is the plaque count for the jth replicate of the ith treatment. Under the model $Y_{ij} \sim \text{Poisson}(\lambda_{ij})$, the pdf of Y_{ij} is

$$f(y_{ij}; \lambda_{ij}) = \frac{e^{-\lambda_{ij}} \lambda_{ij}^{y_{ij}}}{y_{ij}!} = e^{y_{ij} \ln(\lambda_{ij}) - \lambda_{ij} - \ln(y_{ij}!)} = e^{y_{ij}\theta_{ij} - e^{\theta_{ij}} - \ln(y_{ij}!)}.$$

Thus, $Y_{ij} \sim \text{EF}(e^{\theta_{ij}}, 1)$ with canonical parameter $\theta_{ij} = \ln(\lambda_{ij})$ and canonical link function is $g(\lambda_{ij}) = \ln(\lambda_{ij})$. A generalized linear model with the link function $g(\lambda_{ij}) = \ln(\lambda_{ij})$ and linear predictor $\ln(\mu_{ij}) = \alpha + \tau_i$, where τ_i is the ith treatment effect, was fit to the experimental data resulting from each experiment; however, because the design matrix is nonsingular, the model was constrained by setting $\tau_1 = 0$, where τ_1 is the control treatment effect. This model is called a *Poisson regression model*.

7.6.2 Statistical Methods

Maximum likelihood estimation was used to estimate the model parameters, and the log-likelihood function for each experiment is

$$\ell(\alpha, \vec{\tau}) = \sum_{i=1}^{t} \sum_{j=1}^{r} [y_{ij}\theta_{ij} - e^{\theta_{ij}} - \ln(y_{ij}!)]$$

$$= \sum_{i=1}^{t} \sum_{j=1}^{r} [y_{ij}(\alpha + \tau_i) - e^{\alpha + \tau_i} - \ln(y_{ij}!)],$$

where t is the number of treatments in the experiment and r is the number of times each treatment was replicated. The partial derivatives of $\ell(\alpha, \vec{\tau})$ making up the Score function are

$$\frac{\partial \ell(\alpha, \vec{\tau})}{\partial \alpha} = \sum_{i=1}^{t} \sum_{j=1}^{r} [y_{ij} - e^{\alpha + \tau_i}]$$

$$\frac{\partial \ell(\alpha, \vec{\tau})}{\partial \tau_k} = \sum_{j=1}^{r} y_{kj} - e^{\alpha + \tau_k}; k = 1, \ldots, t.$$

The MLEs of the parameters in the model must be found numerically since there are no analytic solutions to $\vec{\ell}(\alpha, \vec{\tau}) = \vec{0}$; however, the MLE of the kth treatment mean $\mu + \tau_k$ is $\hat{\mu}_k = e^{\hat{\alpha} + \hat{\tau}_k} = \bar{y}_{k\bullet}$, where $\bar{y}_{k\bullet} = \frac{1}{r} \sum_{j=1}^{r} y_{kj}$.

The difference between the MLEs of two treatment effects in this model is

$$\hat{\tau}_k - \hat{\tau}_j = \hat{\alpha} + \hat{\tau}_k - (\hat{\alpha} + \hat{\tau}_j) = \ln(\hat{\alpha}_k) - \ln(\hat{\alpha}_j) = \ln\left(\frac{\hat{\mu}_k}{\hat{\mu}_j}\right)$$

Thus, exponentiating difference in two treatment effects is equal to ratio of their corresponding treatment means. That is, $e^{\hat{\tau}_k - \hat{\tau}_j} = \frac{\hat{\mu}_k}{\hat{\mu}_j} = \frac{\bar{y}_{k\bullet}}{\bar{y}_{j\bullet}}$.

The residual deviance for a Poisson regression model with logarithmic canonical link function is

$$\mathcal{D}_{Res} = 2 \sum_{i=1}^{t} \sum_{j=1}^{r} \left[y_{ij} \ln\left(\frac{y_{ij}}{\hat{\mu}_{ij}}\right) - (y_{ij} - \hat{\mu}_{ij}) \right].$$

For the model $\ln(\mu_i) = \mu + \tau_i$, the residual deviance is

$$\mathcal{D}_{Res} = 2 \sum_{i=1}^{t} \sum_{j=1}^{r} \left[y_{ij} \ln\left(\frac{y_{ij}}{e^{\hat{\alpha} + \hat{\tau}_i}}\right) - (y_{ij} - e^{\hat{\alpha} + \hat{\tau}_i}) \right].$$

The residual deviance is used to test the goodness of fit for the model and will also be used in the model utility test.

The deviance residuals and standardized deviance residuals are used to identify any points with unusually large residuals. The ith deviance residual is

$$d_{ij} = \text{sign}(y_{ij} - \hat{\mu}_{ij}) \sqrt{2 \left[y_{ij} \ln\left(\frac{y_{ij}}{e^{\hat{\alpha} + \hat{\tau}_i}}\right) - (y_{ij} - e^{\hat{\alpha} + \hat{\tau}_i}) \right]},$$

and the ith standardize deviance residual is

$$d_{sij} = \frac{d_{ij}}{\sqrt{1 - H_i}},$$

where the hat matrix is $H = V^{1/2} X (X^T V X)^{-1} X^T V^{1/2}$ and V is a diagonal matrix with $\widehat{\text{Var}}(Y_{ij}) = e^{\hat{\alpha} + \hat{\tau}_i}$ on the diagonal.

The asymptotic properties of the MLEs and likelihood function are used for testing the goodness of the model, the model utility, and for pairwise comparisons between the treatment effects, which were used to determine which treatments are significantly different from one another. The asymptotic standard errors of the MLEs are based on the estimated Fisher's Information matrix evaluated at $\hat{\alpha}$ and $\hat{\tau}$. For example, the asymptotic standard error of $\hat{\tau}_k$ is

$$\text{ase}(\hat{\tau}_k) = \sqrt{I_n(\hat{\alpha}, \hat{\tau}_1, \dots, \hat{\tau}_t)_{kk}^{-1}}.$$

The goodness-of-fit test is based on the residual deviance, and the large sample size α critical region for testing the goodness of fit of the model is

$$\mathcal{C}_{\text{GF}} = \{\vec{y} : \mathcal{D}_{\text{Res}} > \chi^2_{K-p,1-\alpha}\},$$

where K is the degrees of freedom of the saturated model and p is the degrees of freedom associated with the fitted model.

The model utility test is based on the difference between the null deviance and the residual deviance, and the large sample size α critical region for testing that all of the treatment effects are equal (i.e. $H_0 : \tau_1 = \cdots = \tau_t$) is

$$\mathcal{C}_{\text{MU}} = \{\vec{y} : \mathcal{D}_{\text{Null}} - \mathcal{D}_{\text{res}} > \chi^2_{p,1-\alpha}\}.$$

Confidence intervals and hypothesis tests for $\tau_1, \dots, \tau_t$ are based on the asymptotic distribution of $\frac{\hat{\tau}_i - \tau_i}{\text{ase}(\hat{\tau}_i)}$, which is approximately distributed as $N(0, 1)$ when n is sufficiently large.

Pairwise comparisons among the treatment effects are based on the *Bonferroni multiple comparison procedure*. A multiple comparison procedure controls for *experimentwise error rate* rather than the *per-comparisonwise error rate* to minimize the chance of making one or more Type I errors.

Definition 7.3 In a multiple comparison procedure, the experimentwise error rate α_{EW} is the probability that at least one of the significance tests results in a Type I error, and the per-comparisonwise error rate α_{PC} is the probability that a Type I error is made in an individual significance test.

The Bonferroni multiple comparison procedure is based on Boole's Inequality, Theorem 1.7, which states that $P(\cup_{i=1}^n A_i) \leq \sum_{i=1}^n P(A_i)$. Thus, if A_i is the event that a Type I error is made on the ith test, then $P(\cup_{i=1}^n A_i)$ is the experimentwise error rate and $P(A_i)$ is the per-comparisonwise error rate. Hence, when k comparisons are being made, by choosing $\alpha_{\text{PC}} = \frac{\alpha_{\text{EW}}}{k}$, it follows that

$$P(\cup_{i=1}^k A_i) \leq \sum_{i=1}^k P(A_i) = \sum_{i=1}^k \frac{\alpha_{\text{EW}}}{k} = \alpha_{\text{EW}}.$$

The Bonferroni multiple comparison procedure is based on a per-comparisonwise error rate of $\frac{\alpha_{\text{EW}}}{k}$ to control the experimentwise error rate.

For example, if 10 comparisons are to be made and the experimentwise error rate should be no larger than $\alpha_{EW} = 0.05$, then the per-comparisonwise error rate in the Bonferroni procedure is $\alpha_{PC} = \frac{0.05}{10} = 0.005$.

7.6.3 Results of the First Experiment

The first IDANP experiment was an exploratory experiment investigating the nine treatments given in Table 7.13. The nine treatments were replicated three times, and the generalized linear model fit to the data is $\ln(\mu_i) = \alpha + \tau_i$ for $i = 1, \ldots, 9$, where the control treatment is treatment 1 and the model is constrained so that $\tau_1 = 0$. The R command $\texttt{glm(model, family=poisson)}$ was used to fit the model, and the output for the fitted model is given in Table 7.13.

There were no unusually large deviance residuals, and the test statistic for the goodness-of-fit test is the residual deviance, which is 7.4797 with 18 degrees of freedom. The p-value for the goodness-of-fit test is $p = 0.0.9854$, and therefore, there is no apparent lack of fit between the model and the observed data.

The model utility test of $H_0 : \tau_1 = \cdots = \tau_9 = 0$ is the difference in the null and residual deviances. The value of the test statistic for the model utility test is

$$\mathcal{D}_{Null} - \mathcal{D}_{Res} = 44.7732 - 7.4797 = 37.2935$$

with 8 degrees of freedom. The p-value for the model utility test is $p = 0.00001$, and therefore, there appear to be at least two treatment effects that are not equal.

The treatment effects were separated with the Bonferroni multiple comparison procedure and pairwise tests of $H_0 : \tau_i = \tau_j$ for $i \neq j$. Since there were nine

Table 7.13 Coefficient summary for the model $\ln(\mu_i) = \alpha + \tau_i$ for the first experiment.

	Estimate	Std. Error	z value	Pr(>|z|)
Intercept	3.2708	0.1125	29.072	< 2e-16
25°C.Cit0%Fe	0.5650	0.1409	4.010	6.07e-05
25°C.Cit20%Fe	0.3668	0.1464	2.505	0.012235
25°C.Cit30%Fe	0.6931	0.1378	5.030	4.90e-07
25°C.Cit40%Fe	0.4012	0.1454	2.760	0.005779
45°C.2XCit30%Fe	0.6142	0.1397	4.397	1.10e-05
45°C.Cit30%Fe	0.5578	0.1411	3.954	7.69e-05
45°C.NoCit30%Fe	0.3219	0.1478	2.179	0.029365
65°C.Cit30%Fe	0.4826	0.1431	3.373	0.000744

Null deviance: 44.7732 on 26 degrees of freedom
Residual deviance: 7.4797 on 18 degrees of freedom

treatments, there are $\binom{9}{2} = 36$ pairwise comparisons to be made, and using per-comparisonwise error rate of $\alpha_{PC} = \frac{0.10}{36} = 0.0028$, the Bonferroni experimentwise error rate is no more than $\alpha_{PC} = 0.10$.

In particular, the pairwise comparison between the ith and jth treatments is based on

$$\frac{\hat{\tau}_i - \hat{\tau}_j}{\text{ase}(\hat{\tau}_i - \hat{\tau}_j)},$$

where

$$\text{ase}(\hat{\tau}_i - \hat{\tau}_j) = \mathbf{c}_{ij}^T I_n(\hat{\alpha}, \hat{\tau})^{-1} \mathbf{c}_{ij}$$

and $\mathbf{c}_{ij}^T$ is a vector with 1 in the τ_i component and a -1 in the τ_j component; for example, with $\vec{\beta} = c(\mu, \tau_1, \ldots, \tau_9)^T$, the contrast $\mathbf{c}_{12}^T = (0, 1, -1, 0, 0, 0, 0, 0, 0, 0)$ is used in a comparison of τ_1 and τ_2 since $\mathbf{c}_{12}^T \vec{\beta} = \tau_1 - \tau_2$.

The comparisons of treatment effects that were significant according to the Bonferroni multiple comparison procedure with $\alpha_{EW} = 0.10$ are given in Table 7.14. Note that the only significant differences are between the control treatment and the IDANP treatments 25C Cit 0%Fe, 25°C Cit 30%Fe, 45°C 2XCit 30%Fe, 45°C Cit 30%Fe, and 65C Cit 30%Fe.

Confidence intervals for the ratio of the means for the control treatment and the treatments 25C Cit 0%Fe, 25°C Cit 30%Fe, 45°C 2XCit 30%Fe, 45°C Cit 30%Fe, and 65C Cit 30%Fe are based on confidence intervals for the difference in treatment effects. In particular, since $\tau_i - \tau_j = \mu + \tau_i - (\mu + \tau_j) = \ln(\mu_i) - \ln(\mu_j)$, it follows that exponentiating the limits of the resulting confidence interval produces a confidence interval for the ratio $\frac{\mu_i}{\mu_j}$. For example, based on the information in Table 7.14, an approximate 95% confidence interval for the ratio of the 25C Cit 0%Fe mean and the control treatment mean is $e^{0.56503 - 1.96 \times 0.14090} = 1.335$ to $e^{0.56503 + 1.96 \times 0.14090} = 2.320$. Thus, based on the experimental data, it appears that the 25C Cit 0%Fe mean is at least 133.5% of the control mean and no more than 232% of the

Table 7.14 A summary of the significant results after performing the Bonferroni multiple comparison procedure with $\alpha_{EW} = 0.10$.

Comparison	Estimate	Std. Error	z value	Pr(>|z|)
CO - 25°C.Cit0%Fe == 0	-0.56503	0.14090	-4.010	0.0001
CO - 25°C.Cit30%Fe == 0	-0.69315	0.13780	-5.030	0.0000
CO - 45°C.2XCit30%Fe == 0	-0.61416	0.13967	-4.397	0.0000
CO - 45°C.Cit30%Fe == 0	-0.55781	0.14108	-3.954	0.0001
CO - 65°C.Cit30%Fe == 0	-0.48258	0.14308	-3.373	0.0007

Table 7.15

Treatment compared with control	Ratio of sample means sample means	95% confidence interval for ratio of means
25C Cit 0%Fe	1.760	1.335–2.320
25°C Cit 30%Fe	2.000	1.527–2.618
45°C 2XCit 30%Fe	1.848	1.406–2.427
45°C Cit 30%Fe	1.748	1.325–2.304
65C Cit 30%Fe	1.621	1.224–2.146

control mean. The confidence intervals for the ratios of the means for the significant comparisons are given in Table 7.15.

The four treatments with the largest means are 25C Cit 0%Fe, 25°C Cit 30%Fe, 45°C 2XCit 30%Fe, and 45°C Cit 30%Fe, and therefore, these four treatments were studied in a second experiment run as a completely randomized design with eight replicates of these treatments along with a control treatment. A Poisson regression model was also used for the second experiment, and the analysis of the resulting data followed a similar analysis to the one used for experiment 1.

Problems

7.6.1 Let $Y \sim \text{Pois}(\lambda) \sim \text{EF}(e^{\theta}, 1)$, where $\theta = \ln(\lambda)$. Show that,
   a) $E(Y) = \lambda$.
   b) $\text{Var}(Y) = \lambda$.

7.6.2 Let $Y_1, \ldots, Y_n$ be independent $\text{Pois}(\lambda_i)$ random variables, and let $\ln(\lambda_i) = \beta x_i$ for known constants $x_1, \ldots, x_n$. Determine
   a) the log-likelihood function as a function of $\vec{\beta}$.
   b) $\frac{\partial \ell(\vec{\beta})}{\partial \beta}$.
   c) the residual deviance as a function of β.
   d) d_i as a function of $\hat{\beta}$.

7.6.3 Using the information in Table 7.13, determine the
   a) the mean for the 25C Cit 0%Fe treatment, $\hat{\mu}_2 = e^{\widehat{\alpha + \tau_2}}$.
   b) deviance residual for $y_{23} = 52$.
   c) standardized deviance residual for $y_{23} = 52$ and $H_{23} = 0.33$

7.6.4 Determine the design matrix for the first experiment
   a) without any constraints on the parameters in the model.
   b) with the parameters in the model constrained so that $\tau_1 = 0$.

7.6.5 Show that when the k comparisons in a multiple comparison problem are independent, the experimentwise error rate is $\alpha_{EW} = 1 - (1 - \alpha_{PC})^k$, where α_{EW} is the experimentwise error rate and α_{PC} is the common per-comparisonwise error rate.

7.6.6 Using the information in Table 7.14, compute a 95% confidence interval for the difference of the treatment effects of the control and 25C Cit 0%Fe treatments.

7.6.7 For the Bonferroni multiple comparison procedure with $\alpha_{PC} = 0.001$, determine the upper bound on the experimentwise error rate when there are
a) $k = 25$ pairwise comparisons.
b) $k = 10$ pairwise comparisons.

7.6.8 Determine the per-comparisonwise error rate for the Bonferroni multiple comparison procedure when
a) $\alpha_{EW} = 0.05$ and $k = 15$
b) $\alpha_{EW} = 0.05$ and there are eight treatments to be compared in a pairwise fashion.

7.6.9 Show that $\text{asyvar}(\mathbf{c}^T \widehat{\beta}) = \mathbf{c}^T I_n(\vec{\beta})^{-1} \mathbf{c}$ for any vector of appropriate dimensions $\mathbf{c}$.

References

1 Kolmogorov, A. (1933). *Grundbegriffe der Wahrscheinlichkeitsrechnung.* Berlin: Julius Springer.
2 Fisher, R.A. (1922). On the mathematical foundations of theoretical statistics. *Philosophical Transactions of the Royal Society of London, Series A: Mathematical, Physical and Engineering Sciences* 222: 309–368.
3 Neyman, J. (1935). Sur un teorema concernente le cosidette statistiche sufficient. *Giorn. Ist. Ital. Att.* 6: 320–334.
4 Lehmann, E. and Scheffé, H. (1950). Completeness, similar regions, and unbiased estimation. *Sankhya* 10: 305–340.
5 Rao, C.R. (1945). Information and accuracy attainable in the estimation of statistical parameters. *Bulletin of the Calcutta Mathematical Society* 37 (3): 81–91.
6 Blackwell, D. (1947). Conditional expectation and unbiased sequential estimation. *Annals of Mathematical Statistics* 18 (1): 105–110.
7 Darmois, G. (1935). Sur les lois de probabilites a estimation exhaustive. *C.R. Acad. Sci. Paris* 200: 1265–1266.
8 Pitman, E. and Wishart, J. (1936). Sufficient statistics and intrinsic accuracy. *Mathematical Proceedings of the Cambridge Philosophical Society* 32 (4): 567–579.
9 Koopman, B. (1936). On distribution admitting a sufficient statistic. *Transactions of the American Mathematical Society* 39 (3): 399–409.
10 Cramér, H. (1946). *Mathematical Methods of Statistics.* Hoboken, NJ: Princeton University Press.
11 Darmois, G. (1945). Sur les limites de la dispersion de certaines estimations. *Rev. Int. Stat. Inst.* 13 (1/4): 9.
12 Fréchet, M. (1943). Sur l'extension de certaines évaluations statistiques au cas de petits echantillons. *Int. Stat. Rev.* 11: 182–205.
13 Casella, G. and Berger, R.L. (2002). *Statistical Inference*, 2e. Duxbury Press.
14 Wald, A. and Wolfowitz, J. (1939). Confidence limits for continuous distribution functions. *The Annals of Mathematical Statistics* 10 (2): 105–118.

Mathematical Statistics: An Introduction to Likelihood Based Inference, First Edition. Richard J. Rossi.
© 2018 John Wiley & Sons, Inc. Published 2018 by John Wiley & Sons, Inc.

15 Adams, K.D., Locke, W.W., and Rossi, R. (1992). Obsidian-hydration dating of fluvially reworked sediments in the West Yellowstone region, Montana. *Quaternary Research* 38: 180–195.

16 Silverman, B.W. (1986). *Density Estimation for Statistics and Data Analysis,* Monographs on Statistics and Applied Probability. Chapman and Hall.

17 McLachlan, G. and Peel, D. (2000). *Finite Mixture Models.* New York: Wiley.

18 Efron, B. and Tibshirani, R. (1993). *An Introduction to the Bootstrap.* Boca Raton, FL: Chapman & Hall/CRC.

19 Neyman, J. and Pearson, E.S. (1933). On the problem of the most efficient tests of statistical hypotheses. *Philosophical Transactions of the Royal Society of London, Series A: Mathematical, Physical and Engineering Sciences* 231: 289–337.

20 Lehmnan, E.L. (1959). *Testing Statistical Hypotheses.* Wiley.

21 Wald, A. (1943). Tests of statistical hypotheses concerning several parameters when the number of observations is large. *Transactions of the American Mathematical Society* 54: 426–482.

22 Rao, C.R. (1948). Large sample tests of statistical hypotheses concerning several parameters with applications to problems of estimation. *Proceedings of the Cambridge Philosophical Society* 44: 50–57.

23 Nelder, J.A. and Wedderburn, R.W.M. (1972). Generalized linear models. *Journal of the Royal Statistical Society A* 135: 370–384.

24 Charnes, A., Frome, E.L., and Yu, P.L. (1976). The equivalence of generalized least squares and maximum likelihood estimates in the exponential family. *Journal of the American Statistical Association* 71: 169–171.

25 McCullagh, P. and Nelder, J.A. (1989). *Generalized Linear Models,* 2e. Boca Raton, FL: Chapman & Hall/CRC.

26 Graybill, F.A. (1976). *Theory and Application of the Linear Model.* Belmont, CA: Wadsworth Publishing Company.

27 Chatterjee, S. and Hadi, A.S. (2012). *Regression Analysis by Example.* Hoboken, NJ: Wiley.

28 Rossi, R.J. (2010). *Applied Biostatistics for the Health Sciences.* Hoboken, NJ: Wiley.

29 Andriolo, J.M., Rossi, R.J., McConnell, C.A. et al. (2016). Influence of iron-doped apatite nanoparticles on viral infection examined in bacterial versus algal systems. *IEEE Transactions on NanoBioscience* 15 (8): 908–916.

A

Probability Models

Table A.1 The discrete probability models.

Distribution	Parameters	Support	PDF	Mean	Variance	MGF
$Bin(1,p)$	$p \in [0,1]$ $q=1-p$	$x=0,1$	$p^x q^{1-x}$	p	pq	(pe^t+q)
$Bin(n,p)$	$p \in [0,1]$ $q=1-p$ $n \in \mathbb{N}$	$x=0,1,\ldots,n$	$\binom{n}{x} p^x q^{n-x}$	np	npq	$(pe^t+q)^n$
$Pois(\lambda)$	$\lambda > 0$	$x \in \mathbb{W}$	$\frac{e^{-\lambda}\lambda^x}{x!}$	λ	λ	$e^{\lambda(e^t-1)}$
$Geo(p)$	$p \in [0,1]$ $q=1-p$	$x \in \mathbb{W}$	pq^x	$\frac{q}{p}$	$\frac{q}{p^2}$	$\frac{p}{1-qe^t}$
$NegBin(r,p)$	$p \in [0,1]$ $q=1-p$ $r \in \mathbb{N}$	$x \in \mathbb{W}$	$\binom{r+x-1}{x} p^r q^x$	$\frac{rq}{p}$	$\frac{rq}{p^2}$	$\frac{p^r}{(1-qe^t)^r}$
$Hyper(M,N,n)$	$M \in \mathbb{N}$ $N \in \mathbb{N}$ $n \in \mathbb{N}$	$x=0,1,\ldots,m^{\star}$ $m^{\star}=\min(M,n)$	$\frac{\binom{M}{x}\binom{N}{n-x}}{\binom{M+N}{n}}$	$\frac{nM}{M+N}$	$\frac{nMN(M-N)}{(M+N)^2(M+N-1)}$	Not useful
$MNom_k(n,\vec{p})$	$\vec{p} \in [0,1]^k$ $\sum p_i = 1$ $n \in \mathbb{N}$	$n_i \in \mathbb{W}$ $\sum n_i = n$	$\binom{n}{n_1,\ldots,n_k} p_1^{n_1} \cdots p_k^{n_k}$	$\mu_i = n_i p_i$	$\sigma_i^2 = np_i q_i$	Not useful

Mathematical Statistics: An Introduction to Likelihood Based Inference, First Edition. Richard J. Rossi.
© 2018 John Wiley & Sons, Inc. Published 2018 by John Wiley & Sons, Inc.

Table A.2 The continuous probability models.

Distribution	Parameters	Support	PDF	Mean	Variance	MGF
$U(\alpha, \beta)$	$\alpha, \beta \in \mathbb{R}$ $\alpha < \beta$	$x \in (\alpha, \beta)$	$\frac{1}{\beta - \alpha}$	$\frac{\alpha + \beta}{2}$	$\frac{(\beta - \alpha)^2}{12}$	$\frac{e^{\beta t} - e^{\alpha t}}{t(\beta - \alpha)}$
$\text{Exp}(\beta)$	$\beta \in \mathbb{R}^+$	$x \in \mathbb{R}^+$	$\frac{1}{\beta} e^{-\frac{x}{\beta}}$	β	β^2	$\frac{1}{1 - \beta t}$
$\text{Exp}(\beta, \eta)$	$\beta \in \mathbb{R}^+$ $\eta \in \mathbb{R}$	$x > \eta$	$\frac{1}{\beta} e^{-\frac{(x - \eta)}{\beta}}$	$\beta + \eta$	β^2	$\frac{e^{\eta t}}{1 - \beta t}$
$\text{Gamma}(\alpha, \beta)$	$\alpha \in \mathbb{R}^+$ $\beta \in \mathbb{R}^+$	$x \in \mathbb{R}^+$	$\frac{1}{\Gamma(\alpha) \beta^\alpha} x^{\alpha - 1} e^{-\frac{x}{\beta}}$	$\alpha \beta$	$\alpha \beta^2$	$\frac{1}{(1 - \beta t)^\alpha}$
$N(\mu, \sigma^2)$	$\mu \in \mathbb{R}$ $\sigma \in \mathbb{R}^+$	$x \in \mathbb{R}$	$\frac{1}{\sqrt{2\pi\sigma^2}} e^{-\frac{1}{2\sigma^2}(x - \mu)^2}$	μ	σ^2	$e^{\mu t + \frac{t^2 \sigma^2}{2}}$
$LN(\mu, \sigma^2)$	$\mu \in \mathbb{R}$ $\sigma \in \mathbb{R}^+$	$x \in \mathbb{R}$	$\frac{1}{x\sqrt{2\pi\sigma^2}} e^{-\frac{1}{2\sigma^2}(\ln(x) - \mu)^2}$	$e^{\mu + \frac{\sigma^2}{2}}$	$e^{2\mu + 2\sigma^2} - e^{2\mu + \sigma^2}$	Not useful
$\text{Beta}(\alpha, \beta)$	$\alpha \in \mathbb{R}^+$ $\beta \in \mathbb{R}^+$	$x \in (0, 1)$	$\frac{\Gamma(\alpha + \beta)}{\Gamma(\alpha)\Gamma(\beta)} x^{\alpha - 1}(1 - x)^{\beta - 1}$	$\frac{\alpha}{\alpha + \beta}$	$\frac{\alpha \beta}{(\alpha + \beta)^2 (\alpha + \beta + 1)}$	Not useful
χ_r^2	$r \in \mathbb{R}^+$	$x \in \mathbb{R}^+$	$\frac{1}{\Gamma\left(\frac{r}{2}\right) 2^{\frac{r}{2}}} x^{\frac{r}{2} - 1} e^{-\frac{x}{2}}$	r	$2r$	$\frac{1}{(1 - 2t)^{\frac{r}{2}}}$
t_ν	$\nu \in \mathbb{R}^+$	$x \in \mathbb{R}^+$	$\frac{\Gamma\left(\frac{\nu + 1}{2}\right)}{\sqrt{\nu \pi}\,\Gamma\left(\frac{\nu}{2}\right)} \frac{1}{\left(1 + \frac{x^2}{\nu}\right)^{\frac{\nu + 1}{2}}}$	0	$\frac{\nu}{\nu - 2}$	DNE
F_{ν_1, ν_2}	$\nu_1 \in \mathbb{N}$ $\nu_2 \in \mathbb{N}$	$x \in \mathbb{R}^+$	$\frac{\Gamma\left(\frac{\nu_1 + \nu_2}{2}\right)\left(\frac{\nu_1}{\nu_2}\right)^{\frac{\nu_1}{2}}}{\Gamma\left(\frac{\nu_1}{2}\right)\Gamma\left(\frac{\nu_2}{2}\right)} \frac{x^{\frac{\nu_1}{2} - 1}}{\left(1 + \frac{\nu_1}{\nu_2} x\right)^{\frac{\nu_1 + \nu_2}{2}}}$	$\frac{\nu_2}{\nu_2 - 2}$ $\nu_2 > 2$	$\frac{2\nu_2^2(\nu_1 + \nu_2 - 2)}{\nu_1(\nu_2 - 2)^2(\nu_2 - 4)}$ $\nu_2 > 4$	Not useful

B

Data Sets

The data sets used in the Chapter 5 and Chapter 7 case studies are given in Tables B.1 and B.2. These data sets, the Donner data set used in the Chapter 6 case study, and the data sets used in the Chapter 5 problems can be found as text files on the website.[1]

www.mtech.edu/academics/clsps/statistics/faculty/
rick-rossi/

Table B.1 The terrace data for the case study in Chapter 5.

1.86	2.34	2.40	2.51	2.63	2.91	2.91	3.14	3.20	3.43
3.54	3.66	3.71	3.77	4.00	4.00	4.06	4.11	4.11	4.23
4.23	4.34	4.34	4.51	4.57	4.63	4.69	4.69	4.74	4.74
4.80	4.80	4.91	4.91	4.97	4.97	5.03	5.09	5.14	5.20
5.20	5.26	5.26	5.26	5.43	5.49	5.60	5.66	5.66	5.66
5.71	5.83	5.89	5.89	5.94	5.94	5.94	5.94	6.00	6.06
6.06	6.06	6.17	6.17	6.17	6.17	6.23	6.23	6.23	6.23
6.29	6.29	6.34	6.34	6.51	6.57	6.57	6.68	6.74	6.74
6.74	6.86	6.86	6.91	6.97	7.14	7.20	7.26	7.26	7.26
7.37	7.37	7.43	7.43	7.43	7.49	7.54	7.54	7.60	7.66
7.71	7.77	7.83	7.89	7.94	7.94	8.06	8.06	8.06	8.06
8.06	8.11	8.17	8.34	8.40	8.57	8.57	8.57	8.69	8.69
8.91	8.97	8.97	9.03	9.03	9.03	9.14	9.14	9.14	9.31

(Continued)

1 Website addresses frequently change at Montana Tech. If this address is no longer valid, go to the Montana Tech website wwww.mtech.edu and search for Rick Rossi.

Mathematical Statistics: An Introduction to Likelihood Based Inference, First Edition. Richard J. Rossi.
© 2018 John Wiley & Sons, Inc. Published 2018 by John Wiley & Sons, Inc.

Table B.1 (Continued)

9.43	9.44	9.49	9.77	10.23	10.29	10.40	10.63	10.74	10.86
11.09	11.09	11.09	11.14	11.26	11.31	11.31	11.49	11.54	11.54
11.66	11.66	11.89	12.00	12.06	12.17	12.23	12.23	12.29	12.34
12.34	12.46	12.57	12.57	12.57	12.80	12.91	12.91	13.09	13.26
13.37	13.37	13.49	13.54	13.89	13.94	13.94	14.17	14.29	14.29
14.34	14.40	14.51	14.51	14.57	14.69	14.97	14.97	15.09	15.14
15.14	15.43	15.43	15.54	15.60	15.66	16.06	16.11	16.74	16.80
16.86	17.03	17.37	17.49	17.60	18.00	18.00	18.00	18.29	18.51
18.69	19.09	19.43	19.54	19.94	20.23	20.86	20.91	20.97	21.20
21.26	21.83	22.11	22.74	22.80	22.86	23.03	23.14	23.26	23.77
23.77	23.83	24.46	26.06	26.63	28.74				

Table B.2 The count data for the case study in Chapter 7.

Treatment	Count	Treatment	Count	Treatment	Count
25°C Cit 0%Fe	48	25°C Cit 40%Fe	39	45°C NoCit 30%Fe	32
25°C Cit 0%Fe	47	25°C Cit 40%Fe	37	45°C NoCit 30%Fe	39
25°C Cit 0%Fe	44	25°C Cit 40%Fe	42	45°C NoCit 30%Fe	38
25°C Cit 20%Fe	43	45°C 2XCit 30%Fe	47	65°C Cit 30%Fe	39
25°C Cit 20%Fe	37	45°C 2XCit 30%Fe	47	65°C Cit 30%Fe	42
25°C Cit 20%Fe	34	45°C 2XCit 30%Fe	52	65°C Cit 30%Fe	47
25°C Cit 30%Fe	53	45°C Cit 30%Fe	38	CO	26
25°C Cit 30%Fe	55	45°C Cit 30%Fe	46	CO	30
25°C Cit 30%Fe	50	45°C Cit 30%Fe	54	CO	23

Problem Solutions

Solutions for Chapter 1

1.1.1 (a) $[0, 4]$; (b) $(0, \infty)$; (c) $(0, \infty)$;
(d) {brown, blonde, black, red, gray, white, other}; (e) $\{2, 3, \ldots, 12\}$;
(f) {Jan 1, $\ldots$, Jan 31, Feb 1, $\ldots$, Feb 29, $\ldots$, Dec 31}; (g) $\{0, 1, 2, \ldots\}$

1.1.2 (a) $\{12, 13, 14, 15, 21, 23, 24, 25, 31, 32, 34, 35, 41, 42, 43,$
$45, 51, 52, 53, 54\}$;
(b) $\{12, 14, 21, 23, 24, 25, 32, 34, 41, 42, 43, 45, 52, 54\}$;
(c) $\{13, 31\}$;
(d) $\emptyset$;
(e) $\{12, 13, 14, 21, 23, 24, 25, 31, 32, 34, 41, 42, 43, 45, 52, 54\}$;
(f) $\{13, 31\}$

1.1.4 (a) $\{AS, 2S, \ldots, KS, AC, 2C, \ldots, KC, AD, 2D, \ldots, KD, AH, 2H, \ldots, KH\}$;
(b) $\{AS, AD, AC, AH\}$; (c) $\{AD, 2D, \ldots, KD\}$;
(d) $\{AS, 2S, \ldots, KS, AC, 2C, \ldots, KC, AH, 2H, \ldots, KH\}$;
(e) $\{AD\}$; (f) $\{AS, AC, AH, AD, 2D, \ldots, KD\}$; (g) $\{AS, AC, AH\}$;
(h) $\{2D, 3D, \ldots, KD\}$

1.1.5 $\{TTTH\}$; (b) $\{H, TH, TTH, TTTH\}$; (c) $\{TTTH\}$;
(d) $\{H, TH, TTH, TTTH\}$

1.1.12 (a) $[-1, 10)$; (b) $[-0.1, 1)$; (c) $[-1, \infty)$; (d) $[0, 1)$

1.1.13 (a) $(0.05, 2)$; (b) $(1, 1.05)$; (c) $(0, 2)$; (d) $\emptyset$

1.2.1 (a) 0.4; (b) 0.25; (c) 0.80; (d) 0.05; (e) 0.20; (f) 0.45; (g) 0.95; (h) 0.25

1.2.3 (a) 0; (b) 0.85; (c) 0.35; (d) 0.5; (e) 0.15; (f) 0.15

1.2.5 (a) $\frac{30}{55}$; (b) $\frac{18}{55}$; (c) $\frac{10}{55}$; (d) $\frac{17}{55}$

1.2.7 (a) 0.97; (b) 0.69; (c) 0.03; (d) 0.22; (e) 0.47

1.2.9 (a) 0.87; (b) 0.13; (c) 0.12

1.2.11 (a) 0.10; (b) 0.60; (c) 0.5; (d) 0.35

1.2.16 (a) $P(A) = P(A \cap B) + P(A \cap B^c)$. Thus, $P(A \cap B^c) = P(A) - P(A \cap B)$, and therefore, $P(A - B) = P(A) - P(A \cap B)$.

1.2.17 (a) $k = 2$; (b) $\frac{242}{243}$

1.2.18 (a) $\frac{1}{2^j} - \frac{1}{2^i}$; (b) $\frac{1}{2^i}$

1.3.1 (a) $\frac{2}{52}$; (b) $\frac{28}{52}$; (c) $\frac{1}{52}$; (d) $\frac{8}{52}$; (e) $\frac{2}{52}$; (f) $\frac{30}{52}$

1.3.3 (a) $\frac{3}{24}$; (b) $\frac{1}{24}$; (c) $\frac{3}{24}$

1.3.5 (a) 32; (b) $\frac{10}{32}$; (c) $\frac{6}{32}$; (d) $\frac{26}{32}$

1.3.7 (a) 1320; (b) 1728

1.4.1 (a) $\frac{10}{14}$; (b) $\frac{50}{210}$; (c) $\frac{10}{15}$; (d) $\frac{100}{210}$; (e) $\frac{110}{210}$

1.4.3 (a) 400; (b) $\frac{192}{400}$; (c) $\frac{208}{400}$

1.4.5 (a) $\frac{4}{51}$; (b) $\frac{16}{2652}$; (c) $\frac{13}{52}$; (d) $\frac{26}{51}$

1.4.7 $\frac{1}{30300}$

1.4.8 (a) 0.9704; (b) 0.8483

1.4.11 (a) 0.6585; (b) 0.7107; (c) 0.1260; (d) Captain I

1.4.13 (a) 0.25; (b) 0.24; (c) 0.36; (d) 0.40

1.4.14 3

1.4.15 $P(A|B) = \frac{P(A \cap B)}{P(B)} > P(A)$. Thus, $P(A \cap B) > P(A)P(B)$, and therefore,

$P(B|A) = \frac{P(A \cap B)}{P(A)} > \frac{P(A)P(B)}{P(A)} = P(B)$

1.5.1 (a) 0.58; (b) 0.18; (c) 0.88; (d) 0.46

1.5.3 (a) 0.38; (b) 0.50; (c) 0.88

1.5.5 (a) 0.14^5; (b) 0.530; (c) $0.86^{k-1} \times 0.14$; (d) 0.779

1.5.6 0.358

1.5.9 (a) 0.71; (b) 0.559

1.5.11 0.064

1.5.13 (a) $\frac{6}{11}$; (b) $\frac{5}{11}$; (c) $\frac{1}{2}$

1.5.15 (a) $P(A \cap B^c) = P(A) - P(A \cap B) = P(A) - P(A)P(B) = P(A)P(B^c)$;
(b) $P(A^c \cap B^c) = 1 - P(A \cup B) = 1 - P(A) - P(B) + P(A)P(B) =$
$[1 - P(A)][1 - P(B)] = P(A^c)P(B^c)$

1.5.17 Since $A \subset B$, $P(A \cap B) = P(A)$. Then, $P(A) = P(A \cap B) = P(A)P(B)$
when $P(A) = 0$ or $P(B) = 0$

1.6.1 (a) $3 \times 7 \times 10^4$; (b) $3 \times 1 \times 10^4$; (c) $3 \times 7 \times (10^4 - 1)$; (d) $3 \times 7 \times 10$

1.6.3 (a) 60; (b) 270; (c) 6144

1.6.5 (a) $\frac{1}{13}$; (b) $\frac{12}{13}$; (c) $\frac{188}{455}$; (d) $\frac{279}{455}$; (e) $\frac{1}{5}$; (f) $\frac{176}{455}$

1.6.7 (a) $\begin{pmatrix} 52 \\ 5,5,5,5,32 \end{pmatrix}$; (b) $\dfrac{\begin{pmatrix} 4 \\ 1,1,1,1,0 \end{pmatrix}\begin{pmatrix} 48 \\ 4,4,4,32 \end{pmatrix}}{\begin{pmatrix} 52 \\ 5,5,5,5,32 \end{pmatrix}}$; (c) $\dfrac{4\begin{pmatrix} 4 \\ 4,0,0,0 \end{pmatrix}\begin{pmatrix} 48 \\ 1,5,5,5,32 \end{pmatrix}}{\begin{pmatrix} 52 \\ 5,5,5,5,32 \end{pmatrix}}$

1.6.9 (a) 1260; (b) 756; (c) 168

1.6.11 (a) $\frac{15}{16}$; (b) $\frac{1}{16}$; (c) $\frac{1}{2}$; (d) $\frac{9^4}{10^4}$

1.6.13 (a) 0.4523; (b) 0.4070; (c) 0.0166; (d) 0.8593

1.6.16 $2^n = (1+1)^n = \sum_{k=0}^{n} \begin{pmatrix} n \\ k \end{pmatrix} 1^k 1^{n-k} = \sum_{k=0}^{n} \begin{pmatrix} n \\ k \end{pmatrix}$

1.6.19 (a) 0; (b) 3^n; (c) α^n

1.7.1 0.8912

1.7.3 (a) $1 - \frac{\frac{52!}{(52-n)!}}{52^n}$; (b) 0.6029; (c) 0.8934; (d) 0.9991

1.7.5 (a) $1 - \frac{\frac{100!}{(100-n)!}}{100^n}$; (b) 0.3718; (c) 0.8696

Solutions for Chapter 2

2.1.1 (a) $\mathcal{S} = \{0, 1, 2, 3, 4\}$, $f_H(h) = \binom{4}{h}\left(\frac{1}{2}\right)^4$;

(b) $\mathcal{S} = \{-4, -2, 0, 2, 4\}$, $f_D(-4) = \frac{1}{16}$, $f_D(-2) = \frac{4}{16}$, $f_D(0) = \frac{6}{16}$, $f_D(2) = \frac{4}{16}$, $f(4) = \frac{1}{16}$;

(c) $\mathcal{S} = \{0, 2, 4\}$, $f_{|D|}(0) = \frac{6}{16}$, $f(2) = \frac{8}{16}$, $f(4) = \frac{2}{16}$

2.1.3 (a) 21; (b) $F(x) = \begin{cases} 0 & x < 1 \\ \frac{j(j+1)}{42} & j \le x < j, \ j = 1, 2, 3, 4, 5, 6; \\ 1 & x \ge 6 \end{cases}$ (c) $\frac{18}{21}$; (d) $\frac{9}{21}$

2.1.5 (a) $\frac{1}{21}$; (b) $F(x) = \begin{cases} 0 & x < 0 \\ \frac{(y+1)(y+2)}{42} & j \le x < j+1, \ j = 0, 1, 2, 3, 4 \\ 1 & x \ge 5 \end{cases}$; (c) $\frac{11}{21}$;

(d) $\frac{12}{21}$

2.1.7 (a) 2; (b) $F(x) = \begin{cases} 0 & x < 1 \\ 2(1 - \frac{1}{x}) & 1 \le x < 2; \\ 1 & x \ge 2 \end{cases}$ (c) $\frac{2}{3}$; (d) $\frac{4}{3}$

2.1.9 (a) 6; (b) $F(x) = \begin{cases} 0 & x < 0 \\ x^2(3 - 2x) & 0 \le x < 1; \\ 1 & x \ge 1 \end{cases}$ (c) 0.648; (d) 0.944

2.1.11 (a) $f(x) = \frac{1}{x^2} I_{[1,\infty)}(x)$; (b) $F(x) = \begin{cases} 0 & x < 1 \\ 1 - \frac{1}{x} & x \ge 1 \end{cases}$; (c) 0.3

2.1.13 (a) $\mathcal{S}_X = [1, \infty)$; (b) $f(x) = \frac{4}{x^5}$; (c) 1.0746; (d) 1.1892

2.1.15 (a) $\mathcal{S}_X = [0, \infty)$; (b) $f(x) = \frac{3x^2}{64} e^{-\left(\frac{x}{4}\right)^3}$; (c) $1 - e^{-\frac{1}{8}}$; (d) 0.5140; (e) 1.4862; (f) 3.53999

2.2.1 (a) 0.32; (b) $f_1(1) = 0.26, f_1(2) = 0.40, f_1(3) = 0.34; \mathcal{S}_1 = \{1, 2, 3\}$;
(c) $f_2(0) = 0.6, f_2(1) = 0.22, f_2(2) = 0.14, f_2(3) = 0.0.4; \mathcal{S}_1 = \{0, 1, 2, 3\}$;
(d) $f(x_1 = 1|x_2 = 2) = \frac{5}{14}, f(x_1 = 2|x_2 = 2) = \frac{8}{14}$,
$f(x_1 = 3|x_2 = 2) = \frac{1}{14}; \mathcal{S}_{X_1|X_2=2} = \{1, 2, 3\}$

2.2.3 (a) $f_1(x_1) = (1 - p)p^{x_1}, \mathcal{S}_{X_1} = \{0, 1, 2, \dots\}$;
(b) $f_2(x_2) = (x_2 + 1)(1 - p)^2 p^{x_2}, \mathcal{S}_{X_2} = \{0, 1, 2, \dots\}$;
(c) $f(x_1|x_2) = \frac{1}{x_2+1}, \mathcal{S}_{X_1|X_2} = \{0, 1, 2, \dots, x_2\}$;
(d) $f(x_2|x_1) = (1 - p)p^{x_2-x_1}, \mathcal{S}_{X_2|X_1} = \{x_1, x_1 + 1, \dots\}$

2.2.5 (a) 6; (b) 0.66; (c) $f_1(x_1) = 3x_1^2, \mathcal{S}_{X_1} = (0, 1)$;
(d) $f_2(x_2) = 3(1 - x_2)^2, \mathcal{S}_{X_2} = (0, 1)$;
(e) $f(x_2|x_1) = \frac{2(x_1-x_2)}{x_1^2}, \mathcal{S}_{X_2|X_2} = (0, x_1)$;
(f) $F(x_2|x_1) = \begin{cases} 0 & x_2 < 0 \\ \frac{2x_1x_2-x_2^2}{x_1^2} & 0 \leq x_2 < x_1 \\ 1 & x_2 \geq x_1 \end{cases}$; (g) 0.25

2.2.7 (a) 0.0137; (b) $f_1(x_1) = 2e^{-2x_1}, \mathcal{S}_{X_1} = (0, \infty)$;
(c) $f_2(x_2) = 2e^{-x_2}(1 - e^{-x_2}), \mathcal{S}_{X_2} = (0, \infty)$;
(d) $f(x_1|x_2) = \frac{e^{-x_1}}{1-e^{-x_2}}, \mathcal{S}_{X_1|X_2} = (0, x_2)$;
(e) $f(x_2|x_1) = e^{-(x_2-x_1)}, \mathcal{S}_{X_2|X_1} = (x_1, \infty)$

2.2.9 (a) $f_1(x_1) = 3(1 - x_1)^2, \mathcal{S}_{X_1} = (0, 1)$;
(b) $f_2(x_2) = 6x_2(1 - x_2), \mathcal{S}_{X_2} = (0, 1)$;
(c) $f_3(x_3) = 3x_3^2, \mathcal{S}_{X_3} = (0, 1)$;
(d) $f(x_1, x_2|x_3) = \frac{2}{x_3^2}, \mathcal{S}_{X_1, X_2|X_3} = \{(x_1, x_2) : 0 < x_1 < x_2 < x_3\}$

2.2.11 $f_2(x_2) = \left(\frac{2}{3}\right)^{x_2} - \left(\frac{1}{2}\right)^{x_2}, \mathcal{S}_{X_2} = \{1, 2, 3, \dots\}$

2.2.13 $f_2(x_2) = 4x_2^3, \mathcal{S}_{X_2} = (0, 1)$

2.2.15 (a) $f_1(x_1) = \frac{1}{(1+x_1)^2}, \mathcal{S}_{X_1} = (0, \infty)$; (b) $F_1(x_1) = \begin{cases} 0 & x_1 < 0 \\ 1 - \frac{1}{1+x_1} & x_1 \geq 0 \end{cases}$

2.3.1 (a) $f_1(0) = 0.4, f_1(1) = 0.4, f_1(2) = 0.2, \mathcal{S}_{X_1} = \{0, 1, 2\}$;
(b) $f_2(1) = 0.3, f_2(2) = 0.45, f_2(3) = 0.25, \mathcal{S}_{X_2} = \{1, 2, 3\}$;
(c) X_1 and X_2 are not independent since $f(0, 1) = 0.1 \neq f_1(0)f_2(1) = 0.12$

2.3.3 (a) $f_1(x_1) = \frac{1}{2^{x_1}}, \mathcal{S}_{X_1} = \{1, 2, 3, \dots\}$;

(b) $f_2(x_2) = \frac{e^{-1}}{x_2!}, \mathcal{S}_{X_2} = \{0, 1, 2, \dots\}$;

(c) X_1 and X_2 are independent since $f(x_1, x_2) = f_1(x_1)f_2(x_2), \forall x_1, x_2$

2.3.5 (a) $f_1(x_1) = 2x_1, \mathcal{S}_{X_1} = (0, 1)$;

(b) $f_2(x_2) = 1, \mathcal{S}_{X_2} = (0, 1)$;

(c) X_1 and X_2 are independent since $f(x_1, x_2) = f_1(x_1)f_2(x_2), \forall x_1, x_2$

2.3.7 (a) $f_1(x_1) = 6x_1(1 - x_1^2)^2, \mathcal{S}_{X_1} = (0, 1)$;

(b) $f_2(x_2) = 12x_2^3(1 - x_2)(1 + x_2), \mathcal{S}_{X_2} = (0, 1)$;

(c) $f_3(x_3) = 6x_3^5, \mathcal{S}_{X_3} = (0, 1)$;

(d) X_1, X_2, and X_3 are not independent since $f(x_1, x_2, x_3) \neq f(x_1)f_2(x_2)f_3(x_3)$

2.3.9 (a) Yes; (b) No; (c) Yes

2.3.11 (a) $f(x_1, \dots, x_n) = \left(\prod a_i \right)^{-1}$; (b) $F(x_1, \dots, x_n) = \prod \frac{x_i}{a_i}$

2.3.13 (a) $f(x_1, \dots, x_n) = \theta^n e^{-\theta \sum x_i}$; (b) $F(x_1, \dots, x_n) = \prod [1 - e^{-\theta x_i}]$

2.4.1 (a) $f_{X+1}(0) = \frac{1}{3}, f_{X+1}(1) = \frac{1}{6}, f_{X+1}(2) = \frac{1}{2}, \mathcal{S}_{X+1} = \{0, 1, 2\}$;

(b) $f_{|X|}(0) = \frac{1}{6}, f_{|X|}(1) = \frac{5}{6}, \mathcal{S}_{|X| 1} = \{0, 1\}$;

(c) $f_{X^2}(0) = \frac{1}{6}, f_{X^2}(1) = \frac{5}{6}, \mathcal{S}_{X^2} = \{0, 1\}$

2.4.3 (a) $f_{X/2}(y) = 1, \mathcal{S}_{X/2} = (0, 1)$; (b) $f_{-\ln(X)}(y) = e^{-y}, \mathcal{S}_{-\ln(X)} = (0, \infty)$;

(c) $f_{1/X}(y) = \frac{1}{2y^2}, \mathcal{S}_{1/X} = (1/2, \infty)$

2.4.5 (a) $f_{X^4}(y) = 1, \mathcal{S}_{X^4} = (0, 1)$; (b) $f_{-\ln(X)}(y) = 4e^{-4y}, \mathcal{S}_{-\ln(X)} = (0, \infty)$;

(c) $f_{1/X}(y) = \frac{4}{y^5}, \mathcal{S}_{1/X} = (1, \infty)$

2.4.7 (a) $f_{1-X}(y) = 6y(1 - y), \mathcal{S}_{1-X} = (0, 1)$;

(b) $f_{4X}(y) = \frac{3}{8}y \left(1 - \frac{y}{4} \right), \mathcal{S}_{4x} = (0, 4)$;

(c) $f_{-\ln(X)}(y) = 6e^{-2y}(1 - e^{-y}), \mathcal{S}_{-\ln(X)} = (0, \infty)$

2.4.9 (a) $f_{X/2}(y) = 2y, \mathcal{S}_{X/2} = (0, 1)$; (b) $f_{X^3}(y) = \frac{1}{6}y^{-\frac{1}{3}}, \mathcal{S}_{X^3} = (0, 8)$;

(c) $f_{(X-1)^2}(y) = \frac{1}{2\sqrt{y}}, \mathcal{S}_{(X-1)^2} = (0, 1)$

2.4.11 (a) $f_{X+2}(y) = \theta(y - 2)^{\theta-1}, \mathcal{S}_{X+2} = (2, 3)$;

(b) $f_{-\ln(x)}(y) = \theta e^{-\theta y}, \mathcal{S}_{-\ln(X)} = (0, \infty)$;

(c) $f_{\sqrt{X}}(y) = 2\theta y^{2\theta-1}, \mathcal{S}_{\sqrt{X}} = (0, 1)$

2.4.13 (a) $|J| = 1$; (b) $f_Y(y_1, y_2) = e^{-y_1}, 0 < y_2 < y_1 < \infty$;
(c) $f_1(y_1) = y_1 e^{-y_1}, y_1 > 0$; (d) $f_2(y_2) = e^{-y_2}, y_2 > 0$

2.4.15 (a) $|J| = \frac{1}{2}$; (b) $f_Y(y_1, y_2) = \frac{1}{4\pi} e^{-\frac{1}{4}(y_1^2 + y_2^2)}, -\infty < y_1, y_2 < \infty$;
(c) $f_1(y_1) = \frac{1}{2\sqrt{\pi}} e^{-\frac{y_1^2}{4}}, -\infty < y_1 < \infty$; (d) $f_2(y_2) = \frac{1}{2\sqrt{\pi}} e^{-\frac{y_2^2}{4}}, -\infty < y_2 < \infty$

2.4.17 (a) $f(x_1, x_2) = 16x_1^3 x_2^3, 0 < x_1, x_2 < 1$;
(b) $|J| = e^{-y_1}$; (c) $f_1(y_1) = 16y_1 e^{-4y_1}, 0 < y_1 < \infty$;
(d) $f_2(y_2) = 4e^{-4y_2}, 0 < y_2 < \infty$

2.4.19 $Y \sim U(0, 1)$

2.5.1 (a) 3; (b) 18; (c) 12; (d) 10

2.5.3 (a) $2p$; (b) $2p(1 + p)$; (c) $2p(1 - p)$; (d) $8p(2 - p) + 2$

2.5.5 3

2.5.7 $p_1 = 0.1, p_2 = 0.3, p_3 = 0.6$

2.5.9 (a) $\frac{5}{3}$; (b) $-\frac{1}{3}$; (c) $\frac{19}{6}$; (d) $\frac{5}{9}$

2.5.11 (a) $\frac{5}{6}$; (b) $\frac{5}{7}$; (c) $\frac{5}{5+n}$; (d) $\frac{1}{5}$

2.5.13 (a) $\frac{1}{\theta}$; (b) $\frac{\theta}{(\theta+1)(\theta+2)}$

2.5.15 (a) 5; (b) 0

2.5.17 (a) $\frac{2e^t}{3-e^t}$; (b) $\frac{3}{2}$; (c) 3; (d) $\frac{3}{4}$

2.5.19 (a) $\frac{1}{(1-t)^2}$; (b) 2; (c) 6; (d) 2

2.5.21 $\frac{1}{\sqrt{1-2t}}$

2.5.23 (a) $\left(\frac{1}{4} + \frac{3e^t}{4}\right)^{10}$; (b) 7.5; (c) 58.125; (d) 1.875

2.5.25 (a) $e^{5t + \frac{16t^2}{2}}$; (b) 5; (c) 41; (d) 16

2.5.27 e^{3t} for $t \in (-\infty, \infty)$

2.5.29 $\int_{-\infty}^{\infty} x \cdot \frac{1}{\pi(1+x^2)} \, dx$ is a divergent integral

2.6.1 (a) $\frac{7}{3}$; (b) 3; (c) 6; (d) 7; (e) $\frac{5}{9}$; (f) 0

2.6.3 (a) $\frac{2}{5}$; (b) $\frac{2}{5}$; (c) $\frac{1}{5}$; (d) $\frac{2}{15}$; (e) $\frac{1}{25}$; (f) $-\frac{2}{75}$

2.6.5 (a) $\frac{5}{8}$; (b) $\frac{5}{8}$; (c) $\frac{7}{15}$; (d) $\frac{3}{8}$; (e) $\frac{73}{960}$; (f) $-\frac{1}{64}$

2.6.7 (a) 0.5; (b) 1.5; (c) 0.5; (d) 3.5; (e) 0.25; (f) 1.25; (g) 1; (h) 0.25;
(i) $\Sigma = \begin{pmatrix} 0.25 & 0.25 \\ 0.25 & 1.25 \end{pmatrix}$

2.6.9 (a) 6; (b) 27; (c) 2; (d) 4

2.6.11 (a) 5; (b) $\frac{13}{4}$

2.6.13 $\text{Var}(X_1 - X_2) = \text{Var}(X_1) + (-1)^2\text{Var}(X_2) = \text{Var}(X_1) + \text{Var}(X_2)$.

2.6.23 (a) $\text{Cor}(X_1, X_2)$; (b) $\sigma_1^2 - \sigma_2^2$

2.6.24 $a_1 = a_2 = \frac{1}{2}$

2.6.27 (a) 63; (b) 149

2.6.29 (a) $\frac{x_2}{2}$; (b) $\frac{1}{2}$; (c) $\frac{x_2^2}{3}$; (d) $\frac{2}{3}$; (e) $\frac{5}{12}$; (f) 2.5

2.6.31 (a) e^{t^2} for $t \in (-\infty, \infty)$; (b) e^{t^2} for $t \in (-\infty, \infty)$

2.6.33 (a) $(0.2e^t + 0.8)^3$ for $t \in (-\infty, \infty)$; (b) 0.6

2.7.1 (a) 24; (b) 123; (c) -1; (d) 21

2.7.3 (a) μ; (b) $29\sigma^2$

2.7.5 (a) 0; (b) 1

2.7.7 μ

2.7.9 (a) 0; (b) $\frac{1}{n}$

2.7.11 (a) $E(T) = \begin{cases} \mu & \text{when } n \text{ is odd} \\ 0 & \text{when } n \text{ is even} \end{cases}$; (b) $n\sigma^2$

2.7.13 $a_i = \frac{1}{n}$ (Use the method of Lagrange Multipliers)

2.7.17 $M_T(t) = (0.6e^t + 0.4)^n$ and $f_T(t) = \binom{n}{t} 0.6^t 0.4^{n-t}$

2.7.19 (a) $M_T(t) = e^{t \sum \mu_i + \frac{t^2}{2} \sum \sigma_i^2}$; (b) $\sum \mu_i$; (c) $\sum \sigma_i^2$; (d) $\sum \sigma_i^2 + \left(\sum \mu_i \right)^2$

2.7.21 (a) $\theta^k k!$; (b) $2\theta^2$; (c) $24\theta^4$; (d) θ^4

2.7.24 (a) 9; (b) 3

2.8.1 (a) $n = 9$ and $n = 10$ are equally most likely; (b) $n = 14$ and $n = 15$ are equally most likely

2.8.2 (a) 10; (b) 15

Solutions for Chapter 3

3.1.1 (a) 1.92; (b) 0.9984; (c) 0.2300; (d) 0.0731; (e) 0.9269; (f) 0.9228

3.1.3 (a) 0.0543; (b) 0.0697; (c) 0.8385; (d) $\frac{10}{6}$; (e) $\frac{50}{36}$

3.1.5 (a) 0.2061; (b) 0.2969; (c) 0.2921

3.1.7 0.9994

3.1.9 (a) $(0.25e^t + 0.75)^{12}$; (b) 11.25

3.1.11 $\sum_{x=0}^{n} \binom{n}{x} p^x (1-p)^{n-x} = [p + (1-p)]^n = 1$

3.1.15 (a) $M_{X_1 + X_2}(t) = (pe^t + q)^{n_1 + n_2}$; (b) $X_1 + X_2 \sim \text{Bin}(n_1 + n_2, p)$

3.1.16 (a) 0.0565; (b) 0.3134; (c) 0.3012; (d) 2

3.1.19 (a) 0.0144; (b) 0.5088; (c) 0.2487

3.1.21 $n(n-1)\frac{M}{M+N}\frac{M-1}{M+N-1}$

3.1.23 (a) 0.0916; (b) 0.9997; (c) 8

3.1.25 (a) 0.0839; (b) 0.9799

3.1.27 $\sum_{x=0}^{\infty} \frac{e^{-\lambda} \lambda^x}{x!} = e^{-\lambda} \sum_{x=0}^{\infty} \frac{\lambda^x}{x!} = e^{-\lambda} e^{\lambda} = 1$

3.1.31 (a) $\lambda = 2$; (b) 0.5940

3.1.33 $e^{-2\lambda}$

3.1.35 (a) $M_{X_1+X_2}(t) = e^{2\lambda(e^t-1)}$; (b) $X_1 + X_2 \sim \text{Pois}(2\lambda)$

3.1.37 (a) 0.4213; (b) 0.1615; (c) 5; (d) 30

3.1.41 (a) $\frac{p^2}{(1-qe^t)^2}$; (b) $X_1 + X_2 \sim \text{NegBin}(2,p)$

3.1.43 (a) 0.6826; (b) 0.1673

3.1.47 (a) $\frac{p^{r_1+r_2}}{(1-qe^t)^{r_1+r_2}}$; (b) $X_1 + X_2 \sim \text{NegBin}(r_1 + r_2, p)$

3.1.49 (a) 0.00756; (b) 0.0158; (c) 2; (d) 5

3.1.51 (a) 0.0020; (b) 0.2642; (c) 0.9844

3.2.1 (a) 0.7; (b) 0.425; (c) 1; (d) $\frac{1}{3}$

3.2.3 (a) 0.5; (b) 0.0675; (c) $f_Y(y) = \frac{1}{2\sqrt{y}}$, $\mathcal{S}_Y = [0, 1)$; (d) $\frac{4}{45}$

3.2.5 $E(X^n) = \int_\alpha^\beta \frac{x^n}{\beta-\alpha}\,dx = \frac{\beta^{n+1}-\alpha^{n+1}}{(\beta-\alpha)\cdot(n+1)}$

3.2.7 (a) 3.75; (b) 5.625; (c) $0.\overline{4}$; (d) 1.843

3.2.11 (a) $M_Y(t) = (1 - \beta t)^{-(\alpha_1+\alpha_2)}$; (b) $Y \sim \text{Gamma}(\alpha_1 + \alpha_2, \beta)$

3.2.13 (a) 10; (b) 100; (c) e^{-2}; (d) 0.0878; (e) 29.957

3.2.15 (a) $-\beta \ln(0.5)$; (b) $-\beta \ln(1 - p)$; (c) $2\beta^2$; (d) $n!\beta^n$

3.2.17 (a) $f_Y(y) = e^{-y}$, $\mathcal{S}_Y = (0, \infty)$; (b) 1

3.2.19 (a) $f_Y(y) = \frac{1}{\beta}e^{-\frac{(y-\tau)}{\beta}}$, $\mathcal{S}_Y = (\tau, \infty)$; (b) $F_Y(y) = \begin{cases} 0 & y < \tau \\ 1 - e^{-\frac{(y-\tau)}{\beta}} & y \geq \tau \end{cases}$;
(c) $\beta + \tau$; (d) β^2

3.2.21 (a) $(1 - 2t)^{-\frac{r+s}{2}}$; (b) $Y \sim \chi^2_{r+s}$

3.2.23 (a) 6; (b) 5040; (c) $\frac{15}{8}\sqrt{\pi}$; (d) $\frac{(2n-1)(2n-3)\cdots 3\cdot 1}{2^n}\sqrt{\pi}$

3.2.25 (a) 0.0062; (b) 0.2023; (b) 157.69

3.2.27 $\mu = 12, \sigma^2 = 100$

3.2.28 $\mu = 100, \sigma = 25$

3.2.32 (a) $e^{2\mu t + t^2 \sigma^2}$; (b) $N(2\mu, 2\sigma^2)$; (c) $e^{t^2 \sigma^2}$; (d) $N(0, 2\sigma^2)$

3.2.38 $\ln(X_i) \sim N(\mu_i, \sigma_i^2)$ and $\ln\left(\frac{X_1}{X_2}\right) = \ln(X_1) - \ln(X_2) \sim N(\mu_1 - \mu_2, \sigma_1^2 + \sigma_2^2)$

3.2.40 $B(m, n) = \frac{\Gamma(m)\Gamma(n)}{\Gamma(n+m)} = \frac{(m-1)!(n-1)!}{(m+n-1)!}$

3.2.42 (a) $F(x) = \begin{cases} 0 & x < 0 \\ 2x(1 - \frac{x}{2}) & 0 \le x < 1 \\ 1 & x \ge 1 \end{cases}$; (b) 0.2929

3.2.44 (a) 0.2898; (b) 0.4102; (c) $\frac{1}{2}$; (d) $\frac{1}{36}$

3.2.46 (a) $\frac{\Gamma(n+\alpha)\Gamma(\alpha+\beta)}{\Gamma(\alpha)\Gamma(\alpha+\beta+n)}$; (b) $\frac{\Gamma(\alpha+\beta)\Gamma(\alpha+n)\Gamma(\beta+m)}{\Gamma(\alpha)\Gamma(\beta)\Gamma(\alpha+\beta+n+m)}$

3.2.50 $f(x; 1, 1) = \frac{\Gamma(2)}{\Gamma(1)\Gamma(1)} x^{1-1}(1 - x)^{1-1} = 1$, for $x \in (0, 1)$

3.3.1 $M_S(t) = M_{X_1}(t)^n = (pe^t + q)^n$

3.3.3 $M_S(t) = M_{X_1}(t)^n = [e^{\lambda(e^t - 1)}]^n = e^{n\lambda(e^t - 1)}$

3.3.5 $M_S(t) = M_{X_1}(t)^n = \left(\frac{p}{1 - qe^t}\right)^n = \frac{p^n}{(1 - qe^t)^n}$

3.3.7 $M_S(t) = M_{X_1}(t)^n = [(1 - \beta t)^{-1}]^n = (1 - \beta t)^{-n}$

3.3.9 $M_S(t) = \prod M_{X_i}(t) = \prod (1 - \beta t)^{-\alpha_i} = (1 - \beta t)^{-\sum \alpha_i}$

3.3.11 $M_S(t) = \prod M_{X_i}(t) = \prod (1 - 2t)^{-\frac{k_i}{2}} = (1 - 2t)^{-\frac{\sum k_i}{2}}$

3.3.13 $M_S(t) = \prod M_{X_i}(a_i t) = \prod e^{\mu_i a_i t + \frac{a_i^2 t^2 \sigma_i^2}{2}} = e^{t \sum a_i \mu_i + \frac{t^2}{2} \sum a_i^2 \sigma_i^2}$

3.3.15 Let $Z_i = \frac{X_i - \mu_i}{\sigma_i}$. Then, $Z_i \sim N(0, 1)$, $S = \sum Z_i^2$, and $M_{Z^2}(t) = (1 - 2t)^{-\frac{1}{2}}$. Thus, $M_S(t) = M_{Z_i^2}(t)^n = (1 - 2t)^{-\frac{n}{2}}$

3.3.17 (a) First, $\sum\left(\frac{X_i - \mu}{\sigma}\right)^2 \sim \chi_n^2$. Thus, $S \sim \sigma^2 \chi_n^2$, and therefore, $M_S(t) = M_{\chi_n^2}(\sigma^2 t) = (1 - 2\sigma^2 t)^{-\frac{n}{2}}$. (b) $E(S) = \frac{n}{2} 2\sigma^2 = n\sigma^2$

3.4.1   (a) 0.5205; (b) 0.5422

3.4.3   First, $S = \sum X_i \sim$ Gamma(25, 10) and $M_S(t) = (1 - 10t)^{-25}$. Hence, $M_{\bar{X}}(t) = M_S\left(\frac{t}{25}\right) = (1 - \frac{10}{25}t)^{-25}$

3.4.5   (a) 0.6827; (b) 0.6908

Solutions for Chapter 4

4.1.1   (a) θ; (b) 0; (c) $\frac{\theta^2}{3n}$; (d) T is MSE-consistent

4.1.3   (a) θ; (b) 0; (c) $T \sim \frac{\theta}{n}\chi_n^2$; (d) $\frac{2\theta^2}{n}$; (e) T is MSE-consistent

4.1.5   (a) Gamma $\left(n, \frac{1}{\theta}\right)$; (b) $\frac{1}{\theta}$; (c) $\frac{1}{\theta}$; (d) $\frac{n}{n-1}\theta$; (e) $\frac{\theta}{n-1}$; (f) $U = \frac{n-1}{n\bar{X}}$; (g) U is MSE-consistent

4.1.9   (a) $\mu^2 + \frac{\sigma^2}{n}$; (b) $\frac{\sigma^2}{n}$; (c) μ^2; (d) 0; (e) 0

4.1.11   (a) μ; (b) $\frac{2(2n+1)}{3n(n+1)}\sigma^2$; (c) 0; (d) $\frac{2(2n+1)}{3n(n+1)}\sigma^2$; (e) $\frac{2(2n+1)}{3(n+1)}$; (f) $\frac{4}{3}$; (g) T is MSE-consistent

4.1.15   (a) μ; (b) $\frac{\sigma^2}{n+m}$; (c) $\frac{\sigma^2}{n+m}$; (d) 0

4.1.17 $\frac{3e^{-\frac{3}{\beta}}}{\beta\sqrt{n}}$

4.1.19 $e^{-\lambda}\sqrt{\frac{\lambda}{n}}$

4.2.1   (a) $L(\theta) = \theta^n(1 - \theta)^{\sum x_i}$; (b) $\sum X_i$

4.2.3   (a) $L(\theta) = \frac{1}{\Gamma(4)^n\theta^n}\prod x_i^3 e^{-\frac{\sum x_i}{\theta}}$; (b) $\sum X_i$

4.2.5   (a) $\sum X_i^2$; (b) $n\theta$; (c) $\frac{1}{n}\sum X_i^2$

4.2.7   (a) $L(\theta) = \frac{1}{(2\pi\theta_2)^{\frac{n}{2}}}e^{-\frac{1}{2\theta_2}\sum(x_i-\theta_1)^2}$; (b) $(\sum X_i, \sum X_i^2)$

4.2.9 $(\sum X_i, \sum X_i^2)$

4.2.11 $\sum |X_i|$

4.2.13 $\prod X_i$

4.2.15 (a) $f(x;\theta) = \underbrace{e^{-\theta}}_{c(\theta)}\ \underbrace{e^{x\ln(\theta)}}_{e^{T(x)q(\theta)}}\ \underbrace{\dfrac{1}{x!}}_{h(x)}$; (b) $\sum X_i$

4.2.17 (a) $f(x;\theta) = \underbrace{\theta}_{c(\theta)}\ \underbrace{e^{(\theta-1)\ln(x)}}_{e^{T(x)q(\theta)}} \cdot \underbrace{1}_{h(x)}$; (b) $\sum \ln(x_i)$

4.2.19 (a) $f(x;\theta_1,\theta_2) = \underbrace{\dfrac{\Gamma(\theta_1+\theta_2)}{\Gamma(\theta_1)\Gamma(\theta_2)}}_{c(\theta_1,\theta_2)}\underbrace{e^{(\theta_1-1)\ln(x)+(\theta_2-1)\ln(1-x)}}_{e^{T_1(x)q_1(\theta_1,\theta_2)+T_2(x)q_2(\theta_1,\theta_2)}} \cdot \underbrace{1}_{h(x)}$;

   (b) $(\sum \ln(X_i), \sum \ln(1-X_i))$

4.2.21 (a) $f(x;\theta_1,\theta_2) = \underbrace{\dfrac{1}{\Gamma(\theta_1)\theta_2^{\theta_1}}}_{c(\theta)}\underbrace{e^{(\theta_1-1)\ln(x)-\frac{x}{\theta_2}}}_{e^{T_1(x)q_1(\theta_1,\theta_2)+T_2(x)q_2(\theta_1,\theta_2)}} \cdot \underbrace{1}_{h(x)}$; (b) $(\sum \ln(X_i), \sum X_i)$

4.2.23 S_X depends on θ

4.3.1 (a) $\frac{1}{\theta}$; (b) $\frac{\theta}{n}$; (c) $\overline{X}$ is UMVUE of θ; (d) $e^{-2\theta}\frac{\theta}{n}$

4.3.3 (a) $\frac{1}{\theta^2}$; (b) $\frac{\theta^2}{n}$; (c) $\overline{X}$ is UMVUE for θ; (d) First, $\overline{X} \sim \text{Gamma}\left(n, \frac{\theta}{n}\right)$. Then, $E(\overline{X}^2) = \frac{n^2+1}{n^2}\theta^2$. Thus, $E\left[\frac{n^2}{n^2+1}\overline{X}^2\right] = \theta^2$, and $\frac{n}{n+1}\overline{X}^2$ is entirely based on a complete sufficient statistic.; (e) No. $\text{Var}\left[\frac{n^2}{n^2+1}\overline{X}^2\right] \neq \text{CRLB}_{\theta^2}$

4.3.5 (a) $\frac{1}{2\theta^2}$; (b) $\frac{2\theta^2}{n}$; (c) $\frac{1}{n}\sum X_i^2$ is an unbiased estimator of θ based on a complete sufficient statistic

4.3.7 (a) $\sum \ln(X_i)$; (b) $\frac{1}{\theta^2}$; (c) $\frac{\theta^2}{n}$; (d) $-4\frac{1}{n}\sum \ln(X_i)$

4.3.9 (a) $\sum \ln(1-X_i)$; (b) $\frac{2\theta^2+2\theta+1}{\theta^2(1+\theta)^2}$; (c) $\frac{\theta^2(1+\theta)^2}{n(2\theta^2+2\theta+1)}$; (d) $-\frac{1}{n}\sum \ln(1-X_i)$

4.3.11 (a) $(\sum \ln(X_i), \sum X_i)$; (b) $\begin{pmatrix} \Psi_1 & \frac{1}{\theta_2} \\ \frac{1}{\theta_2} & \frac{\theta_1}{\theta_2^2} \end{pmatrix}$ where $\Psi_1 = \frac{\Gamma''(\theta_1)\Gamma(\theta_1)-\Gamma'(\theta_1)^2}{\Gamma(\theta_1)^2}$; (c) $\frac{\theta_1}{\theta_1\Psi_1-1}$;

   (d) $\frac{\Psi_1\theta_2^2}{\theta_1\Psi_1-1}$

4.4.1 (a) $f_{(1)}(x) = \frac{n}{\theta}e^{-\frac{nx}{\theta}}, x \in (0, \infty)$; (b) $\frac{\theta}{n}$; (c) $\frac{\theta^2}{n^2}$;

   (d) $M_T(t) = M_{X_{(1)}}(nt) = (1 - \frac{\theta}{n}nt)^{-1} = (1 - \theta t)^{-1}$; (e) $E(T) = \theta$ and $\text{Var}(T) = \theta^2$. Thus, $\lim_{n\to\infty}\text{MSE}(T;\theta) = \lim_{n\to\infty}\theta^2 \neq 0$

4.4.3 (a) $f(x_1, \ldots, x_n; \theta) = (1-\theta)^{-n} \prod I_{(\theta,1)}(x_i) = \underbrace{(1-\theta)^{-n} I_{(-\infty, x_{(1)}]}(\theta)}_{g(x_{(1)};\theta)} \cdot$

$\underbrace{1}_{h(x_1,\ldots,x_n)}$;

(b) $f_{(1)}(x) = \frac{n}{(1-\theta)^n}(1-x)^{n-1}, x \in (\theta, 1)$; (c) $\frac{n\theta+1}{n+1}$; (d) $\frac{n(1-\theta)^2}{(n+1)^2(n+2)}$;

(e) $\frac{n(1-\theta)^2}{(n+1)^2(n+2)} + \frac{(1-\theta)^2}{(n+1)^2}$; (f) $X_{(1)}$ is MSE-consistent;

(g) Suppose $h(x)$ is a function such that $E[h(X_{(1)})] = 0, \quad \forall \theta$. Then,

$$0 = \frac{\partial E[h(X_{(1)})]}{\partial \theta} = -\frac{nh(\theta)}{(1-\theta)^n}(1-\theta)^{n-1} = -\frac{nh(\theta)}{(1-\theta)}, \quad \forall \theta.$$

Thus, $h(\theta) = 0, \forall \theta$, and therefore, $X_{(1)}$ is a complete sufficient statistic for θ.;

(h) $\frac{(n+1)X_{(1)}-1}{n}$ is a UMVUE of θ

4.4.5 (a) $F(x) = \begin{cases} 0 & x < 0 \\ \frac{3x^2\theta - 2x^3}{\theta^3} & 0 \le x < \theta; \\ 1 & x \ge \theta \end{cases}$

(b) $f_{(1)}(x) = \frac{6n}{\theta^{3n}}x^{2(n-1)}(\theta-x)[3\theta - 2x]^{n-1}, x \in (0, \theta)$

4.4.7 (a) $X_{(n)}$; (b) $F(x) = \begin{cases} e^{x-\theta} & -\infty < x < \theta; \\ 1 & x \ge \theta \end{cases}$;

(c) $f_{(n)}(x) = ne^{n(x-\theta)}, x \in (-\infty, \theta)$; (d) $\frac{n\theta-1}{n}$; (e) $\frac{nX_{(n)}+1}{n}$

Solutions for Chapter 5

5.1.1 $\frac{-n}{\sum \ln(x_i)}$

5.1.3 $\left(\frac{\sum x_i^3}{n}\right)^{\frac{1}{3}}$

5.1.5 (a) $\sqrt{\frac{1}{n}\sum x_i^2}$; (b) $\frac{\theta^2}{2n}$; (c) $\frac{1}{n}\sum x_i^2$

5.1.7 (a) $\frac{\bar{X}}{m}$; (b) θ; (c) $\frac{\theta(1-\theta)}{mn}$; (d) $\frac{\theta(1-\theta)}{mn}$; (e) $\hat{\theta}$ is a UMVUE of θ since $\text{Var}(\hat{\theta}) = $ CRLB

5.1.9 (a) $\frac{n}{\sum \ln(x_i)}$; (b) $E(\hat{\theta}) = \frac{n\theta}{n-1}$, $\text{Var}(\hat{\theta}) = \frac{n^2\theta^2}{(n-2)(n-1)^2}$; (c) $\frac{\theta^2}{n}$

5.1.11 (a) $\frac{1}{\sigma^2}\sum(x_i - \theta)$; (b) $\frac{n}{\sigma^2}$

5.1.13 Using out5113=fitdist(x,dgamma,start=list
(shape=3),fix.arg=list(scale=2.5))
(a) 4.02; (b) 0.266; (c) 10.05; (d) 25.125; (e) Use the command
plot(out5113)

5.1.15 Using out5115=fitdist(x,dbeta,start=list
(shape2=2),fix.arg=list(shape1=1.2))
(a) 3.532; (b) 0.467; (c) 0.252; (d) 0.033

5.1.17 (a) 9 and 10 are MLEs of θ; (b) $2x$ and $2x - 1$ are MLEs of θ

5.1.19 (a) $e^{75}0.4^{\Sigma x_i}$; (b) 2; (c) 5

5.1.21 (a) $X_{(n)}$; (b) $\frac{n}{n+1}\theta$; (c) $\frac{n\theta^2}{(n+1)^2(n+2)}$

5.1.23 (a) $\sum X_i$; (b) $-\ln\left(\frac{\overline{X}}{1-\overline{X}}\right)$; (c) $\overline{X}$; (d) $\frac{(1+e^{-\theta})^2}{ne^{-\theta}}$

5.1.25 $\widehat{\theta}_i = \frac{x_i}{n}, i = 1, \ldots, k$

5.1.27 $\widehat{\theta} = -\frac{-\sum x_i + \sqrt{4n\sum x_i^2 + (\sum x_i)^2}}{2n}$

5.1.29 (a) $\frac{n-1}{n}\theta_1$; (b) $\theta_2 + \frac{\theta_1}{n}$; (c) $\frac{n}{n-1}\widehat{\theta}_1$; (d) $\widehat{\theta}_2 - \frac{\widehat{\theta}_1}{n-1}$

5.1.30 (a) $\theta_1^n e^{-\theta_1 \sum(x_i - \theta_2)} \prod I_{(\theta_2,\infty)}(x_i)$; (b) $\widehat{\theta}_1 = \frac{n}{\sum(x_i - x_{(1)})}, \widehat{\theta}_2 = x_{(1)}$; (c) $\theta_2 + \frac{1}{n\theta_1}$

5.1.32 Using the command out5132=fitdist(x,dgamma,start=
list(shape=2,scale=2))
(a) $\widehat{\theta}_1 = 6.573, \widehat{\theta}_2 = 0.400$; (b) $se(\widehat{\theta}_1) = 1.656, se(\widehat{\theta}_2) = 0.105$; (c) 2.632;
(d) 1.054
(e) Use the command plot(out5132)

5.1.34 Using the command out5134=fitdist(x5134,dweibull,
start=list(shape=2,scale=3))
(a) $\widehat{\theta}_1 = 2.552, \widehat{\theta}_2 = 1.756$; (b) $se(\widehat{\theta}_1) = 0.288, se(\widehat{\theta}_2) = 0.102$; (c) Use
the command plot(out5134)

5.1.35 Using the command out5135=fitdist(x5135,dlogis,
start=list(location=30,scale=3))
(a) $\widehat{\theta}_1 = 50.606, \widehat{\theta}_2 = 2.077$; (b) $se(\widehat{\theta}_1) = 0.512, se(\widehat{\theta}_2) = 0.243$; (c) Use
the command plot(out5135)

5.1.36 $\widehat{\theta} = \begin{cases} \overline{x} & \overline{x} > 5 \\ 5 & \overline{x} \leq 5 \end{cases}$

5.1.38 $\hat{\theta} = \begin{cases} \frac{1}{n}\sum x_i^2 & \frac{1}{n}\sum x_i^2 > 1 \\ 1 & \frac{1}{n}\sum x_i^2 \leq 1 \end{cases}$

5.2.1 (a) $f(\theta|\vec{x}) = \begin{cases} \dfrac{0.25*0.3^{\sum x_i}0.7^{n-\sum x_i}}{0.25*0.3^{\sum x_i}0.7^{n-\sum x_i}+0.75(0.5)^n} & \theta = 0.3 \\[2mm] \dfrac{0.750.5^n}{0.25*0.3^{\sum x_i}0.7^{n-\sum x_i}+0.75(0.5)^n} & \theta = 0.5 \end{cases}$;

(b) $f(\theta|\vec{x}) = \begin{cases} 0.241 & \theta = 0.3 \\ 0.759 & \theta = 0.5 \end{cases}$;

(c) 0.452

5.2.3 (a) Beta$(n+1, \sum x_i + 1)$; (b) $\frac{n+1}{n+\sum x_i+2}$

5.2.5 (a) $N\left(\frac{\sum x_i+\mu}{n+1}, \frac{\sigma^2}{n+1}\right)$; (b) $\frac{\sum x_i+\mu}{n+1}$; (c) $\frac{\sum x_i+\mu}{n+1}$

5.2.6 (a) 0.783; (b) 0.785

5.2.9 (a) Gamma$(\sum x_i + \alpha, (30+\frac{1}{\beta})^{-1})$; (b) $\frac{\sum x_i+\alpha}{30+\frac{1}{\beta}}$; (c) 5.803; (d) 5.792

5.2.11 (a) $N\left(\frac{\sum x_i+\frac{\nu}{\tau^2}}{n+\frac{1}{\tau^2}}, (n+\frac{1}{\tau^2})^{-1}\right)$; (b) $\frac{\sum x_i+\frac{\nu}{\tau^2}}{n+\frac{1}{\tau^2}}$; (c) $\frac{\sum x_i+\frac{\nu}{\tau^2}}{n+\frac{1}{\tau^2}}$; (d) 8.796

5.2.13 (a) $\frac{e^{-\frac{\lambda}{2}}(0.5\lambda)^{\theta-x}}{(\theta-x)!}, \theta = x, x+1, \ldots$; (b) $x+\frac{\lambda}{2}$; (c) 45

5.3.1 (a) $T \sim N\left(0, \frac{1}{n}\right)$; (b) $\bar{x} \pm z_{1-\frac{\alpha}{2}}\frac{1}{\sqrt{n}}$;

(c) $\theta \geq \bar{x} - z_{1-\alpha}\frac{1}{\sqrt{n}}$

5.3.3 (a) For $Y = \frac{X_{(n)}}{\theta}, f_Y(y) = ny^{n-1}, y \in (0,1)$; (b) $\left(\frac{x_{(n)}}{\left(1-\frac{\alpha}{2}\right)^{\frac{1}{n}}}, \frac{x_{(n)}}{\left(\frac{\alpha}{2}\right)^{\frac{1}{n}}}\right)$;

(c) $\theta \geq \frac{x_{(n)}}{(1-\alpha)^{\frac{1}{n}}}$

5.3.5 (a) For $Y = X_{(1)} - \theta, f_Y(y) = ne^{-ny}, y \in (0,\infty)$;

(b) $\left(x_{(1)} - \frac{\ln\left(1-\frac{\alpha}{2}\right)}{n}, x_{(1)} - \frac{\ln\left(\frac{\alpha}{2}\right)}{n}\right)$;

(c) $\theta \leq x_{(1)} - \frac{\ln(1-\alpha)}{n}$

5.3.7 (a) $\frac{(n-1)S^2}{\theta_2} \sim \chi_{n-1}^2$; (b) $\left(\frac{(n-1)s^2}{\chi_{n-1,1-\frac{\alpha}{2}}^2}, \frac{(n-1)s^2}{\chi_{n-1,\frac{\alpha}{2}}^2}\right)$; (c) $\left(\sqrt{\frac{(n-1)s^2}{\chi_{n-1,1-\frac{\alpha}{2}}^2}}, \sqrt{\frac{(n-1)s^2}{\chi_{n-1,\frac{\alpha}{2}}^2}}\right)$

5.3.9 (a) $\frac{\theta(1-\theta)}{n}$; (b) $\left(\bar{x} - z_{1-\frac{\alpha}{2}}\sqrt{\frac{\bar{x}(1-\bar{x})}{n}}, \bar{x} + z_{1-\frac{\alpha}{2}}\sqrt{\frac{\bar{x}(1-\bar{x})}{n}}\right)$; (c) $(0.361, 0.499)$

5.3.11 (a) $\frac{-n}{\sum \ln(x_i)}$; (b) $\frac{\theta^2}{n}$; (c) $\frac{-n}{\sum \ln(x_i)} \pm z_{1-\frac{\alpha}{2}}\sqrt{\frac{n}{(\sum \ln(x_i))^2}}$

5.3.13 (a) (3.130, 6.762); (b) (1.439, 3.251)

5.3.14 (a) (2.787, 6.275); (b) (0.708, 1.384)

5.3.17 (a) (0.383, 0.734); (b) $\theta \geq 0.608$; (c) $\theta \leq 0.427$

5.3.19 (a) (0.208, 0.320); (b) $\theta \geq 0.182$; (c) $\theta \leq 0.115$

5.4.1 (a) 14; (b) 11

5.4.3 (a) Using $\texttt{lambda=c(0.6,0.4)}$, $\hat{p}_1 = 0.462$, $\hat{p}_2 = 0.538$, $\hat{\mu}_1 = 6.130$, $\hat{\mu}_2 = 14.166$, $\hat{\sigma}_1 = 1.807$, $\hat{\sigma}_2 = 5.397$
(b) Using $\texttt{lambda=c(0.4,0.3,0.2,0.1)}$, $\hat{p}_1 = 0.454$
$\hat{p}_2 = 0.088$, $\hat{p}_3 = 0.297$, $\hat{p}_4 = 0.162$, $\hat{\mu}_1 = 5.733$, $\hat{\mu}_2 = 8.340$,
$\hat{\mu}_3 = 12.817$, $\hat{\mu}_4 = 20.508$, $\hat{\sigma}_1 = 1.641$, $\hat{\sigma}_2 = 0.722$, $\hat{\sigma}_3 = 2.258$,
$\hat{\sigma}_4 = 3.476$

5.4.4 The solutions are dependent on the data generated by $\texttt{rnormmix}$

5.4.5 (a) $\text{se}(\widehat{\text{age}}) \approx 946 \times |\hat{\mu}_1| \times \text{se}(\hat{\mu}_1)$; (b) 1738.9

5.4.8 First, the continuity of an M-component continuous finite mixture model follows from being sum of M continuous pdfs. The nonnegativity of $f_M(x)$ follows from $p_1, \ldots, p_m$ and $f_1(x, \vec{\theta}_1), \ldots, f_M(x; \vec{\theta}_M)$ being nonnegative, and

$$\int_{-\infty}^{\infty} f_M(x)\, dx = \int_{-\infty}^{\infty} \sum p_i f_i(x; \vec{\theta}_i)\, dx$$

$$= \sum p_i \int_{-\infty}^{\infty} f_i(x; \vec{\theta}_i)\, dx = \sum p_i = 1$$

Solutions for Chapter 6

6.1.1 (a) composite; (b) simple; (c) composite; (d) composite

6.1.3 (a) 0.016; (b) 0.868

6.1.5 (a) 0.049; (b) 0.898; (c) The solution to Problem 6.1.5c is given in Figure S6.1.
(d) 1.343

6.1.7 (a) 12; (b) 0.596

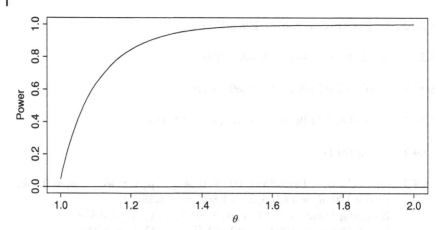

Figure S6.1 The solution to Problem 6.1.5c.

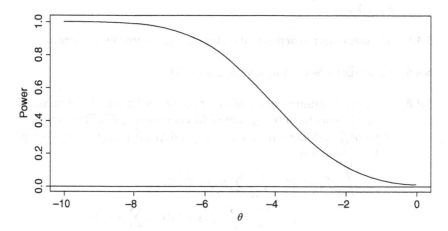

Figure S6.2 The solution to Problem 6.1.9c.

6.1.9 (a) −4.029; (b) 0.712; (c) The solution to Problem 6.1.9c is given in Figure S6.2

6.1.11 (a) Gamma(20, 1); (b) 0.04; (c) 0.897; (d) 0.023

6.2.1 (a) 54; (b) 0.304

6.2.3 (a) 39.403; (b) 0.360

6.2.5 $C = \{\vec{x} : \sum x_i \geq k\}$; (b) Bin$(2n, \theta)$; (c) 30; (d) 0.839

6.2.7 (a) $C = \{\vec{x} : \sum x_i \leq k\}$; (b) Gamma$(n, \theta)$; (c) 77.575; (d) 0.475

6.2.9 (a) $C = \{\vec{x} : \sum x_i \geq k\}$; (b) Gamma$(2n, \theta)$; (c) 151.090; (d) 0.869

6.2.11 (a) $C = \{\vec{x} : \sum \ln(x_i) \geq k\}$; (b) -5.208

6.2.13 $C = \{\vec{x} : \sum \ln(x_i) \geq k\}$

6.2.15 $C = \{\vec{x} : \sum x_i^2 \geq k\}$

6.3.1 $f(x; \theta) = \theta e^{x \ln(1-\theta)}$ so $f(x; \theta)$ is in the exponential family and by Theorem 6.3, $L(\theta)$ has MLR in $\sum x_i$

6.3.3 $f(x; \theta) = \theta^{-\frac{1}{2}} e^{-\frac{1}{2\theta} x^2} h(x)$ so $f(x; \theta)$ is in the exponential family and by Theorem 6.3, $L(\theta)$ has MLR in $\sum x_i^2$

6.3.5 For $\theta_1 < \theta_2$, $\Lambda(\theta_1, \theta_2) = \left(\frac{\theta_2 - 1}{\theta_1 - 1}\right)^n I_{(x_{(n)}, \infty)}(\theta_1) I_{(x_{(n)}, \infty)}(\theta_2)$, which is nonincreasing in $x_{(n)}$. Thus, $L(\theta)$ has MLR in $X_{(n)}$

6.3.7 For $\theta_1 < \theta_2$, $\Lambda(\theta_1, \theta_2) = \left(\frac{1 - \theta_2}{1 - \theta_1}\right)^n I_{(-\infty, x_{(1)})}(\theta_1) I_{(-\infty, x_{(1)})}(\theta_2)$, which is nondecreasing in $x_{(1)}$. Thus, $L(\theta)$ has MLR in $X_{(n)}$

6.3.9 (a) $X_i \sim$ Gamma$(2, \theta^{-1})$, so $L(\theta)$ has MLR in $\sum t(x_i) = \sum x_i$; (b) $C = \{\vec{x} : \sum x_i \geq k\}$; (c) Gamma$(2n, \theta^{-1})$; (d) 10.615; (e) 0.818

6.3.11 (a) $C = \{\vec{x} : \sum x_i \geq k\}$; (b) $C = \{\vec{x} : \sum x_i \leq k\}$; (c) Pois$(n\theta)$

6.3.13 (a) $C = \{\vec{x} : \sum x_i \leq k\}$; (b) $C = \{\vec{x} : \sum x_i \geq k\}$; (c) NegBin$(n, \theta)$

6.3.15 (a) $C = \{\vec{x} : \sum (x_i - \mu)^2 \geq k\}$; (b) $C = \{\vec{x} : \sum (x_i - \mu)^2 \leq k\}$; (c) $\theta \chi_n^2 =$ Gamma$(\frac{n}{2}, 2\theta)$

6.3.17 (a) $C = \{\vec{x} : \sum x_i \geq k\}$; (b) $C = \{\vec{x} : \sum x_i \leq k\}$; (c) Gamma$(n\alpha, \theta)$

6.3.19 (a) $C = \{\vec{x} : \sum \ln(x_i) \geq k\}$; (b) $C = \{\vec{x} : \sum \ln(x_i) \leq k\}$

6.3.21 (a) $C = \{\vec{x} : \sum \ln(1 - x_i) \geq k\}$; (b) $C = \{\vec{x} : \sum \ln(1 - x_i) \leq k\}$

6.3.23 (a) $C = \{\vec{x} : x_{(1)} \geq k\}$; (b) $C = \{\vec{x} : x_{(1)} \leq k\}$

6.3.25 $C = \{\vec{x} : \sum x_i \leq k_1 \quad \text{or} \quad \sum x_i \geq k_2\}$

6.3.27 $C = \{\vec{x} : \sum \ln(x_i) \leq k_1 \quad \text{or} \quad \sum \ln(x_i) \geq k_2\}$

6.3.29 (a) $C = \{\vec{x} : \sum x_i^3 \geq k\}$; (b) $C = \{\vec{x} : \sum x_i^3 \leq k\}$

6.4.1 (a) $\bar{x}$; (b) $\widehat{\theta}_0 = 0.25$; (c) $\lambda(\vec{x}) = \left(\frac{1-\bar{x}}{3\bar{x}}\right)^{\sum x_i}\left(\frac{.75}{1-\bar{x}}\right)^n$

6.4.3 (a) $\frac{1}{n}\sum x^2$; (b) $\widehat{\theta}_0 = \begin{cases} 1 & \text{when } \frac{1}{n}\sum x_i^2 < 1 \\ \frac{1}{n}\sum x_i^2 & \text{when } \frac{1}{n}\sum x_i^2 \geq 1 \end{cases}$;

(c) $\lambda(\vec{x}) = \begin{cases} 1 & \text{when } \frac{1}{n}\sum x_i^2 \geq 1 \\ \left(\frac{1}{n}\sum x_i^2\right)^{\frac{n}{2}}e^{\frac{n}{2}-\frac{1}{2}\sum x_i^2} & \text{when } \frac{1}{n}\sum x_i^2 < 1 \end{cases}$;

(d) $\{\vec{x} : \lambda(\vec{x}) \leq k\} = \{\vec{x} : \frac{1}{n}\sum x_i^2 \leq k\}$

6.4.5 (a) $\widehat{\theta}_1 = \bar{x}, \widehat{\theta}_2 = \frac{1}{n}\sum (x_i - \bar{x})^2$; (b) $\widehat{\theta}_{10} = \bar{x}, \widehat{\theta}_{20} = 1$;

(c) $\lambda(\vec{x}) = \sqrt{\frac{1}{n}\sum (x_i - \bar{x})^2}^n e^{\frac{n}{2}-\frac{1}{2}\sum (x_i-\bar{x})^2}$;

(d) $\{\vec{x} : \lambda(\vec{x}) \leq k\} = \{\vec{x} : \sum (x_i - \bar{x})^2 \leq k_1 \quad \text{or} \quad \sum (x_i - \bar{x})^2 \geq k_2\}$

6.4.7 (a) $\frac{1}{n}\sum(x_i - x_{(1)})$;

(b) $\widehat{\theta}_{10} = \begin{cases} 1 & \text{when } \frac{1}{n}\sum(x_i - x_{(1)}) < 1 \\ \left[\frac{1}{n}\sum(x_i - x_{(1)})\right]^n e^{n-\sum(x_i-x_{(1)})} & \text{when } \frac{1}{n}\sum(x_i - x_{(1)}) \geq 1 \end{cases}$;

(c) $(\bar{x} - x_{(1)})^n e^{n-\sum(x_i-x_{(1)})}$ when $\frac{1}{n}\sum(x_i - x_{(1)}) < 1$ (d) $\{\vec{x} : \lambda(\vec{x}) \leq k\} = \{\vec{x} : \sum(x_i - x_{(1)}) \leq k\}$

6.4.9 (a) 5.797; (b) 0.176; (c) $C = \{\vec{x} : -2\ln[\lambda(\vec{x})] \geq 3.841\}$; (d) Fail to Reject H_0; (e) 0.062

6.4.11 (a) $\ell(\theta) = -n\ln[\Gamma(\theta)] - n\theta \ln(10) + (\theta - 1)\sum \ln(x_i) - \frac{\sum x_i}{10}$;
(b) -1151.39; (c) $C = \{\vec{x} : -2\ln[\lambda(\vec{x})] \geq 6.635\}$; (d) 30.54; (e) Reject H_0

6.5.1 (a) 1.58; (b) $\frac{1}{\theta}$; (c) $C = \left\{\vec{x} : \frac{\bar{x}-2}{\sqrt{\frac{\bar{x}}{100}}} \leq -1.645\right\}$; (d) Reject H_0

6.5.3 (a) 1.504; (b) $\frac{1}{\theta^2}$; (c) $C = \left\{\vec{x} : \frac{\widehat{\theta}-1}{\frac{\widehat{\theta}}{\sqrt{100}}} \geq 2.326\right\}$; (d) Reject H_0

6.5.5 (a) $C = \left\{\vec{x} : \frac{\widehat{\theta}-1}{se(\widehat{\theta})} \geq 1.751\right\}$; (b) Reject H_0; (c) 0.000

6.5.7 (a) $C = \left\{ \vec{x} : \frac{\hat{\theta}_1 - 6}{se(\hat{\theta}_1)} < -1.881 \right\}$; (b) Fail to Reject H_0; (c) 0.067;

(d) $C = \left\{ \vec{x} : \left| \frac{\hat{\theta}_2 - 2}{se(\hat{\theta}_2)} \right| > 1.96 \right\}$; (e) Reject H_0; (f) 0.040

6.5.9 (a) $\dot{\ell}(\theta) = -\frac{n}{\theta} + \frac{\sum x_i}{\theta^2}$; (b) 1.106; (c) Fail to Reject H_0; (d) 0.211

6.5.11 (a) $\dot{\ell}(\theta) = -\frac{n}{2\theta} + \frac{\sum x_i^2}{2\theta^2}$; (b) −1.603; (c) Fail to Reject H_0; (d) 0.257

6.5.13 (a) $\frac{\sum(x_i - \theta)}{25}$; (b) 5.144; (c) Reject H_0; (d) 0.010

6.6.1 (a) Fail to Reject H_0; (b) Reject H_0; (c) 0.262; (d) 0.724; (e) 54.066

6.6.3 (a) $L(\beta) = e^{\beta \sum y_i x_i} \prod (1 + e^{\beta x_i})^{-1}$; (b) $\ell(\theta) = \beta \sum y_i x_i - \sum \ln(1 + e^{\beta x_i})$;
(c) $\dot{\ell}(\theta) = \sum y_i x_i - \sum \frac{x_i e^{\beta x_i}}{1 + e^{\beta x_i}}$

Solutions for Chapter 7

7.1.1 (a) $f(y; \lambda) = e^{y\theta - e^\theta - \ln(y!)}$; $\theta = \ln(\lambda), b(\theta) = e^\theta, a(\phi) = 1$;
(b) $E(Y) = b'(\theta) = \lambda$;
(c) $\mathrm{Var}(Y) = a(\phi)b''(\theta) = \lambda$; (d) $\theta = \ln(\lambda)$

7.1.3 (a) $f(y; \eta) = e^{y\theta + \ln(-\theta)}$; $\theta = -\eta, b(\theta) = \ln(-\theta), a(\phi) = 1$;
(b) $E(Y) = b'(\theta) = \frac{1}{\eta}$; (c) $\mathrm{Var}(Y) = a(\phi)b''(\theta) = \frac{1}{\eta^2}$; (d) $\theta = -\eta$

7.1.5 (a) $f(y; \beta) = e^{y\theta - \ln\left(\frac{-1}{\theta}\right)}$; $\theta = \frac{-1}{\beta}, b(\theta) = \ln\left(\frac{-1}{\beta}\right), a(\phi) = 1$;
(b) $E(Y) = b'(\theta) = \beta$;
(c) $\mathrm{Var}(Y) = a(\phi)b''(\theta) = \beta^2$; (d) $\theta = \frac{-1}{\beta}$

7.1.7 (a) $f(y; p_i) = e^{y_i \theta_i - \ln(1 + e_i^\theta)}$; $\theta_i = \ln\left(\frac{p_i}{1 - p_i}\right), b(\theta) = \ln(1 + e^\theta), a(\phi) = 1$;
(b) $V(\mu) = \mu(1 - \mu)$

7.1.9 $f(y; \mu_i) = e^{-\frac{y_i \mu_i - \frac{\mu_i^2}{2}}{\sigma^2} - \frac{y_i^2}{2\sigma^2} - \frac{1}{2}\ln(2\pi\sigma^2)}$; $\theta = \mu_i, b(\theta) = \frac{\mu_i^2}{2}, a(\phi) = \sigma^2$

7.2.1 (a) $L(\vec{\mu}) = e^{-\sum \frac{y_i}{\mu_i}} \prod \frac{1}{\mu_i}$; (b) $\ell(\vec{\mu}) = -\sum \frac{y_i}{\mu_i} - \sum \ln(\mu_i)$; (c) $\frac{\partial l}{\partial \mu_i} = -\frac{1}{\mu_i} + \frac{y_i}{\mu_i^2}$; (d) $D = 2\sum \left[\ln\left(\frac{\hat{\mu}_i}{y_i}\right) - \left(1 - \frac{y_i}{\hat{\mu}_i}\right) \right]$

7.2.3 (a) $L(\vec{\mu}) = \Gamma(4)^{-n} e^{-4\sum \frac{y_i}{\mu_i}} \prod \left(\frac{\mu_i}{4}\right)^{-4} \prod y_i^3;$

(b) $\ell(\vec{\mu}) = -4\sum \frac{y_i}{\mu_i} - n\ln[\Gamma(4)] - 4\sum \ln\left(\frac{\mu_i}{4}\right) + 3\sum \ln(y_i);$

(c) $\frac{\partial l}{\partial \mu_i} = -\frac{4}{\mu_i} + 4\frac{y_i}{\mu_i^2};$

(d) $D = 8\sum \left[\ln\left(\frac{\hat{\mu}_i}{y_i}\right) - \left(1 - \frac{y_i}{\hat{\mu}_i}\right)\right]$

7.2.5 (a) $L(\beta) = (2\pi)^{-\frac{n}{2}} e^{-\frac{1}{2}\sum(y_i - \beta x_i)^2};$ (b) $\ell(\beta) = -\frac{n}{2}\ln(2\pi) - \frac{1}{2}\sum(y_i - \beta x_i)^2;$

(c) $\hat{\beta} = \frac{\sum y_i x_i}{\sum x_i^2};$ (d) $D = \sum(y_i - \hat{\beta}x_i)^2$

7.2.7 $d_i = 1.39$

7.3.1 (a) $p = 0.0264;$ (b) $(-0.397, 0.009)$

7.3.3 (a) $p = 0.9306;$ (b) $p = 0.0034$

7.4.2 (a) $\frac{\sum y_i x_i}{\sum x_j^2};$ (b) $\beta;$ (c) $\frac{\sigma^2}{\sum x_j^2};$ (d) $\frac{x^{*2}\sigma^2}{\sum x_j^2}$

7.4.4 (a)

$$\sum y_i x_{1i} = \beta_1 \sum x_{1i}^2 + \beta_2 \sum x_{1i}x_{2i}$$
$$\sum y_i x_{2i} = \beta_1 \sum x_{1i}x_{2i} + \beta_2 \sum x_{2i}^2 \; ;$$

(b) $\hat{\beta}_1 = 0.25, \hat{\beta}_2 = 1.25$

7.4.8 (a) $(137.7375, 2.1507);$

(b)

SOV	df	SS	MS	F	P
Regr	1	81309	81309	150.03	0.0000
Error	250	135485	542		
Total	251	216794			

(c) $p = 0.0000;$ (d) $0.176;$ (e) $t = 12.2481, p = 0.0000$

7.5.1 (a) $\sum[y_i x_i \beta - \ln(1 + e^{\beta x_i})];$ (b) $\sum \left[y_i x_i - \frac{x_i e^{\beta x_i}}{1 + e^{\beta x_i}}\right];$ (c) $I(\beta) = \sum \frac{x_i^2 e^{\beta x_i}}{(1 + e^{\beta x_i})^2};$

(d) $\sqrt{I(\beta)^{-1}}$

7.5.3 (a) $p = 1.000;$ (b) $p = 0.000;$ (c) $p = 0.5449;$ (d) 0.197

7.5.5 (a) $p_i = 1 - e^{-e^{x_i \beta}};$ (b) 0.365

7.5.7   (a) $d_1 = -1.641$; (b) $d_{s1} = -1.873$

7.6.1   (a) $b'(\theta) = e^\theta = e^{\ln(\lambda)} = \lambda$; (b) $a(\phi)b''(\theta) = e^\theta = e^{\ln(\lambda)} = \lambda$

7.6.3   (a) $\hat{\mu}_2 = 46.33$; (b) $d_{23} = 1.276$; (c) $d_{s23} = 1.559$

7.6.5 $P(\cup_{i=1}^k A_i) = 1 - P(\cap_{i=1}^k A_i^c) = 1 - \prod_{i=1}^k P(A_i^c) = 1 - (1 - \alpha_{\text{PC}})^k$

7.6.7   (a) 0.025; (b) 0.010

Index

Mathematical Statistics: An Introduction to Likelihood Based Inference, First Edition. Richard J. Rossi.
© 2018 John Wiley & Sons, Inc. Published 2018 by John Wiley & Sons, Inc.